DIESEL MECHANICS

ERICH J. SCHULZ

Department Head
Mechanical Division

Pacific Vocational Institute—Burnaby Campus
Burnaby, British Columbia

Gregg Division
McGraw-Hill Book Company

New York   San Francisco   Düsseldorf   Mexico   Paris   Sydney
St. Louis   Auckland   Johannesburg   Montreal   São Paulo   Tokyo
Dallas   Bogotá   London   New Delhi   Singapore   Toronto
                 Madrid   Panama

Library of Congress Cataloging in Publication Data

Schulz, Erich J
  Diesel mechanics.

  Includes index.
  1. Diesel motor—Maintenance and repair. I. Title.
TJ799.S38   621.43'6   76-15386
ISBN 0-07-055664-4

Diesel Mechanics

34567890 BABA 783210987

The editors for this book were Ardelle Cleverdon and Susan Berkowitz, the designer was Tracy A. Glasner, the art supervisor was George T. Resch, and the production supervisor was Iris A. Levy. It was set in Melior by Typographic Sales, Inc.

Cover photo: Krider Studios Inc.

CONTENTS

FOREWORD

Vast research and development money is now being spent by industrialized nations around the world to design new internal-combustion engines that will meet the challenges of the future, particularly with regard to clean emissions and fuel economy. Therefore, today's diesel engine mechanics must maintain these engines at peak operating efficiency so that the full benefit of this research and development may be obtained by their operators.

The author wrote this textbook for the diesel engine mechanic with this in mind. It is neither too elementary nor too highly engineering-oriented. Its basic concept is to pass on the author's many years of experience in maintaining large diesel-engine-powered fleets of on- and off-highway vehicles. It is a textbook that shows the operating requirements of the equipment. Design concepts are discussed where necessary. Particular emphasis is placed on the diagnosis, repair, and service of specific malfunctions.

After many years of instructing students in the classroom, the author has put together this textbook from class lecture notes, student questions, and on-the-job observation of the students' requirements in actual applications. It is hoped that the realistic, practical style of presenting the information will enable diesel engine mechanics to avoid many of the mistakes that can occur in diagnosing and servicing diesel engine equipment.

Diesel Mechanics should be used in conjunction with the various manufacturers' service manuals so that the manufacturers' particular specifications are strictly adhered to. The author shows students how to practice preventive maintenance and constantly reminds them of the need for a high quality of workmanship. These fundamentals are essential to prevent expensive downtime on diesel equipment.

The text includes those engines and related equipment in use at the time the manuscript was prepared and deals specifically with diesel engines that the students may work on upon completion of their training. However, during the preparation of the manuscript, the author constantly referred to the latest catalogs and service manuals to ensure that the information was current and up to date.

Derek Bland

Vice Principal
British Columbia
Vocational School

PREFACE

Great progress has been made in mechanical engineering since the time when wind and water were harnessed as the sources of mechanical power. The extent of this progress is particularly noticeable when one considers today's gas turbine, jet propulsion, and atomic power units.

The internal-combustion piston engine would now appear to have lost its leading position as the source of mechanical power. A new era of engine development seems to be taking its place. In some cases this has already occurred. In the aircraft industry, for example, the gas turbine engine has almost replaced all piston-type engines and, especially in the larger power ranges, is now beginning to replace the diesel or diesel electric locomotive engines. Tests are also being made with on- and off-highway trucks and tractors to replace their piston engines with gas turbines. The rotary engine is seriously being considered as a replacement for the reciprocating piston-type engine.

It would be presumptuous to forecast at this time what types of engines will be used in the next 20 to 30 years, but at present it appears that the internal-combustion diesel engine, and perhaps the rotary diesel engine, will be the main sources of power for the on- and off-highway trucks, large buses, and tractors.

Because the diesel engine has not yet reached its maximum power and efficiency potential, it will be with us for many more years to come, and in some areas, may take precedence over other types of engines. With advanced technology, its efficiency may increase from its present 42 percent to 50 percent or perhaps even higher. Furthermore, its operational cost is low, it is reliable, and it has a long service life. Another inherent advantage of the diesel engine in today's pollution-conscious society is its relatively clean emission.

Credit is due to Dr. Rudolf Diesel, a German engineer, for crystallizing engineering thought in 1892 on the importance of using high-compression air to attain the self-ignition of fuel oil. He obtained the financial backing of Baron Friedrich Alfred von Krupp and the giant Maschine-Fabrick Augsburg Nurnberg. Strangely enough, the engine got off to a very shaky start. Experiments that began with coal dust as fuel were dropped. Modifications were made, and in 1897 a successful compression ignition engine was built by Dr. Lauster and the engineering staff of the M. A. N. Company in collaboration with Dr. Diesel. A number of manufacturers were licensed to build similar engines. Adolphus Bush

was one of those who bought a license to manufacture and sell compression ignition engines in the United States and Canada. A diesel engine built for his brewery was the first to go into service anywhere in the world.

The original oil-burning engine injected fuel by a high-pressure air blast. This high-pressure air blast atomized the fuel, and the resulting turbulence facilitated the mixing of the air and fuel with the compressed air in the cylinder. Because of Dr. Diesel's insistence that this engine operate at an almost constant temperature cycle, the engines were cumbersome and rotated at a slow speed.

With the lifting of this restriction and with the greater advancements in metallurgical and manufacturing technologies, the diesel engine and its fuel-injection system have been developed to make it the versatile engine that it is today. The introduction of a more refined fuel oil has also contributed to its success. Further metallurgical research into light-weight/high-strength materials, combined with refined engineering design concepts, will enable the diesel engine to meet the challenges of the energy and pollution crises and place it in the forefront of the world's industrial power units for many years to come.

The prime purpose of this textbook is to train diesel mechanics who will be able to diagnose, repair, and service today's diesel engines. The *Workbook for Diesel Mechanics* has been prepared to further this training by providing hands-on experience in the shop. The *Instructor's Guide for Diesel Mechanics* is also available to give the instructor additional suggestions for lectures and shop activities as well as answers to the workbook questions.

The textbook, workbook, and instructor's guide have been constructed so that the instructor can begin with any unit and then progress from one unit to another. With a fairly advanced class, the instructor may wish to use the preliminary units simply as review material.

Suggestions for equipment, tools, training aids, and references have been included in the instructor's guide. Dual dimensioning has been used throughout this program. Metric measurements are given in brackets following the U.S. Customary measurements. Metric conversion tables and a glossary of diesel terms are included at the back of the textbook.

It is hoped that this program will ensure that the students achieve sufficient knowledge and practical experience to make them employable in today's growing diesel field.

Erich J. Schulz

ACKNOWLEDGMENTS

No mechanics' textbook would be considered complete, and few would be intelligible to the novice, without graphs, photographs, and other illustrative material pertinent to the trade. To the numerous companies, therefore, which have consented to allow reproductions for this text from their service manuals and other literature, I wish, on behalf of my readers and myself, to express our appreciation for their contributions. I refer particularly to those companies listed below.

Aeroquip Corporation
Allis-Chalmers Engine Division
American Bosch-Ambac Industries Inc.
J I Case Company Components Division
Caterpillar Tractor Co.
CAV Limited
Chrysler Canada Ltd.
Chevron Research Company
Cummins Engine Company, Inc.
Dana Corporation, Spicer Transmission Division
Eaton Corporation Transmission Division
Federal-Mogul Replacement Sales Division
GMC Truck and Coach Division
GMC Detroit Diesel Allison Division
Greenfield Tap and Die Corporation
International Harvester Company
KHD, Klockner-Humboldt-Deutz AG
Mack Trucks Canada Limited
Owatonna Tool Company (Tools and Equipment Division)
Robert Bosch GmbH, Stuttgart, West Germany
Roosa Master Stanadyne/Hartford Division
Sperry Rand Canada Limited Vickers Division
Stratoflex of Canada Inc.
Sunnen Products Company
Sun Oil Company
The L. S. Starrett Company
United Delco-AC Products General Motors of Canada Limited
The Weatherhead Co. of Canada Ltd.

To my friend and colleague, Derek Bland, who, despite pressure from other commitments, freely gave his time to proofread the major draft, I am deeply indebted. And to my wife I gratefully acknowledge the help she gave me from start to finish.

Last but not least I wish to thank my publishers, who recognized the need for this book by those training to become diesel mechanics.

Erich J. Schulz

SECTION 1
Introduction to the Diesel Engine

UNIT 1

Basic Components

What a Diesel Engine Is The simplest way to describe a diesel engine is to compare it with an ordinary gasoline engine such as the one you have in your car. Both engines are of the internal-combustion design because they burn the fuel within the engine. The major components of both engines are the same. However, the components of a diesel engine with the same horsepower as a gasoline engine are heavier because they must withstand greater dynamic force and more concentrated stress and load put on them due to the greater pressure (Fig. 1–1).

The greater pressure is the result of the higher compression ratio. In a gasoline engine the compression ratio (which controls the compression temperature) is limited by the detonation and preignition quality of the air fuel mixture.

In the diesel engine the compression ratio can be as high as desirable, say, 24:1, or as low as 14:1 because it compresses only air. This is one factor contributing to the high efficiency of the diesel engine.

Gasoline engines are self-speed limiting. Engine speed is controlled by the butterfly valve in the intake manifold, which limits the air supply as well as the air/fuel mixture. Limiting the air/fuel mixture taken in for combustion limits the engine speed.

Diesel engines are not self-speed limiting. Engine air intake for combustion is not limited; the cylinders always have more air than is needed to support combustion. Since the amount of fuel injected into the cylinders controls the engine speed, a diesel engine requires a speed limiter (the governor). A manual control would be near impossible because a diesel engine can accelerate at a rate of more than 2000 revolutions per second. Furthermore the diesel engine requires no ignition system because the fuel is injected as the piston is near the top of its stroke. The fuel vaporizes and ignites as it comes in contact with the hot air which has been compressed by the piston. The injection pump and governor of a diesel engine control the quantity of fuel injected by the fuel nozzle.

Cylinder Block, Crankcase, and Oil Pan The cylinder block and crankcase comprise the framework of a liquid-cooled diesel engine. It is generally a single unit made from cast iron. The air-cooled diesel engine usually has a separate cast-iron crankcase and individual cylinder blocks. The cylinder block has openings for the cylinder sleeve (cylinder liner) and oil and water passages, and bores for the crankshaft and camshaft bearings.

The upper half of the cylinder block contains the water jackets. The lower half of the cylinder block where the crankshaft, camshaft followers, and pushrod are located is called the *crankcase*. An oil pan, which is bolted to the crankcase forms the oil reservoir for the lubrication system.

1

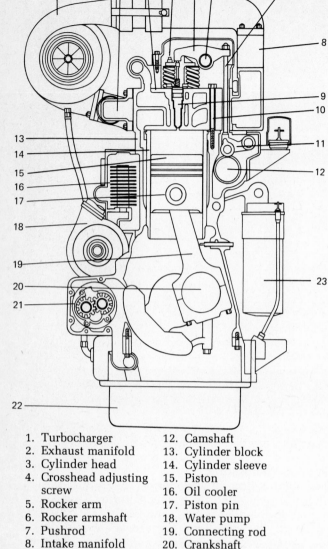

1. Turbocharger
2. Exhaust manifold
3. Cylinder head
4. Crosshead adjusting screw
5. Rocker arm
6. Rocker armshaft
7. Pushrod
8. Intake manifold
9. Injector
10. Cylinder-head bolt
11. Cam follower
12. Camshaft
13. Cylinder block
14. Cylinder sleeve
15. Piston
16. Oil cooler
17. Piston pin
18. Water pump
19. Connecting rod
20. Crankshaft
21. Oil plunger
22. Oil pan
23. Oil filter

Fig. 1–1 Sectional view of a liquid-cooled engine.

Piston and Piston Rings The piston with piston rings acts as a piston pump while moving up and down in the cylinder sleeve. Pistons are made from aluminum or cast-iron alloy. Piston rings are made from cast-iron alloy, and compression rings are commonly chrome-plated.

The two main functions of the piston and piston rings are to seal the lower side of the combustion chamber and to transmit the pressure of the combustion gases via the piston pin and connection rod to the crankshaft.

Connecting Rod The connecting rod is made from drop-forged, heat-treated steel and is the link between the crank and the piston. It has a bore on each end. The upper bearing bore (*piston-pin bore*) has a bushing pressed into it in which the piston pin oscillates. The lower bearing bore (*crankpin bore*), is split in half by the connecting-rod cap. The two

halves of the connecting-rod bearings fit tightly into the rod cap and connecting rod. When the connecting rod is bolted to the connecting-rod journal, it follows a rotational pattern, causing the connecting rod to move up and down.

Cylinder Sleeve The cylinder sleeve (cylinder liner) forms the combustion chamber walls. When the cylinder sleeve is in direct contact with the coolant it is referred to as *wet sleeve*. When the cylinder sleeve is indirectly in contact with the coolant (that is, the sleeve is enclosed in a cylinder), it is referred to as *dry sleeve*. In either case, it is through the contact of the cylinder sleeve with the coolant or cylinder block that efficient cooling is achieved. Wet sleeves have special sleeve seals that seal the coolant at the lower end of the cylinder sleeve and block. The accurately machined surface of the liner flange, cylinder block, and cylinder-head gaskets create a seal at the block surface.

Crankshaft The crankshaft, made of forged steel, has accurately machined and hardened main bearings and connecting-rod journals (Fig. 1–2). The offset cranks of the crankshaft are balanced for proper weight distribution to ensure even force during rotation. Some crankshafts use counterbalance weights (or a gear train) to achieve balancing.

The shaft rotates in its main bearings and lubricating oil from the drilled passages within the crankcase feeds the main bearing. Drilled passages in the crankshaft pass lubricating oil to the connecting-rod journals. A thrust bearing is used to prevent excessive end movement.

Flywheel The flywheel serves three purposes. First, through its inertia, it reduces vibration by smoothing out the power stroke of the cylinders. Second, it is the mounting surface of the clutch pressure plate and the friction surface for the clutch. (When a fluid clutch is used, the impeller is splined or bolted to the flywheel.) Third, the "shrunk on" flywheel ring gear is used for transmitting the power of the cranking motor to the crankshaft.

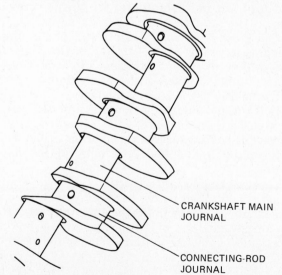

CRANKSHAFT MAIN JOURNAL

CONNECTING-ROD JOURNAL

Fig. 1–2 Typical crankshaft.

Vibration Damper (Balancer) A vibration damper is a unit which counteracts the twisting or torsional vibration caused by force variations (usually from about 3 to 10 tons [2724 to 9080 kg (kilograms)]) on the piston and subsequently the crank. Torsional vibration is an oscillation (rhythm) which occurs within every power stroke. The application of force, and its removal a split second later, cause the crank to be alternately twisted out of alignment and snapped back into place. If a preventive measure is not taken against this action, the engine will run rough and the crankshaft may crystallize and break.

Vibration dampers of the viscous or rubber-element design are fastened to the front of the crankshaft (Fig. 1–3). Since torsion vibration differs with engine design, vibration dampers are constructed to suit specific engines.

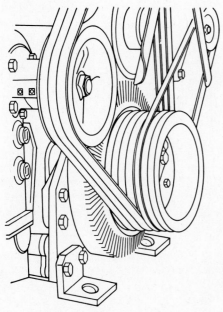

Fig. 1–3 Pulley and vibration-damper mounting.

Cylinder Head and Valves The cylinder head is cast as a one-piece unit. It is the upper sealing surface of the combustion chamber. It may serve one, two, or even six cylinders. The valve guides, which direct the valve stem during the opening and closing of the valve, are pressed into the cylinder head (see Fig. 1–4). The intake valve and seat control the entry of air into the combustion chamber via the intake manifold. The exhaust valve and seat control and release the combustion pressure in the combustion chamber.

Timing Gears, Camshaft, and Valve Mechanism The timing gears (see Fig. 1–5) transmit rotary motion to the camshaft. The camshaft with the cam lobes then rotates on friction-type bearings in the crankcase and converts the rotary motion to reciprocating motion. This motion is transmitted to the follower, pushrod, rocker arm, and valves. On some engines, the camshaft is located above the valve stem. In that case the cam lobes open and close the valves either directly or indirectly.

Timing-Gear Cover and Valve Cover The timing-gear cover encloses the gear train and seals the crankshaft and sometimes the external drive shafts. On some engines it has bearings or support shafts for the timing gear, idler gear, and fuel-injection-pump drive gear. The valve cover encloses the upper part of the cylinder head and the valve mechanism.

Gaskets and Seals Gaskets and seals are used to seal engine components which are bolted to the subframe, the cylinder block, and the crankcase.

Diesel-Engine Systems To operate, a diesel engine requires four supporting systems: the cooling system, the lubrication system, the fuel-injection system, and the air-intake system. The function of each system is equally important to the engine as a whole.

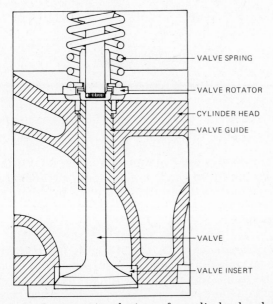

Fig. 1–4 Cross-sectional view of a cylinder head and valve.

VALVE SPRING

VALVE ROTATOR

CYLINDER HEAD

VALVE GUIDE

VALVE

VALVE INSERT

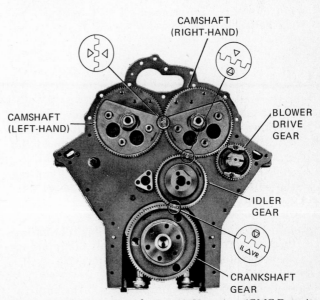

CAMSHAFT (RIGHT-HAND)

CAMSHAFT (LEFT-HAND)

BLOWER DRIVE GEAR

IDLER GEAR

CRANKSHAFT GEAR

Fig. 1–5 Timing gears of a Detroit V engine. (GMC Detroit Diesel Allison Div.)

4

1. Cylinder head
2. Aftercooler
3. Water-temperature regulator housing
4. Coolant outlet from radiator
5. Air compressor
6. Bypass water line
7. Coolant outlet to aftercooler
8. Water manifold
9. Coolant outlet to water manifold
10. Cylinder block
11. Oil cooler
12. Coolant inlet to water pump
13. Water pump
14. Shunt line
15. Radiator
16. Vent line
17. Upper chamber
18. Bleed tube
19. Lower chamber

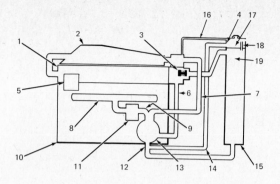

Fig. 1–6 Cooling-system flow diagram. *(Caterpillar Tractor Co.)*

1. Injector
2. Rocker lever
3. Rocker-lever shaft
4. Push tube
5. Cam follower
6. Cam-follower shaft
7. Camshaft
8. Lubricating oil filters
9. Main oil passage
10. Piston-cooling oil passage
11. Piston-cooling nozzle
12. Connecting rod
13. Piston pin
14. Crankshaft
15. Oil flow to gear train

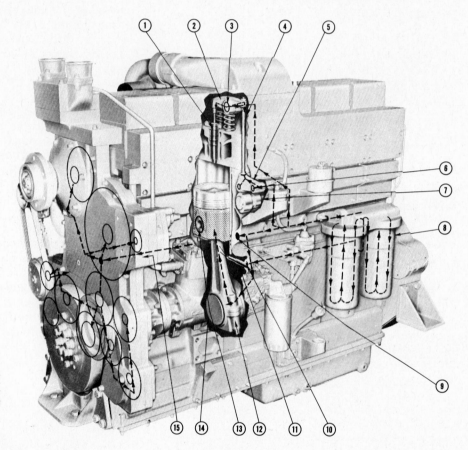

Fig. 1–7 Typical pressure lubrication system (Cummins K model). (Internal arrows show the oil flow.) *(Cummins Engine Co., Inc.)*

Cooling-System Components The water pump, in conjunction with the thermostat, internal cooling passages in the cylinder block and cylinder head, the radiator, and the fan, is responsible for maintaining an even cooling temperature of about 190°F (degrees Fahrenheit) [88°C (degrees Celsius)]. See Fig. 1–6.

Lubrication-System Components The oil pump, through the internal passages, supplies lubricating oil to the bearings, gears, and other components which need lubrication and cooling (see Fig. 1–7).

Fuel-Injection-System Components The fuel-injection pump and injectors are responsible for supplying and injecting the required amount of fuel to

each cylinder at the right time. Fuel filters are necessary to keep the fuel clean.

Air-Intake-System Components The air cleaner, intake manifold, and sometimes a turbocharger and/or aftercooler supply clean and cool air to the cylinders, supply air for scavenging, and reduce the airflow noise.

Other Components A diesel engine has many additional components, and all of these will be covered in later units. Before attempting to analyze individual components, however, you should have at least a superficial knowledge of the mechanical terminology that relates to your job as a diesel mechanic.

Engine Performance Terminology

Bore and Stroke Bore and stroke refer to the cylinder size of an engine. The *bore* is the diameter of the cylinder. The *stroke* is the distance that the piston travels from bottom dead center (BDC) to top dead center (TDC).

Engine Displacement Engine displacement is the volume that is displaced by all the pistons in making one upward stroke each. To find the engine displacement, multiply the volume of one cylinder by the number of engine cylinders:

$$R^2 \times \pi \times \text{stroke} \times \text{no. of cylinders}$$

$$= \text{engine displacement}$$

Suppose that the bore is 3.500 in (inch) (bore radius R is 1.750 in), that the stroke is 4.000 in, and that there are four cylinders. The engine displacement is $1.750^2 \times 3.14 \times 4.000 \times 4 = 153.86$ in³ (cubic inches).

In the metric system of measurement, engine displacement is expressed in liters (l). 1 l = 61.02 in³. The engine in our example, therefore, has a displacement of 2.521 l (153.86/61.02 = 2.521).

Ratio Ratio is the relationship, or proportion, that one number bears to another. For example, if the distance from the fulcrum to the lever on one side is three times greater than the distance from the fulcrum to the lever on the other side, then the ratio is 3:1. Likewise, when one gear has three times as many teeth as another, the ratio is 3:1.

Compression Ratio and Clearance Volume The amount of compression developed by a cylinder is expressed as the *compression ratio*. The volume of this ratio is determined to a large extent by the efficiency with which the engine will convert the heat energy contained in the fuel to useful mechanical energy. The compression ratio compares the cylinder volume when the piston is at the bottom of its stroke (BDC) and the volume when the piston is at the top of its stroke (TDC) (Fig. 2–1).

The volume to which the air is compressed is the *clearance volume* (the space remaining in the combustion chamber). Clearance volume in the combustion chamber is almost impossible to calculate because it is usually irregular in shape. A common method of measuring clearance volume is to fill the combustion chamber with oil. This can done by filling a measuring glass (known as a *graduate*) with oil and recording the indicated cubic centimeters (cm³). To make this measurement, remove the injector and bring the piston to precisely top dead cen-

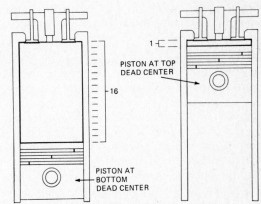

Fig. 2–1 Compression ratio. Here, it is 16:1.

ter. Pour into the combustion chamber the amount of oil required to fill the space.

Let us assume your graduate has a capacity of 100 cm³ and after pouring the oil into the combustion chamber, 58.525 cm³ still remains in the glass. It is evident that 41.475 cm³ (100 − 58.525 cm³) of oil is required to fill the combustion chamber. By using a conversion scale to convert this volume (41.475 cm³) to its equivalent expressed in cubic inches (in³), it is only necessary to multiply 41.475 by 0.061. This will equal 2.53 in³.

To calculate compression ratio, you need to know the displacement volume, the clearance volume, and the following formula:

Compression ratio =

$$\frac{\text{displacement volume} + \text{clearance volume}}{\text{clearance volume}}$$

Suppose that the engine displacement for a four-cylinder engine is 153.86 in³. The displacement volume of one cylinder is then 153.86 divided by 4, or 38.465 in³. If you use the clearance volume as calculated above (2.53 in³), the compression ratio is (38.465 + 2.53)/2.53 = 16.2:1.

Hydraulics The word *hydraulic* comes from the Greek words *hydros*, meaning water, and *aulis*, meaning tube or pipe. Originally the study of hydraulics included only the behavior of water at rest and in motion. Now, hydraulics covers the behavior of all liquids.

All liquids, such as fuel, oil, and water, possess characteristics of both solids and gases. The molecules of liquids are flexible yet rigid enough to maintain the fluid distance between each; that is, they can change shape, divide into parts, and reunite. In other words, given certain conditions, liquids will move.

How this relates to diesel engines is apparent to the mechanic who must understand fluids and fluid flow in order to service the various hydraulic systems—the fuel, lubricating, and cooling systems.

Many factors affect hydraulic action, so it is important for the diesel mechanic to understand such concepts as force, pressure, specific gravity, atmospheric pressure, work, and power (horsepower), and to apply these concepts to the behavior of fluids.

Force   When you use muscles to push or pull things you are using force. Force, however, is not always correlated with motion. For instance, if two equal and opposite forces act on a body, there is no movement; or if there is no resistance, there is also no force. Weight is a force caused by gravity (gravitational force) acting on any object. For example, if an object is weighed on a scale at sea level, the scale will register more than it would if the same object were weighed on the same scale at the top of a mountain. (The standard unit of measurement for force is pounds or ounces [kilograms or grams]).

Pressure   Pressure P is defined as force per unit area. For example, you can easily hold a crowbar in your hand, but if you try to balance it with the point on your hand, you will find the experience a painful one. The force on your hand is the same in both cases because it is equal to the weight of the crowbar. The difference in how it feels is caused by the pressure. When calculating the force of a unit area 1 in^2 (square inch) or 1 cm^2 (square centimeter), you are calculating the pressure, that is, the pounds per square inch (psi) [or the kilograms per square centimeter (kg/cm^2)]. The formula to calculate pressure is area divided into force equals pressure, or $F/A = P$.

Specific Gravity   If a stone is thrown into a lake it will sink. It sinks because it is heavier than the water. Specific gravity is a comparison of the weight of an object with the weight of an equal volume of water. To compute specific gravity, divide the weight of an object by the weight of an equal volume of water. Because it is a ratio, specific gravity has no units.

Atmospheric Pressure   The weight of the atmosphere also plays an important role in hydraulic systems. Its applied force acting over unit area introduces pressure. If a column of air 1 in^2 extends upward as high as the atmosphere, this column of air would weigh, on the average, about 14.7 lb (pounds) at sea level, or 1 atmosphere (atm). This means that at sea level the force of atmospheric pressure on 1 ft^2 (square foot) would be well over a ton. At a higher elevation, the pressure is less.

When listening to a weather report you may hear, "barometric reading is 29.9 inches and rising." When the barometer reads 29.9 in [75.9 cm], the air is exerting a pressure of 14.7 psi [1.03 kg/cm^2] on the earth's surface. In other words, 14.7 psi of atmospheric pressure forces the mercury 29.9 in up a tube. Changes in the weight of the surrounding air (air pressure) will change the barometric reading.

Work   Work has a technical meaning as well as its common meaning of manual labor. In each case, however, a force is exerted which produces motion. Physicists interpret work as force acting through distance. The work done is expressed in foot-pounds (ft·lb) [meter-kilograms (m·kg)]. The formula is force times distance equals work, or $F \times D = W$.

Torque   Torque is the twisting effort applied to a crank at a 90° angle. The twisting effort on the shaft is equal to the distance from the center of the shaft to the point at which the force is applied. The standard unit of measurement for torque is the pound-foot [lb·ft] or pound-inch [lb·in] [kilogram-meter (kg·m)].

Torque can only produce work when the applied torque is greater than the resistance. Torque can be increased by increasing the distance, the weight, or both.

Power (Horsepower)   Power is the work done in a given time, or the rate of doing work. Power may be expressed in terms of work per minute or work per second. However, in the English system of measurement it is expressed in horsepower (hp). This term was first used by James Watt as he tried to compare the power of a steam engine with the power of a horse. He found that a horse could lift a 150-lb load $3\frac{2}{3}$ ft/s (feet per second) doing 550 ft·lb of work every second, or 33,000 ft·lb of work every minute. From these observations came the horsepower (hp) unit, the unit of power in the English system of measurement.

In the metric system horsepower is measured in watts (W) or in kilogram-meters per second (kg·m/s). (1 hp = 745.7 W = 76.04 kg·m/s.)

Indicated Horsepower   Engines are rated according to their power output and torque. Indicated horsepower (ihp) is the power transmitted to the pistons by the gas in the cylinders. When designing an engine, designers mathematically calculate the indicated horsepower. This calculation is about 15 percent more than the actual horsepower at the flywheel. Some latitude is allowed when calculating the dimension (the length, height, and width) of an engine. In order to determine the dimension of an engine, and thereby its mean effective pressure (mep), designers must consider: the length of the stroke, the rotation speed and the cross-sectional area of the cylinder, the number of cylinders, the compression ratio, and the volumetric efficiency.

Every factor is easy to calculate or to measure except the mean effective pressure because the pressure varies throughout the power stroke. Therefore, the effective pressure has to be measured at several points throughout the power stroke and then calculated to the mean effective pressure. As an example: 10 measurements were taken, totaling 1200 psi. This would give you the mep of 1200/10 = 120 psi. On a naturally aspirated engine the pressure difference is between 95 and 140 psi, and it is up to 300 psi on a turbocharged engine.

The formula used to calculate ihp for a four-cycle engine is

$$ihp = \frac{R^2 \times \pi \times L/12 \times N \times P \times n}{33,000}$$

where $R^2 \times \pi$ = area of a cylinder, in^2
$L/12$ = length of stroke, in
N = engine speed, rpm
P = mean effective pressure, psi
n = number of cylinders

For a two-cycle engine the formula is

$$ihp = \frac{\left(\dfrac{R^2 \times \pi}{4}\right) \times L/12 \times N \times P \times n}{33,000} \times 2$$

Brake horsepower (bhp) is the rated horsepower not the indicated horsepower, and it refers to the mechanical efficiency of the engine. Brake horsepower is the usable power of the engine.

Mechanical efficiency is the difference between indicated horsepower and brake horsepower. This difference is due to the friction within the engine and its associated components. The mechanical efficiency of a four-cycle engine is about 82 to 90 percent. This is slightly lower than the efficiency of the two-cycle engine.

The piston and its rings, the pumping of air in and out of the cylinders, the friction of the bearing surface, and the weight of the moving components are the major contributors to decreased mechanical efficiency. The water pump, oil pump, and fuel-injection pump, can also add measurably to mechanical efficiency or inefficiency.

A *dynamometer* is an instrument which measures power or torque of an engine by means of a mechanical, hydraulic, or electrical device. When the dynamometer is scaled in pound-feet (lb·ft), the following formula is used to arrive at brake horsepower:

$$bhp = \frac{2\pi \times F \times R \times N}{33,000} = \frac{F \times R \times N}{5252}$$

where F = force, lb
R = brake arm radius, ft
N = engine speed, rpm

When the dynamometer used is a generator, the scale is in watts (W) or in kilowatts (kW). Since 1 hp = 746 W, the following formula is used to arrive at horsepower:

$$hp = \frac{total\ electrical\ watt\ power}{746}$$

Heat Heat is a form of energy which can raise the temperature of a body, substance, or any material. It may be generated by combustion, friction, chemical action, radiation, conduction, or convection, and it can be converted into other forms of energy. The total absence of heat is known as *absolute zero* and corresponds to $-459.6\,°F$ [$-273\,°C$].

EFFECT OF HEAT As the temperature changes, a substance will expand or contract. Some substances will expand to their limit, vaporize and then burn, while others will just vaporize. When heated or cooled beyond the point where their molecular structure is stable, some substances will destroy themselves; others actually change their state.

Temperature Temperature refers to the degree of heat in a substance as measured by a thermometer. Temperature is measured in degrees Fahrenheit (°F) or, in the metric system, degrees Celsius (°C). It is essential that you understand the difference between the Fahrenheit and Celsius (or centigrade) scales. Both are based on the freezing and boiling points of water at a pressure of 14.7 psi (1 atm), but the number of graduations between these points differs. The formula for converting degrees Fahrenheit to degrees Celsius is $°C = 5/9\ (°F - 32)$.

The temperature ranges to be measured on diesel engines vary from ambient temperature to exhaust temperature. (Ambient temperature is the temperature of the surrounding air. Exhaust temperature is the temperature which exists at the exhaust manifold.) Thermometers are therefore made of materials having different properties which change as temperature changes. The most commonly used thermometers are thermocouples and liquid tube thermometers.

Thermocouple A thermocouple is a pair of dissimilar metals joined so as to produce a thermoelectric effect when the contact surfaces are at different temperatures. The thermocouple illustrated in Fig. 2–2 shows two twisted wires made from dissimilar materials connected to a galvanometer. A *galvanometer* is an instrument which detects and measures a small electric current. As the heat of one junction exceeds that of the other, a small current will flow. The voltage of this current will be indicated by the galvanometer. The scale used, however, can be calibrated in degrees thus permitting the needle to directly show the degree of temperature.

The amount of voltage produced depends on the difference in temperature between the opposite ends of the junction. The greater the difference in temperature, the higher the voltage.

Liquid Tube Thermometer The common liquid thermometer consists of a graduated glass capillary tube or stem with a bulb containing mercury or an-

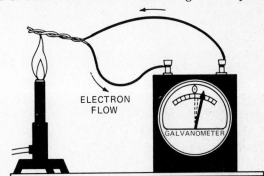

Fig. 2–2 Thermocouple principle. (*Ford Motor Company of Canada, Ltd.*)

other type of liquid which will expand or contract as temperature rises or falls.

Expansion Ratio   Expansion ratio is the ratio of the total volume at the end of the power stroke to the total volume at the beginning of the power stroke. An increase of the expansion ratio depends on the following factors: a greater volumetric efficiency, a higher compression ratio, a higher air swirl, a uniform size and distribution of fuel droplets, a higher fuel volatility, and the bore and stroke ratio.

In order to make any comparisons, you must know the heat value of the fuel and the formula to determine thermal efficiency (see the following section, British Thermal Unit).

British Thermal Unit   The British thermal unit (Btu) is used to measure the heat value of a fuel, or the amount of heat transferred from one object to another. It is derived from the amount of energy required to heat 1 lb of water 1°F at a barometric pressure of 14.7 psi.

Suppose that an engine on an engine dynamometer uses 4 gal (gallons) of fuel in 1 hour. [During this hour the continuous engine output (brake horsepower) is 100 hp, or 100 hp • h (horsepower-hours).] Also, let us assume that the heat value of 1 gal (U.S.) of fuel is 135,000 Btu. We can say, therefore, that the engine uses 4 × 135,000, or 540,000 Btu in 1 hour.

Thermal efficiency is the total efficiency of an engine compared with the heat value of the fuel and the brake horsepower in British thermal units. If we continue with the example cited above, since 1 hp equals 2547 Btu, then 100 hp equal 254,700 Btu. Thermal efficiency may then be calculated as 254,700/540,000 = 47.16 percent.

Service manuals show fuel consumption in pounds per brake horsepower (lb/bhp). In the case illustrated above this would be the weight of fuel per gallon multiplied by the number of gallons used, divided by the brake horsepower. Since 1 gal (U.S.) of fuel weighs 6.8 lb, from the example above, (6.8 × 4/100) = 0.272 lb/bhp.

Boyle's Law and Charles' Law   Boyle's law states that the pressure exerted by a body of gas is inversely proportional to its volume if the temperature is held constant. In effect, when the volume is halved, the pressure will be doubled.

Charles' law relates to volume and temperature. It states that the volume of a gas is directly proportional to its absolute temperature if the pressure remains constant. The compression of a gas causes it to become heated, thereby causing an increase in pressure. Conversely, expansion causes cooling and a reduction in pressure. Why are these laws important to a diesel mechanic? If you understand them they will help you to understand what takes place inside a cylinder when the air is being compressed, and what forces operate against the diesel engine components. Such general information points out the importance of bearings, crankshaft, connecting rod, pistons and rings, sleeves, valves, and cylinder-head gaskets as they relate to proper fit, clearance, alignment, and torque.

Questions

1. What is the engine displacement of a six-cylinder engine when the bore is 4 in and the stroke is 5 in?

2. What is the compression ratio of the engine in Question 1 when the clearance volume is 5.5 in³?

3. Define *brake horsepower.*

4. Define *thermal efficiency.*

5. Convert the exhaust temperature of 1100°F to Celsius.

6. Convert the coolant temperature of 112°C to Fahrenheit.

7. What is the brake horsepower of an engine when it develops 650 lb-ft at 1825 rpm in horsepower? What is the bhp reading in watts?

8. What is the indicated horsepower of a six-cylinder four-cycle engine that has a bore of 5 in, a stroke of 5.6 in, an engine speed of 2300 rpm, and an mep of 140 psi?

9. What is the thermal efficiency of an engine being tested on a dynamometer which has a fuel consumption of 97.6 lb/h, an engine output of 225 bhp during the hour tested, and a fuel heat value of 19,800 Btu/lb?

Cycle Operation

Cycle The word *cycle* refers to a series of events that repeat themselves. Cycle, in relation to a four-cycle engine indicates that it requires four strokes (intake, compression, power, and exhaust) and therefore 720° of crankshaft rotation (two full crankshaft rotations) to accomplish one working cycle (Fig. 3–1). With a two-cycle engine the entire event (that is, power, exhaust, intake, and compression) requires only two strokes and it is accomplished with one crankshaft rotation (360°) (Fig. 3–2).

HOW A FOUR-CYCLE DIESEL ENGINE OPERATES

Assume you have a four-cycle, four-cylinder diesel engine with a bore of 3.5 in (inches)[88.9 mm (millimeters)], a stroke of 4 in [101.6 mm], and a compression ratio of 16:1. The firing order is 1-4-3-2, and the valve timing is as shown in Fig. 3-3.

Intake Stroke As you (1) press the starter button or turn the key switch to start the rotation of the electric

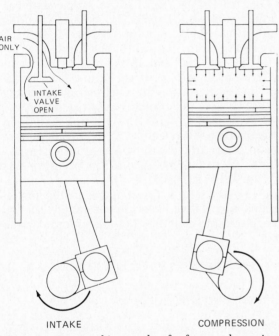
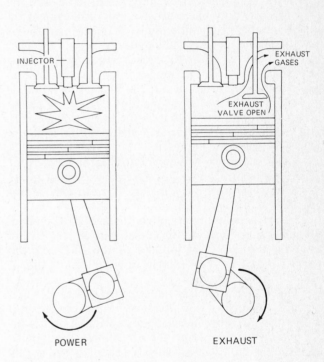

Fig. 3–1 One working cycle of a four-cycle engine.

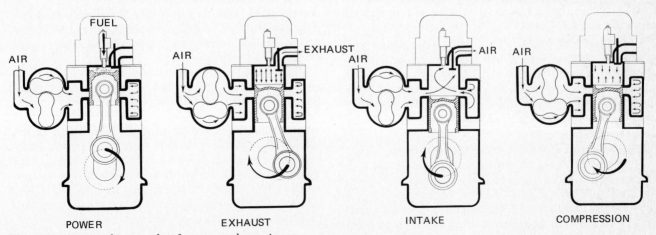

Fig. 3–2 One working cycle of a two-cycle engine.

cranking motor, or (2) pull the control valve of a hydro starter, or (3) activate the air valve of an air starter, the pinion of the cranking motor engages the ring gear and turns the crankshaft.

As the No. 1 and No. 3 pistons are moved upward, and No. 2 and No. 4 pistons are moved downward by the rotation of the crank and connecting rods, the timing gear on the crankshaft turns the timing gear on the camshaft at a ratio of 2:1; that is, the crankshaft turns twice for every single revolution of the camshaft.

Using only the No. 1 cylinder, let us follow the cycles from beginning to end. As piston No. 1 nears top dead center (TDC), precisely 28° before top dead center (BTDC) (Fig. 3–3a), the camshaft lobe starts to lift the follower of the intake valve. The pushrod moves upward and pivots the rocker arm on the rocker-arm shaft. As the valve lash is taken up, the rocker arm pushes the valve inward, opening the combustion chamber to atmospheric pressure.

At this time the exhaust valve is still open and the exhaust gas flow has created a low pressure within the combustion chamber and the intake manifold. Atmospheric air enters the cylinder.

At about 12° after top dead center (ATDC), the exhaust valve follower moves from the cam nose to the closing flank and then to the base circle which is 23° ATDC. At the same time the valve spring closes the valve and transmits its force to the rocker arm, pushrod, and cam follower to maintain the follower on the camshaft cam.

A 51° duration (valve overlapping) is needed for this engine design to force out the exhaust gases (scavenging) and draw in the oxygen-rich atmosphere.

The efficiency of an engine depends on the effectiveness of the scavenging and air intake. (Note that during the 51° duration the piston has traveled about 0.310 in [7.874 mm] upward and 0.250 in [6.350 mm] downward.

As the piston reaches bottom dead center (BDC), it creates no further low pressure. However, air still enters the cylinder because the restriction on the intake side has increased the air velocity, causing a pressure drop. The intake valve is not closed at this point in order to allow the full amount of air to enter the cylinder.

NOTE The piston travel in relation to crankshaft rotation is very small.

Volumetric Efficiency This is the ratio of the volume of atmospheric air at standard temperature (60°F) and pressure (14.7) that actually enters the cylinder compared with the volume that could theoretically enter the cylinder. volumetric efficiency = $\dfrac{\text{volume entering (60°F 14.7 psi)}}{\text{piston displacement}} \times 100 = \%$

Compression Stroke At 35° after bottom dead center (ABDC) (Fig. 3–3b) the intake valve starts to close in a manner similar to that of the exhaust valve.

At 43° ABDC (or 137° BTDC) the intake valve is on its seat; the piston at this point has traveled about 0.250 in [6.350 mm] upward. A pressure of 14.7 psi [1.033 kg/cm²] at a temperature of 80°F [26.7°C] air is trapped in the cylinder (see Table 3–1).

Table 3–1 IDEAL TEMPERATURE AND PRESSURE IN RELATION TO PISTON TRAVEL

Piston position, °BTDC	Distance piston travels, in [mm]	Temperature, °F [°C]	Pressure, psi [kg/cm²]
180	BDC	80 [26.7]	14.7 [1.033]
137	0.250 [6.350]	80 [26.7]	14.7 [1.033]
70	2.125 [53.975] *	160 [71]	34.2 [2.404]
43	3.062 [77.775] *	360 [182]	85.2 [5.989]
28	3.530 [89.662] *	720 [382]	277.1 [19.48]
20	3.757 [95.428] *	1280 [694]	742.0 [52.16]
0	TDC		

* Volume in the cylinder is halved.

At about 70° before top dead center (BTDC) the piston has traveled 2.125 in [53.975 mm], or about half of its stroke. This action reduces the volume in the cylinder by half. In accordance with gas laws, the temperature has now doubled to 160°F [71°C] and the pressure has reached 34.2 psi [2.404 kg/cm²].

At about 43° BTDC the piston has traveled upward 3.062 in [77.775 mm] of its stroke and has halved the volume once again. The result is that the temperature has doubled to 360°F [182°C] and the pressure has risen to 85.2 psi [5.989 kg/cm²].

As the piston has traveled upward 3.530 in [89.662 mm] of its stroke, it has halved the volume again. The temperature doubles again, reaching 720°F [382°C] and the pressure has reached 277.1 psi [19.48 kg/cm²]. When the volume is halved again the piston has traveled 3.757 in [95.428 mm] of its stroke. At this point the temperature has climbed to 1280°F [694°C] and the pressure has reached 742.0 psi [52.16 kg/cm²]. This is approximately 7135 lb [3232.1 kg] of force on the piston because the area of the piston in your engine is 9.616 in². The formula to calculate pounds of force on the piston is

$$R^2 \times \pi \times psi = 1.750 \times 1.750 \times 3.14 \times 742 = 7135.26 \text{ lb}$$

where R = radius of bore, in²

These calculations would be true if, during the compression stroke, heat were not lost to the pistons, cylinder sleeves, and cylinder head. Furthermore, heat is reduced due to lower compression pressure. The compression pressure is lower than calculated because of engine breathing inefficiency, inefficient piston sealing as well as lower temperature caused by the heat losses.

In actual operation the engine reaches a compression pressure of 410 psi [28.82 kg/cm²] and a combustion pressure of 925 psi [64.827 kg/cm²] and a combustion temperature of 2410°F [1280°C]. This means that the force previously calculated is exceeded by 1759 lb [797.706 kg]. The discrepancy between the actual fact and the calculated figures should be borne in mind when servicing components. The *actual force* on the pistons, wrist pin, connecting rod, bearings, crankshaft, crankcase, and cylinder head should not be overlooked.

NOTE Some of the major causes of cooling-system inefficiency are: loss of coolant, inadequate coolant, worn water pump, loose drive belts, restricted radiator and loss in coolant pressure. Malfunctions which cause increased temperature and pressure will destroy the molecular structure of the piston rings, piston, and piston sleeve. It is essential, therefore, that the maintenance and service of cooling-system components be equal to that of the engine components.

Combustion Fuel in a liquid state is injected into the cylinder at a precise rate to ensure that the combustion pressure is forced on the piston neither too early nor too late, that is, between 10 and 15° ATDC. The fuel enters the cylinder where the heated compressed air is present.

Fuel will burn only when it is in a vaporized state (attained through addition of heat) and intimately mixed with a supply of oxygen. All these condtions are present in the cylinder. When the first minute droplet of fuel enters the combustion chamber, it is

quickly surrounded by its own vapor because the compression temperature at this point is about 650°F [343°C]. Heat is withdrawn from the air surrounding the droplet causing it to vaporize. It takes time for the heat to build up again and cause the vapor to ignite since the core of the droplet is still liquid and relatively cold. Once ignition has started and a flame is present, the heat required for continuous vaporization is supplied from that released by combustion and the higher compression temperature. The liquid droplet, surrounded by its own vapor, burns as fast as fresh oxygen is supplied. This process continues unchanged until the fuel oil is burned or the oxygen is used up.

Injection starts at 28° BTDC and ends at 3° ATDC (Fig. 3–3c). The fuel therefore is injected over a period of 31°. The piston progresses upward and causes a rise in temperature. This creates quicker ignition (almost as the droplets leave the fuel nozzle) and instantly increases combustion pressure. At the

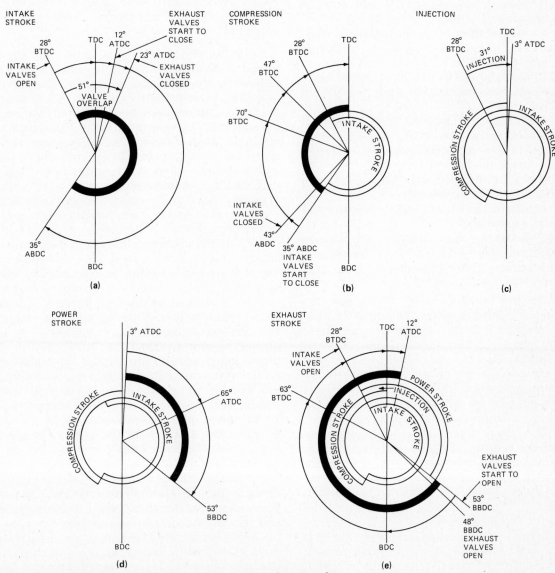

Fig. 3–3 Valve-timing diagram of a four-cycle engine, showing the progressive steps in one working cycle. (a) Intake stroke, (b) compression stroke, (c) injection, (d) power stroke, (e) exhaust stroke.

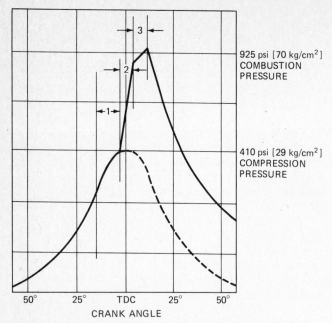

925 psi [70 kg/cm²] COMBUSTION PRESSURE

410 psi [29 kg/cm²] COMPRESSION PRESSURE

CRANK ANGLE

Fig. 3–4 The three phases in the combustion process.

end of injection, the oxygen is reduced and the process slows down.

Figure 3–4 illustrates the three phases of the combustion process. These overlap during injection and complete combustion.

The first phase shown (1 in Fig. 3–4) is the delay period in which some fuel has been injected but not yet ignited.

The second phase (2) shows the period where ignition starts. There is a rapid rise in temperature and pressure since the droplets have already started to vaporize during the first phase.

The third phase (3 in Fig. 3–4) is the period where the last fuel enters the combustion chamber. The fuel burns almost instantly and a rise in pressure and temperature is noticeable.

Some engines have an undesirable fourth phase in which the unburned fuel mixes with the unused oxygen and burns. Other engines may have only two phases.

Factors Affecting Delay Periods Phase one, the delay period, is needed to achieve complete combustion. However, it must be as short as possible to control the burning at the second phase. A short delay period results in a controlled pressure and temperature change which gives a smooth, efficiently running engine. Because of many variable factors, it is not easy for a designer to reduce the delay period and maintain constant speed and power.

The ignition delay period depends, among other things, on the motion of air in the combustion chamber at the time of injection. If the air were motionless, only a certain number of fuel droplets would find the oxygen they need for combustion. On the other hand, if the droplets and air traveled the same velocity, the droplets would also lack oxygen and be smothered. A controlled airflow (air swirl) is needed to bring a steady supply of fresh air to each vaporized or burning droplet. The means by which engine manufacturers have achieved a controlled air swirl is outlined in Combustion Chamber Design (Unit 4).

Other factors which affect the delay periods are: the droplet size, velocity, injection pressure, and spray pattern. An ideal condition would be to have relatively large droplets (0.1mm) injected at fairly low pressure and equally distributed over the total area of the piston. This would give the second phase of combustion a lower pressure rise and would cause a higher pressure in the third phase. This would result in higher efficiency and a smooth running engine. To date, this ideal condition has not been realized completely. The droplets are not uniform, but are smaller on the outside of the spray cone than on the inside. The smaller droplets have less velocity and soon travel the same speed as the air. They are suffocated by their own combustion. Injection pressure is also dissimilar during injection, therefore, the droplet size varies as does the area which the droplets cover. Manufacturers presently compromise so that the majority of droplets are large enough, uniform, and evenly distributed to give a fairly low-pressure rise in the second phase of combustion.

NOTE When servicing an injector do not forget the importance of the pressure, the hole size, and the spray pattern.

Volatility will also increase or decrease the delay period as will the self-ignition temperature (cetane number) of the fuel (refer to Unit 24, "Diesel Fuel").

A fuel of low volatility and high cetane should be used when the engine speed is high and the load and the ambient temperature are low. This aids in maintaining a constant time in the delay period since the compression temperature under such circumstances would be low. Conversely, with a low speed, high load, and high ambient temperature, a fuel with high volatility and a low cetane number should be used.

Although at first it may seem possible to maintain an even temperature or higher compression temperature by simply increasing the compression ratio, this actually only solves part of the problem. With a higher compression ratio the compression space is very small and dead-end pockets are formed which cannot be reached by the fuel droplets. Obviously these pockets do not utilize the air in the second and third phases of combustion, so there is a power loss during the power (expansion) stroke. Also, with a higher compression ratio, precise tolerances are essential since the clearance between the valves, piston, and piston cylinder head is reduced. For this reason and because a higher compression ratio would lower the volumetric and the mechanical efficiency and also would increase pollution problems, manufacturers keep the compression ratio within a range of 15:1 to 18:1.

To increase the compression temperature various designs of precombustion chambers are used. These chambers store heat, stabilize the temperature, and shorten the delay period. There is very little uncontrolled burning; the heat which is released, proceeds smoothly but rapidly, although relatively late in the cycle. It is important to note that engines with precombustion chambers are low in hydrocarbon (HC) and nitrogen oxide (NO_x) emissions.

Production of Combustion The burning of hydrocarbon fuel is an oxidation process. The speed with which it burns depends on the temperature. At this stage you need not be concerned with the chemical formula involved in this process. Rather, you should concern yourself with the volume and density of the emission from the exhaust stack as it relates to pollution standards. Pollution standards are among the factors which govern the need for an engine tuneup or service.

Power (Expansion) Stroke After the piston passes TDC, heat is rapidly released causing a rise in pressure. This rise in pressure forces the piston downward and increases the force on the crankshaft. During this time (to about 65° ATDC) (Fig. 3–3d) the force on the crank remains nearly constant because the expansion of gases causes a counteracting pressure and temperature loss. The crank, in relation to the force applied, gains advantage because of an increasing angle between the connecting rod and the crankshaft. From that point on the pressure decreases as the temperature of the expanding gases decreases.

Exhaust Stroke At 53° before bottom dead center (BBDC) the cam of the exhaust valve forces the follower upward. At about 48° BBDC the exhaust valve lifts off its seat and exhaust gases enter the exhaust manifold and pass into the atmosphere. (**NOTE** The exhaust temperature changes with engine speed and load.)

After stopping at BDC, the piston moves upward and accelerates to its maximum at 297° or 63° BTDC and from there on the piston decelerates. As the piston speed slows down, the velocity of the gases creates a pressure slightly lower than atmospheric pressure. At 28° BTDC the inlet valve opens. This causes a scavenging effect on the remaining gases and marks the start of a new working cycle.

Relation between Time and Crankshaft Rotation Time is required to move exhaust gas out and air into the cylinder, to compress the air, and to inject and burn the fuel. Although this time is extremely short, each of the events just mentioned must be timed to the crankshaft rotation. The position of the crankshaft governs the movement of the pistons, valves, and injection.

Refer to the piston-travel diagram (Fig. 3–5) to trace the movement of the piston and valve action. Let us take the same engine, for example, which is governed at 2100 rpm high idle. In this case it would mean that in 1 second the crankshaft has turned 35 rotations (12,600°) or that the crank has advanced 1° in 0.00008 second. This leaves only 0.03288 second to remove the exhaust gases and bring fresh air in, and about 0.00248 second to inject the fuel; in other words, the piston has to travel in its stroke a distance of 4 in in about 0.01440 second.

Rotation of Crankshaft and Piston Travel Crankshaft rpm is controlled by the governor and fuel-injection pump. The governor, however, does not control piston travel. The speed of the piston varies from no movement at TDC to its maximum speed at a point before the halfway mark of its intake stroke (about 63°) (see Fig. 3–6). In other words, when the centerline of the connecting rod and centerline of the crank form a 90° angle, the piston speed is at its maximum. Piston speed then decelerates to a complete stop at BDC. With further rotation of the crank, the piston moves upward and accelerates in speed. Acceleration reaches its maximum at about 297°. From this point on, although the piston decelerates, it travels a greater distance (about 1.500 in [38.100 mm]) with respect to the crankshaft degree. (This causes a quick rise in compression pressure.) When the crank and connecting-rod centerline are at a 90° angle, the piston has traveled about 2.515 in [63.881 mm]. This means that at half-stroke the piston has traveled 2 in + 0.515 in.

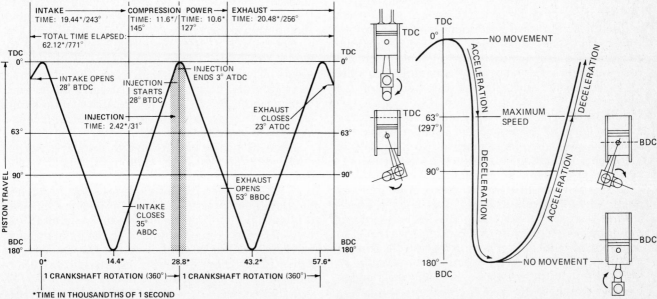

Fig. 3–5 Piston-travel diagram when engine speed is 2100 rpm.

Fig. 3–6 Piston travel related to piston speed and crankshaft angle.

14

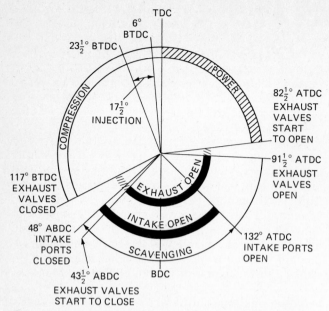

Fig. 3–7 Valve-timing diagram of a two-cycle engine.

Relation between Force and Crankshaft Position When the piston is at TDC with a force applied on the piston and crank, there will be no rotation, but there will be a great force placed on the piston, connecting rod, bearings, crankshaft, and engine crankcase. As the crankshaft rotates to 20° ATDC, 30 percent of the applied force is transmitted to the crank thus causing rotation. At about 63° ATDC the centerline of the connecting rod and crank form a 90° angle giving the greatest torque advantage. As the crank angle decreases, so does the force advantage.

HOW A TWO-CYCLE DIESEL ENGINE OPERATES

The occurrence of intake, compression, expansion (power), and exhaust of a two-cycle diesel engine is confined to two strokes (one complete crankshaft revolution or 360°) (see Fig. 3–7). With each downward movement of the piston there is a power cycle and with each upward piston movement a compression cycle. The intake and exhaust cycle may be considered a unit which is part of the power and compression cycle. It begins after completion of the power stroke and ends after the piston closes off the inlet ports.

About 44 percent (or 160°) of the total working cycle is needed to remove the exhaust gases and bring in fresh air. In comparison to a four-cycle diesel engine this time is relatively short. The four-cycle requires about 57 percent (or 411°) of crankshaft rotation; in other words, nearly three times as much as a two-cycle engine. A two-cycle diesel engine requires an air pump (blower) capable of pumping air at a pressure of about 2 to 7 psi [0.141 to 0.492 kg/cm²] into the cylinder to replace the exhaust gases with fresh air. The volume of air is about 40 times the cylinder volume. You should also remember that on a two-cycle engine the reciprocating components (pistons, connecting rods, and crank-

shaft) are devoted solely to generating power and therefore have less friction losses than they do in the four-cycle engine. The reciprocating components of a four-cycle engine require 411° of a crankshaft rotation to pump exhaust gases out of the cylinders and draw fresh air in. The four-cycle engine gives very little supporting power to scavenging and charging of the cylinders.

You may conclude, when comparing a two-cycle engine with a four-cycle engine having the same displacement, stroke, compression ratio, and rpm, that the two-cycle engine would have twice the horsepower. This is incorrect because the power stroke has to be shortened by about 20 percent and the compression stroke by about 12 percent to gain time for scavenging and charging the cylinder. Also, the blower on the two-cycle engine requires engine power to pump the air into the cylinder. However, the blower of a two-cycle engine increases the volume within the cylinder, and as a result, the temperature and pressure during compression are nearly the same as those of a four-cycle engine. The exhaust temperature, on the other hand, is slightly higher than that of a four-cycle engine because of the shorter expansion stroke.

Although power of a four-cycle engine usually is related to breathing efficiency and the area of the cylinder, the power of a two-cycle engine depends on cylinder volume. (A reminder here—do not confuse power with torque. Torque is governed by the length of the stroke and the time in degrees of the power stroke.)

The stroke of a two-cycle engine is limited, since the lower part of the piston (piston skirt) must cover and seal the inlet port when the piston is at TDC. As a result, the piston of a two-cycle engine must be approximately 10 percent longer than its stroke.

The Two Strokes and Action of a Diesel Engine Let us assume that you have an engine which has been cranked over, started, and is running at 2100 rpm. Let us then follow one of its pistons for one complete revolution (one working cycle); refer to Fig. 3–7 as you read. Do not forget that the crankshaft and camshaft turn at the same speed.

At 82½° ATDC, with the piston near the end of its power stroke, the exhaust cam begins to lift the follower. The valve lash is then taken up and 9° later, the rocker arm forces the exhaust valve off its seat. Exhaust gases escape into the exhaust manifold and cause the cylinder pressure to decrease. This is the first phase of scavenging.

After the piston has traveled three-quarters of its stroke (132° of crankshaft rotation), the piston opens the inlet ports to begin the second phase of scavenging. The blower pressure, being slightly higher than cylinder pressure, creates an airflow through the open ports. Consequently, any remaining exhaust gases are forced out through the open exhaust valve. Detroit diesel engine manufacturers have established for these engines a maximum time allotment of 96° of crankshaft rotation.

At 43½° ABDC the camshaft has rotated sufficiently so that the follower is resting on the closing flank of the cam. Spring pressure forces the exhaust valve to close and at 117½° BTDC the valve is on its seat. In the same period of time, the piston starts to move upward, and at 48° ABDC the piston closes the inlet ports.

NOTE This is 4½° after the exhaust valve starts to close. Also note that the exhaust valves of some engines close before the inlet ports close. Cylinder pressure and volume then increase, as does the power. The engine is then supercharged.

Let us stop for a moment and analyze the foregoing as it relates to service. Consider first the timing of the crankshaft gears to the idler and camshaft gears. If the camshaft gears are out of time, the relationship between exhaust valve and port opening changes. If the ports open too early or too late in relation to the exhaust valves, the air volume and power output are affected. When the exhaust valve lash is too small, the power stroke is reduced. When it is too large, the scavenging phase is reduced, and again a loss in power is noticeable.

COMPRESSION STROKE OF A TWO-CYCLE ENGINE After the exhaust valve is seated, the compression stroke begins. The temperature and pressure rise is nearly the same as that of a four-cycle engine. At 23½° BTDC the injector cam on the camshaft starts to lift its follower and pushrod. As the rocker arm pivots, the injector is forced downward, causing fuel to be injected. After 17½° (6° BTDC) of camshaft rotation, the fuel injection ends. The fuel injection does not end by the cam action of the unit injector. It ends as the helix passes the lower port. (See Unit 31, "Detroit Diesel Fuel-Injection System").

Let us review the beginning of injection in relation to engine performance. If the unit injector is adjusted too high or too low the injection occurs too late or too early, resulting in a loss of power.

POWER STROKE OF A TWO-CYCLE ENGINE Combustion pressure forces the piston downward and at 91½° ATDC the exhaust valve is open and a new working cycle begins.

OTHER FACTORS AFFECTING CYCLE OPERATION

Blowers The process of driving exhaust gases out of the cylinder and replacing them with fresh air is called *scavenging*. Two-cycle engine scavenging is achieved through an air pump (blower). Its effectiveness depends on pumping a capacity of air through the cylinder 30 to 50 times the volume of the cylinder. This excessive volume of air is necessary because the exhaust gases, to some extent, resist motion and therefore mix with the incoming air. However, it also has the beneficial effect of cooling the internal components.

There are many types of air pumps in use but perhaps the most common is the Roots-type blower. Detroit Diesel engines use this blower exclusively. It has the advantage of being a positive displacement

blower with little mechanical friction. It operates reasonably well at low speed, requires no internal lubrication, and therefore delivers oil-free air.

Supercharging Some of the objectives of diesel engine manufacturers are to increase engine power output (hp), increase thermal efficiency, improve reliability, and hold down maintenance costs while keeping within imposed emission standards. These objectives have been met by modifying air motion, fuel spray characteristics, combustion chamber configuration, compression ratio, injection timing, fuel-injection rate, and, by supercharging the engine. (An engine is referred to as *supercharged* when the manifold pressure exceeds atmospheric pressure.)

While at first it may seem feasible to increase the piston diameter and the stroke or to increase the number of cylinders to augment the power, size, and weight of an engine, neither of these changes increase the mean effective pressure (mep) or engine efficiency. It is possible to increase engine speed or piston speed, but this would raise the inertia force. Furthermore, inertia force demands greater strength; therefore, a heavier engine would be necessary. Greater engine speed also increases friction losses, produces lubrication and wear problems, and greatly increases maintenance problems and costs. It also increases the airflow rate. However, there is a limit to which the manifold, valves, and cylinders can take in the airflow without the necessity of either supercharging or redesigning the engine.

The most effective gain in mep, causing a corresponding gain in power output and thermal efficiency, results from supercharging, since this increases the mass of air handled by a given swept volume. Supercharging is considered the most efficient method; experimental engines with an mep of 400 psi [28.12 kg/cm²] have been developed.

Mechanically driven air pumps such as the positive displacement Roots blowers or the nonpositive centrifugal blowers are not commonly used on modern four-stroke diesel engines. This is because Roots blowers are limited to a rotating speed of about 6000 rpm and a pressure of about 20 inHg (inches of mercury) [50.8 cmHg (centimeters of mercury)].

Centrifugal blowers are very efficient and have a relatively high ratio (about 3:1). However, they require a complicated drive arrangement to achieve the compressor speed of 20,000 to 60,000 rpm. Since both types require power from the engine to drive the air pump, the total power output of the engine is reduced. Nearly all modern high-output engines are now supercharged, by using exhaust-gas-driven turbines to drive the compressor (turbocharger).

Engine Heat Balance A diagram of a typical engine heat balance of a two- and a four-cycle engine is illustrated in Fig. 3–8. You will notice that 33 to 45 percent of the total heat of the fuel oil is converted into usable power (bhp) and 28 to 33 percent of the heat is given up during the exhaust stroke. During the periods of combustion, expansion, and exhaust, 27 to 33 percent of the heat, plus the heat generated by friction and the rings, is given up by conduction,

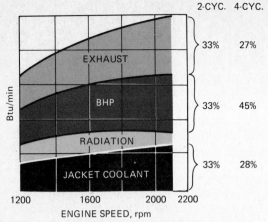

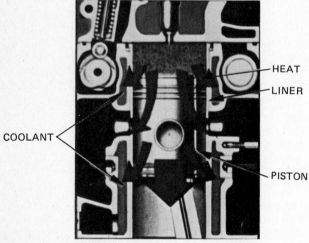

Fig. 3–8 Typical engine heat balance of two- and four-cycle engines. (*GMC Detroit Diesel Allison Div.*)

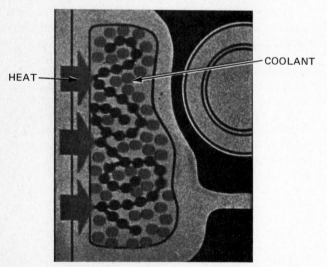

Fig. 3–9 Principle of conduction. (*GMC Detroit Diesel Allison Div.*)

Fig. 3–10 Principle of convection. (*GMC Detroit Diesel Allison Div.*)

convection, and radiation. Let us define these three terms:

- *Conduction* is the transmission of heat through matter without motion of the conducting body (Fig. 3–9).
- *Convection* is the transer of heat from one body to another through a liquid or gas by motion of its parts (Fig. 3–10).
- *Radiation* is a transmission of heat in the absence of a gas, liquid, or physical conductor, and by the energy of molecules and atoms undergoing internal changes.

Note that the heat balance diagram (Fig. 3–8) does not include the heat carried away by the lubricating oil or that given up by radiation or convection through the external walls of the engine or its components.

As a rule, the heat balance figures given relate to engines at full load. On a four-cycle engine at reduced load the flow of heat through the cooling medium is less than that of a two-cycle engine because heat is given up or transferred from one object to another by way of conduction, convection, or radiation (or all three) until both bodies reach an equal temperature.

Questions

1. What is the difference between a two-cycle and a four-cycle engine?

2. Draw a valve-timing diagram patterned after Fig. 3–7 for an engine which has the following specifications: intake valve opens at 36°, intake valve closes at 48°, exhaust valve opens at 53°, exhaust valve closes at 23°, injection starts at 15°.

3. Explain why a piston travels a greater distance on the first 90° of its down-stroke travel of crankshaft rotation than on the remaining 90°.

4. What factors affect the delay period?

5. List four reasons why a two-cycle engine with the same displacement, compression ratio, and speed as that of a four-cycle engine does not have twice the power of the four-cycle engine.

6. Why do two-cycle engines require a blower?

7. List several components of an engine which dissipate heat through conduction and through convection.

8. List several reasons why some diesel engines have a greater valve overlap than others.

Combustion Chambers

Combustion Chamber Design   To achieve complete combustion, gain power and thermal efficiency, and stay below the emission limits, a maximum of inhaled air (or forced air) must be brought into contact with the injected fuel during the combustion process.

In order to achieve this objective, a relatively high velocity is needed between the fuel droplets and the air.   The air inlet passages to the cylinder, the combustion chambers, or special chambers, are specially designed to control the airflow, velocity, and direction.   Also, pressure at the beginning of combustion is used to control flow and velocity (squish).   Fuel velocity is controlled by locating the fuel nozzle in a position that allows the fuel spray to cover the total area of the airflow.   There must be no interference with cooling or valve size since this can reduce the volumetric efficiency.

To achieve a given total fuel velocity and direction of fuel spray, the velocity of the fuel droplets (pressure) and droplet size must be considered.

Classification of Combustion Chambers   The various combustion chamber designs may be classified as: (1) open combustion (direct-injection) chambers, (2) precombustion chambers, (3) turbulence combustion chambers, (4) air cell type (power cells or energy cells).

Open Combustion Chamber   A typical open or direct-injection combustion chamber is shown in Fig. 4–1. The combustion chamber is formed in the pistonhead and the injector is positioned over the center of the piston so that the spray from the multiorifice nozzle will be distributed evenly.   This spray must match the air and droplet movement in order to eliminate the dead air space within the combustion chamber.

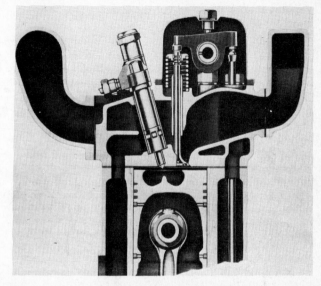

Fig. 4–1   Cross-sectional view of a Hercules-engine combustion chamber.

Action within an Open Combustion Chamber   As the piston moves downward, atmospheric pressure forces air through the scroll-shaped intake passage and promotes a circular air motion.   This circular airflow then passes through the intake valve opening(s) into the cylinder (Fig. 4–2a).   This motion helps to fully charge the cylinders.

As the piston comes up on compression stroke, it accelerates air motion by reducing the area in which the air is rotating (Fig. 4–2b).   As the piston approaches the top of its stroke, the confined swirl of air is forced into an even smaller area in the pistonhead (Fig. 4–2c).   This area has an inverted cone-shaped center (sometimes called *Mexican head*) which adds an even further rolling motion to the air. Into this bowl of accelerated swirl, fuel is injected evenly in four, or as many as eight directions (Fig.

Fig. 4–2   Air action in an open combustion chamber: (a) intake, (b) compression, (c) extreme compression, (d) power. (J. I. Case Company Components Div.)

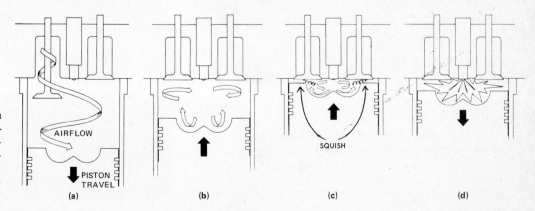

4–2*d*). The tip of the fuel spray hits the rim of the bowl rather than the outer circle of the piston or the comparatively cool cylinder walls.

The open combustion chamber design used on some engines from Maschinenfabrik Augsburg–Nurnberg A. G. and Kloeckner-Humboldt Deutz A. G., MANN, and International-Harvester consists of a special spherical chamber in the piston (a deep bowl) and uses an intake duct of a special configuration for greater air motion in the cylinder during the intake stroke. A pintaux nozzle is used to spray the fuel against the wall of the spherical combustion chamber. As the fuel hits the wall, it forms a thin coat over the hot chamber wall, ensuring favorable conditions for vaporization. The fiery vortex spinning causes the vaporized fuel to come off the walls of the combustion chamber in layers, thus bringing about complete combustion.

Advantages of Open Combustion Chamber

1. It has high brake thermal efficiency.
2. All energy produced by the fuel acts directly on the pistonhead.
3. The cylinder and piston suffer minimum effects of combustion due to the low temperature.

Disadvantages of Open Combustion Chamber

1. Emission control is difficult.
2. The engine tends to run rough due to a shorter delay period. This delay causes a high and rapid pressure rise.
3. It is sensitive to fuel and timing.
4. The multiorifice nozzles and high injection pressure tends to increase fuel-injection problems.

Specially Designed Combustion Chamber All specially designed combustion chambers are separate units which are bolted or screwed into the cylinder head. They connect with either a small or large passage to the cylinder above the piston. Fuel is injected directly into the chamber by a single hole, pintle, or delay nozzle. A portion of the clearance volume is inhaled into the chamber. During the last few years many companies have dropped the specially designed chambers from their newer engine designs in favor of direct injection. This trend is promoted by the generally lower thermal efficiency in the specially designed chambers due to a greater heat loss and lower brake mean effective pressure (6 mep). Some companies, however, still produce these specially designed chambers because of their great success in controlling hydrocarbon (HC) and nitrogen oxide (NO_x) emissions. Carbon monoxide (CO) emissions levels from diesel engines remain at relatively low levels as long as the engine operates at an air/fuel (A/F) ratio which satisfactorily controls exhaust smoke levels.

Precombustion and Turbulence Chambers The injector nozzle of a precombustion chamber engine is located in a specially formed chamber which contains about 30 percent of the total clearance volume. Most chamber openings point toward the center of the piston.

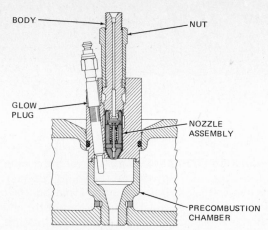

Fig. 4–3 Cross-sectional view of a precombustion chamber and fuel-injection valve. (*Caterpillar Tractor Co.*)

The Caterpillar Tractor Company uses precombustion chambers on their larger engines (see Fig. 4–3).

A turbulence chamber engine is one which has the injector nozzle located in a specially formed chamber containing about 80 percent of the clearance volume. The chamber passage to the cylinder is large and the chamber is located on one side of the cylinder (Fig. 4–4). Since the injector is located on one side of the cylinder, it will not interfere with larger valves. The volumetric efficiency and mean effective pressure are, therefore, higher than those of the precombustion chamber.

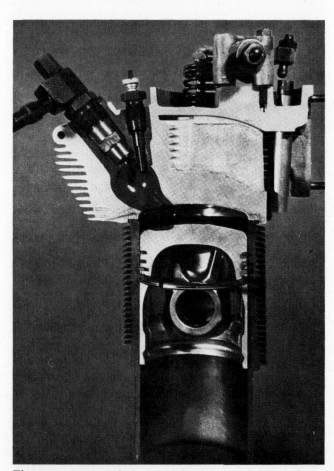

Fig. 4–4 Deutz turbulence chamber.

Action within Precombustion and Turbulence Chambers The action within the precombustion and turbulence chambers is very similar. On the compression stroke, air is pumped at high velocity through the restricted passage into the chamber. This causes a high turbulence within the chamber.

As the piston approaches TDC the injector discharges fuel into the chamber at low pressure. Combustion then causes the pressure to rise, forcing the fiery droplets, air and gases, at high velocity, into the main combustion area. Although this causes a high turbulence within the main combustion chamber, burning nevertheless is controlled by the dependency of combustion on injection.

Energy and Power Cells This type of combustion chamber has the energy or power cell located on one side of the cylinder head and the injector positioned across from the main combustion area.

With this design, manufacturers generally use a flat-head piston with the oval cylindrical main combustion chamber located in the cylinder head.

These types of cells have the same advantage (over the direct-injection engine) as the precombustion and turbulence chambers, that is, a continuous mixing of fuel and air. This gives a controlled combustion rate and a lower peak pressure than the direct-injection engine. The engine runs smoothly with less noise and with generally good fuel economy.

Questions

1. Why is the (atomized) injector spray directed into the piston bowl instead of the outer circle of the piston?

2. What factors control the fuel velocity during compression and ignition?

3. List the advantages and disadvantages of an open combustion chamber.

4. What are the main differences between a precombustion chamber and an energy cell?

5. List the main advantages of an energy cell over the open combustion chamber.

UNIT 5

Engine Disassembly

Purpose of This Unit This unit introduces you to the working parts of a diesel engine by explaining how to remove the engine from the vehicle and how to disassemble its components. (Later units describe operation and service procedures in detail.) You may wish to refer to Section 6, "Shop Equipment," while reading this and following units to find out how to use various tools and equipment.

CAUTION When working with machines having hydraulically, mechanically, and/or cable controlled equipment—such as shovels, loaders, dozers, scrapers, etc.—be certain the equipment is lowered to the ground before servicing, adjusting, and/or repairing. If it becomes necessary to have the equipment partially or fully raised to gain access to the engine, be *sure* the equipment is properly supported by means other than the hydraulic lift cylinders, cable, and/or mechanical devices used for controlling the equipment.

Starting and Stopping a Diesel Engine Before attempting to disassemble an engine you must, of course, get the vehicle into your shop. If the machine is one that you have not previously driven, you should familiarize yourself with the location of the instruments, throttle, brake system, hydraulic control, and transmission shift mechanism *before*

starting the engine. If you have not been informed about the engine's present condition, you should also check the coolant, the oil level and, if applicable, the hydraulic and transmission oil levels *before* moving the vehicle. Make certain that no other person has previously worked on the engine and removed parts that might prevent the engine from starting or damage the engine or its accessories.

You can then place the transmission in neutral or in the safety position. Move the throttle to half engine speed and, if a master switch is used, close the switch. If an engine preheater (glow plug) is used, turn it to the "on" position if necessary. After about 15 seconds, engage the electric, hydraulic, or air starter; the engine should start after about three full revolutions.

Once the engine is running, check the oil and fuel pressure. Watch closely the coolant temperature and air-pressure gauge. If the air-pressure gauge does not come up to "set pressure," the air tank drain may not be closed. Close it.

NOTE: Operate the engine at low idle for 3 to 5 min then increase the engine speed to about 1000 rpm. When the engine has reached an even temperature, you can move the vehicle into the shop.

Before you stop the engine, reduce the engine temperature to about 160°F [71°C] by reducing the engine speed to around 1000 rpm. Reduce the engine speed to low idle and then use the fuel stop mechanism to prevent fuel flow to the injectors. Use the decompressor or the emergency shutdown device only in an emergency.

Before leaving the vehicle, lower the hydraulic equipment and apply the parking brakes. You are now ready to remove the engine.

Removal of Engine   It is not possible, for obvious reasons, to list a common set of procedural steps to remove an engine from a truck, tractor, crane, loader, or boat. The service manual for each particular vehicle should always be referred to first, but if a service manual is not available, the following rules will serve as a general guide.

1. Before steam-cleaning the engine and its surrounding area, visually inspect and make a record of the overall condition of the engine and its components. This will reveal a great deal about the service it requires.

2. Secure or remove all hydraulic equipment which may interfere with the engine removal. Remove electrical components (including the battery), and tag the electrical wires to speed up reassembly.

3. Drain the complete system of lubricating oil, fuel, and coolant. Bleed the air system if an air system is used.

4. If electrical components are not to be removed, cover them to protect them from direct steam.

5. Steam-clean the engine first, to remove accumulated road grime, grease, and oil. This will permit a superficial visual inspection and will also speed up disassembly.

6. Clean the area around the engine so that the lifting device will not be obstructed when the engine is being lifted from its mounting.

7. Before you lift the engine from its mounting, remove radiator and hose connections, power steering and hoses (if used). Disconnect oil, coolant gauges, and connections to the pyrometer, if used.

8. Remove coolant, oil, fuel filters and connections, mechanical controls, tubing or hoses, turbocharger connections, air cleaner, and exhaust pipes. Remove transmission (if necessary).

9. Cover all openings, hoses, and tube ends to prevent the entry of dirt.

10. Attach lifting eyebolts or brackets at specified mounting points to assure the proper balance point.

11. When possible, use an adjustable lifting beam with an adjustable pull point. If a chain or cable is used, it must be parallel and as near perpendicular as possible to the object to be lifted. Be sure that the eyebolts or bracket bolts are threaded into the hole a distance of at least 1½ times their diameter to give proper thread strength.

12. After positioning a hoist (or other lifting device) properly and taking up the slack, remove the engine mounting bolts. Begin lifting the engine. Check that the load is in balance and again check for obstructions which could interfere during lifting.

CAUTION   Stay clear of the load and lift only to the height required to swing the engine free. With a rope, guide the engine clear during the lifting. Lower the engine to within a few inches off the ground, then guide it to your work area.

Mounting Engine to Engine Stand   Before mounting the engine on the engine stand you may have to remove components and parts (mounting brackets, clamps, belts, etc.). Do this in a systematic order. Identify mounting components with the location from which they were removed. Place parts and components as units on a rack or in a tray. Remove the turbocharger, exhaust and intake manifolds, water pump, water manifold and thermostat housing, oil cooler and connections. If the fuel-injection pump is located so that it could be damaged, it should be removed also.

As each component is removed, check it visually for distortion of flanges and brackets, cracked or broken flanges, discoloration, and damaged threads. Examine gaskets and seals to determine if they have been the cause of previous air, oil, or fuel leakage. Bolt the recommended mounting adapter plate securely to the engine and position the engine on the stand before releasing the hoist or lifting the vise.

Disassembly Procedures   Because there are so many engine designs, it is only possible to recommend basic disassembly procedures. They are as follows:

1. Always use the correct hand or power tools. Do not use a punch and hammer where a puller should be used.

2. When lifting heavy components, make sure that the hoist has sufficient lifting capacity and is secure. When disassembling, take care not to drop components.

3. Store subcomponents separately and as units.

4. Identify all removed parts and mark their original location on the engine so that they can be replaced quickly after inspection and/or diagnosis.

5. It will save time in the long run if you inspect each component or part during disassembly and make a record of the parts which need to be replaced.

Intake Manifold, Exhaust Manifold, and Turbocharger   The *intake manifold* connects the intake ports of a row of cylinders and the inlet port with the air cleaner and aftercooler, or to the compressor side of the turbocharger. It is designed to aid scavenging. It is a one-piece cast-iron alloy or aluminum-alloy casting (Fig. 5–1).

The exhaust manifold is also manufactured from a cast-iron alloy so that it will withstand temperature changes without cracking or distorting. It is designed to aid scavenging and may be built as a one-piece unit or in sections. It is the function of the manifold to deliver exhaust gases from the exhaust ports of the cylinders to pipes vented to atmosphere. When a turbocharger is used, the exhaust manifold is connected to the turbine of the turbocharger and from there exhaust gases are directed through pipes into the atmosphere.

The mounting flanges of the exhaust manifold are machined to seal (with asbestos or steel gaskets) the

INTAKE MANIFOLD

EXHAUST MANIFOLD

Fig. 5–1 Intake and exhaust manifolds on the vehicle.

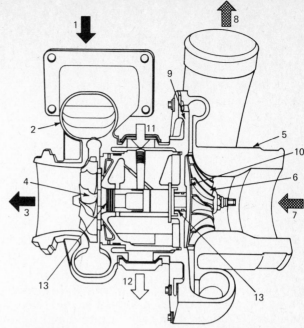

1. Exhaust in	7. Air in
2. Turbine housing	8. Air to engine
3. Exhaust out	9. Diffuser plate
4. Turbine wheel	10. Oil-seal assembly
and shaft	11. Oil in
5. Compressor housing	12. Oil out
6. Impeller	13. Piston ring seal

Fig. 5–2 Schematic view of a turbocharger.

exhaust ports to the cylinder head and the turbine or to the exhaust pipe flange. When multisectional manifolds are used, leak-free slip joints connect the ends to the center section. The manifolds of truck and tractor engines are air-cooled but marine manifolds are cast with integral water jackets surrounding the exhaust manifold. A bypass from the coolant system is passed through the water manifold to aid in cooling.

A *turbocharger* is an air pump which utilizes exhaust-gas energy to drive a centrifugal blower (compressor). The turbine side of the turbocharger is connected directly to the exhaust manifold and the compressor is connected between the air cleaner and the intake manifold.

The horsepower gain when using a turbocharger may range from 50 to 150 percent depending on the boost pressure and the efficiency of the aftercooler. The average exhaust temperature at lower engine load is about 500°F [351°C], and the boost pressure is about 5 inHg [12.7 cmHg]. At maximum engine load (maximum torque), the exhaust temperature may reach 1000 to 1200°F [577.7 to 740°C]. At this stage the turbine speed is at its maximum designed speed and has reached its maximum designed or controlled boost pressure. Torque reduction, however, causes reduced exhaust pressure and rapidly decreases turbine speed and boost pressure.

To balance the quantity of fuel being injected with the boost pressure, various control devices are used. A fuel ratio control or an aneroid is used to control the fuel-rack movement. An aneroid is used to control the fuel pressure (Cummins), and an air ratio control is used to control the boost pressure. Each of these devices reduces exhaust smoke during acceleration and deceleration. A throttle-delay mechanism (Detroit Diesel) is used to postpone fuel-rack movement when the engine is accelerating. (See Unit 26, "Antipollution Control Devices.")

Action in Turbocharger The turbocharger is bolted by its turbine housing to the outlet of the

exhaust manifold. Either the compressor housing or the compressor extension is connected to the inlet manifold.

Refer to Fig. 5–2 as you read the following description of the action in the turbocharger. After the engine is started, the exhaust gases leave the exhaust manifold and enter the turbine housing. The exhaust gases flow with high velocity into the volute-shaped (cork screw) turbine housing. The housing gradually decreases in area, causing a further increase in velocity. The high-velocity air is directed through the nozzle onto the turbine and from there discharges through the exhaust pipes to the atmosphere.

The exhaust gases force the turbine and the compressor wheel to rotate which in turn creates a low pressure at the inlet of the compressor housing. Atmospheric pressure forces air at high velocity into the inlet opening of the compressor housing or compressor extension. The continued increasing rotational speed of the impeller increases the air velocity. As the air is forced through the diffuser, it gradually slows down, converting the kinetic energy into pressure. The diffuser may be in the form of an open passage with a cross-sectional area which gradually increases toward the outer circumference or it may be in the form of blades on an older type of turbine. The purpose of the diffuser is to decrease the velocity and raise the pressure in the chamber formed between the compressor housing cover and the inlet manifold.

Removal of Intake and Exhaust Manifolds First remove the air cleaner connecting hose and tubings

Fig. 5–3 Cutaway impeller-design pump. (The arrows show the fluid flow.)

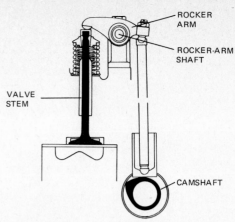

Fig. 5–4 Rocker-arm action. (*Cummins Engine Co., Inc.*)

to the turbocharger. Next, remove oil inlet and outlet hose connections from the turbocharger. Then remove the turbocharger.

NOTE Cover any turbocharger openings to prevent the entrance of dirt or other foreign matter. Remove the nuts and bolts from each manifold and remove each manifold, as a unit, from the cylinder head(s).

Nonpositive Displacement Coolant Pumps Nonpositive displacement pumps are those which supply a continuous flow until the resistance in the discharge line equals the designed capacity of the pump.

Impeller-Design Pump Nonpositive displacement pumps of the centrifugal design, that is, the impeller design, are used as coolant pumps (Fig. 5–3). Basically they are designed with an impeller which has straight, curved, or coned-shaped blades. The impeller is positioned on a shaft in a spiral- or volute-shaped housing. The fluid inlet is near the center of the housing and the outlet is on the outer circumference of the housing.

When the impeller is driven, the fluid starts to revolve and the centrifugal force moves the liquid out between the blades to a greater and greater radius. This increases the tangential velocity and centrifugal force. The liquid thrown off the tip of the blades follows the contour of the housing. From Fig. 5–3 you will see that the pump chamber (in which the impeller rotates) gradually increases in area toward the outlet port.

This design allows the velocity head to be converted gradually to a pressure head. When the resistance in the discharge line and the liquid force of the pump are equal, there is no flow. However, the impeller still moves the liquid within the housing and produces heat. The design and the rpm determine the maximum pressure and volume. No pressure relief valve is needed.

Removal of Coolant Pump Remove the necessary coolant hose, the transfer, and the coolant bypass tube. Remove the mounting bolts and lift the pump from the engine.

Removal of Fuel-Injection Pump Remove the high-pressure injection lines. Cover the openings of the injectors, injection-pump discharger fittings, and both sides of the injection lines.

CAUTION Use two wrenches when loosening the nuts and do not bend or kink the fuel lines.

Remove the fuel lines leading from the second-stage filters to the injection pump and cover the openings. Remove the mounting bolts, lift the injection pump from the engine, and store it in a safe place.

Removal of Valve Cover The valve mechanism cover is made from cast iron, cast aluminium, or is metal-stamped. Simply remove the necessary bolts and lift the cover from the cylinder head.

Decompression Lever Assembly The decompression mechanism is an assembly of levers and shafts that open either the intake or exhaust valve(s). When actuated, the engine can turn freely since it does not have to compress the air.

Rocker Arm and Shaft The rocker arm pivots around the rocker-arm shaft to transmit motion from the camshaft to the valve stem (see Fig. 5–4). On Detroit Diesel and Cummins diesel engines, the pushrod also activates the unit injector or injector.

Rocker arms are made of forged cast alloy or laminated steel. One end has a ball socket adjustment mechanism and the other end, which contacts the valve stem, is especially hardened or has an insert of high-quality steel. The contour of the contact area is ground so that a minimum of side thrust is exerted on the valve stem or bridge as the rocker arm pivots. When an overhead camshaft is used, the rocker arm has a cam roller on one end, or the camshaft lobe may act directly on the follower assembly (Fig. 5–5). Some rocker arms are cross-drilled to supply lubrication to both ends. A bushing is pressed into the bore of the rocker arm to reduce friction and to increase service life. The pivot point is closer to the adjusted end to multiply the camshaft lift to a ratio of about 1:1.5. The hollow or solid rocker-arm shaft is made of hardened steel. It is supported either in a separate rocker-arm housing or by two or more rocker-arm-shaft brackets. Holes are drilled into the shaft at precise locations to lubricate the rocker-arm

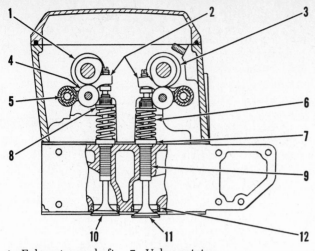

1. Exhaust camshaft
2. Rocker-arm assemblies
3. Inlet camshaft
4. Roller
5. Rocker-arm shaft
6. Valve spring
7. Valve rotator
8. Valve retainer and lock
9. Valve guide
10. Exhaust valve
11. Inlet valve
12. Valve-seat insert

Fig. 5–5 Valve and valve mechanism (Caterpillar 1674 truck engine). *(Caterpillar Tractor Co.)*

bushing. Lubrication commonly is supplied to the rocker-arm shaft through the rocker-arm bracket or through a separate connection (Fig. 5–6).

The decompression lever assembly usually is removed before the rocker-arm assembly. When overhead cams are used, remove the camshaft assembly or lever housing.

CAUTION It is essential to release tension evenly because the valve springs force the assembly upward. Hex bolts and nuts must, therefore, be loosened one after another, a little at a time.

If special hex bolts are used to pass oil from the cylinder block to the cylinder head and thereby lubricate the rocker-arm assembly, be sure to mark their location. When valve bridges (crossheads) are used, remove them before removing pushrods and followers and place them in a holder in the order of their removal.

Cylinder Head All cylinder heads are made of a special iron alloy casting containing carbon, silicon,

and copper. This mixture provides elasticity, good thermal conductivity, and a low thermal expansion rate. The size of the cylinder head is not determined by the number of cylinders but rather by such factors as the cylinder block design, the number of bearings, the expected thermal stress, and the anticipated cooling and sealing difficulties (of the cylinder head). Factory complications in casting or foundry work can also affect cylinder-head size.

Whether an individual cylinder head is used for each cylinder (Fig. 5–7) or whether the cylinder head covers two, three, four, or six cylinders, it must nevertheless have adequate strength and stiffness. It must act as a sealing surface between the cylinder sleeve, cylinder block, and oil and cooling passages, without causing distortion of the sleeve or valves. The cylinder head must be sufficiently strong so that it does not crack between the cylinder-head bolts (studs), between the intake and exhaust valve, or between the valves and injector sleeve.

The internal cooling passages must be located to ensure that the coolant flow has a high velocity at and around the valves and injector tubes. It must remove heat (steam bubbles) and prevent the accumulation of deposit or scale. The passages should have no dead ends. The external openings must prevent turbulence and permit unrestricted circulation from the cylinder block to the cylinder head and from the cylinder head to the radiator.

The valves must be located so that the fuel spray can reach the total combustion area, but they must be far enough apart so that the coolant can circulate freely between them, thereby preventing the cylinder head from cracking between the valve seats.

The location of the valve passages and openings as well as the valve throats must assist rather than reduce breathing capacity (refer to Combustion Chamber Design in Unit 4).

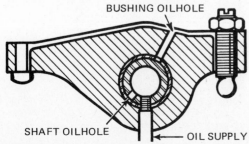

Fig. 5–6 Stamped-steel rocker arm. *(J. I. Case Company Components Div.)*

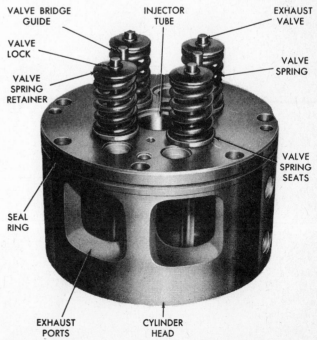

Fig. 5–7 Cylinder head (Detroit 149 Series). *(GMC Detroit Diesel Allison Div.)*

24

The location of the injector, precombustion chamber, turbulence chamber, and the air cells must ensure good precipitation of the fuel droplets to aid in combustion and cooling.

Removal of Cylinder Head Care should be taken when removing the cylinder-head bolts or nuts. Also, the reversed tightening sequence should be used to prevent distortion of the cylinder head.

CAUTION Never remove the cylinder head when it is hot because it will become distorted (warped).

If the cylinder head is very heavy, use a hoist to lift it from the cylinder block. If the cylinder head is small, screw the lift handles into the cylinder head to lift it from the block. If a cylinder head becomes stuck, do not drive a chisel or screwdriver between the cylinder block and head because this will cause damage to both surfaces. Lightly tap the cylinder head with a bronze or lead hammer or use a block of wood to break it loose.

Carefully inspect the combustion chamber once it is exposed. Close scrutiny might reveal the cause of high oil consumption, overfueling, water leakage, or overheating. Damage to pistons, cylinder sleeves, and cylinder block can also be seen.

CAUTION When removing the cylinder head, take care not to damage it or the cylinder block surface or threads. If studs are used, take care not to bend them. After removal, place the cylinder head in a holding fixture, or if it happens to be square, you may place it on a workbench.

Valves Valves are exposed to extreme temperature changes, to being hammered against the valve seat, to corrosion from combustion, and to constant changes in direction. The composition of the intake valves may range from a low-alloy steel to an alloy of chromium with carbon added. Exhaust valves may have the same alloy base or they may have nickel, manganese, or nitrogen added.

The external appearance of valves is reasonably similar (see Figs. 5–8 and 5–9). There are major differences, however, in the dimensions of the valve stem and head, the collet grooves (valve keeper grooves), and the seat angle. The *seat angle* is the angle of the valve face surface against the horizontal surface of the insert and is either 30° or 45°. The angle is governed by such factors as the valve throat, the valve lift, and whether it is a naturally aspirated or a supercharged engine.

A 45° valve has about 20 percent more seating force than a 30° valve and it produces a better cleaning action against accumulated carbon deposit. However, a 45° valve tends to deform the valve seat faster because of the greater force of the valve spring. Intake valves are generally about 40 percent larger in area than exhaust valves in order to increase breathing capacity.

There are other differences in the valves which you cannot detect, for instance, the valve may be welded or part of the valve stem may be made of one alloy while the rest of the stem and the valve head may be made of another alloy to improve its durability against corrosion and high temperatures. Some valves are faced and/or the head is coated, or the ends of the stem may be hardened to increase corrosion and wear resistance. Other valves are hollow and filled with metallic sodium to reduce valve-head temperature by convection (see Fig. 5–9). Valve-head temperature can be reduced about 200°F [109°C] when this type of valve is used.

Removal of Valves and Injectors Use a valve-spring compression tool as shown in Fig. 5–10, to remove, in sequence, the valve keeper, valve-spring cap or valve rotator, valve spring, valve-spring seat,

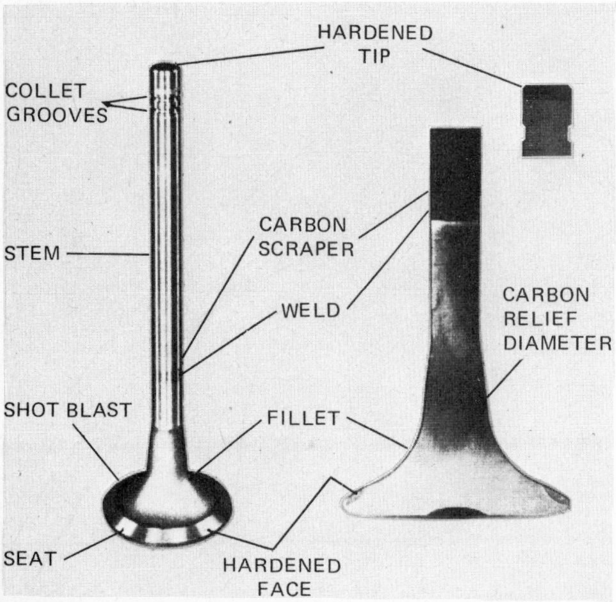

Fig. 5–8 Typical valve design and construction. *(Cummins Engine Co., Inc.)*

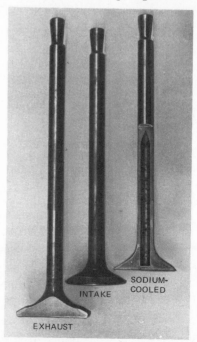

Fig. 5–9 Typical valve types.

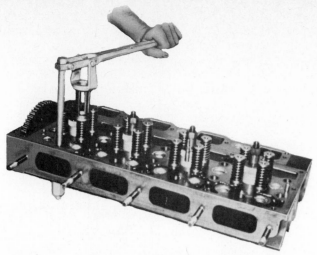

Fig. 5–10 Removing a valve-spring retainer.

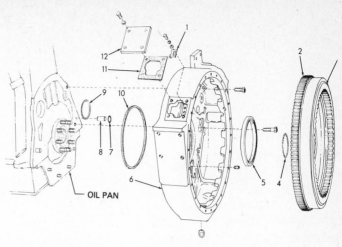

1. Timing pointer	8. Housing dowel
2. Ring gear	9. Flywheel housing-to-camshaft O ring
3. Flywheel	
4. O ring	10. Flywheel housing-to-block O ring
5. Crankshaft rear oil seal	
	11. Cover gasket
6. Flywheel housing	12. Timing-hole cover
7. Dowel O ring	

Fig. 5–11 Flywheel and flywheel housing. *(Allis-Chalmers Engine Div.)*

and valve. Throughout removal, place each valve in a holding fixture.

Remove bolts or nuts from injector hold-down studs and then carefully remove the injectors. When removed, the injectors must be kept in numerical order and placed where they cannot be damaged. If an injector is stuck in the head, use two prybars (injector bars) to loosen it. Grip close to the injector body and ease out evenly so that the injector comes straight out of the hole. If the injector is exceptionally tight due to carbon formation or corrosion build-up between the injector hole and the sleeve, you can use penetrating oil on the stud and body to help in the removal.

Flywheel Housing The flywheel housing is a one-piece casting, ribbed for strength, positioned with dowel pins, and secured with hex bolts to the cylinder block. The flywheel housing covers the flywheel and is the mounting surface for the transmission and the cranking motor. It is usually the mounting surface for the rear engine mount, and it can also act as the timing-gear housing. Flywheel-housing size as well as the diameters of the bores, the spacing of bolt holes, and the size of the holes vary with engine design. Accordingly, the Society of Automotive Engineers (SAE) recommends nine designs from SAE00 to SAE6. The crankshaft rear oil seal is pressed into the flywheel housing (see Fig. 5–11). When a rear seal cover is used, no seal is required in the flywheel housing.

Flywheel and Ring Gear The *flywheel* is one-piece cast iron. Its size and weight are determined by the engine speed, the number of cylinders, and whether it is a two- or four-cycle engine. The flywheel is machined to provide a true fit with the crankshaft flange, pressure plate surface and clutch friction surface, or with the clutch drive ring gear, or torque converter impeller mounting ring.

A flywheel used for a double-disk clutch is machined on the inner circumference and has four drive pins (drive locks) to hold and drive the intermediate plate into position. The center of the fly-

wheel has either a machine bore into which the pilot bearing or bushing is press-fitted, or it has a spline drive flange bolted to it to drive an auxiliary shaft. The flywheel bolt holes are precise in dimension and usually one hole is offset to ensure that the flywheel will fit onto the crankshaft in only one position.

Part of the outer circumference is machined to provide the seating area of the ring gear.

A *ring gear* is a steel ring with external teeth which is shrunk to the outer circumference of the flywheel (see Fig. 5–11). It is the connecting link between the cranking motor pinion and the engine. Ring gears are designed to provide smooth engagement and disengagement of the cranking motor drive. The average gear ratio between the pinion and the ring gear is about 25:1.

Removal of Flywheel and Flywheel Housing To remove the flywheel, bend the lock plates free from the flywheel hex bolts. Install the lift hook as shown in Fig. 5–12, and then remove the flywheel hex bolt. When necessary use two hex screws threaded their entire length to act as jack screws to pull the flywheel from the crankshaft. Remove the hex bolts which hold the flywheel housing to the cylinder block. Using a sling and lift, raise the flywheel housing from the cylinder block.

Vibration Damper Diesel engines which are required to operate throughout a wide speed range, are protected against normal operation frequencies, vibration (roughness), and stress, by means of a vibration damper. Two types of vibration dampers are used: rubber element and viscous element.

The *rubber-element damper* is made of an inner iron-alloy steel flange or has a mounting facility, and

Fig. 5–12 Removing a flywheel. *(GMC Detroit Diesel Allison Div.)*

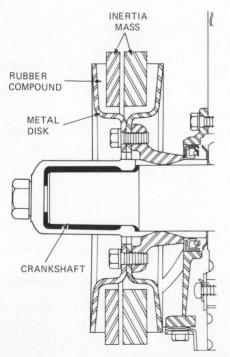

Fig. 5–13 Cross-sectional view of an installed rubber-element vibration damper.

INERTIA MASS

RUBBER COMPOUND

METAL DISK

CRANKSHAFT

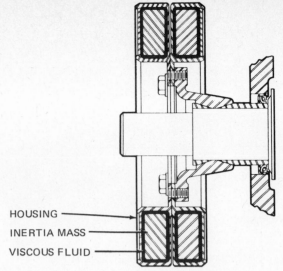

HOUSING

INERTIA MASS

VISCOUS FLUID

Fig. 5–14 Cross-sectional view of an installed viscous vibration damper.

an outer cast-iron alloy weight assembly (inertia mass). The inner flange is bonded to the outer weight by a rubber compound (Fig. 5–13).

The *viscous damper* forms a two-piece housing. The weight (inertia mass) and a special greaselike fluid is placed between the two pieces which then are welded together. The clearance between the housing and the weight is about 0.010 in [0.25 mm] on all sides. It is this space which is filled by the viscous fluid (Fig. 5–14).

Removal of Crankshaft Pulley, Vibration Damper, and Front-Mounting Bracket Some crankshaft pulleys and vibration dampers are bolted together, press-fitted and keyed to the crankshaft; some are separately keyed and are pressed onto the crankshaft; others are bolted and doweled to a hub which is keyed and pressed onto the crankshaft. Regardless of the crankshaft pulley and vibration damper arrangement, always use a puller similar to the one shown in Fig. 5–15. *Never drive the vibration*

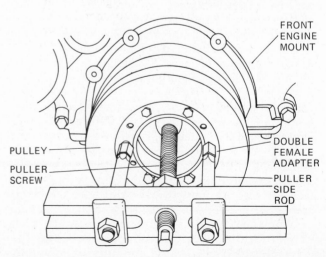

FRONT ENGINE MOUNT

PULLEY

PULLER SCREW

DOUBLE FEMALE ADAPTER

PULLER SIDE ROD

Fig. 5–15 Removing a viscous vibration damper.

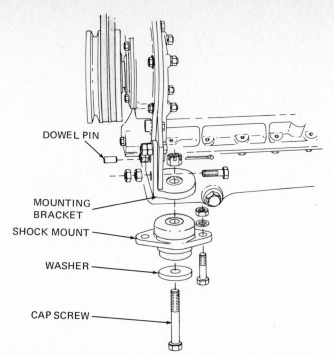

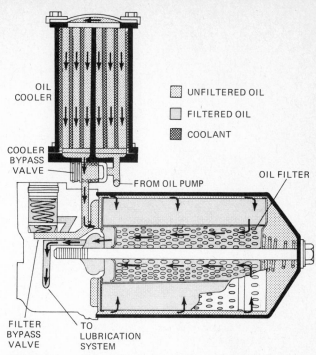

Fig. 5–16 Engine front support. *(Allis-Chalmers Engine Div.)*

Fig. 5–17 Oil flow through a cooler and filter. *(J. I. Case Company, Components Div.)*

damper assembly from the crankshaft assembly with a hammer. Improper removal damages the vibration dampers as they are precision-built to a close fit.

It is common practice to replace the viscous vibration damper during major engine overhauls. After removal of the hex bolts from the front-mounting bracket, remove and discard the synthetic liners but retain the mounting bracket and cap as a unit. One of many engine front-mounting brackets is shown in Fig. 5–16.

Oil Pan The oil pan is the reservoir for the lubricating system. It allows sediment to settle at the bottom. In addition it dissipates heat by radiation from the external walls or by use of internal or external radiation devices.

The oil pans used on diesel engines are machined cast iron, cast aluminum alloy, or metal-stamped. To strengthen the oil pan, internal and external reinforcement struts are integrally cast or pressed. The oil pan usually is constructed as a single unit. Less common are the two-piece oil pans (designed for special applications) which have an upper and lower section. Most oil pans have a large rear sump. When the engine position or the operating condition does not exceed a 10° longitudinal incline, flat oil pans are used. Some pans have large inspection covers and clean-out covers with drain plugs, while others have only a drain plug. Some have magnetic chip detectors. All oil pans have oil baffles (separator plates) to prevent the lubricating oil from splashing onto the connecting rods. Some engines use a steel ribbon-type oil level indicator (dipstick) whereas others have high and low oil-level check point plugs.

Oil Cooler The lubricant must be kept within the temperature of 180 to 250°F [82 to 121°C] if it is to maintain viscosity and absorb and dissipate heat. Although a safe operating temperature can be achieved by increasing oil capacity, under certain operating conditions this can cause excessive heat to develop, and thereby decrease oil control.

An oil cooler can more effectively control oil temperature than the system previously mentioned. As illustrated in Fig. 5–17, oil is first pumped through the oil cooler, where it is cooled. From there it passes through the oil filter, and finally into the lubrication system.

Some oil coolers have many copper tubes welded or soldered to a seal ring. The assembly is sealed in a shell-type housing. Coolant is passed from one end to the other through the copper tubes. Lubricant enters the shell at one end, surrounds the tubes and leaves the shell at the other end. The coolant usually enters in the direction opposite to the lubricant. (This is a very effective cooling method.) With other types of oil coolers, the cooler is connected directly to the lubrication pump and main oil galleries by an opening in the cylinder block. Both types of oil coolers use a relief valve to supply the engine with oil in case of an oil-flow blockage.

Lubrication Pump Most diesel engines use gear-type pumps to circulate and distribute sufficient oil to lubricate, cool, clean, and seal the components. Engines which use in-track or wheel-type loaders or dozers rely on an additional scavenging gear pump or a tube scavenging system to maintain a constant supply to the lubrication pump when the machine is working on a grade (see Fig. 5–18).

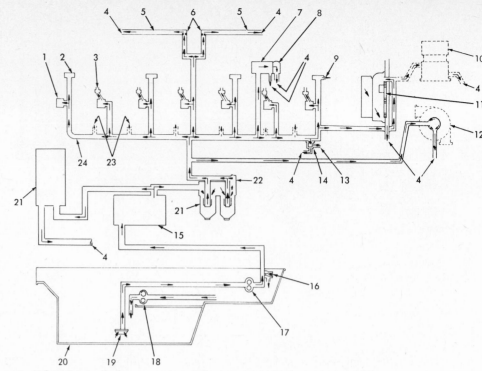

1. Main bearings
2. Camshaft bearings
3. Connecting-rod bearings and to piston pin
4. Return to oil pan
5. Rocker-arm shaft
6. Orifice in rocker-arm shafts (located at Nos. 3 and 4 rocker-arm-bracket positions)
7. Fuel-injection pump
8. Governor
9. To camshaft gear
10. Air compressor
11. Accessory drive gears
12. Turbocharger
13. Oil restrictor hole (oil bleeds to timing-gear housing)
14. Oil-pressure regulating valve
15. Engine oil cooler
16. Oil-pump pressure relief valve
17. Oil-pressure pump
18. Oil scavenging pump
19. Suction screen
20. Oil pan
21. Oil filter
22. Oil-filter bypass valve
23. Piston-cooling oil jets
24. Main oil gallery

Fig. 5–18 Engine lubricating oil-flow diagram. (*Allis-Chalmers Engine Div.*)

The lubrication pumps are driven directly from the crankshaft timing gear, through an idler, or from a gear on the camshaft. The oil pressure is maintained within the engine by restrictive oil passages and drillings. The system is protected from over-pressurization by a relief valve.

All oil pumps are of positive displacement design. A valve is required to relieve displacement once maximum operating pressure has been reached.

Since all pumps are designed with a positive seal, they have very little volumetric slip when the pressure increases at the discharge port. The efficiency of any pump depends on the design of its rotating members and its ability to maintain its sealing effectiveness at various pressures and temperatures.

Pump Rating   Whether a pump is positive or nonpositive it is rated according to its performance. Performance is judged by the specific pump volume at a specified rpm against a fixed resistance measured in psi [kg/cm²].

Rotary Pumps   Classification of the various types of rotary pumps generally is made according to the type of rotating members and their construction. The three most common rotary pumps used as lubrication pumps are: the gear-type pump, the internal gear pump, and the gerotor pump.

Gear-Type Pump   In the gear-type pump, fluid under atmospheric pressure exists at the inlet (Fig. 5–19). As the power source rotates the drive gear (which is in mesh with the driven gear), a partial vacuum is created at the inlet side of the pump. When the gear teeth unmesh, fluid flows to fill in the space. This fluid is carried in an opposite direction around the pockets formed between the gear teeth, the wear plates, and the center section (the housing). These components and the oil form a seal, therefore, fluid cannot pass to the inlet side of the pump. The gear teeth mesh on the outlet side, and thereby force the fluid from both pockets out against system pressure.

Internal Gear Pump   The drive gear in an internal gear pump is an inner gear with outward projecting teeth (Fig. 5–20). The teeth of the outer gear (driven gear) project inward and are in mesh with the inner gear between the outlet and inlet side of the pump.

The drive shaft of the pump is set off-center in a circular chamber. This chamber is the bearing and support of the outer gear. Opposite to the inlet and outlet port there is a crescent-shaped seal which is machined into the pump housing. This seal is positioned between the gears and provides a close clearance within them.

The pumping principle of the internal gear pump will be familiar to you since it is patterned after all

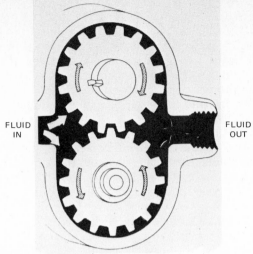

Fig. 5–19 Schematic view of a gear-type pump.

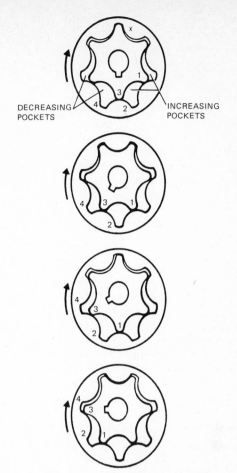

Fig. 5–21 Schematic views of a gerotor pump.

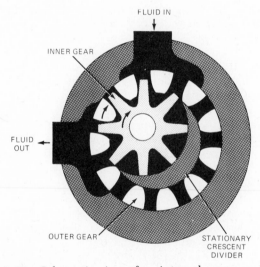

Fig. 5–20 Schematic view of an internal gear pump.

other gear pumps. The shaft rotates the inner gear which, in turn, causes the outer gear to rotate. Since the two gears are in mesh, everything within the chamber rotates (except the crescent). This motion traps liquid in the gear space as it passes the crescent. It is then carried by the gear spaces from the inlet side to the outlet side. From there it is forced out of the pump by the meshing of the gears. As the fluid is carried away from the intake side of the pump, pressure is reduced, allowing more fluid to enter.

Gerotor Pump   The gerotor pumping element consists of a pair of strange-looking gears, one within the other, in a housing (see Fig. 5–21). The inner gear, which has always one tooth less than the outer gear, is the drive gear. At least four teeth at a time are in sliding mesh with the outer gear ring. At all times one pair of teeth is in full mesh position.

In Fig. 5–21 the full mesh position is shown at the top (X). Note that when the inner gear is driven, the inlet pockets (1–2) increase in size while the discharge pockets (3–4) decrease in size. By reading the diagram from top to bottom you will see that the

pockets on the right-hand side increase in size while those on the left-hand side decrease in size. From this you can determine that the intake side of the pump is on the right side and the discharge side of the pump is on the left side.

The advantage of this type of pump design is that the pumping elements and gears rotate at a greatly reduced speed in comparison to that of the drive shaft. This reduces component wear.

Removal of Oil Pan and Oil Pump   To remove the oil pan, turn the engine over and remove the hex bolts which hold the oil pan to the crankcase. If the oil pan is dowel fitted, remove the dowel bolts with a soft-faced hammer. Do not drive a chisel or screwdriver between the oil pan and crankcase because this will damage the mating surfaces. Use a lead hammer and gently tap the oil pan loose.

Remove the necessary tube connections. Loosen and remove the oil pump from its mounting.

NOTE   Some oil pumps are doweled. Take extra care when loosening them from their mountings.

Timing-Gear Cover   The timing-gear cover is usually a one-piece casting mounted to the cylinder block which provides a cover for the timing gears. In some applications the timing cover and the flywheel housing are one unit.

Removal of Timing-Gear Cover The timing-gear cover can be removed easily from small engines. However, when the timing gears are at the rear of the engine and the engine is large, the following procedures should be followed:

1. Remove the oil pan and attach a suitable hoist and sling.
2. Remove the bolts which secure the timing-gear housing to the crankcase and cylinder block.
3. Remove the timing-gear housing, taking care not to damage the inner bearings which support the various drives and idler gears. When the timing gears are at the rear, only a front cover which has the crankshaft seal and bearing support for the front-mounting brackets is used.

Camshaft Camshafts are made from special high-tensile-strength iron alloy. The bearing journals and cam lobes are heat-treated and precision-ground. Usually one bearing journal is positioned between each cylinder to support the camshaft. The contour of the lobes varies with the engine design to conform to predetermined valve duration, valve lift height, and lift velocity (the speed with which the valves open and close). Cummins engines and Detroit Diesel engines use one extra lobe for each cylinder to actuate the injector-operating mechanism.

V engines may have two camshafts which rotate in opposite directions (Fig. 5–22), or they may have only one camshaft (Fig. 5–23) to actuate the valve mechanism of the left and right cylinder banks. Some engines have two camshafts which rotate in opposite directions; one camshaft actuates the intake valve mechanism and the other camshaft actuates the exhaust valve mechanism.

Thrust bearings, thrust plates, or thrust washers, located between the camshaft gear and the shoulder of the first journal, position the camshaft in the cylinder block. Some engines are designed with a spring-loaded plunger to hold the camshaft in place, others have two thrust washers to take up the camshaft thrust. (It is the helical gear that causes the thrust on the camshaft.) One washer is located between the rear thrust surface and the other between the bearing and camshaft gear. In the latter design, the camshaft is held in place by the camshaft gear.

Removal of Camshaft, Timing Gears, and Accessory Drive Gears It is wise to measure the backlash before removing the camshaft or the accessory gears to determine if they are reusable. However, this is only possible when the wear on bearings and bushings does not exceed service manual specifications.

In most cases the camshaft and gear can be removed as a unit once the thrust plate bolts are removed (see Fig. 5–24). When it is necessary to remove the camshaft timing gear, always use a suitable puller. Take care when removing the idler accessory drive gears to arrange them in a systematic order.

Connecting Rod Connecting rods for the trunk-type piston are made of drop-forged steel, often in I design, with one small and one large closed hub (see Fig. 5–25a). The two connecting-rod ends are first bored, then the large end is cut in half, holes are drilled and tapped and its surface machined. Then they are bolted together with specially designed bolts or with stud bolts and nuts. They are again machined to a precise dimension, heat-treated, and electronically balanced.

Some connecting rods are center drilled to provide lubricant to the piston pin. Some have an orifice at the lower end which is press-fitted into the counterbore to control the oil flow. These connect-

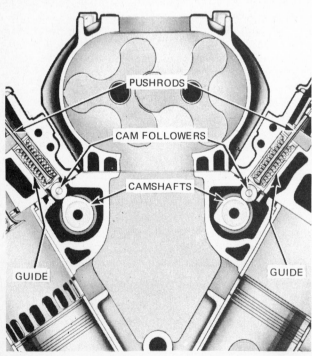

Fig. 5–22 Detroit Diesel V engine having two camshafts. *(GMC Detroit Diesel Allison Div.)*

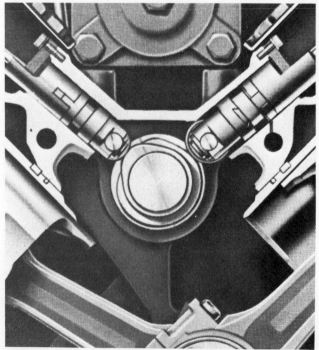

Fig. 5–23 Cummins diesel V engine having one camshaft.

ing rods require helical grooved bushings with a space between them to assure good lubrication of the piston-pin bushings. In addition, many engine manufacturers use a controlled oil flow (from the spray nozzle) to cool the underside of the piston head.

Another connecting-rod design has a funnel shape cut in the upper end of the rod to store the oil. This oil is used to supply lubricant to the single- or double-radial or spiral piston-pin bushing. Yet another connecting rod design is one which has serrated mating surfaces to increase rigidity and to place the load uniformly on both the connecting-rod cap and the connecting rod. When the (larger) end of the connecting rod will not pass through the cylinder sleeve because of its size in relation to the bore size, manufacturers simply change the separation point of the cap. This changes the connecting-rod bolt position and reduces the outside dimension.

The connecting rod shown in Fig. 5–25b is a crosshead piston type used with the Detroit Diesel 71, 92, and 149 Series engines. You will notice that the piston-pin bore and bushings have been replaced with a saddle. The piston pin is held rigidly by two bolts which reach from below through the saddle and into a special nut inserted in the piston pin. Note the bending stresses imposed on the piston pin in a conventional trunk-type piston (Fig. 5–25a) as opposed to the central force on the crosshead piston (Fig. 5–25b).

All connecting-rod bolts, or stud bolts, are made from special steel and are specially designed to assure proper alignment between the rod and the cap and to prevent any spreading of the radius of the connecting-rod bore under maximum combustion pressure.

Piston Today's diesel engine must operate at a much higher speed and, at the same time, produce more power than in previous years. The pistons (balanced to close limits) and the piston rings must meet this challenge. Most pistons, therefore, are manufactured of an aluminum alloy made chiefly of magnesium and copper or of magnesium and nickel. This alloy is light in comparison to its size, yet it is strong enough to withstand pressure, speed, and quick changes in direction. More important, pis-

tons of this composition are able to transmit heat three times faster than malleable iron pistons. However, because of the thermal expansion rate, piston dimensions must be precise in order to ensure the exact clearance. (Silicon is added to the aluminum alloy to reduce thermal expansion.)

There are many different types of pistons in use today, the most dissimilar being the trunk and crosshead designs.

Trunk Piston Most truck engines, earth-moving machinery, light marine vessels, and farm tractor engines use trunk-type pistons. Exceptions are some Detroit Diesel 71 Series, all 92 Series, and all 149 Series engines, and the Mack 300 Series Maxidyne. These engines use crosshead pistons.

Trunk-type pistons are one-piece, precision-machined, internally reinforced, and designed to transfer heat from the piston crown to the ring groove and skirt area (see Fig. 5–26). The pistons generally are cooled by an oil spray or splash at the inner side of the piston crown and/or indirectly through conduction and convection to the cooling system. Some are tin-plated to reduce scuffing and to permit a closer fit.

The area above the rings is known as the piston crown. The crown diameter is slightly smaller than the diameter of the ring area and is tapered to overcome distortion when the piston becomes hot. The top of the crown is irregularly shaped and forms the combustion chamber bowl. High compression ratio engines using direct-injection or combustion chamber design require valve reliefs to prevent the valves from interfering with the piston when in TDC position.

Caterpillar engines which use precombustion chambers have a stainless steel heat plug cast into the top of the crown. The heat plug prevents the high pressure and high heat which escape from the combustion chamber orifice from damaging the aluminum surface. This plug absorbs the heat, then

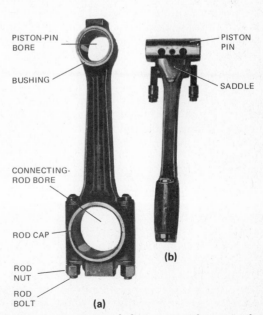

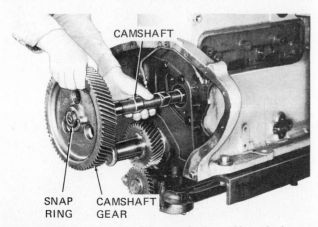

Fig. 5–24 Removing the camshaft. (*Allis-Chalmers Engine Div.*)

Fig. 5–25 Connecting-rod designs: (*a*) for a trunk-type piston, (*b*) for a crosshead piston.

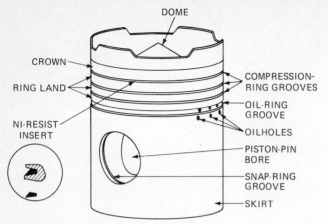

Fig. 5–26 Cummins trunk-type piston.

Fig. 5–27 Cummins insert on the piston skirt. (*Cummins Engine Co., Inc.*)

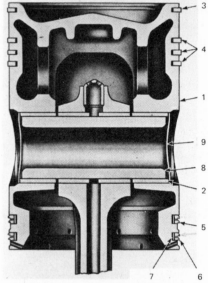

1. Piston
2. Piston-pin bushing
3. Compression ring (top)
4. Compression rings
5. Oil-control ring (upper half)
6. Oil-control ring (lower half)
7. Oil-control-ring expander
8. Piston pin
9. Piston-pin retainer

Fig. 5–28 Cross-sectional view of a Detroit trunk-type piston. (*GMC Detroit Diesel Allison Div.*)

transfers it to the underside of the piston crown where the heat dissipates.

The ring groove area is located below the crown. The ring grooves, which hold and guide the rings, are diverse in shape and spacing. The surface between them is called the *ring land*.

Some pistons are equipped with a cast-iron band or BOND-O-LOC insert to reduce the wear rate of the ring grooves. Behind and slightly below the lower ring groove, oil holes are drilled through the piston wall to allow the oil to escape after it is wiped off the cylinder wall by the oil-control rings. Some engines are designed to utilize this escaped oil to cool the crown of the piston and to lubricate the piston-pin bushings.

The region below the ring grooves is the piston skirt area. It may be divided into three sections: the area below the ring grooves, the pin bore, and the skirt area.

The large skirt area spreads the force evenly against the cylinder walls thereby preventing the piston from cocking as it moves up and down. The piston may have either a full skirt or a semislipper-type skirt. When the stroke equals the bore dimension, the engine is said to be square. When the stroke is shorter than the diameter of the bore, it is said to be under-square. In either case, the connecting rod or the crankshaft could interfere with the skirt when the crankpin is at a 90° angle; for this reason the skirt is partly removed.

The piston skirt's surface finish (which reduces friction and scuffing) is not always the same. A popular finish is the interrupted surface. It has high points and plateaus parallel to the top of the piston. Another popular finish is the diamond-shaped raised surface.

Cummins Engine Company, Inc., uses Teflon inserts on the piston skirts of some of their pistons to reduce cylinder sleeve vibration and piston and sleeve pitting (see Fig. 5–27).

Larger aluminum pistons usually are *cam-ground*. This means that the piston is oval with the narrowest diameter across the piston-pin bore. This feature is necessary because the added material used to strengthen the piston-pin area does not allow the piston to expand uniformly when it becomes hot. However, when the piston approaches operating temperature, it expands to become round.

Detroit Diesel engines have oil-ring grooves in the lower part of the skirt which seal (block) the scavenging air to the crankcase (Fig. 5-28). Holes drilled through the walls of the piston, below and through the lower oil groove, allow any excess oil which has collected to return to the crankcase.

The reinforced piston-pin bore (*boss*) into which the piston pin is fitted, is precision honed or machined to accommodate the two helical oil groove bushings, which provide bearings for the floating piston pin. The piston pin is the link between the piston and the connecting rod.

Piston Pin The hollow piston pins are made from special quality high-carbon steel to withstand fatigue failure. The pin is hardened and ground, then its surface is lapped and polished to extremely close tolerances to reduce friction and wear.

Some piston-pin bores are grooved on both ends. Snap rings fit into these grooves and hold the piston pin in place. Some piston-pin bores are slightly enlarged to accept the piston-pin retainers. On many pistons, the piston-pin boss area is relieved to allow the piston pin to flex. This technique reduces pin seizure and prevents the pin bore from cracking.

Crosshead Piston The crosshead piston used in Detroit Diesel 71, 92, and 149T1 Series engines and Mack 300 Series Maxidyne engines consists of two major pieces: the piston crown, which has the compression-ring grooves, and the skirt. These are held together as a unit by the piston pin.

Detroit Diesel 71, 72, and 149T1 Series engines also employ a piston ring as a seal between the piston crown and the skirt. The skirt has oil-ring grooves.

The two major pieces of the floating piston in the 149 Series consist of the carrier and the piston. These are held together by a snap ring.

Detroit pistons are designed with a slipper bearing which is locked in the upper half of the piston-pin bore (see Fig. 5-29). The connecting rod is bolted to the piston pin. The combustion force of the floating piston is transmitted from the piston to the thrust washer onto the carrier, slipper bearing, piston pin, connecting rod, and crankshaft. In the crosshead design the force of combustion from the piston crown goes directly onto the slipper bearing and the piston pin. The skirt, being separate, is free from vertical load distortion. It is also free from thermal distortion since the hotter crown expands during engine operation.

In the Detroit Diesel 71, 92, and 149T1 Series engines, oil under pressure is forced through a drilled passage in the connecting rod, through the piston pin and then into the cocktail shaker. This increases lubrication and reduces piston temperature.

In the Detroit Diesel 149 Series floating piston design, oil under pressure is forced along the groove of the slipper bearing insert to the area between the carrier and the piston where it is trapped during piston reciprocation. A percentage of oil drains through two holes in the carrier into the crankcase.

Removal of Piston and Connecting-Rod Assembly Turn the engine right side up and clean all carbon from the upper inside of the cylinder sleeve. If the ring travel ridge is too deep and the removal of the piston would be difficult, the ridge must be removed. Use a hone or a ridge reaming tool similar to that shown in Fig. 5-30. Install the ridge removal tool so that the cutting guide is below the ridge and the lower part of the cutting tool is tight against the cylinder sleeve. A clamp must be used to hold the cylinder sleeve in order to prevent it from turning in the cylinder block. Release the spring force. This will force the cutting blades outward so that there is a clean cut when turning the tool.

CAUTION Take care not to damage the connecting-rod journals and the finish of the cylinder sleeves when removing the piston assembly.

Fig. 5–29 Detroit crosshead piston. (GMC Detroit Diesel Allison Div.)

Turn the engine 90° and remove the hex bolts or cotter key and nuts from the connecting rods. Remove the caps. If the caps are excessively tight, tap them gently with a lead hammer to break them loose. Mark each connecting rod and cap as soon as you remove it if it is not identified by the manufacturer. Tape mating bearing shells together and record the connecting rod or cylinder from which you have taken the bearings so that you can correctly diagnose the cause of failure. Place the components in a safe place if they can be reused.

Removal of Piston Rings and Connecting Rods from Piston Hold the connecting rod in a vise fitted with protecting covers over the steel jaws. Use a piston ring installing tool, as shown in Fig. 5-31, to remove the rings. Do not break the rings to remove them as this could cause damage to either the ring lands or the ring grooves.

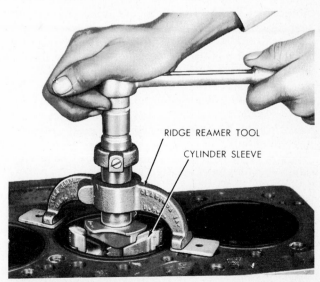

Fig. 5–30 Ridge reaming.

Fig. 5–31 Removing piston rings. *(GMC Detroit Diesel Allison Div.)*

Most diesel engines use a full-floating piston pin. This means that the piston pin is free to move in the connecting-rod bushing and in the piston-pin bushing. The pistion pin usually is retained by two snap rings. Some manufacturers, however, provide a clamp bolt to hold the piston pin in the connecting rod while others have the connecting rod bolted directly to the piston pin.

Cylinder Sleeve Two types of cylinder sleeves are used: the *wet* and the *dry* cylinder sleeve (Fig. 5-32). Both types are made from cast-iron alloy and machined to dimension, then surface-hardened, ground, and honed, and sometimes, chemically treated. The degree of hardness of the cylinder surface varies among manufacturers. Piston rings must, therefore, be selected from among those recommended by the individual manufacturer, if the best service life is to be anticipated.

The removable wet-type or dry-type cylinder sleeve is inserted into the cylinder bore from the top and is held at the top by a flange which fits into a machined counterbore in the cylinder block. The cylinder-head gasket, or individual-head gasket, is compressed between the flange and the cylinder head, holding the cylinder sleeve in place and sealing the coolant and lubricating oil, as well as the pressure at the upper end of the sleeve. The wet-type cylinder sleeve uses silicon or Buna seal rings fitted into grooves, either in the lower part of the cylinder bore or in the lower outside circumference of the cylinder sleeve (see Fig. 5-32a). This prevents coolant leakage into the oil pan.

In the Detroit Diesel 53 Series engine the sealing arrangement differs since the seal is placed between the flange and the water jacket. Additionally, the upper half of the cylinder sleeve is surrounded by

the water jacket. The lower half, including the oval-shaped inlet ports, is surrounded by the scavenging air.

The wet-type cylinder sleeve of a four-cycle engine is surrounded completely by the water jacket. It extends from the sealing ring to the flange of the cylinder sleeve. The dry-type cylinder sleeve used on either two- or four-cycle engines is thinner, however, the total material thickness of the cylinder bore and cylinder sleeve is nearly the same as that of the wet-type cylinder sleeve.

The Detroit two-cycle engine, which uses dry-type cylinder sleeves, has an upper and lower water jacket divided at the inlet port, but connected by hollow struts. Coolant from the water pump enters the lower water jacket and passes through the hollow struts into the upper water jacket, then into the cylinder head.

Removal of Cylinder Sleeves The following rules are applicable to the removal of all cylinder sleeves:

1. Do not use a punch and hammer or a prybar to remove a cylinder sleeve because they can damage the cylinder block or the sleeve. Use only the correct pullers or clamps (Fig. 5-33).
2. To remove a sleeve using a slide hammer, insert the lower puller clamp by twisting it on the tapered seat so that it will slide down the sleeve. As the clamp clears the bottom of the sleeve it will fall back on its tapered seat.
3. Slide the upper clamp into place against the top end of the sleeve. Hold the puller rod, and thereby the lower clamp, against the cylinder sleeve. Using sharp blows, slide the puller against its stop to release the sleeve from its bore.
4. When an aluminum block is used, immerse the cylinder block in water heated to 180°F [82°C] or force hot water through the cylinder block. Allow sufficient time for the bore to expand so that less effort will be required to remove the sleeve.

NOTE If for some reason the cylinder sleeve is to be replaced but the crankshaft is to remain in place, you should protect (with cardboard or gasket material)

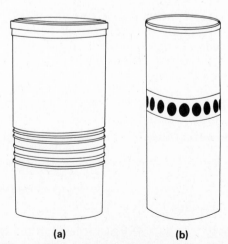

(a)   (b)

Fig. 5–32 *(a)* Wet cylinder sleeve and *(b)* dry cylinder sleeve.

the crankshaft and the inside of the crankcase from contamination and from coolant fluid.

Crankshaft The crankshaft of almost all diesel engines is made from drop-forged and heat-treated steel, to assure the utmost strength and durability (see Fig. 5-34). The accurately ground main and connecting-rod journals are induction hardened. Counterweights are forged integrally with the crank or bolted to the crankshaft. All crankshafts are dynamically balanced for proper weight distribution to assure even force during rotation. They are T-drilled to pass lubricating oil (under pressure) from the main bearings to the connecting-rod bearings.

Fig. 5–33 Removing a cylinder sleeve. (*Owatonna Tool Co.*)

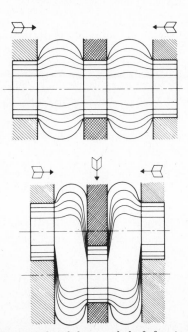

Fig. 5–34 Principle of the crankshaft forging process.

One main journal is especially machined with two thrust surfaces to control the end thrust in conjunction with thrust washers. Sometimes the end play is controlled by a set of special main-bearing shells.

The number of main journals and connecting-rod journals depends on the number of cylinders and the design (in-line or V block). The surface area of the journals is governed by the potential engine speed and combustion pressure. The crankshaft-throws arrangement determines the firing order.

Removal of Crankshaft Main Bearing
1. Turn the engine upside down and remove cotter keys and nuts (or hex bolts) from the main-bearing caps.
2. Make sure that all bearing caps are marked so that they can be installed later in their original location.
3. Remove the main-bearing caps from the dowels. Some manufacturers recommend the use of a special main-bearing-cap puller.
4. Using protective hooks, remove the crankshaft. Alternately, wrap two throws of the crankshaft with cloth to protect the bearing surface and then use a sling, as shown in Fig. 5-35, to hoist the crankshaft out of the upper bearings. Take care not to damage the threads or bend the studs (if used).
5. Remove the bearing shells from the cap and crankcase. Tape the mating halves of the bearing shells together after identifying their original posi-
6. Store the crankshaft in a protected place.

Cylinder Block Once you have removed all the components, you are left with the cylinder block. This main structural part of the engine is a one-piece casting, made of cast-iron alloy. When the cylinder block is cored to receive wet-type cylinder sleeves, or when the manufacturers prefer a multicylinder-head style, the cylinder block has additional transverse members cast integrally to provide rigidity and strength. This ensures accurate alignment of the crankshaft bearings and cylinder sleeves and also prevents distortion of the cylinder block. The walls of all bearing bores (cylinder, upper half of the

Fig. 5–35 Removing the crankshaft. (*Cummins Engine Co., Inc.*)

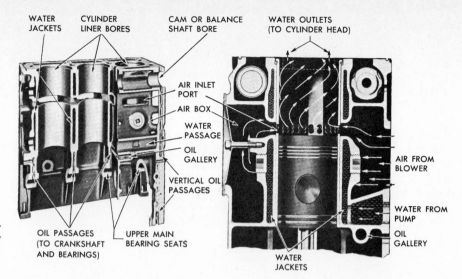

Fig. 5–36 Cutaway views of a Detroit Diesel cylinder block using dry cylinder liners (sleeves). *(GMC Detroit Diesel Allison Div.)*

crankshaft, camshaft, etc.) and all openings and passages for the coolant and lubrication system, being part of the cylinder block, are integrally cast. Detroit Diesel also has air passages and openings within the cylinder block (see Fig. 5-36). The main-bearing supports and the camshaft bores, are line-bored to assure longitudinal and horizontal alignment. Most cylinder blocks have one main oil gallery which extends lengthwise through the cylinder block. Internal passages direct oil from the main oil gallery to the crankshaft main bearings and through vertical passages, to the rocker-arm assemblies, timing gears, and fuel-injection pump. All recently designed engines use a full-flow lubrication system with an oil cooler (Fig. 5-18). Here, a vertical passage, from the bottom of the crankcase, extends to the filter mounting, oil cooler, and turbocharger. Oil from the filters flows through a passage into the main gallery and from there to the various components which require lubrication.

Cleaning Components Some components can be inspected for excessive wear or cause of failure before cleaning. Others need to be cleaned first and then followed through by inspection. Typical component cleaning procedure is as follows:

1. Remove all seals, gaskets, and plugs from the disassembled components.
2. Steam-clean and dry the disassembled components with compressed air. Units such as oil pans, cylinder block and heads, oil cooler, heat exchanger, valve covers and rocker-arm assemblies, should be cleaned immediately after they are disassembled to prevent hardening of accumulated grime and other foreign substances.
3. Use a glass-bead cleanser for components such as valves, pistons, cylinder heads, water pump housings, and impellers, etc. Bead size U.S. Sieve No. 70 is recommended with an operating pressure of not more than 90 psi [6.32 kg/cm²].
4. Do not expose the components to the bead blast any longer than is absolutely necessary. This is especially important when cleaning soft material. Be sure to remove all glass-bead substance and other

foreign material by washing the parts in solvent and then drying them with compressed air.
5. It is wise to clean all other components in a hot tank either as a unit or, if disassembled, in a wire-mesh basket. **(NOTE** Never clean soft metal parts in hot tanks.) Check manufacturer's recommendations as to solution concentration, etc. Most solutions should be heated within 180 to 200°F [82 to 93°C].
6. Make certain that all coolant passages and oil galleries have been cleaned thoroughly and that the coolant jackets are free of scale.

Questions

1. List the steps to be taken when removing an engine.

2. What poor work habits, when removing an engine, could cause damage to components and create removal difficulties?

3. Why should all components be promptly examined when they have been removed from the engine?

4. Select a service manual for a particular engine. From the manual list the steps to be followed when disassembling this engine.

5. Outline the procedures to remove a piston assembly from the engine, the piston rings, and the piston from the connecting rod.

6. How would you remove a cylinder sleeve using the tools shown in Fig. 5-33?

7. Why should you tape and identify the connecting-rod shells and the main-bearing shells upon disassembly?

8. When components have seals, plugs, etc., why should they be removed before cleaning the component?

9. Good work habits require conforming to given safety rules. List some of the rules you must obey when cleaning components using solvent, hot tank, glass beads, and compressed air.

SECTION

Engine Components Service

2

UNIT 6

Cylinder Block

Inspecting Cylinder Block   After the cylinder block has been properly cleaned and dried with compressed air, it should be moved to a flat surface for inspection.   A thorough inspection will reveal whether it can be refinished or whether it should be discarded.

After you have cleaned the cylinder block, make sure that all solvent or hot tank solutions are removed.   Recheck to determine if there is any loose scale present in the water jacket or passages.   Make sure that the openings have not been reduced by corrosion.   At the same time check for eroded water holes which could prevent proper sealing of gaskets or O rings (grommets).   Check for cracks, porosity, and leaks.   Using dye penetrant over areas where flaws are suspected, apply developer over the dried surface.   Flaws will be detectable by the appearance of solid or dotted lines of dye penetrant at the surface.   Flaws will also be detectable by pressure testing the cylinder block.

Pressure-Testing Cylinder Block   A cylinder block can crack open between the cylinder bores, the water jacket and crankcase, the water jacket and oil passages, the water jacket and air box, and in many other places.   It may crack from overheating due to insufficient coolant and/or oil; from an improperly functioning water pump, fan, or shutter; from extremely high friction of one or more moving components within the engine; from inadequate cy-

linder-head torque; from vibration; or from internal or external damage of the casting due to mechanical failure of one or more components.

Many manufacturers recommend that the cylinder block be pressure-tested before any servicing takes place.   Others do not give explicit instructions, relying, instead, on the mechanic's ingenuity in locating cracks or flaws.   Lack of instructions may also be attributed to the fact that blocks have chronic locations where cracking takes place but which are known to all seasoned mechanics.

When a pressure test of the cylinder block is warranted, insert coolant, coring, and drill plugs.   Seal all coolant openings with plates, gaskets, or O rings.   When wet-type sleeves are used, insert and position the cylinder sleeves also.

Fill the coolant passage with an antifreeze solution.   Not only is its surface tension lower than that of water, permitting it to penetrate small cracks, but its color will also help you to locate any cracks.   Pressurize the coolant passages to about 80 psi [5.62 kg/cm²] and maintain the pressure for about 2 hours.

Another method (actually the most effective one in determining cylinder block cracks) is the use of a hot water tank.   The same preparation just described is required with the exception that no antifreeze solution is necessary.

After you have pressurized the coolant passage to about 80 psi [5.63 kg/cm²], immerse the cylinder block in the solution which has been heated to about

180°F [82°C]. After the cylinder block has reached the same temperature as the solution, again check the coolant-passage air pressure. Also watch the solution for bubbles, another indication of leaks.

CAUTION Make certain that the bubbles are not caused by poor cylinder seals, sealing plates, O rings, gaskets, or plugs.

Check all mounting surfaces for flatness with a straightedge. Check for any raised areas at the stud or bolt holes. If required, refinish surfaces according to the manufacturer's recommended procedures.

Check all threaded holes and, when necessary, restore any damaged threads. If this is not possible, drill and tap the hole for a helical thread insert (coil).

Check the top surface of the cylinder block for flatness with a straightedge, both transversely and longitudinally. Compare the results with the specifications. Should the tolerance exceed the specifications, the surface must be refinished; otherwise, it will be difficult to maintain compression, water, and oil sealing.

If it is necessary to refinish the top surface of a cylinder block (using either a milling machine or a large surface grinder), do not refinish the block to a height less than the manufacturer's specification.

Check the cylinder-block bores, the mounting surfaces of components, and sleeve counterbores for corrosion and erosion. Discard the cylinder block if the areas cannot be cleaned, refinished, or resleeved.

Engines using dry-type cylinder sleeves depend on good cylinder sleeve to cylinder bore contact since cooling is accomplished through conduction. Excessive clearance or cylinder-bore distortion reduces heat transfer. Therefore, before measuring the cylinder-block bores for out-of-round or taper, clean the bores using an adjustable rigid hone having a 120-grit stone.

Honing Cylinder Bore (Dry Sleeve) The cylinder bore must be honed or bored to receive an oversize cylinder sleeve. Honing the bore should be done with an adjustable rigid hone (Fig. 6-1) because it will not follow the irregularities of the bore as does the spring-loaded deglazer hone.

A ½- or ¾-inch electric drill will provide the right speed and torque to drive the hone. Use either wet or dry 120-grit stones. Install the honing mast so that the center of the hone is in the center of the cylinder bore and so that the spring-tension position of the hone is about ½ in [12.7 mm] above the cylinder block top surface. Adjust the stones so that they lie firmly against the smallest measured section. This will ensure good honing and prevent the hone from fluttering.

Start the drill. Irregularities will be felt by an increased drag on the stones. Move the drill up and down with short strokes concentrating on the high spots first. Readjust the hone as necessary to assure firm surface and stone contact. When using dry stones, frequently clean them with a wire brush to prevent stone loading.

CAUTION On two-cycle engines, do not hone for as long at the air inlet port area. The hone cuts more

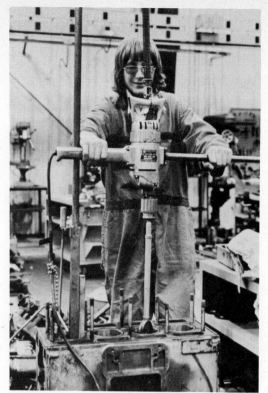

Fig. 6-1 Honing the cylinder bore.

swiftly (where there are holes) than in the rest of the bore.

When 70 percent of the bore is honed, move the hone a full stroke over the full length of the bore.

CAUTION Do not allow the stones to protrude more than 1 in [25.4 mm] above or below the cylinder bore. Do not pull the hone out of the cylinder bore when the hone is spinning.

Continual measuring and inspecting of the bore and cleaning of the stone is essential if a perfect bore is to be achieved. A *perfect bore* means that the dimension over the total area is within the minimum allowable tolerance. The unhoned area must not be larger than 1 in [25.4 mm] in diameter.

Measuring Cylinder Bore First, clean the cylinder bore with soap and hot water. Dry it with compressed air, and coat it thoroughly with clean lubricating oil. Use white paper towels to wipe the lubricant from the walls of the cylinder bore. Repeat the procedure until there is no evidence of residue on the white paper towel.

To determine the diameter of the cylinder bore, measure it with an inside micrometer or use a telescopic gauge. When you use the telescopic gauge to determine the diameter of the cylinder bore, measure the distance of the telescopic gauge with an outside micrometer. (Refer to Unit 48, "Measuring Tools," for procedures on using these as well as other hand measuring tools mentioned later.)

Usually neither method will produce a precise measurement because tool and surface are difficult to align. A cylinder or an out-of-round gauge which

Fig. 6–2 Measuring the cylinder bore.

have a dial indicator calibrated in thousandths of an inch [hundredths of a millimeter] are more suitable for measuring the cylinder bores. To adjust the dial indicator of either tool to zero, place the gauge in the bore of a *new* cylinder block, in a master ring, or between the anvils of an outside micrometer (adjusted to specification). Set the dial indicator to zero. Insert the tool into the cylinder bore, as illustrated in Fig. 6-2, and record your measurement. If the pointer moves to the left from the zero mark, the diameter of the cylinder bore is large. If the pointer moves to the right of one zero mark, the diameter of the cylinder bore is smaller. The measurement is determined by the millimeter or fraction of a millimeter that the pointer has moved from zero.

Boring Cylinder Bore When the cylinder bore does not meet specification or does not clean up through honing, the dimension of the bore must be increased to accommodate an oversized cylinder sleeve. This can be done with a special precision cylinder honing machine or with a portable boring bar.

When you use a Sunnen honing machine or boring bar to oversize the cylinder bore, always follow the manufacturer's instructions. Make certain the main-bearing caps are torqued to specification. Be sure that the cutting tool is sharpened properly and positioned tightly in the holding fixtures. Be sure that the boring bar is centered with the counterbore and that the boring bar is clamped securely. Cut the bore 0.001 in [0.025 mm] smaller than the intended completion size. When the hole is bored, back off the cutter and then manually remove the bar. Always sharpen the cutter after each hole is bored. Use a hone to remove the tool marks and to bring the bore to the specified dimension. (See Honing)

CAUTION Remember that boring causes a heat buildup. Therefore, allow the block to cool before measuring the cylinder bore. Again, because of heat, it is advisable to bore alternate cylinder holes to reduce the chance of hole distortion.

Checking and Measuring Cylinder-Sleeve Protrusion Check the counter bore surface for distortion because the cylinder sleeve may have moved due to incorrect cylinder-head torque or insufficient

Fig. 6–3 Measuring the depth of the counterbore. (*Allis-Chalmers Engine Div.*)

cylinder-sleeve protrusion. Make certain that the flange of the cylinder sleeve fully contacts the counter bore.

There are several methods for measuring or checking sleeve protrusion. One method is to measure the width of the cylinder-sleeve flange and the depth of the counterbore at four or more locations (see Fig. 6-3). Subtract the counterbore depth from the width of the cylinder-sleeve flange. The result is the sleeve protrusion.

Another method is to insert the cylinder sleeve into the cylinder bore (without any shims). Hold the sleeve firmly in place. Hold a straightedge across the sleeve flange. Use a feeler gauge to measure the flange protrusion or use a depth micrometer or a dial indicator. **NOTE** When you use a straightedge, make certain that the straightedge is on the cylinder-sleeve flange and not on the cylinder-sleeve firing ring.

If the protrusion is not within specification, select the shim with the thickness to achieve specified protrusion and place it in the counterbore. Mark each sleeve to identify it with the bore to which it must be installed.

If shimming does not correct the cylinder-sleeve protrusion, if the counterbore is distorted, or if the cylinder-sleeve protrusion between adjacent sleeves is more than specified, the counterbore must be machined.

Reseating Cylinder-Sleeve Counterbore Regardless of the reseating tool used, the manufacturer's procedures must be followed exactly. Following are some procedures which usually are included but often are overlooked by the mechanic:

1. Make certain that the lower and upper cylinder bores are free of carbon, scale, or corrosion. Failure to clean adequately can cause the tool to misalign and therefore cut an untrue seat.
2. Secure adapter plate. An improperly secured adapter plate can cause an untrue seat and/or enlarge the diameter.
3. Adjust the cutting tool against the shoulder. Improper diameter adjustment can cause an oversize diameter or a step shoulder in the counterbore.

4. Use a sharp tool bit. A loose or improperly sharpened tool bit can cause an untrue surface.

5. Do not exceed a cutting rate of 0.001 in [0.025 mm] per revolution. To do so can cause an untrue surface.

6. Measure depth frequently. Neglecting to do so may result in the counterbore being machined too deep. This causes intolerance with the adjacent cylinder.

7. Thoroughly clean the cylinder block. Imperfect cleaning of the cylinder block after machining can destroy the total engine.

Checking Main-Bearing Caps and Bores Check the mating surface of each cap, the upper main-bearing support, and the bearing surface, for wear, nicks, and burrs.

Make certain that the correctly numbered (or otherwise identified) cap is in the proper position and fits with no perceptible clearance. If there is no clearance, the cap will not rock. The machined surface of the cap must rest on the mating surface portion of the block, otherwise, when tightening, distortion will result.

Install caps and tighten hex bolts or nuts to specification. Measure the main-bearing bores horizontally, vertically, and diagonally with an inside micrometer or with a dial gauge which has been properly zeroed to an outside micrometer or ring gauge.

When the main-bearing bores are within specification, check them for alignment. Misalignment may be caused by overheating, overspeeding, vibration, broken crankshaft, or other severe stress. The alignment of main bearings can be checked with a master bar or with the boring tool. When a master bar (which is about 0.001 in [0.025 mm] less than the main-bearing bore) is used, place the alignment shaft in the cleaned and lubricated upper bearing support and install the caps and torque to the specified torque. The method most commonly followed is to install a new or reground crankshaft, using new bearing shells of the correct size. Apply adequate lubrication and make sure that the bearing caps are torqued to specification. The alignment is correct when the crankshaft can be rotated freely by hand.

When the main-bearing bore(s) is (are) out of alignment, damaged, distorted, or too large in dimension, the cylinder block must be discarded. When it is possible to salvage the block by using replaceable main-bearing caps, the main-bearing bores must be serviced.

Replacing Main-Bearing Caps Most manufacturers produce semifinished or finished main-bearing caps for the purpose of restoring the cylinder block. However, it is not unusual to have to try several finished replacement caps before one will fit and align correctly.

When it is necessary to use semifinished replacement caps, restoring the serviceability of the cylinder block can be somewhat difficult. Machining must reduce the ends of the caps equally when decreasing the diameter of the bearing bore. In many

cases the replacement caps have to be redoweled and bearing cap(s) rebored to specification.

NOTE When resizing the main bores, the cylinder block must rest on a flat surface. Before making any adjustment or starting to ream or hone the main-bearing bores, the cylinder block should be allowed to stabilize to room temperature.

Checking and Measuring Camshaft Bores Camshaft bores and/or balance shaft bores of a cylinder block are rarely distorted, worn, or damaged. Nevertheless the bores must be checked for nicks and burrs that may have occurred due to improper removal of the camshaft bearing. Bearing bores should be cleaned thoroughly and the oil passages checked. When it is doubtful that the replacement bearings will press-fit into the bores, measure the bores with an inside micrometer or with a telescopic gauge (Fig. 6-4).

Removing and Installing Camshaft Bearings (Bushings) The precision replacement camshaft bearings (bushings) are available in standard sizes and in 0.010 in [0.25 mm] undersizes. The location of the bearings differs with the engine design. They could be in the crankcase, in the upper part of the cylinder block or, when overhead camshafts are used, in a special housing.

The average clearance between the camshaft journals and bearings is 0.002 in [0.50 mm]. When the clearance is 0.008 in [0.203 mm], a replacement is needed. To determine the running clearance, with the crankshaft removed, measure the inside diameter of each bearing and compare it with the manufacturer's specification. Then measure each camshaft journal and subtract it from the bearing measurement. The result will be the running clearance.

Camshaft bearings should be replaced whenever the engine requires a major overhaul. Use the same

Fig. 6–4 Measuring the camshaft-bearing bore with a telescopic gauge.

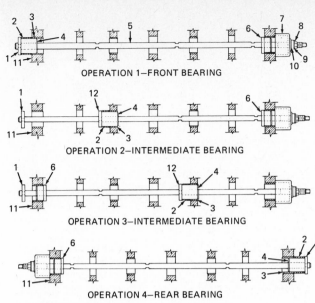

OPERATION 1—FRONT BEARING

OPERATION 2—INTERMEDIATE BEARING

OPERATION 3—INTERMEDIATE BEARING

OPERATION 4—REAR BEARING

1. C washer (stop plate)	8. Hex nut
2. New camshaft bearing	9. Flat washer
3. Old camshaft bearing	10. Thrust washer
4. Installing pilot	11. Front machined surface
5. Puller shaft	of cylinder block
6. Shaft pilot	12. C washer
7. Collar	

Fig. 6–5 Sequence of operations for removing and installing camshaft bearings. *(Allis-Chalmers Engine Div.)*

Fig. 6–6 View of camshaft-bearing caps and thrust plates. *(GMC Detroit Diesel Allison Div.)*

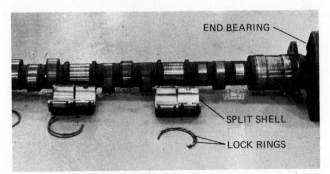

Fig. 6–7 Camshaft and camshaft bearings.

tools with which you removed the camshaft bearings to install the replacement bearings. Visually check the camshaft bores, particularly for nicks and burrs which could damage the bearings during installa-

tion. Lay out the bearings on the cylinder block in the order in which they are to be installed; camshaft bearings are not all of the same width nor are the oil supply holes situated in corresponding locations.

To assure precision hole alignment, mark the location of the oilhole on the cylinder block. Lubricate the inside and outside of the bearing and install it as shown in Fig. 6-5. Draw a line on the sleeve or washer to indicate the oilhole. When pulling the bearing into place, be sure that it does not turn out of position. Take care that the oilholes of the bearing and block do not restrict one another. Since there are different tools available for removal and installation of bearings, make sure you use a tool which "pulls" the bearing in and out of the bore to prevent damage to the bearing or bore.

It is good procedure to check the dimension of newly installed bearings with an inside micrometer or telescopic gauge for the possibility of out-of-roundness.

Detroit Diesel 149 Series engines use bearing shells and caps to support the camshaft, and a thrust plate to take up the thrust (Fig. 6-6). Part of the camshaft cap is used for the rocker-arm-shaft support, and the cam lobes act directly on the rocker arm.

Caterpillar overhead-camshaft engines use one-piece bearings in the supports.

The Detroit Diesel 53, 71, and 92 Series engines, differ from the 149 Series with respect to camshaft bearings and their location (see Fig. 6-7). The two end bearings form one unit which is bolted to the front and the rear cylinder block. The remaining camshaft bearings are made up of split shells, which are held together by two lock rings and positioned in their bores by a setscrew.

The camshaft bearing checks and measurement procedures for the 53, 71, and 92 Series are the same as for the conventional camshaft bearing models.

The removal and installation checks and measurements of overhead-camshaft bearings are also the same as those for conventional camshaft bearings.

Questions

1. List the visual checks you must make when examining the cylinder block.

2. Detail the steps you must take when preparing to hone a cylinder bore.

3. What is the importance of having a good contact between the cylinder bore and the outside diameter of the cylinder sleeve?

4. Why is sleeve protrusion necessary?

5. List the steps to prepare a cylinder block for the pressure test.

6. Crankshaft bore alignment must not exceed *(a)* 0.015 in, *(b)* 0.005 in, *(c)* 0.0015 in, or *(d)* 0.0035 in.

Camshaft

Fig. 7–1 Measuring camshaft runout.

Checks and Measurements   Wear of camshaft journals and camshaft cam lobes is minimal if the engine always is operated with clean oil. Nevertheless, checks and measurements are still required.

Some camshafts have oil galleries. When you service these camshafts, clean the galleries with a wire brush and compressed air. Check the surface condition of the journals and cam lobes for roughness and scoring and check the thrust surfaces.

Inspect the keyway and threads for damage. When the cam lobes (or the keyways) are damaged, the camshaft must be replaced. None of the major engine manufacturers recommend regrinding camshaft lobes or building the cam lobes up by welding or spraying.

When the keyway, threads, and cam lobes are in good condition, place the camshaft on V blocks and use a dial indicator to measure the runout (Fig. 7-1). The runout at the center bearing should not exceed 0.002 in [0.050 mm].

Measure the camshaft-bearing journal; the average maximum wear limit is 0.004 in [0.101 mm]. Should the measurement exceed the manufacturer's wear limit, the camshaft-bearing journals must be reground and undersized camshaft bearings installed. To grind the bearing journals, a crankshaft grinding machine is commonly used.

Measure the intake and exhaust lobes (Fig. 7-2) by taking micrometer readings at $A-C$ and $B-D$; then subtract $B-D$ from $A-C$. This will give you the lobe lift. Some manufacturers specify maximum lobe and lift wear, others do not.

Check the thrust-bearing plate washer for wear. Replace it if the wear area is rough or if the wear exceeds specification.

Installing Camshaft Gear and Camshaft   Before you install the camshaft gear, check it for wear, nicks, and scored or broken teeth. If the bore is so enlarged that a press-fit, say 0.002 in [0.050 mm], is no longer possible, or if the keyway is damaged, the gear must be replaced.

To install the camshaft gear, follow the same procedure applicable to installing the crankshaft gear

[see Installing Crankshaft (Unit 9)]. In addition, use a split yoke or parallel bar to support the camshaft in the press at its first bearing journal so that it rests parallel to the base of the press. Be certain that the camshaft is lubricated at the camshaft gear location, that the thrust plate or washers are in place, and that a new key is installed.

Using a new keyway, heat the gear evenly to 400°F [204°C]. Position and align the gear keyway to the key, then push the gear straight on.

Check the clearance between the thrust face of the camshaft (or bearing) and the thrust plate (or camshaft gear). The average clearance should measure about 0.004 in [0.101 mm]. It is wise to recheck the running clearance between camshaft journal and bearings, especially if you were not the only mechanic to work on the engine.

After lubricating the camshaft bearings and journals, carefully insert the camshaft. Do not scratch or

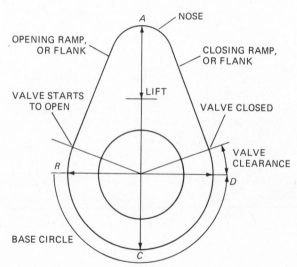

Fig. 7–2 Determining lobe lift.

Fig. 7–3 Checking camshaft-gear backlash using a dial indicator. (*Cummins Engine Co., Inc.*)

burr the bearings. Before the camshaft gear comes in mesh with the crankshaft gear, align the thrust plate or washer and the timing marks of mating gears. Slide the camshaft into place, and recheck that the timing marks of mating gears are aligned.

When working on Detroit Diesel engines you will find that some require standard timing while others require one-tooth advanced timing on the timing gear.

CAUTION Never drive the camshaft gear onto the shaft. To do so may push out the Welch or Cup plug or camshaft-bearing support at the rear of the cylinder block.

Your final check should be the gear backlash. Using a dial indicator and mount, as shown in Fig. 7-3, make certain that the dial stem rests squarely on the helical tooth before zeroing the dial.

The average backlash is from 0.002 to 0.006 in [0.050 to 0.152 mm]. Should the maximum backlash exceed specification, one or both gears must be replaced. It is wise now to recheck the end play.

NOTE If a Welch or Cup plug is used, install it at the rear camshaft opening.

Measuring Camshaft Wear and Lobe Lift with Camshaft Installed On occasion it becomes necessary to check camshaft wear and lift to locate engine trouble involving valve action or reduced compression. To check the lobe lift, install a dial indicator as shown in Fig. 7-4.

Turn the engine until the needle of the dial indicator no longer moves. The cam follower will then rest on the cam base circle. At this point, zero the dial indicator, rotate the engine until the needle of the dial indicator starts to move in the opposite direction. The follower will then leave the cam nose.

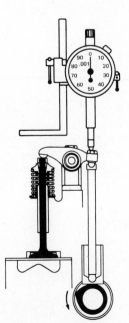

Fig. 7–4 Measuring cam lobe lift with the camshaft installed. *(Cummins Engine Co., Inc.)*

The difference between zero and the indicator reading is the lift for one valve only. Record the reading and compare it with the manufacturer's specification.

To check camshaft and bearing wear, remove whatever components or parts are necessary to pry the camshaft up. Install a dial indicator, either directly or indirectly, to the camshaft. Zero the dial and pry the camshaft up.

The graduation of needle movement indicates the total wear of the journal and bearing.

Idler and Auxiliary Gears Many engines, because of the camshaft location, require an idler gear between the camshaft and crankshaft timing gears. Most idler gears have an odd number of teeth to prevent meshing of identical teeth with each revolution.

As a result, tooth life is increased and gear noise reduced. Some engines require additional drive gears to drive the fuel-injection pump, the compressor, or the blower. The idler or auxiliary gears rotate on bushings or bearings on a dummy shaft and are pressure-lubricated.

Inspecting and Servicing Idler or Auxiliary Gear After all parts are cleaned and dried with compressed air, examine the dummy shaft for wear or grooving. When the shaft also acts as the bearing surface of the idler or auxiliary gear bushing, measure the shaft with a micrometer, and compare the diameter size with specifications. Measure the bushing and check for grooving and pitting. On engines which have a bushed dummy shaft, the idler or auxiliary-gear bore is also the bearing surface. Check the gear teeth for scoring, pitting, and chipping. To replace the bushing, firmly support the gear before pressing the old bushing out and the new bushing in, using the correct size bushing installer. Make certain that the oilholes are in line with the supply holes.

When tapered roller bearings are used, match-mark the bearing cup and cone, then, if reusable, the bearing cup and cone can be assembled as a unit. Always examine the bearing cups and cones carefully. If they show an uneven wear pattern, the bearings are running too loose. Also check the bearings and cups for pitting, scoring, and flat spots.

If a replacement is necessary, place the gear squarely on the bed of the press ensuring that it is well supported. To press the cup out, use a bushing adapter which covers the bearing cup but does not interfere with the bore, then press the inner bearing cup onto the idler gear until it bottoms against the shoulder of the gear. Place the outer spacer against the cup and press the outer bearing cup lightly against the spacer. Press the inner bearing cone onto the idler gear hub. When positioning the inner spacing ring, allow a gap of 180° from the oilhole.

Place the gear and the bearing onto the press so that the inner bearing is supported. To seat the bearings, press the outer bearing against the inner spacer while rotating the gear.

Checking Bearing Preload Tapered roller bearings require either a preload or end play (depending on application) to ensure long bearing life and to provide a rigid support for the rotating gear [see Tapered Roller Bearings (Unit 17)].

To check the preload for this type of mounting, mount the idler gear in a vise. Attach a small cord to a short piece of welding rod and lay the rod between the teeth of the gear, then wrap the cord a couple of times around the gear. Attach a spring scale, calibrated in pounds [kilograms] to the other end of the cord. Pull steadily on the spring scale. The idler gear should rotate freely and produce an even scale reading. The preload is correct when the pull required to produce an even scale reading is not lower than 1¼ lb [0.51 kg] and not higher than 6¾ lb [3.06 kg]. Pull variation should not exceed 2¾ lb [1.25 kg].

Questions

1. Why is it important to measure the camshaft journals, the lobes, and the runout?

2. Worn camshaft lobes of an engine could cause (a) valve noise, (b) valve "float" at higher engine speeds, (c) loss of engine power, or (d) camshaft follower breakage.

3. Camshafts are timed to (a) the valves, (b) the pistons, (c) the crankshaft, (d) engine firing order, (e) only have one valve open at a time, or (f) (e) is correct and (a), (b), (c), and (d) are wrong.

UNIT 8

Cylinder Sleeve

Checks and Measurements Check the cylinder sleeves, *when installed*, for vertical cracks, scoring, sleeve protrusion, and for dimension.

Since it is not possible to detect, externally, the condition of the lower or upper sealing surface, the cylinder sleeve should be removed. After removal, inspect the water jacket, counterbore, and (lower) cylinder bore. This will give you a clearer picture of possible malfunctioning components.

The cylinder sleeves shown in Fig. 8-1 are comparable to those you may be called upon to service. You should be able to recognize the cause of their condition and know how to correct it. The illustrations shown are divided into two groups: Group 1 failures are caused either by their not being serviced to specification or by careless installation, handling, or maintenance. Group 2 failures are caused by an unbalanced condition between coolant and temperature, by chemical or mineral content in the coolant and iron, or by improper maintenance of the coolant system (see Unit 20, "Cooling System").

After the cylinder sleeve has passed superficial inspection, it should be measured for out-of-roundness and taper with a bore gauge or inside micrometer. The average allowable taper and out-of-roundness for a 4- to 6-in [101.6 to 152.4 4-mm] cylinder bore is about 0.002 in [0.050 mm].

The cylinder sleeve should be measured in two directions for wear. First, parallel to the crankshaft and secondly, at a 90° angle to the cranksahft. Measurement should be taken just below the top of ring travel and also at several locations within the area of piston travel.

The normal wear pattern will show maximum wear at the top tapering off at about three-quarters ring travel (see Fig. 8-2). From the measurements taken you will know whether to hone or deglaze the cylinder sleeve to specification. (It may be that honing will enlarge the bore beyond the specified limit.) The average wear limit is about 0.008 in [0.2 mm] for a 4- to 6-in [101.6- to 152.4-mm] cylinder bore. Most manufacturers do not supply oversized pistons, therefore, when the sleeve bore exceeds specification it must be replaced.

Honing or Deglazing Cylinder Sleeve When the cylinder sleeve can be reused, the glazed surface area of the piston-ring travel must be refinished. Deglazing of the walls assures the shortest break-in period. By honing or deglazing the bore, the high spots of the cylinder-sleeve surface are removed, the bore is made round, and the taper is removed. At the same time, the surface is roughened to a precise microfinish. If the surface finish is too rough, the ring wear will be too great. Too rough a surface finish could also break the piston ring or the piston land. If the surface is too smooth or if the crosspattern is not within 25 to 35° (Fig. 8-2), the piston rings will not seat within 200 to 300 hours of operation. This prolonged break-in period will cause high oil consumption, lack of power, hard starting, and possibly will prevent the ring from seating perfectly.

Whether the cylinder sleeve should be honed or deglazed will depend on the variation of the sleeve bore measurements. If the variation of out-of-roundness and/or taper is small, say, 0.001 in [0.025 mm], but it is within specification, a rigid adjustable hone should be used to restore the cylinder-sleeve bore.

When honing or deglazing a cylinder sleeve, place the sleeve in a holding fixture or in an old cylinder block. Under no circumstances should a vise

GROUP 1 | GROUP 2

SCORING

POSSIBLE CAUSES:

Rings scuffing, piston scoring, improper cold start, honing debris, dirt in intake air, broken snap ring or ring land (insert), improperly installed cylinder sleeve or sealing (packing) rings, distortion of cylinder-block bore

CORROSION

POSSIBLE CAUSES:

Lack of water treatment, long storage without coolant drain, improperly maintained corrosion resistor, high mineral and/or chemical content in coolant

ABRASIVES IN COOLANT

POSSIBLE CAUSES:

Core sand or cleaning abrasives, sand or grit in coolant

CRACKING

POSSIBLE CAUSES:

Improper machining of flange counterbore, improper press fit, improper shimming of flange counterbore area, overtorqued head cap screws, improper protrusion, upper block distortion. Cracks at lower liner due to packing-ring problem, overheating or hot spot, corrosion, scoring.

CAVITATION

POSSIBLE CAUSES:

Aerated coolant, liner movement, high coolant temperature, inadequate coolant treatment, low coolant flow

SCALE

POSSIBLE CAUSES:

Inadequate coolant treatment, high mineral and/or chemical content in coolant

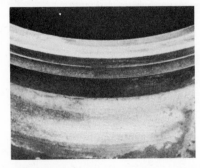

FRETTING

POSSIBLE CAUSES:

Improperly torqued head cap screws, uneven liner protrusion, loose liner press fit, damaged or worn head gasket

VERTICAL BREAKS

POSSIBLE CAUSES:

Handling damage, severe cavitation, result of piston seizure

DISCOLORATION OF LINER OD

POSSIBLE CAUSES:

Light or medium gray normal discoloration due to chromates in coolant system. Beige or brown usually denotes overheating

Fig. 8–1 Cylinder-sleeve-failure conditions and possible causes. (*Cummins Engine Co., Inc.*)

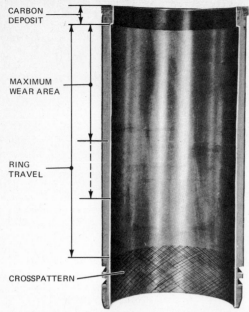

CARBON DEPOSIT

MAXIMUM WEAR AREA

RING TRAVEL

CROSSPATTERN

Fig. 8–2 Normal wear pattern on a cylinder sleeve. *(J. I. Case Co. Components Div.)*

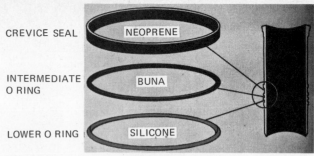

CREVICE SEAL — NEOPRENE

INTERMEDIATE O RING — BUNA

LOWER O RING — SILICONE

Fig. 8–3 Cummins cylinder-sleeve packing rings. *(Cummins Engine Co., Inc.)*

1. Dial indicator
2. Indicator contact point
3. Cylinder-sleeve tool
4. V section (cut from sleeve)
5. Mounting bar

Fig. 8–4 Measuring the out-of-square relation of the cylinder-sleeve counterbore with the centerline of the cylinder. *(Allis-Chalmers Engine Div.)*

of any kind be used. The sleeve must be held only at its flange to prevent distortion.

Deglazing Cylinder Sleeve Select a ½- or ¾-in electric drill for correct speed and power. Select glazing stones with a grit size of 180 to 220. These stones produce a surface finish of about 25 μm (microns) (25 millionths of a millimeter). Coat the surface generously with honing oil or kerosene. Insert the deglazer and adjust spring tension so that the stone rests with a light-to-medium pressure on the surface of the sleeve wall.

Start the drill and feel out the bore for high spots which cause the stone to grab. Deglaze the high spots first. Be careful not to allow the stone to protrude more than 1 in [25.4 mm] above the lower or upper sleeve surfaces. Move the drill fast enough over the total sleeve area to achieve the desired crosspattern.

After about 15 strokes, stop the drill and remove deglazer while holding the stones compressed. Check the bore for low spots and crosspattern. Measure for taper and out-of-roundness. Repeat the procedure until the low spots are removed and the bore is within specification.

After you have deglazed the cylinder walls, glass-bead (peen) the area of the sleeve flange and sealing rings. Wipe as much abrasive as possible from the walls and clean the surface with hot water and soap. Dry with compressed air and generously coat the surface with lubricating oil, then using white paper towels, wipe off the lubricating oil. Repeat this procedure until there is no evidence of residue on the white paper towels. It is also advisable to clean new cylinder sleeves to ensure that no abrasives are present.

Installing Wet-Type Cylinder Sleeve Be certain that the counterbore and the cylinder bore are clean. When reinserting cylinder sleeves, make sure that they are deposited into the original hole. The sleeve should fit in place without force and be turnable in the bore by hand. A sleeve which has been forced into place will not turn freely and will therefore require additional cleaning or an alignment check of the counterbore to the lower cylinder bore.

To install the seal ring, stand the sleeve on a clean workbench. Install new seal rings in the grooves on the sleeves or cylinder bore. The seal must not be twisted during assembly as leakage may result.

To prevent any twist, slip a finger or a pencil under the sealing rings and run it around the sleeve under the sealing rings two or three times, allowing the sealing rings to slide back into grooves without a twist.

Some manufacturers use three types of seal rings (see Fig. 8-3). It is therefore essential that you follow the manufacturer's instructions regarding position and lubricant. If there are not any instructions available concerning lubricant, use liquid soap or petroleum jelly. Otherwise apply a light coat of the recommended lubricant on the lower cylinder bore and sealing ring.

With extreme care slide the cylinder sleeve through the upper cylinder bore into the lower bore. With firm force, place the cylinder sleeve in position. (Some manufacturers recommend the use of a seal-protecting tool to ensure that the seal rings are not cut on the sharp edges of the bores when being installed.) After the cylinder sleeve is installed, check the protrusion and the out-of-square relationship of the cylinder-sleeve counterbore with the centerline of the cylinder (Fig. 8-4).

Installing Dry-Type Cylinder Sleeve There is no difference in service or installation procedure between a Detroit Diesel dry-type and wet-type cylinder sleeve. The dry-type sleeve should push-slide into position without the use of excessive force and without turning. Excessive force will cause cylinder-sleeve distortion and may damage the rings or piston. If the sleeve does not slide into position, remove it, turn it 90°, and try again. If it still will not slide into position, remove it once more, and measure the cylinder bore. It may need rehoning.

If a shim is required to obtain proper cylinder-sleeve protrusion (or retraction), install the shim between the insert and the cylinder block. Mark the sleeve and cylinder bore so that the sleeve can be returned to the same bore and position. When removing the sleeve, do not remove the shim or the insert.

NOTE Dry sleeves having no counterbore flanges are press-fitted into the cylinder bore. This type of sleeve may be placed in dry ice or alcohol for 25 minutes and then pressed into position. Later, if necessary, it should be honed to specification to assure proper piston-to-sleeve clearance.

Questions

1. Explain in detail how to determine the serviceability of a cylinder sleeve?

2. When is it necessary to deglaze a cylinder sleeve?

3. When is it necessary to hone a cylinder sleeve?

4. Why should the cylinder sleeve and the cylinder bore be marked after being fitted?

5. Explain in detail how to deglaze a cylinder-sleeve bore.

6. Before installing a wet cylinder sleeve what two important checks must be made to the outer surface of the liner which is exposed to the water?

7. When installing a wet cylinder sleeve into a cylinder block, what four checks must be made prior to installation?

8. When installing a wet cylinder sleeve, what precautions must be taken with regard to the lower cylinder sleeve seals?

9. When installing a dry cylinder sleeve into the cylinder bore (of a two-cycle engine) what four checks must be made *prior* to installation?

10. What is the difference between a clearance fit and an interference fit (dry sleeve)?

11. Dry sleeves which are fitted too loosely in the cylinder bore could cause (a) poor cooling, (b) head-gasket failure, or (c) too tight a piston fit.

12. A wet-type cylinder sleeve must (a) fit tight at the bottom so that it will not leak, (b) be a loose fit so that it will not distort under heat, (c) be sealed only at the bottom, (d) not protrude above the cylinder block, (e) have the same fit as a dry liner, or (f) have clearance at the top and bottom.

UNIT 9

Crankshaft

Preliminary Inspection Before the crankshaft can be inspected remove all plugs and clean the crankshaft, preferably in a hot tank. Alternatively, you can clean the shaft and oil passages with solvent or fuel, using a wire brush. Blow-dry the crankshaft with compressed air.

Visually inspect the journals for scoring, chipping, grooving, excessive wear, or signs of overheating. Overheating will usually be indicated by discolored or blue bearing journals.

Check the oil-seal surface for wear or grooving. If it cannot be restored by polishing with emery cloth

or crocus cloth, you may have to use a spacer or a sleeve. Check the keyways for cracks and wear.

Locating Fine Cracks After the crankshaft passes a superficial visual inspection, it must be checked for further cracks which are not immediately visible. Several methods are used: dye penetrants, magnetic field and iron particles, magnetic field fluorescent, and x-ray. The magnetic field fluorescent method is most commonly used since it is the least expensive and yet detects very fine cracks. The other methods (with the exception of x-ray) are not as dependable.

If you use the magnetic field fluorescent method to test the crankshaft for flaws, first spray the shaft with the special solution containing magnetic fluorescent particles. These will glow under black light (invisible ultraviolet rays). Slide the powerful electric magnetic ring slowly over the crankshaft with the light pointing at the shaft. Magnetic particles are attracted to the edges of any cracks by the magnetic force. Cracks or flaws are visible under the black light as a white line, whereas the undamaged shaft shows up as dark blue.

Causes of Crankshaft Failure The crankshaft seldom fails (breaks) when operated under normal conditions but should this occur, the cause must be determined promptly. Some conditions which will promote failure are:

1. Improper storage or handling.
2. Overspeeding the engine. This can produce crankshaft vibration that exceeds vibration damper control.
3. Incorrect radii at journal fillets and oilholes. Either can create fatigue cracks (Fig. 9-1).

Causes of crankshaft failure over which you as a mechanic have control are as follows:

1. A vibration damper that is loose due to insufficient torque or that has been damaged during storage or installation. Under such circumstances the crankshaft vibration cannot be controlled and the result is torsional stress at the connecting-rod journal areas.
2. Improper bearing cap fit, loose bearing cap, or an obstruction between the bearing cap and the upper cap surface. This leaves the crankshaft unsupported causing it to bend with each revolution.
3. Misalignment of the main-bearing bores, too high a runout of the shaft, or worn bearings. Any of these can cause the crankshaft to bend in two directions.
4. Misalignment of the torque converter, transmission, generator set, etc., to the flywheel housing. Misalignment causes excessive side load on the rear main-bearing and connecting-rod journals.
5. Excessive end thrust or improper end clearance. In either case the result is a lack of oil between the thrust wear surfaces causing fatigue cracks, wear, and overheating. The combination of abrasion and heat can wipe out a main bearing, leaving the crankshaft without support. This imposes a bending stress on the shaft.
6. Inadequate oil improper grade of oil, or oil that is contaminated due to careless engine service. (See Unit 12, "Engine Oil.")

Measuring Crankshaft Each bearing journal should be measured with a micrometer, as shown in Fig. 9-2. Start with the main journal. On each journal, write the out-of-roundness measurement on the balance weight (plus or minus), so that when measuring the runout (alignment) the out-of-roundness is taken into consideration. Failure to take the out-of-roundness measurement into your calculation will affect the dial indicator runout reading.

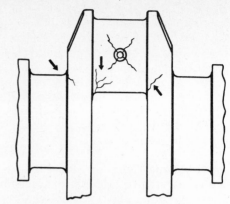

Fig. 9—1 Common areas of fatigue cracks on a crankshaft. (Cummins Engine Co., Inc.)

Fig. 9—2 Measuring the main crankshaft journal.

Continue to measure each journal until the smallest diameter is obtained. Take three measurements at the smallest location on the journal—one on the left, one at the center, and one on the right side—and record the readings. Take a separate set of measurements 90° to the first set and record them. The difference between the two sets of recorded figures will give you the out-of-roundness and the taper of the journals. Do not forget to measure the thrust flange for wear (Fig. 9-3).

When measuring the connecting-rod journals, you will notice that the wear, out-of-roundness, and also perhaps the taper, is greater than on the main journals. The reason for this is that the connecting-rod journals have less oil pressure and volume plus the additional factor of fluctuation.

To measure the crankshaft runout, support the crank on its front and rear main journals in lubricated V blocks or between the centers in a lathe or crankshaft grinder. For large models, it may be necessary to support the the crankshaft on its center journal to prevent sagging.

Place the dial indicator on the center main journal first. Set the dial to zero, slowly turn the crankshaft, and then write down the reading. Be sure that the dial indicator is resting on a smooth surface and that the base is securely mounted so that it cannot move. The average allowable runout reading is about 0.0025 in [0.0635 mm].

Check all main-bearing journal readings. Say, for example, one journal reading is +0.001 in [0.025 mm] and the adjacent journal reading is 0.002 in [0.050 mm]. In this case the runout would be 0.003

Fig. 9–3 Measuring the thrust flange for wear. (*Cummins Engine Co., Inc.*)

in [0.076 mm]. If the runout exceeds specification the shaft must be straightened. **CAUTION** Do not forget to take the out-of-roundness into consideration when measuring runout.

When all journals are within specification and in a serviceable condition and the runout is within the specified limit, the crankshaft may then be re-installed. If one journal is more than the allowable out-of-roundness or taper tolerance, or if the surface shows groove or score marks greater than 0.0005 in [0.0127 mm], then the crankshaft must be reground. Sometimes the shaft will not sustain further regrinding and must be replaced.

Polishing Crankshaft If the journals are within the out-of-roundess or taper tolerance, and if the surfaces show only minimum roughness, grooving or scoring, say 0.0005 in [0.0127 mm], the crankshaft should nevertheless be polished before being installed.

To polish the crankshaft, mount it in a lathe and rotate the shaft about 100 rpm. Wrap wet and dry 600-grit emery cloth or crocus cloth around the journals. Apply moderate force in a shoeshine motion. After polishing the shaft, thoroughly clean its surface and oil passages to remove any abrasive residue.

Selection of Bearings Most manufacturers furnish replacement bearing shells in undersizes of 0.002, 0.010, 0.020, and 0.030 in [0.050, 0.254, 0.508, and 0.762mm] for service purposes. To determine whether to use standard or 0.002-in [0.50-mm] undersize bearing shells, measure the journal wear and compare your measurement with the specification. Be sure that minimum bearing clearance is maintained. The average main-bearing clearance is about 0.004 in [0.101 mm] for a 3-in [76.2-mm] journal, and about 0.006 in [0.152 mm] for a journal above 3 in. If the minimum clearance is not obtained, the crankshaft bearing may seize, ruining both the bearing and the shaft.

Grinding Crankshaft Crankshaft grinding is a special field which requires precision grinding equipment and a specialized operation. It is not practical for most service shops to maintain such equipment

since only a few crankshafts or camshafts need to be ground. It is cheaper to sent the camshaft or crankshaft to a specialty shop which keeps these grinding tools and a qualified machinist available.

Before grinding the crankshaft, your first step should be to measure the shaft for runout. If the measurement of the runout is more than specified, the crankshaft must be straightened before being ground.

First, the 80-grit grinding wheel is dressed, and then the shaft is mounted in the cross-sliding chucks and adjusted to the grinding wheel. Next, the shaft journals are measured. The journal with the lowest measurement is used to determine the undersize of all main or connecting-rod journals.

The crankshaft is driven in one direction and the grinding wheel in the opposite direction. Large shafts must be supported by a steady bearing. The operator must proceed with caution to avoid localized heating, which may produce cracks, and to prevent wheel chatter marks. (Coolant should be used generously.) Oilholes and sharp edges must be removed within the radius specified by the engine company.

The journals then are polished to an undersize diameter of approximately 0.0002 in [0.005 mm]. This removes any grinding marks and permits a finer microfinish when polishing the shaft. At this time the fillet radius is also polished. The shaft is again checked for cracks, after which the crankshaft is thoroughly cleaned, lubricated, and stored. After cleaning (and before storing), the machinist usually installs all drill plugs.

Installing Crankshaft Seal Sleeve When a sleeve (wear ring) is used and the seal surface of the front or rear seal area is damaged, the sleeve must be replaced (Fig. 9-4).

To install a sleeve, first heat it in an oven or heat it with a heating torch (not with a cutting torch) to about 400°F [205°C]. Then slide the sleeve in place. The temperature can be checked with a heat stick. If you do not heat the sleeve, use the correct sleeve driver and drive the sleeve in place.

Installing Crankshaft Whenever a new or reground shaft is to be installed, new bearings and thrust washers must be used. **NOTE** Always check the components for the correct part number.

Fig. 9–4 Installing a sleeve (wear ring) on the crankshaft. (*Mack Trucks Canada, Ltd.*)

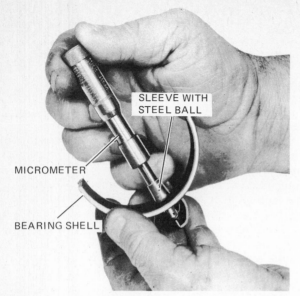

Fig. 9–5 Measuring bearing-shell thickness. (Allis-Chalmers Engine Div.)

Fig. 9–6 Installing the upper bearing shell. (Cummins Engine Co., Inc.)

The bearing size must coincide with the main and connecting-rod bearing sizes and with the size of the thrust washers which is stamped on the crankshaft. The most accurate way to measure the bearing size or to determine the amount of wear, is to measure each shell with a micrometer, as shown in Fig. 9-5, and compare the reading with the manufacturer's specification.

Check that all plugs are correctly installed.

Before installing the crankshaft gear, coat the shaft at the gear location with high-pressure lubricant. Install the gear key and heat the crankshaft gear to about 400°F [205°C]. Position the gear keyway to the key. Using the correct driver, work as quickly as possible while driving the gear onto the shaft in order to prevent gear seizure due to cooling.

To install the main bearings, remove the main-bearing caps and place them, in sequence, on a clean surface. Recheck the bearing surfaces of the caps and the upper bearing supports for nicks and burrs. Remember that 15 percent of all bearing failures are caused by improper installation or careless assembly.

Remove the new bearing shells from the packaging material one at a time. Before installing them,

make sure that the backs of the bearing shells are completely free from dirt, grit particles, and oil. Correctly install the upper bearing shell (with oil-hole and groove) in the upper bearing support (Fig. 9-6). Install the lower bearing shell (without oilhole or groove) in its bearing cap.

NOTE Make sure that all oilholes are in line because not all oilholes are drilled at the same angle. Do not use a lubricant of any kind between the bearing shell and bearing support.

Make certain that the bearing shells snap into place and that the lock tang (of the shell) is in the groove of the cap or in the upper bearing support. Also, when using special thrust main-bearing shells, be certain that they are fitted snugly into the bearing seats.

Check the crankshaft for identification marks. If the thrust surfaces are refinished, oversize washers which coincide with the markings should be used, if applicable.

Install the upper half of the crankshaft thrust washers and then install the lower half into the cap.

NOTE The grooved sides of the crankshaft thrust washers must face the thrust surface of the crankshaft to assure their adequate lubrication.

Clean the crankshaft main journals and lubricate the journals, bearing surfaces, and thrust washers with the lubricant recommended by the manufacturer. (Some manufacturers recommend special grease while others recommend engine oil.)

Lift the crankshaft into place and install the bearing caps. Lubricate all hex or stud bolt threads and bearing surfaces if so recommended, then screw the bolts snugly. Then check the shaft to see that it rotates freely.

Starting with the center-cap hex bolt, screw each bolt uniformly tight to its specified torque. After the center-cap hex bolt, tighten the hex bolt of the nearest main-bearing cap to the right. Next, the bolts of the main-bearing cap to the left are tightened. Continue in this sequence until all the bolts are uniformly tightened to specification.

Some manufacturers recommend the torque-turn method to tighten cap bolts. This method places an additional tension on the hex bolt or stud. When you have to use the torque-turn method, the first step is to tighten the hex bolts to the recommended torque value, using the same sequence just outlined.

With a washable marker, identify each hex bolt or nut to indicate its position on the main-bearing cap. Turn the nut or bolt an additional turn as suggested in the service manuals. This will stretch the hex bolt and place additional tension on it, holding the bearing cap firmly in place. Regardless of the method used, if the main-bearing caps and bearing shells are correctly installed, the crankshaft bore, the runout, the end play, and the bearing clearances are within specification, the crankshaft will turn freely.

How to Measure or Check Bearing Clearance The bearing clearance may be measured by using a plastic strip, such as Plastigage, or virgin lead wire manufactured for this purpose. With either method, you must remove the main-bearing or connecting-rod

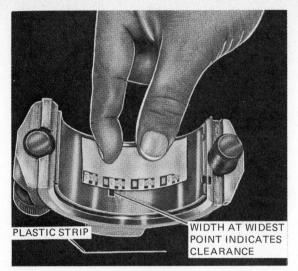

PLASTIC STRIP

WIDTH AT WIDEST POINT INDICATES CLEARANCE

Fig. 9–7 Checking bearing clearance. *(Mack Trucks Canada, Ltd.)*

cap and clean any lubricant from the bearing shells and exposed journals. When using virgin lead wire to check the bearing clearance, make sure that you relubricate the bearing journals and bearing shell. Do not, however, use any type of lubrication when using plastic strip because the plastic material is soluble in oil.

After placing a suitable length of virgin lead wire or a plastic strip across the bearing surface, install the cap and bolt, then torque the bolts or nuts to specification. On installation, the lead wire or plastic strip will crush (flatten) to the thickness of the bearing clearance between the bearing shells and the journal. When using a virgin lead wire, remove the bearing, gently remove the virgin lead and measure it with a micrometer. If a plastic strip is used, however, it is not removed. Instead, the graduation scale from the strip's original envelope is used to measure the width at the widest point of the flattened plastic material (Fig. 9-7).

Crankshaft End Play Crankshaft end play can be checked with a feeler gauge or with a dial indicator. The average end play is around 0.010 in [0.25 mm].

To check the end play, force the crankshaft with a soft-faced hammer or prybar against its thrust surface. Insert various thicknesses of feeler gauge stock between the thrust surface of the crankshaft and thrust washer until one slides in with only light drag. When using the dial indicator method, force the crankshaft to one side, set the dial indicator in place, and zero the dial. With a prybar, force the crankshaft in the opposite direction to obtain the end-play reading.

Replacing Main Bearings with Crankshaft Installed Occasionally main bearings must be replaced without the engine being removed. In such circumstances, the procedure set out below should be followed:

1. Steam-clean the engine thoroughly.
2. Follow all safety rules with regard to hydraulic equipment and accessories.

3. Allow yourself room to work without obstruction.
4. Drain the oil from the crankcase, then remove the oil filters, oil pan, oil pump, and also the oil tubing, if used.
5. To allow the crankshaft to turn freely, remove the injectors from the engine (or release the compression release if one is used). Cap all the fuel lines.
6. Remove one main-bearing cap at a time and check the bearing journals. Measure the bearing wear with a micrometer. If a plastic strip or virgin lead is used, apply a slight force with a jack against the crankshaft to keep the shaft in contact with the upper bearing shell. This assures a true measurement since the weight of the crankshaft is not resting on the measuring material while the main-bearing cap is being torqued. To measure the main-bearing journals for out-of-roundess, take another measurement after the crankshaft has been turned 90°.
7. Evaluate all your measurements, the surface condition of the journals, and the condition of the bearing shells. As a result you will know whether to install new standard shells, or whether, in fact, it is necessary to regrind the shaft. **CAUTION** Do not install a mixture of new and old or standard and undersize bearing shells. To do so will cause an uneven support of the crankshaft and, eventually, its failure.
8. Remove the lower bearing shell from the bearing cap and check the cap for nicks and burrs. Remove the upper bearing shell by turning the crankshaft to a position where you can insert a special removal lug or a cotter pin (with the head flattened) into the oilhole (Fig. 9-8). Turn the crankshaft in a direction that will allow the asserted force to induce the shell to rotate tang first from the bore. Make sure that the insert does not interfere with the bore.
9. If it appears practical to reuse the bearing shell, you should check the bearing spreads against specification to determine if the shell will fit snugly into the bearing bore. If no specifications are available, make sure that the shell snaps into place or can be gently forced into place. It is not always necessary to discard bearing shells because they do not fit snugly into the bearing bore since excessive or insufficient spread can be corrected as shown in Fig. 9-9.

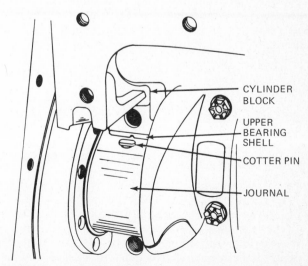

CYLINDER BLOCK

UPPER BEARING SHELL

COTTER PIN

JOURNAL

Fig. 9–8 Removing the rear main-bearing upper shell. *(Allis-Chalmers Engine Div.)*

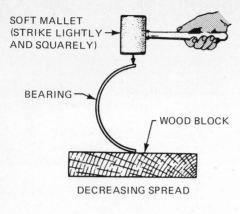

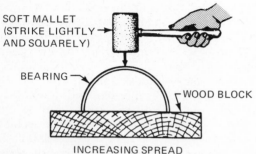

Fig. 9–9 Adjusting bearing spread. *(International Harvester Co.)*

10. Installation of the bearings requires as much attention as installation of the main bearings on a removed crankshaft. Furthermore, it is difficult to see dirt or lint in the upper bearing bore. It should therefore be checked very carefully before rolling the lubricated upper shell around the crankshaft journals into place. Remember, too, that the upper bearing bore must be installed in a direction opposite to that used when it was removed, until the tang on the bearing shell is positioned in the bearing support slot. Lubricate the bearing shell in the main bearing cap and place it on the upper bearing support. Install hex bolts or nuts and tighten them to the specified torque.

NOTE: On Cummins engines, you must place the bearing shell on the main journal first, position it and then place the main bearing cap in position.

11. Installation procedure of the thrust washers with the crankshaft removed is the same as that with the crankshaft in place. **NOTE** Check end play.

12. Always check the operating condition of the lubrication pump and the pressure relief valve before reinstallation.

13. The remainder of the installation is simply a reversal of the removal procedure.

14. Fill the crankcase with the lubricant specified by the engine manufacturer, and replace oil filters. Run the engine up to operating temperature and recheck for oil leaks. Check the oil level.

Engine Counterbalancer Some engines are equipped with an inertia balancer to counteract the secondary forces created by piston acceleration (Fig. 9-10). It should not be confused with a torsional dampener, which is used on some engines to reduce the torsional stress in the crankshaft, and not as a means of reducing secondary inertia force.

Although the two sets of pistons in a four- or six-cylinder in-line engine are equal in weight and move in opposite directions, their vertical inertia forces do not completely neutralize one another. Since the secondary forces, which tend to limit the smoothness of an engine, occur at twice the speed of the engine, these vertical forces can be canceled by rotating two counterweights at twice the speed of the crankshaft.

When the total secondary force is downward, it is counteracted by the two counterweights exerting an equal upward force (Fig. 9-10a). When the total secondary force is upward, it is counteracted by the counterweights exerting an equal downward force (Fig. 9-10d).

When the secondary forces are canceled out (because of piston position), the counterweight forces are also canceled out (note their positions in Fig. 9-10b and c). Only the natural inherent balance of the engine exists.

Questions

1. What is the main difference between the crankshaft of an in-line engine and the crankshaft of a V engine?

2. Draw the crankshaft throws of a six-cylinder in-line engine which has a firing order of 1-5-3-6-4-2.

3. List the visual inspections you must make when checking a crankshaft.

4. What measurements must be taken to determine the condition of a crankshaft? Explain how these would be made.

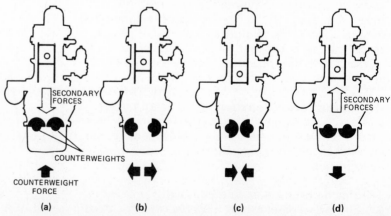

Fig. 9–10 Action of the counterbalancer. *(a)* total secondary force downward, *(b)* and *(c)* secondary forces canceled out, *(d)* total secondary force upward. *(International Harvester Co.)*

5. Describe how to measure one main journal.

6. List several causes of crankshaft failure.

7. What is the reason for measuring crankshaft runout?

8. List the steps to be taken to install a crankshaft.

9. Why is it essential to follow the manufacturer's torque sequence when tightening main-bearing cap screws?

10. Explain how to measure the running clearance of the main bearings using plastic strip.

11. Outline the procedure to replace a set of main bearings with the crankshaft installed.

12. Why do crankshafts require end play? How is end play checked?

13. The average main-bearing clearance for a 76.2-mm journal should be about (a) 0.152 mm, (b) 0.330 mm, (c) 0.008 mm, or (d) 0.076 mm.

UNIT 10

Connecting Rod

Connecting-Rod Failure Although design and construction ensures negligible connecting-rod problems, this does not mean connecting-rod problems do not exist. However, when difficulties arise, they are inevitably due to improper operation or inadequate service and maintenance.

Connecting-rod failure over which you as a mechanic have no control is caused by stress raiser fatigue, overspeeding, seized pistons, foreign material in the cylinder, or hydraulic lock. However, connecting-rod failure due to worn bearings or bushings should not occur because it can be prevented by conducting a good maintenance program. Proper part assembly, torquing to specificaton, and checking and measuring the connecting rod are among the good procedures that will ensure longer service life of the connecting rod.

Inspection Before you make any inspection or perform any service, you should thoroughly clean the connecting rod. Make certain that the rod and cap are kept as an assembly. Before you check the connecting rod for alignment, magnaflux all the connecting rods to reveal cracks. Pay particular attention to the center web above the large connecting-rod bore and below the piston-pin bore because these areas are more prone to cracking.

If you have determined that the connecting rod is serviceable, you should then check that the connecting-rod bolts or nuts rest squarely on the machined surface without interfering with the connecting-rod or cap fillet.

Check that the threads are not distorted and that the bolt diameter is within specification.

Make sure that the cap and the connecting-rod holes are not worn or enlarged. (The connecting-rod bolts must fit tightly in the bores.)

You should then check the piston-pin and crankpin bore and measure them in several places with an inside micrometer, a dial bore gauge, or a telescopic gauge. The average allowable tolerance for the crankpin bore is 0.001 in [0.025 mm] and for the piston-pin bore it is 0.0005 in [0.012 mm]. When either bore is out of specification, the bores must be resized or the connecting rod must be replaced.

When the piston-pin bushings are not removed, check and measure them for wear. The average wear limit is 0.005 in [0.127 mm] and the average clearance (when new) between the piston pin and bushing is 0.001 to 0.002 in [0.025 to 0.050 mm].

Servicing Piston-Pin Bore When it becomes necessary to rebore the piston-pin bore, mount the connecting rod in a holding fixture and bore or ream the hole to specified size. Remove any sharp edges with a taper reamer. (Follow the manufacturer's instructions for using the boring machine.)

Replacing Piston-Pin Bushings Position the connecting rod in a holding fixture or follow the manufacturer's recommendations. With a bushing driver, press or drive out the bushing. Before you reinstall new standard or oversize bushings, lubricate the bore and carefully align the oilhole in the bushing with the hole in the piston-pin bore. Using the same bushing driver, press or drive the bushing flush with the outside of the bore. Do not concern yourself with the location of the bushing split but rather with the oilhole alignment.

Some manufacurers recommend broaching the bushing (that is, pressing a broaching tool into the bushing bore) to ensure that the bushing fits tightly against the bore surface to prevent the bushing from turning and to give a uniform bore size. A tight-fitting bushing also transfers heat well.

Resizing Crankpin Bore If you have to resize the crankpin bore, you first should determine if a resizing of the bore is possible. If the connecting rod is too short, it will be impossible to resize the crankpin bore. Therefore you first should measure the length of the connecting rod with an alignment fixture or with a checking fixture. To set up the connecting rod for measuring, remove the piston-pin bushing, tighten the cap to specification, and position the rod as illustrated in Fig. 10-1.

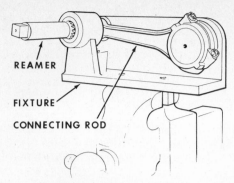

Fig. 10–1 Checking connecting-rod length. *(J. I. Case Co., Components Div.)*

Fig. 10–2 Precision connecting-rod grinding.

Fig. 10–3 Measuring the bore with the special gauge on the honing machine.

Fig. 10–4 Honing the crankpin bore.

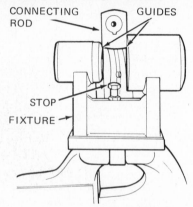

Fig. 10–5 Simple connecting-rod alignment tool. *(J. I. Case Co., Components Div.)*

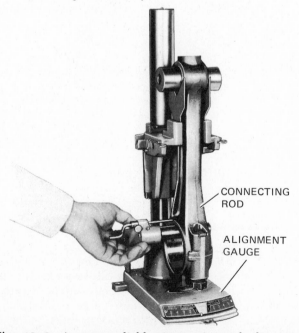

Fig. 10–6 A more reliable connecting-rod alignment tool. *(Allis-Chalmers Engine Div.)*

Check the length of the connecting rod using a master pin instead of the reamer. If the master pin fits into the piston-pin bore, the connecting-rod crankpin bore can be resized. Remove about 0.009 in [0.23 mm] from the connecting-rod and cap surface with a connecting-rod resizing grinder (Fig. 10-2). Make certain that the cap or connecting rod is clamped sufficiently tight so that it cannot move during the grinding operation. This will assure alignment of the connecting-rod bolts with the bores and proper contact of the mating surfaces.

It is usually necessary to lap the mating surfaces to ensure at least an 80 percent contact. The contact area can be checked by bluing the mating surfaces, then rubbing them together. The area of contact should be around the connecting-rod bolts and not toward the center of the crankpin bore.

Bolt the mating cap to the connecting rod. When you torque the bolts to specification, hold the connecting rod in a special vise or hold it in a soft-jaw vise to prevent it from twisting or bending. Measure the bore with an inside micrometer or with the special gauge on the honing machine to determine if it is out-of-round (see Fig. 10-3).

Honing Crankpin Bore   When using the Sunnen honing machine, make sure that you select and install the correct hone. The stones selected should give you a 70-μm finish bore or better. True the stones.

Check the bronze bearing supports. When necessary, use an oil stone to remove high spots to ensure a true bore.

Slide the crankpin bore over the hone, adjust the stone tension and direct the spout of the honing oil onto the lower part of the hone. Support the rod on the support arm of the honing machine. Start the motor and again adjust the stone tension until you feel no chatter. Then move the connecting rod back and forth over the stone while maintaining snug stone tension (Fig. 10-4). Frequently check the inside dimension to prevent an oversize bore. The average allowable tolerance is 0.0005 in [0.012 mm].

Checking Connecting-Rod Alignment   Many different alignment fixtures are used to check connecting-rod alignment; one of the simplest is shown in Fig. 10-5. To check the alignment using this type of tool, place the crankpin bore on the large pin and tighten the connecting-rod bolts to specification. Push the piston-pin bore end down between the guides. If the piston-pin bore end does not fit between the guides, the rod is twisted or bent. The average allowable twist or bend is 0.008 in [0.20 mm].

A more reliable alignment tool is shown in Fig. 10-6. Direct measurements for bend, twist, and connecting-rod length can be taken with this tool.

Reaming Piston-Pin Bushing   The connecting-rod length is very important for establishing the compression ratio and in preventing piston interference with the valves. The piston-pin bushing, therefore, must be reamed in a fixture to assure a precise center-to-center length.

When reaming a bushing, gently insert the reamer into the guide bushings to ensure perfect alignment and then carefully ream the bushing. Do not force (crowd) the reamer or turn it in a counterclockwise direction because this will produce a rough finish. After reaming, check the smoothness of the bushing surfaces for even thickness and then measure the diameter. When the thickness of the bushings varies or the reamer has failed to give a full surface contact, the connecting rod is bent or twisted. It must be straightened or discarded.

Crosshead-Piston Connecting Rod   The checks, tests, and service procedures for a crosshead-piston connecting rod are almost identical to that of the trunk-type-piston connecting rod. But, with the crosshead type you must check saddle alignment.

Questions

1. State the three ways in which alignment is maintained between the connecting rod and the cap.

2. Outline the procedure used to measure the connecting-rod bore when using an inside micrometer.

3. Why is it very important to have the length of the connecting rod within correct specification from the center of the pin bore to the connecting-rod bore?

4. If a connecting rod is installed which is twisted beyond the allowable specification, how would it affect the piston, the connecting rod, the connecting-rod bearing, and the journal?

5. Why should a piston-pin bushing be reamed in a holding fixture rather than honed?

6. Bent connecting rods may be caused by (a) water in the cylinders, (b) preignition, (c) excessively worn connecting-rod bearings, or (d) excessive crankshaft end play.

7. One of the most common connecting-rod problems encountered when replacing bearings is (a) bent rod, (b) twisted rod bolts, (c) rod bore stretched out-of-round beyond specifications, or (d) mismatched bearings.

8. Connecting rods and piston assemblies should (a) be given a complete and careful alignment check with every ring job; (b) be corrected if out of alignment; or (c) be replaced if out of alignment.

Piston and Rings

Maintenance To ensure long service life of the piston, piston rings, and cylinder sleeve, use only those components recommended by the manufacturer when replacements are required. Do not, through careless workmanship or through poor storage, allow dirt to get into the components. You should recheck completed work before you start a new assignment.

A good maintenance program begins with a thorough break-in run [see Break-in Run (Unit 50)]. Give adequate attention to starting, hot shut down, daily checks of coolant, oil level, and the air-intake system.

Tune up the engine regularly, and repair defective components as soon as any malfunction is evident. Stop all oil or coolant losses instantly. Change the oil and filters regularly. Do not adjust the governor or fuel above specifications.

Do not neglect the exhaust system. Replace a damaged muffler or pipe that may cause restrictions and increase combustion temperature.

Inattention to any of the above maintenance practices can lead to early engine failure.

Piston Failure Figure 11-1 illustrates piston failures that result from poor workmanship or poor maintenance.

Figure 11-1a shows a piston with score marks on each side of the piston-pin bore. These marks could have been caused by low coolant, restricted coolant flow, continuous improper cold starting, hot shutdown, or improper piston-pin-to-bore fit.

Figure 11-1b shows a piston with irregular score marks from the skirt area to the crown. These marks could have been caused by low coolant, restricted coolant, a partly plugged radiator, high coolant temperature (due to defective radiator cap, loose fan belt, damaged water pump), continuous improper cold starting, hot shutdown, lack of lubrication, overfueling (due to damaged turbocharger), faulty injector, and/or improper adjustment of the fuel-injection pump.

Figure 11-1c shows a piston with vertical score marks. These marks have developed over a long period of time on the thrust and antithrust side of the piston skirt. They could have originated from scuffed rings, overfueling, piston distortion due to

(a) (b) (c)

(d) (e) (f) (g)

Fig. 11–1 Piston failures due to (a) four-point scoring, (b) irregular scoring, (c) thrust or antithrust scoring on skirt, (d) burning, (e) damaged crown, (f) ring fracture, (g) piston fracture. (*Cummins Engine Co., Inc.*)

thermal expansion, distortion of cylinder-sleeve bore and/or cylinder-sleeve O rings.

Figure 11-1d shows a piston that has started to erode (burn) on the outside of the piston crown. The erosion could have originated through detonation, restricted air supply, excessive use of ether (for starting), overfueling, faulty injector, and/or enlarged nozzle orifice.

Figure 11-1e shows a piston with a damaged crown. The damage may have originated from improper ring installation which caused the ring or the ring land to break, from valve or valve insert failure, from extremely worn connecting-rod bearings, or from foreign particles in the combustion chamber which entered through the intake system.

Figure 11-1f shows a damaged piston due to a broken ring land. This damage could be the result of improper compression-ring installation, ring breakage or ring-land damage prior to installation, or manufacturer's casting porosity. It may also be the result of excessive use of ether for starting, or extreme detonation.

When the piston (including the skirt area) shows black or brown deposits, it is an indication that engine oil or combustion gases have passed by the rings. Combustion gases can pass by the rings if there is a poor ring seal, poor quality oil used, high oil level in the crankcase, high oil temperature, aerated lubricant, or excessive time between oil changes.

Piston fracture, such as that shown in Fig. 11-1g, can be caused by detonation, overfueling, or continuous overloading of the engine. Piston fracture, when it occurs due to defects in the manufacturer's casting, will commonly show on top or below the crown.

Piston-Ring Failure Ring scuffing (Fig. 11-2) is by far the major cause of hard starting, loss of power, and high oil consumption. It can also cause piston and cylinder wall scoring.

Ring scuffing may originate during the assembly by failure to maintain piston and/or ring specifications, by improper surface characteristics, and/or improper maintenance of the lubricant, coolant, and air-intake systems. Other causes of ring scuffing are: improper adjustment of fuel-injection pump, turbocharger failure, thermostat failure, plugged radiator, nonworking radiator shutter, improper engine break-in, or misuse of the engine during the first few hundred miles of operation.

The use of keystone rings virtually eliminates ring sticking. However, should this problem occur, the cause may be traced to operating during cold ambient temperature, running the engine when it is too cold, allowing too long an interval between oil changes, or heating the engine intermittently.

Ring breakage usually arises from a damaged piston crown and/or piston land, or it can occur after excessive detonation. The greatest percentage of ring breakage, however, is due to improper installation procedures, incorrect ring-gap clearance, or damage to the ring land while inserting the piston into the cylinder sleeve. Another cause of ring breakage, though not common, is the failure to remove the ring

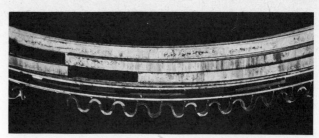

Fig. 11-2 Ring scuffing. (*Cummins Engine Co., Inc.*)

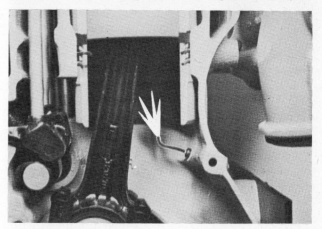

Fig. 11-3 Piston cooling jet, showing coolant flow. (*Cummins Engine Co., Inc.*)

ridge. This can cause the compression ring to break.

Piston-Cooling Jets or Nozzles Many engines are equipped with a jet or a nozzle to direct lubrication oil to the underside of each piston crown (see Fig. 11-3). In addition to lubricating the pin, this oil absorbs the heat from the pistons.

Lubrication oil is supplied either from the main-bearing cap, from a separate oil manifold, or from the camshaft oil gallery.

As you become aware of the importance of piston cooling, you will realize the importance of the position and the condition of the nozzle. When it is obvious that the nozzle is damaged or deformed, do not attempt to realign it; it should be replaced. (A plastic target is sometimes used to check the spray area.)

Piston Inspection and Service Pistons should be replaced if they are fractured or burned, or if they show a damaged crown, a fractured or burned ring land, or damaged piston bores. When a piston has passed inspection, it should be cleaned internally and externally with solvent or fuel to remove any gummy substance or carbon deposits.

CAUTION Be selective when using solvent because some solvents contain a chemical which attacks aluminum alloy. Do not use a wire brush to remove deposits from aluminum pistons because it may damage the surface. Furthermore, if the skirt is coated, a wire brush could remove this coating.

When there is carbon in the ring grooves, remove it with a ring-groove cleaner. Do not cut any material from the ring lands or from the bottom of the groove. Another method of cleaning the ring

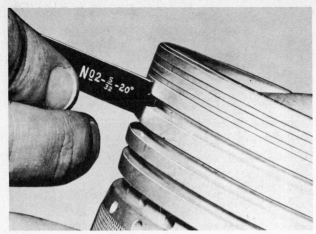

Fig. 11-4 Checking ring-groove wear with a wear gauge. (*Cummins Engine Co., Inc.*)

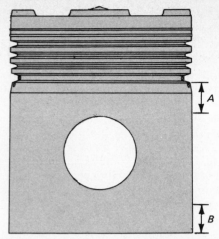

Fig. 11-5 Measuring a straight or tapered piston. (*Cummins Engine Co., Inc.*)

grooves is to break a discarded ring in half. File the butt ends to form a sharp corner which may be used as a scraper to clean the ring grooves. Do not forget to check the oil return holes in the piston wall. If the solvent has not cleared the holes, use a drill (correct size) to remove the deposit.

After the piston has been cleaned and dried with compressed air, reinspect it, particularly in the piston-pin boss area. Cast-iron pistons which have slight score marks on the skirt area need not be replaced. These can be cleaned up by polishing with crocus cloth, or an India stone. Next, check the piston-ring grooves for wear. Use a feeler gauge or wear gauge (Fig. 11-4) at several points around the circumference of the ring grooves. The average maximum wear limit is about 0.006 in [0.152 mm]. If the wear exceeds this, replace the piston.

Pistons are usually replaced because of excessive clearance in the first or second compression-ring grooves (caused by normal wear) and not because of normal wear in the oil-ring groove, piston skirt, or piston bores. No diesel engine manufacturer recommends extending the service life of the piston by restoring the ring grooves. Do not recut the ring grooves with a manual lathe to accommodate a top groove spacer.

Check the piston-to-sleeve clearance (running clearance) by measuring the piston diameter and comparing it with the inside diameter of the cylinder sleeve. The average running clearance is about 0.006 in [0.152 mm] and not less than 0.002 in [0.050 mm]. Insufficient clearance will cause premature piston and/or cylinder-sleeve failure.

The procedure to measure the pistons is not identical for all pistons. You must, therefore, check the appropriate service manual: For instance, the measurement of a cam-ground piston is taken at right angles to the piston-pin bore. For a straight or tapered piston (Fig. 11-5) it is taken at *A* and at *B*.

NOTE All measurements should be taken at 71°F [21.1°C].

For straight pistons, the out-of-roundness should not exceed 0.0005 in [0.001 mm]. The running clearance of all straight pistons can also be measured by the method shown in Fig. 11-6. When using this

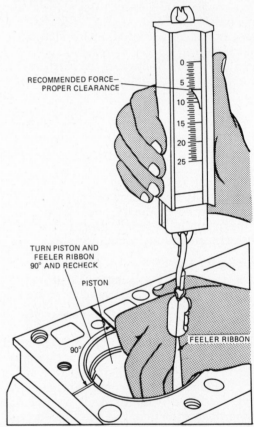

Fig. 11-6 Measuring the running clearance of a straight piston.

method, lubricate the piston (without rings) and cylinder sleeve with a thin film of light oil. Insert a 1/2-in [12.7-mm] feeler gauge in the cylinder sleeve as shown. Insert the piston, bottom up, about 2 in [50 mm] below the block surface with the piston-pin bore in line with the crankshaft. When the spring scale indicates the recommended force in pounds [grams] as the feeler gauge is being withdrawn, the running clearance is correct. (The running clearance is actually about 0.001 in [0.025 mm] greater than that of the feeler gauge.)

Take at least one more measurement, about 90° from the first one, to confirm correct measurement. When the running clearance exceeds 0.010 in [0.254

Fig. 11–7 Measuring the piston-pin bore with an inside micrometer.

mm], the piston, (and at times the piston and cylinder sleeve) must be replaced.

Although not recommended by diesel engine manufacturers, knurling is a means of resizing a piston (increasing its skirt size). *Knurling* is the action of accurately displacing the surface metal. The displaced metal is forced between the teeth of the knurling tool. By controlling the pressure and the position of the knurling tool, a narrow raised pattern is formed on the outer surface of the piston skirt, thereby increasing the skirt diameter.

Measuring Piston Pin and Bore Before you measure the piston-pin bore, check the piston temperature because a temperature variation above or below 70°F [21.1°C] affects the bore size. Measure the bore with an inside micrometer, as shown in Fig. 11-7. You should then inspect and measure the piston pin. If it is out-of-round more than 0.001 in [0.025 mm], or if it shows signs of etching and/or corrosion, or if it is worn to its limit, the piston pin must be replaced.

Rarely should you ream or hone the piston-pin bore and substitute an oversize piston pin. Reaming or honing may result in misalignment and can cause piston seizure or connecting-rod-bearing failure.

Removing and Installing Piston-Pin Bushings It is seldom necessary to replace bushings nor is it usually practical, since in most cases the piston, at this time, is worn and cannot be reused. Furthermore, some manufacturers do not supply replacement bushings.

If the situation exists, however, where there is no alternative but to replace the bushings, take extreme care not to damage the piston when pushing or driving the bushings in or out. Always use the correct fixture to support the piston. Position the joint of the bushing toward the lower end of the skirt to increase bearing support. Do not hone or ream the bushing free hand; use a reaming fixture.

Fig. 11–8 Checking the fit of the piston pin. (*Cummins Engine Co., Inc.*)

Assembling Trunk-Type Piston to Connecting Rod When you assemble the connecting rod, make sure that the rod, cap, and piston identification numbers coincide. If you are reassembling a used connecting rod and piston, check the fit of the piston pin in the connecting rod and piston, as shown in Fig. 11-8. The piston pin should fit snugly. In other words, considerable hand pressure should be applied when inserting the pin into the piston. The fit in the connecting rod, however, should be comparatively loose.

When you assemble the connecting rod to the piston, place the piston in a holding fixture or place the connecting rod in a soft-jaw vise. Apply clean engine oil to the piston-pin bore, connecting-rod bushing, and piston pin.

Install one snap ring into the groove or install a piston-pin retainer into the piston-pin bore. Press the piston pin through the piston-pin bore until a slight protrusion is noticeable, then place the connecting rod with its piston-pin bore onto this protrusion. Force the piston pin through the bore against the snap ring.

Install the other snap ring or piston-pin retainer. Make certain that the recommended clearances exist between the snap ring and piston pin and also between the connecting rod and piston-pin boss. Because of the close tolerance, some pistons must be heated to about 210°F [98.9°C] before installation of the piston pin. This prevents damage to the piston-pin bore.

NOTE When installing pistons with valve reliefs, check the service manual to determine in which direction the relief should be pointing. The manual will also indicate the direction in which the connecting rod must face. If you have no service manual, check the marking on the piston crown for piston direction. As a general rule the connecting-rod number faces toward the camshaft in in-line engines and toward the outside in V engines.

Assembling Crosshead Piston to Connecting Rod (Fig. 11-9) Lay the piston (crown down) on a clean surface and insert the slipper bearing by forcing the half bushing into its seat. One of two types of seal rings are used: a seal ring with an inside bevel or a

Fig. 11–9 Assembling the crosshead piston.

seal ring which is marked "top." Lubricate the seal ring as well as the ring groove. Use a ring-installation tool to install the seal ring so that either the inside bevel or the ring marked "top" faces toward the top of the piston crown. The seal ring must not stick in its groove, and the clearance must not exceed specification. Lubricate the seal-ring surface of the piston skirt. Compress the seal ring with a special ring compressor, and slide the skirt over into position on the crown. The skirt must be able to rotate freely by hand to assure satisfactory piston performance.

Piston Ring Design In an effort to design a multipurpose piston ring that will seal the combustion chamber, control the oil on the cylinder walls, dissipate heat from the piston to the cylinder wall, withstand heat and pressure, and have a long service life, manufacturers have produced a great variety of piston rings. The material from which pistons are made is one of the variables. Piston rings may be made from cast iron, malleable iron, or ductile iron with the addition of other elements such as carbon, silicon, graphite, manganese, phosphorus, chromium, copper, molybdenum, or vanadium. Most compression rings and some oil-control rings are chrome-faced to reduce friction and to ensure longer service life.

All piston rings, regardless of the manufacturer, have a built-in controlled ring tension and are precision-machined and lapped to specific dimension. In a free state, the piston rings are oval. When confined to the cylinder, they fit flat against the piston-groove sides and ring face and conform to the cylinder wall to ensure proper sealing and oil control.

Compression and Oil-Control Rings (Fig. 11-10) All compression rings use the inherent ring tension, the combustion pressure on the power stroke, the exhaust pressure on the exhaust stroke, the crankcase pressure on the intake stroke, and the compression pressure on the compression stroke to seal between the moving piston and the cylinder wall. In addition, they effectively control the oil film on the cylinder walls and thus prevent metal-to-metal contact between the piston rings and the cylinder sleeves. However, too much static pressure causes scuffing, scoring, and excessive wear. Insufficient static pressure results in less oil control and high oil consumption.

When you install piston rings, always use a ring-installation tool. It will prevent the piston ring from becoming distorted or overexpanded during installation.

Oil-control rings are used to limit the oil film on the cylinder walls and to provide adequate lubrication to the compression rings. If the oil film left on the cylinder walls is excessive, it cannot be controlled by the compression rings and it therefore

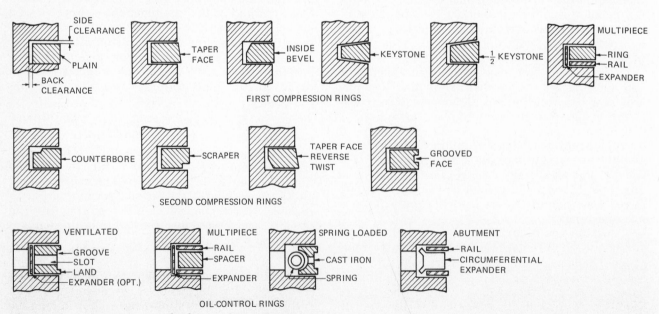

Fig. 11–10 Compression- and oil-control-ring designs.

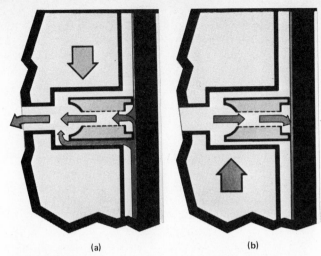

Fig. 11–11 Oil-control ring action during the (a) down and (b) up strokes.

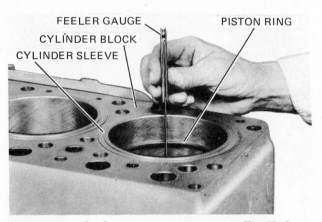

FEELER GAUGE   PISTON RING
CYLINDER BLOCK
CYLINDER SLEEVE

Fig. 11–12 Checking piston-ring gap. (*Allis-Chalmers Engine Div.*)

enters the combustion chamber. Insufficient oil film on the walls will allow a metal-to-metal contact between the compression rings and cylinder walls and will result in scoring and scuffing.

Figure 11-11 shows an installed oil-control ring and the openings or passages from the face to the back of the rings. These passages allow the oil (wiped from cylinder walls as the piston moves downward) to pass into the ring grooves, piston wall openings, and back into the crankcase (Fig. 11-11a). At the same time, this oil dissipates heat from the piston (Fig. 11-11b). On some engines it also lubricates the piston pin.

Most oil-control rings use some type of expander to force them, or the rails, against the cylinder wall (see Fig. 11-10). This force is a major oil-control factor and should never be altered.

Checking Piston-Ring Gap   Before you install the piston ring on the piston, you should check each piston ring for correct ring gap. Insufficient gap will cause damage to the piston, piston ring, and cylinder walls. Excessive ring gap can cause compression losses or a buildup of carbon in combustion gases, or can allow oil to enter the combustion chamber.

To check the ring gap, insert each ring into its cylinder sleeve. With a piston, push the ring down

half the distance of piston travel so that it rests square to the cylinder sleeve. With a feeler gauge, measure the space (the gap) between the ends of the piston ring (Fig. 11-12).

Do not file chrome-plated piston-ring ends because the chrome plating may loosen and, at a later date, damage or score the cylinder sleeve or piston.

When it becomes necessary to increase the ring gap, clamp a mill file or an oil stone in a vise. The width of the file or stone should be the same as that of the ring gap, to assure proper ring-end angle. Hold the ring in alignment when forcing both ring ends against the file or stone. Cut from the outside of the ring surface to the inside only.

Installing Piston Rings   After all pistons rings are checked and, if required, fitted, clean the piston assembly and rings with compressed air, then generously lubricate the piston and piston rings with engine oil. Clamp the connecting rod in a vise so that the piston skirt rests on the vise jaws. Install the lowest piston ring first. Follow the manufacturer's service manual procedure, if a manual is available, or install piston rings as illustrated in Fig. 11-13.

When you install a three-piece oil-control ring, be certain that the butt ends of the expander do not overlap. The ends must be located above the piston-pin bore area and the color indicators must be visible at the expander gap. Install the rail (starting with the gap) about 90° from the piston-pin bore, then roll the rail into the ring groove. It is advisable to hold a thumb over the expander gap to prevent it from overlapping. Use a shim stock under the sharp edge of the rail end to prevent the piston from scoring. The second rail gap, when installed, should be 180° from the first rail gap.

Some Detroit Diesel engine pistons employ two, three-piece oil-control rings. Others have a three-piece oil-control ring in the lower groove and a two-piece oil-control ring in the second groove. When you install either type, make certain that the expander does not overlap, that the scraping edge faces down, and that the ring gaps are properly spaced.

CAUTION   Never shorten an expander. To do so can reduce oil control and possibly result in cylinder damage.

Compression rings marked with a "T" or a dot have an inside bevel or a counterbore. These must be installed toward the crown. The scraper compression ring (outside counterbore) must be installed toward the skirt. Some engine manufacturers use a color code to identify the rings, when the first and second or the second and third compression rings are of the same type and width. If you have any doubt, check your service manual for the proper location of these rings.

After the piston rings are installed, check the rings for freeness in their grooves and make certain that the ring face can retract (about 0.002 in [0.050 mm]) below the piston land.

Ring-Gap Spacing   Always follow the manufacturer's instructions about ring-gap spacing. The

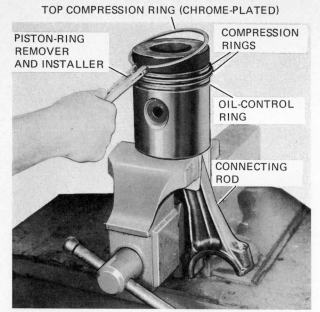

TOP COMPRESSION RING (CHROME-PLATED)

PISTON-RING
REMOVER
AND INSTALLER

COMPRESSION
RINGS

OIL-CONTROL
RING

CONNECTING
ROD

Fig. 11-13 Installing piston rings. (Allis-Chalmers Engine Div.)

Fig. 11-14 Positioning the connecting rod onto the crankshaft journal. (Mack Trucks Canada, Ltd.)

ring gaps should not be in line with each other. They should not be placed over the piston-pin bore or on the thrust or antithrust sides of the piston skirt.

Installing Piston Assembly and Connecting Rod Check and clean the cylinder sleeve, connecting-rod journal, and piston assembly before coating it generously with engine oil. Check and clean the connecting-rod bore and remove the connecting-rod cap.

Position the crankshaft journal at BDC. Recheck the ring spacing and the connecting rod. Make sure that the numbers correspond.

Install the ring compressor or, when a tapered sleeve is used, place the sleeve in the cylinder bore.

Remove the connecting-rod bearing shells from their pack and install them in the connecting rod and cap. **NOTE** Check that the backs of the bearing shells are free from dirt and that they snap into place.

Generously lubricate both halves of the bearing shells. Slide the assembly into the corresponding cylinder, being careful not to damage the walls with the connecting-rod bolts.

Align the piston in position. If you are using piston-cooling nozzles, be careful not to bend the cooling tube. You may have to insert the piston a few degrees out of alignment in order to clear the nozzle.

With a hammer handle, push or tap the piston into its bore until it is flush with the cylinder block. If you have difficulty inserting the piston, the piston rings may be improperly installed, the piston and/or the rings may not have been properly checked for clearance, or the ring gap or the ring grooves may not be clean. In such instances, the assembly must be removed and the problem located and corrected.

Align the lower end of the connecting rod with the crankshaft journal after determining that there is

no dirt on the bearing shell or journal. Pull the connecting rod into place by hand or with a tool similar to that shown in Fig. 11-14.

Check the bearing shell location in the connecting rod, and then install the connecting-rod cap. Identification numbers must not only correspond but must also be positioned on the left or right side of the engine according to the service manual directions.

Some manufacturers recommend inserting two strips of feeler gauge stock (the size of the connecting-rod clearance) at the connecting-rod parting line before tightening connecting-rod bolts to specification. This is to ensure proper connecting-rod-to-cap alignment.

With the recommended lubricant, lubricate the connecting-rod bolts, nuts, or stud threads to ensure proper torque.

Check the running clearance between the connecting-rod bearing and journal by the same method you used to check the running clearance of the main bearings. The average running clearance is about 0.003 in [0.076 mm] when new. The average maximum running clearance is about 0.007 in [0.18 mm]. Check to be sure there is sufficient side clearance between the connecting-rod side face and the journal face. The average side clearance is about 0.005 in [0.13 mm].

Detroit Diesel recommends installing the connecting-rod and piston assembly into the matching cylinder sleeve, externally, as shown in Fig. 11-15. When you do this, position the matching marks of the cylinder sleeve and bore toward one corner of the wooden block. Insert the assembly as shown in operation 2, and align the identification number of the connecting rod toward the corresponding mark on the sleeve. Force the connecting rod downward until the ring compressor is free from the piston. Remove the connecting-rod cap and the ring compressor. Install the upper half of the connecting-rod-bearing shell (the one without oil grooves), and then push the piston down until the compression rings clear the sleeve ports. With the crankshaft

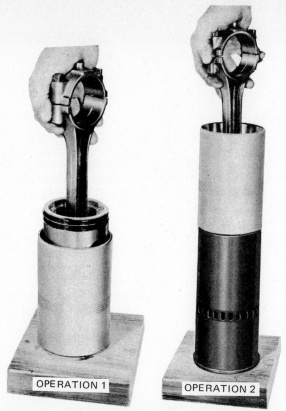

3. Why are holes drilled in the oil-ring groove and below the oil ring (on the thrust and antithrust side)?

4. List several reasons why a piston should not be cleaned with a wire brush.

5. Give several reasons why a singularly designed piston ring (compression ring) and/or oil-control ring cannot be used on all engines.

6. What checks and measurements must be made before installing the piston rings to the piston?

7. Why should a piston-ring installer and removal tool be used to remove or install piston rings?

8. Why is ring-gap spacing so important?

9. Name the three methods used to cool a piston.

10. Failing to remove the cylinder ridge when installing new piston rings could cause (a) compression leakage, (b) broken rings and piston-ring lands, (c) misfiring cylinders, or (d) piston slap.

11. Excessively worn compression rings in an engine with a low total number of operating hours is probably a result of (a) engine overspeed, (b) glazed cylinder liners, (c) improper maintenance of the air-intake system, or (d) mixing the SAE weights of oil in the crankcase.

12. The proper wrist-pin-to-bushing clearance for a 25.4-mm wrist pin is (a) 0.0076 mm, (b) 0.127 mm, (c) 0.076 mm, or (d) 0.381 mm.

13. Low cylinder compression is an indication of (a) compression rings worn, stuck, or broken; (b) partially clogged exhaust parts; (c) excessively worn oil-control rings; or (d) excessive exhaust valve clearance.

14. Scuffing or scoring of pistons, rings, and cylinders can result from insufficient lubrication because of (a) lean air fuel ratio, (b) excessive clearance at camshaft or main bearings, (c) insufficient connecting-rod bearing clearance, (d) overadvanced ignition timing, (e) excessive clearance in oil pump, (f) failure to service air cleaner, or (g) broken oil-pressure relief valve spring.

Fig. 11–15 Installing the piston and connecting-rod assembly in the ring compressor and cylinder sleeve. *(GMC Detroit Diesel Allison Div.)*

journal in its lowest position, carefully guide the assembly into the cylinder bore.

Align the corresponding marks and force the cylinder-sleeve flange against the counterbore. Make sure that the journal and bearing shell are clean. Lubricate both of them thoroughly, then align the connecting rod and pull it onto the crankshaft journal. After the connecting-rod cap bolts are torqued to specification, place a holddown clamp on the cylinder sleeve. This prevents the sleeve from being forced out when the crankshaft is rotated for the installation of the next unit.

Questions

1. What are the main differences between a trunk-type piston and a crosshead piston?

2. Why is the surface of the piston skirt roughened?

Engine Oil

Lubrication System   One of the greatest enemies of an internal combustion engine is the damaging heat created through friction.   To reduce friction by preventing metal-to-metal contact of rotating, sliding, and rocking components, oil is pumped through passages, drillings, and openings to the components by the lubrication pump.   This also cools and cleans the components.   Most diesel engines use gear-type pumps, but in a few cases gerotor pumps are used.

Engine Oil Requirements   Engine oil must be able to withstand extreme temperatures and loads without breaking down.   Although it must form a sliding seal between the piston rings and cylinder walls, the oil film must be thick enough to seal, but thin enough to minimize fluid friction.   It also functions as a semipurifier by holding solid particles in suspension until they are removed by the filter.   Some particles are so fine that they cannot be filtered out, and therefore they remain in the oil until it is changed.

As pointed out, the effectiveness of the engine oil ultimately extends the life of the engine.   To achieve maximum effectiveness the crankcase oil should possess the following qualities: good wetting ability, correct viscosity, minimum evaporation in service, relative nontoxicity, no damaging sediments, and no tendency to deposit varnish, gum, or sludge deposits.

However, unless the air-intake system, the coolant system, and the fuel system are maintained properly and the oil and filters changed regularly, the engine oil will not perform satisfactorily.

Crude Oil   Crude oil is a complex mixture of hydrocarbons and other substances.   The crudes differ in chemical structure and, as a result, in boiling point.

The three main types of hydrocarbons found in crude oil are parraffinic, naphthenic, and aromatic.   Their complex differences in chemical structure dictate the refining processes required for each crude and influence the viscosity, viscosity index (VI), the oxidation stability, the volatility, and the wax content of the final product.   Naphthenic-base crudes produce lubricant of medium viscosity index

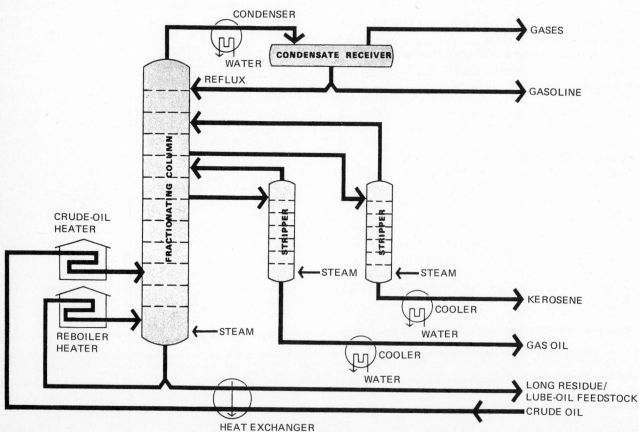

Fig. 12–1   Crude-oil refining process. *(Allis-Chalmers Engine Div.)*

whereas paraffinic-base crudes produce a lubricant with a high viscosity index and are less volatile than naphthenic-base crudes.

REFINING CRUDE OIL   Since the process of refining different crude oils into various petroleum products is highly technical, only a brief description of the basic stages follows.   Refer to Fig. 12-1 as you read.

The first stage of refining of all types of crude oils is distillation, which separates the lighter fractions from the residual (topped crude).   Distillation is carried out at atmospheric pressure in a fractionating column by heating the crude oil rapidly to a constant temperature of 662°F [350°C].   The hydrocarbons which have a lower boiling point vaporize first and, as the temperature of the crude oil rises, other hydrocarbons vaporize.

As the vapor cools it changes back to liquid. An unavoidable overlapping in boiling range occurs between the various products (fractions).

In the second stage of refining, feedstock from the atmospheric still is distilled under vacuum since these stocks cannot be heated under pressure above 662°F [350°C] without rupturing the molecular structure (cracking).   This method requires a temperature of 662°F [350°C] but with lower pressure.   By lowering the pressure, more vaporization occurs and further fractions such as fuel and lubricating oil feedstock are collected.   The remaining residual mixtures from the crude have very high boiling points and therefore cannot be vaporized even under vacuum distillation.   The character of this residue is determined by the original crudes.   From paraffinic-base crudes, comes lubricating oil (residual), from naphthenic-base crudes comes asphalt, and from mixed crudes comes fuel (residual).

From this point lubricating-oil feedstock is processed through a separate though related lubricating-oil refinery where it is transformed into a variety of clean, bright lubricating-oil blending stocks.   Natural contaminants are removed, together with wax and any unstable molecules from the original crude.

Finished oils are then manufactured by blending the base stocks and fortifying them with potent additives in order to produce lubricants best suited to their intended end uses.

Viscosity   Viscosity is the physical property of an engine oil with which the mechanic is most concerned.   Viscosity is a measurement of fluidity or resistance to flow.   Fluidity or resistance to flow changes with temperature; therefore its measurement is always related to temperature.   Of the many instruments developed to measure viscosity, the one most used by technicians and engineers is the Saybolt Universal viscometer.   This viscometer consists of a calibrated reservoir (in which the test oil is placed) with a fixed (universal) orifice at the bottom.   The reservoir is surrounded by a bath, which is heated by heating coils to bring the test oil to the temperature at which the viscosity of the oil is to be measured.   A container marked at the 60-cm³ level is placed under the orifice.   When the test oil

Table 12-1   DISTILLATION VAPORIZATION POINTS

Hydrocarbon	Degrees Farenheit	Degrees Celsius
Gasoline	86–392	30–200 (approx.)
Kerosene	284–536	140–280
Diesel fuel	392–662	200–350
Topped crude	662	350 upwards

SOURCE: Chevron Research Company.

reaches the desired temperature, the orifice is opened.   The number of seconds it takes for the oil to reach the 60-cm³ level is expressed in Saybolt Universal Seconds (SUS), also referred to as Seconds Saybolt Universal (SSU).

A recent method of testing the viscosity of engine oil at 210°F [98.8°C] uses a kinematic viscometer tube.   A measured amount of test oil is drawn into a viscometer tube and placed into a bath heated to 210°F [98.8°C].   The time that it takes for the measured amount of oil to flow through the small-diameter tube (called a capillary tube) is then measured.   The pressure which causes the flow in this instrument is the height of the column of oil above the tube plus the density or gravity of the oil.   The viscosity determined in this way is called *kinematic viscosity*.   It is expressed in Saybolt Universal Seconds (SUS) and is converted into engineering units, that is, centistokes (cSt).   Centistoke viscosity is calculated by multiplying the SUS by the correction factor of the viscometer.

A recent method of testing the viscosity of engine oil at 0°F [-18°C] to simulate cold starting is by measuring, under precise conditions, the shear strength of a film of an oil.   The measurement is taken at 0°F [-18°C] and in centimeters, grams, and seconds.   The unit is called a *poise* (P) or *centipoise* (cP).

The SAE crankcase oil viscosity number chart (Table 12-2) specifies ranges of viscosity at low and high temperatures.   SAE viscosity numbers without the letter W are based on 210°F [98.8°C] viscosity. SAE viscosity numbers with the letter W are based on 0°F [-18°C] viscosity.

MULTIVISCOSITY OIL   A *multiviscosity oil* is one which at a viscosity of 0°F [-18°C] falls within the range of one of the W numbers and at a viscosity of 212°F [100°C] falls within the range of one of the non-W numbers.   Thus an oil with a viscosity in the range of 1200 to 2400 cP at 0°F [-18°C] and in the range of 9.6 to 12.9 cSt at 210°F [98.8°C] is a multiviscosity oil of 10W-30.

NOTE   viscosity is the fluidity of a liquid and is not a classification of the quality of an oil.

Viscosity Index   The *viscosity index* (VI) is a numerical method of indicating the extent to which an oil changes viscosity with changing temperature. The higher the viscosity index, the less the oil changes viscosity as the temperature changes. For example, an oil with a VI of 100 will thin out less than an oil with a VI of 50 when they both are heated.

Table 12-2 SAE VISCOSITY NUMBERS FOR CRANKCASE OILS

Viscosity number	Viscosity units§	Viscosity range‡ at 0°F [−18°C]		Viscosity number	Viscosity units§	Viscosity range‡ at 210°F [98.8°C]	
		Min.	Max.			Min.	Max.
SAE 5W	Centipoises (cP) Centistokes (cSt) Saybolt Universal Seconds (SUS)		Less than 1200 1300 6000	SAE 20	Centistokes (cSt) Saybolt Universal Seconds (SUS)	5.7 45.0	Less than 9.6 58.0
SAE 10W	Centipoises Centistokes Saybolt Universal Seconds	1200* 1300 6000	Less than 2400 2600 12,000	SAE 30	Centistokes Saybolt Universal Seconds	9.6 58.0	Less than 12.9 70.0
SAE 20W	Centipoises Centistokes Saybolt Universal Seconds	2400† 2600 12,000	Less than 9600 10,500 48,000	SAE 40	Centistokes Saybolt Universal Seconds	12.9 70.0	Less than 16.8 85.0
				SAE 50	Centistokes Saybolt Universal Seconds	16.8 85.0	Less than 22.7 110.0

* Minimum viscosity at 0°F may be waived provided viscosity at 210°F is not below 4.2 cSt (40 SUS).
† Minimum viscosity at 0°F may be waived provided viscosity at 210°F is not below 5.7 cSt (45 SUS).
‡ The viscosity of all oils included in this classification shall not be less than 3.9 cSt at 210°F (39 SUS).
§ The official values in this classification are based upon 210°F viscosity in centistokes (ASTM D 445) and 0°F viscosities in centipoise (ASTM D 2602). Approximate values in other units of viscosity are given for information only.

SOURCE: Sun Oil Company.

Pour Point The *pour point* is the point at which an oil solidifies either through the formation of wax crystals or through a gradual increase in viscosity as the ambient temperature drops. Pour point is only of concern when the engine operates where the ambient temperature is lower than 10°F [-12°C].

Flash Point The *flash point* is the temperature at which an oil vaporizes and ignites from a source of ignition. The flash point of oil varies with the viscosity and quality of the oil. It can range from 380 to 500°F [190° to 260°C] or more. When the flash point falls below 320°F [160°C] due to contamination by fuel of lower volatility, a crankcase explosion could occur.

Specific Gravity *Specific gravity* is the comparison of the weight of an oil with the weight of an equal volume of water; it does not relate to the viscosity or quality of an oil. However, the specific gravity of a diesel fuel is very important because it determines the heat volume of the fuel (see Unit 24, "Diesel Fuel").

Engine Oil Additives The success of today's engine oils depends very much on the complex chemicals (additives) which are blended into the crankcase oil base stock. Many of the additives affect more than one of the oil's characteristics.

OXIDATION INHIBITORS To prevent varnish and sludge formation and to prevent corrosion, organic compounds, those containing sulfur, phosphorus, or nitrogen (such as organic amines, sulfides, or phenols), are used. Metals such as tin, zinc, or barium are often incorporated.

ANTICORROSION (RUST) ADDITIVES To prevent failure of alloy bearings and to prevent other metal surface corrosion, metal salts of thiophosphoric acid and sulfurized waxes are added in addition to the organic compounds mentioned previously.

DETERGENT ADDITIVES To keep metal surfaces clean and to prevent any type of deposit formation, metallo-organic compounds such as phosphates, phenolates, sulfonates, alcoholates, or soaps containing metals like magnesium, barium, calcium, or tin are added. These additives are used basically to prevent corrosive acids or sludge particles, produced as a result of combustion, from forming deposits and to hold such particles in suspension.

ANTISCUFF AND ANTIWEAR ADDITIVES These additives reduce friction and prevent galling, scoring, and seizure and also give the oil extreme pressure properties. Organic compounds containing chlorine, phosphorus, and sulfur such as chlorinated waxes, and organic phosphates are used. Phosphates such as tricresyl phosphate and zinc dithiophosphates are used.

FOAM INHIBITORS Due to the operating condition of an engine and the effects of contamination, an engine oil without the additives (silicone polymers) will form small air bubbles. The additive combines the small bubbles to make larger bubbles, which separate faster.

Crankcase Oil Classification For many years the American Petroleum Institute (API) provided a crankcase oil classification system that related to engine operating conditions. However, with the increasing demands placed on crankcase oil, the API, in cooperation with SAE, American Society for Testing and Materials (ASTM), and various car, truck, and engine manufacturers published (in 1970) a new crankcase oil classification system.

This system, based on a simple letter designation, describes precise engine operating conditions as well as the necessary oil performance properties in regard to engine design and construction.

SA (OLD CLASSIFICATION ML) An oil used for gasoline and diesel engine service, under such mild condi-

Table 12-3 PROPERTIES OF CD CRANKCASE OIL

Property	10W	20/20W	30	40	50
Gravity °API*	31.7	29.1	28.8	27.5	27.4
Color ASTM† D1500	L2.0	L3.0	L3.0	3.0	3.0
Flash point, °F [°C]	420 [215.5]	450 [232.2]	470 [243.3]	490 [250.0]	520 [271.1]
Pour point, °F [°C]	−25 [31.7]	−15 [−26.1]	−5 [−20.6]	0 [−17.8]	0 [−17.8]
Viscosity poise at 0°F [−18°C]	18.5	62			
SUS at 100°F [37.8°C]	198.8	346	566	742	1276
SUS at 210°F [98.9°C]	46.8	57.5	67.1	76	100
Viscosity index ASTM D2270	106	112	97	96	94
Sulfated residue, wt. %	0.83	0.83	0.83	0.83	0.83
TBN—AP‡	7.2	6.7	6.4	6.7	6.7
Calcium, wt %	0.23	0.23	0.23	0.23	0.23
Zinc, wt %	0.075	0.075	0.075	0.075	0.075
Copper corrosion at 212°F [100°C]	Negligible	Negligible	Negligible	Negligible	Negligible
ASTM foam test	Passes	Passes	Passes	Passes	Passes

* Degrees American Petroleum Institute.
† American Society for Testing and Materials.
‡ Total Base Number—aromatic paraffinic crude oils.

SOURCE: Sun Oil Company.

tions that the protection afforded by compounded oils is not required.

SB (OLD CLASSIFICATION MM) This oil is designed for medium-duty gasoline engine service. It provides only antiscuff capability and resistance to oil oxidation and bearing corrosion.

SC (OLD CLASSIFICATION MS) An oil designed for use in gasoline engines to provide control of high- and low-temperature deposits and to give protection against wear, rust, and corrosion.

SD (OLD CLASSIFICATION MS) An oil designed for gasoline engine service. It gives more protection against high- and low-temperature engine deposit, wear, rust, and corrosion than an oil of the SC classification.

SE (NO OLD CLASSIFICATION) An oil used for modern gasoline engines. It is designed to provide more protection against oil oxidation deposit, high-temperature engine deposit, rust, and corrosion than oils of the SD or SC classifications.

CA (OLD CLASSIFICATION DG) An oil used for light-duty diesel engine service to provide protection against bearing corrosion and from high-temperature deposits in normally aspirated diesel engines.

CB (OLD CLASSIFICATION DM) An oil used for moderate-duty diesel engine service. It is designed to provide necessary protection from bearing corrosion and from high-temperature deposits in normally aspirated diesel engines with higher sulfur fuels.

CC (OLD CLASSIFICATION DM) An oil used for moderate-to-severe-duty diesel and gasoline engine service. Used in trucks, industrial and construction equipment, and farm tractors. It provides protection from high-temperature deposits in lightly supercharged diesel engines and also from rust, corrosion, and low-temperature deposits in gasoline engines.

CD [OLD CLASSIFICATION DS (SERIES 3 OIL)] An oil used for severe-duty diesel engine service. It provides protection against bearing corrosion and from high-temperature deposits in supercharged engines when fuels of a wide quality range are used.

Table 12-3 is a typical property chart of a CD crankcase oil.

As an example of how the API classifications are used, Cummins Engine Company recommends an SC lubricant which has low-temperature sludge protection for their normally aspirated engines in stop-and-go service. The same engine, when used in heavy-duty operation, requires a lubricant of CC performance level. Cummins turbocharged engines require a CD performance level.

Storing and Handling Crankcase Oils To reduce the risk of contaminating the lubricant, the oil drums should be stored indoors, in a designated area, and placed in an oil-drum rack. Always keep the drum heads, taps, measuring cans, dispensers, and surrounding area clean. Never use empty antifreeze cans for dispensing any oil. Identify each oil clearly to reduce the possibility of dispensing the wrong oil. Use a power dispenser or a separate measuring can for each product. When possible use taps for each drum to prevent damage to the bungs or to prevent lubricant contamination. Bungs and drum heads must be replaced and tightened immediately after use. Use only a bung wrench in order to prevent damage from frequent opening and closing.

Use only lintfree rags when cleaning the dispensers because lint can accumulate and stop oil circulation. If you must store the drums outside, tilt them slightly with the bungs at 3 o'clock and 9 o'clock or lay them on their sides so that water cannot enter. Always wipe up any spilled oil around the drums. Do not use sand or a chemical to soak up the oil. Not only is a chemical a fire hazard, but it may get into the oil.

Questions

1. List the various purposes of an engine oil.

2. Why can't one oil serve all engines?

3. Which type of crude oil is the most suitable for use as an engine oil?

4. Why are viscosity and the viscosity index more important to the mechanic than the pour point and the flash point?

5. What is the difference between an SAE multiviscosity 10-30 oil and an SAE 10 oil?

6. What effect has a low VI oil on the oil pressure when the engine is at operating temperature?

7. What are the major differences between an oil with an SA classification and one with a CC classification?

8. List some changes (in your shop area) which in your opinion would improve the handling and storing of fuel oil and grease.

UNIT 13

Lubrication Pump and Oil Cooler

Disassembling Lubrication Pump Before you disassemble the lubrication pump, clean the pump assembly. Remove the oil screen, drive gear, and pump cover. With a felt pencil, identify the pump gears so that the teeth will be in mesh when reinstalled. Remove the gear along with the drive shaft, idler shaft, or driven shaft. Remove the relief valve assembly and wash component parts in clean fuel or solvent.

Inspecting and Measuring the Lubrication Pump Check the faces and sides of the drive gear and pump gears for burrs, scoring, and grooves. Measure the bushings and shaft for wear. If the clearance is more than 0.003 in [0.076 mm], the bushings and/or shaft must be replaced. Check the pump cover and the sides of the gear housing for wear and scoring. If the wear pattern on the low-pressure side (pump intake) has passed 3 o'clock, the housing must be replaced.

Another method of checking the amount of wear is to measure the clearance between the pump body and the gear teeth by placing the gears in the pump body and using a feeler gauge to measure the wear on the low-pressure side (see Fig. 13-1). When the radial clearance exceeds 0.004 in [0.101 mm], the pump body must be replaced. The pump body and

cover must be smooth and must show no scratches, score marks, or rough spots.

Measure the height of the gears with a micrometer and, with a depth micrometer, measure the depth of the body to determine the running clearance of the gears. **NOTE** When you take this measurement, do not forget to place a new gasket on the housing before measuring the body depth. When the running clearance exceeds 0.005 in [0.127 mm], replace the pump body and/or gears. Lapping the housing to reduce clearance is not recommended.

Another way to measure running clearance is with a plastic strip (Plastigage) (see Fig. 13-2).

Servicing and Reassembling the Lubrication Pump Remove nicks and burrs from gears, cover plate, shafts, keyways, and splines. If bushings need to be replaced, do so with a bushing remover

Fig. 13–1 Measuring pump-gear-to-body wear. (International Harvester Co.)

Fig. 13–2 Measuring pump-gear running clearance. (J. I. Case Co., Components Div.)

Fig. 13–3 Typical oil-pump installation and piping. *(GMC Detroit Diesel Allison Div.)*

adapter. Press the old bushing out and the new one in. Your service manual will indicate the bores into which the bushings must be pressed and to what depth. It will also indicate if the bushings require reaming.

By using a new shaft, check bushing-to-shaft clearance. After you clean the oil screen, check for broken or loose screen wires before you reinstall the screens into or onto the oil-pump cover.

Clean and polish the relief-valve bore and plunger. Measure the relief-valve spring height and, if it is below specification, replace it. Do not stretch the spring.

Lubricate all parts and components with engine oil and reassemble them in precisely the reverse sequence to which they were disassembled.

When new pump gears are used, check their backlash with a feeler gauge. Make certain that the relief-valve plunger will not stick and that the assembled pump can turn without binding. Do not force the drive gear onto the drive shaft. It should be heated and installed as outlined in Installing Crankshaft (Unit 9).

Installing Lubrication Pump to Engine Most lubrication pumps are doweled to their mounting support so that the backlash, when properly adjusted, cannot change. The backlash of some oil-pump drive gears is adjusted by placing (or removing) shims between the mounting surfaces. On other oil-pump drive gears the backlash is adjusted by moving the pump sideways, then drilling dowel holes into the locating plate. Regardless of the installation method, the backlash must be within specification.

When the lubrication pump is bolted in place and torqued to specification, the scavenging pump is connected to it with the drive shaft or with the scavenging tubes. The necessary tubes are then connected to the pump and the cylinder block (Fig. 13-3).

NOTE Use Teflon sealing tape on pipes and plug threads.

Inspecting and Servicing Oil Pan After you have thoroughly cleaned the oil pan and removed the old gasket material from the mounting surfaces, check the oil pan for dents, cracks, or loose baffle plates. Make repairs when possible. Inspect the condition of the flanges. Inspect the bolt holes for elongation, and inspect the condition of the threads.

When necessary, straighten the metal-stamped flanges. Check for obstructions in the oil-pan passages which direct oil to the main oil gallery. Particularly check the threads and the sealing surface of the oil drain bolts and the oil pan. Damaged threads can be repaired with a helicoil or a new plug assembly can be welded or soldered into place.

Installing Oil Pan Make a cursory check to ensure that all assembled components or parts are in place, properly torqued, and secured. Check the oil pan and cylinder block mating surfaces for cleanliness. Make sure the oil baffles are securely in place and that the oil pickup screen and seal (if used) are in place. Use a nonhardening sealing compound to cement, when recommended, a new gasket to the front and side rails of the oil pan. Alternatively, use the method shown in Fig. 13-4 to mount the new gasket. For easier installation, install guide studs to each corner of the cylinder block. If sealing rings are used, hold them in place with shim stocks.

Place the pan in position and install all oil-pan bolts. (Sometimes two or more aligning bolts are used to position the oil pan.) Remove the guide studs and holding device and install the aligning bolts and remaining oil-pan bolts, then tighten, in sequence, all bolts to the recommended torque.

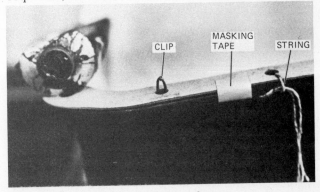

Fig. 13–4 Holding the gasket in place.

NOTE Some oil pans are bolted to the flywheel housing. When this is the case, use a nonhardening sealing compound in the corner of the oil pan and housing to improve sealing. To ensure a leakfree joint, you must be particularly careful to tighten the oil-pan and flywheel-housing bolts.

Finally, install the clean-out cover or inspection cover as well as the oil level plug and drain plug or bolts.

Servicing Oil Cooler Before you disassemble a tube-type oil cooler for service, fill the cooler housing, through its ports, with mineral spirits (or a similar cleaning fluid) to clean it and to loosen the oil residue and contamination within the cooler, otherwise it may be difficult to remove the element.

After a period of time, say 1 hour, remove the plug and drain the fluid. You could, in the meantime, clean the waterside (the inside) of the copper tubing. To do this, immerse the cooler in a solution of 9 parts water to 1 part muriatic acid. Add oxalic acid and pyridene in proportion to the solution (strength) required. For example, when using 9 gal (U.S.) [7.44 gal (Imp.) or 33.82 l] water, and 1 gal (U.S.) [0.83 gal (Imp.) or 3.78 l] muriatic acid, add to this 2 lb [0.9 kg] oxalic acid, and 0.02 gal (U.S.) [0.016 gal (Imp.) or 0.08 l] pyridene.

Leave the oil cooler in the solution until the foaming and bubbling have stopped. Then remove the cooler and flush it with hot or warm water to remove the cleaning solution.

NOTE Because of the multitude of metals used, check with your service manual for the proper solution strength, as well as for the recommended cleaning procedure.

To remove the cooler element, first remove the cover, the gasket retainer, and the O ring. Insert two stud bolts into the cooler element puller hole and secure a suitable puller bar to them. To prevent hardening of the residue, clean the oil side of the cooler as soon as the element is removed. Flush trichloroethylene or an oakite or alkaline solution (depending on the extent of clogging of the cooler) around the cooler tubes or into the inlet of the cooler cores.

When metal particles from broken or worn engine components are present in the lubrication system, a core-type cooler must be replaced because you cannot check the effectiveness of your cleaning. Use hot water to flush the cleaning solution from the cooler.

CAUTION Use rubber gloves and eye protection. Clean only in a well-ventilated room.

Inspecting and Testing Oil Cooler Dry the cooler with compressed air and flush the tubes with light engine oil. Check for damaged tubes or core. Check the flared ends of the tubes for corrosion and for welding or soldering cracks. Check the housing and connections for damaged threads and the flanges for nicks, gouges, or cracks. Check the

Fig. 13–5 Preparing the cooler core for pressure testing. *(GMC Detroit Diesel Allison Div.)*

bypass valve spring for corrosion or a damaged surface.

To check for cooler tube or core leakage, seal both ends of the tubes (or when testing a cooler core, seal the core with a plate, as illustrated in Fig. 13-5). Connect an air hose to the drilled and tapped hole. Using an air-pressure regulator for control, pressurize the core to the recommended pressure.

NOTE In view of the variety of oil coolers, be sure to refer to your applied pressure specification. It varies from 1 to 150 psi [0.07 to 10.54 kg/cm^2].

Immerse the core in water heated to 180°F [82.0°C]. When air bubbles rise, mark the location. In practice, a cooler which has a leak is either sent to a radiator shop for repair or replaced.

To repair damaged soldered cracks, resolder the flared ends of the tubes. Take care not to melt the solder on the adjacent tubes.

To repair damaged tubes, insert a smaller tube into the damaged tube, flare both ends, then solder the new and the old tube to the end plate.

Reassembling and Installing Oil Cooler Reassemble the oil cooler in a sequence precisely the reverse to which it was disassembled (see Servicing Oil Cooler). Make certain that the O rings are lubricated in their grooves and not twisted, that the gaskets are positioned properly, and that all hex bolts or nuts are torqued properly. Install the bypass valve in the correct sequence. When you mount the oil cooler onto the engine, make certain the mounting and sealing surfaces are clean and that the cooler faces in the correct direction in relation to the oil "in" and "out" and the coolant "in" and "out" connections.

When a multioil cooler is used, take care not to interchange the torque converter oil lines with the engine oil lines. Make sure the hex bolts and plugs are torqued properly and that the radiator and block hose connections or tubes are positioned correctly and clamped securely.

Questions

1. What is the purpose of the lubrication pump?

2. What determines the maximum oil pressure?

3. What determines the minimum oil pressure?

4. Describe how you would measure the clearance from the oil-pump gears to the housing using a micrometer.

5. What is the purpose of measuring the backlash of the drive and driven oil-pump gears?

6. What is the purpose of the oil-pan baffles?

7. Why is the engine oil directed to the cooler before passing through the filter to the components?

8. What precautions must you take when connecting the oil and coolant lines to an oil cooler?

UNIT 14

Cylinder Head and Valves

Cylinder-Head Failure Service records indicate that cylinder-head and valve-train failures are usually due to inadequate service, misadjustment, and poor maintenance.

Regarding misadjustment, a point which cannot be overstressed is the necessity to retain correct sequence when torquing bolts. Torquing bolts out of sequence may cause the head gasket to leak or may damage the valve train, which could result in far more serious impairment to the valve, valve seat, and injector.

Improperly torqued cylinder-head bolts can cause the valves to distort or the injector nozzles to stick, resulting in misfiring or timing change. One side of the distorted valve will contact the seat earlier than the other and the contacted side will cool more quickly. Additionally, the seating force is higher on the contacted side. Since the remainder of the seat is open, hot combustion gases escape, the fillet of the valve expands, and the opening increases. It follows that the valve will fail. Prior to valve failure the escaped combustion gases cause turbulence in the exhaust manifold with the serious result of incomplete scavenging.

Fig. 14—2 Left, white ash deposit; right, carbon accumulation. *(Cummins Engine Co., Inc.)*

Cracks between the valve port and the injector or between the valve ports are often due to overheating and/or overfueling, hot shutdown, and loss or reduction in coolant flow. Cracks also can result from defective casting, improper machining, improper installation, or the use of excessive ether for starting.

Damage to the cylinder head shown in Fig. 14—1 is the result of foreign material entering the combustion chamber. The material may have come from a damaged valve, broken valve insert, broken piston, broken ring land and/or piston ring. It is also possible that the foreign material originated elsewhere and was taken into the intake system.

Valve and Valve-Insert Failure Excessive valve-face wear is caused by improper valve-spring tension, loose valve adjustment, high speed, or high engine temperature. When such wear is found only on the intake valves and inserts, it is the result of dirt being taken in on the intake stroke.

A warped valve is caused by improper setting, carbon and/or varnish accumulation, inadequate cooling, air-intake restriction, high exhaust temperature due to overloading the engine, overfueling, turbocharger failure, tight valve adjustment or a weak valve spring.

Deposits of white ash or carbon on valve fillets are shown in Fig. 14—2. This usually results from burned engine oil residue. This oil could come from the crankcase because of a broken piston, worn oil-control ring, damaged sleeve, or simply because there was originally too much oil in the crankcase.

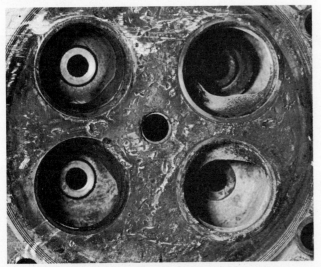

Fig. 14—1 Cylinder-head damage due to foreign material. *(Cummins Engine Co., Inc.)*

Fig. 14–3 Varnish accumulation. *(Cummins Engine Co., Inc.)*

Fig. 14–4 Damage due to excessive heat. *(Cummins Engine Co., Inc.)*

Fig. 14–5 Damage due to contact with the piston. *(Cummins Engine Co., Inc.)*

It can also come from the rocker arm side of the cylinder head due to excessive clearance of rocker-arm bushings, improper positioning of the rocker-arm shaft or worn valve guides and/or valve stems.

The valve shown in Fig. 14–3 has a poor seat and there are varnish deposits on the fillet. These defects could be the result of improper valve adjustment, incomplete combustion, exhaust restriction, long low-idle period at low ambient temperature, extended time between oil and filter changes, or contaminated oil.

Damage to the valve shown in Fig. 14–4 can be due to excessive heat, low coolant and/or reduced coolant flow, overfueling, improper injection timing, overloading the engine, air-intake restriction, or a damaged turbocharger. Continuous operation under any of these conditions can lead to overheating of valve inserts as well as valve burning or channeling. However, valve channeling can also be caused by a broken insert, bent valve, out-of-round valve or valve-seat insert, worn valve guide, damaged injector nozzle, deposits on valve seat and/or misadjustment which prevents the valve from seating properly. When more than one valve insert is cracked, it is usually due to overheating. When only one insert is cracked the cause may be improper installation, an improperly machined insert bore, or an insert loose in its bore.

The defective valve shown in Fig. 14–5 contacted the piston. This does not happen often but when it

does, look for a broken valve spring, a valve sticking in the guide, the possibility of insufficient lubrication, a bent valve stem, a weak valve spring, or carbon deposits on the valve stem or guide. A valve can also come in contact with the piston from overspeeding the engine, a bent valve-bridge guide pin, improper installation, a worn valve keeper, or worn grooves.

Broken valve stems are uncommon. This defect can usually be traced to overspeeding the engine, improper valve adjustment, a valve seat not ground square to the valve guide, excessive engine temperature, or a weak or broken valve spring.

Abnormal valve-guide wear is caused by abrasives in the lubricant, inadequate lubrication, overheated engine, valve or valve insert being out-of-round, a bent valve spring, or an improperly ground rocker-arm arc contacting the valve-stem end. Inadequate lubrication of the valve guide can originate from incorrect clearance between valve guide and valve stem, low oil level, low oil-pump pressure (worn pump), sticking pressure relief valve, or contaminated lubricant.

Valve-bridge failure can be caused by a damaged adjusting screw. Side wear or excessive surface wear of the valve bridge usually is caused by an improperly adjusted or improperly torqued locknut, rocker arm and/or tappet adjustment failure, a bent bridge guide pin, a bent pushrod, or overspeeding the engine.

Rocker-arm breakage, though not common, is usually due to defective casting or careless handling during shipping. The rocker-arm can also break due to failure of some other valve-train component.

Abnormal wear of the rocker-arm bushings or shaft can be caused by inadequate oil supply resulting from incorrect installation, contaminated oil, entrance of dirt during assembly or adjustment, low oil level, or low oil pressure.

The pushrod damage (Fig. 14–6) is infrequent. It is brought about by overspeeding the engine, a damaged follower, a damaged camshaft lobe, or an improperly adjusted rocker arm.

Abnormal socket or ball-end wear (shown in Fig. 14–7) can be caused by insufficient lubricant, plugged oil passages, contaminated lubricant, or dirt left in the socket during assembly or service.

Fig. 14–6 Pushrod damage. *(Cummins Engine Co., Inc.)*

Fig. 14–7 Abnormal ball-end wear. *(Cummins Engine Co., Inc.)*

A broken valve spring, though a relatively uncommon defect, is one of the most damaging to the engine. The spring may originally have had a factory defect but most likely it broke due to flutter or fatigue when overspeeding the engine.

Cam follower failure does not usually arise until after other components of the valve train have been damaged or have failed. A damaged camshaft lobe will cause uneven force on the follower roller and pin, which in turn increases stress and reduces lubrication. Improper installation, improper assembly, or overspeeding the engine can also bring about cam follower failure.

Cylinder-Head Service   After you remove the valves, plugs, cups, and cooling tubes (or nozzles, if used), steam clean the cylinder head. If steam cleaning does not remove the lime or scale formation, the cylinder head must be cleaned by the hot tank method.

After you thoroughly clean the cylinder head, visually inspect it for damage. Pay particular attention to the fuel and lubrication passages and, when necessary, clean them with a suitable wire brush (Fig. 14–8).

Fig. 14–8 Cleaning the cylinder head.

Check the injector tube (sleeve) for evidence of coolant leakage or a damaged nozzle seat. If the injector sleeve is damaged, do not replace it until the cylinder head has been air-tested or water-tested.

Check the cylinder-head surface for scratches and corrosion. Measure the cylinder-head surface for flatness with an accurate straightedge and feeler gauge. Use a power tool to refinish the head surface. Also check it for unevenness at the gasket sealing areas. Compare the measurements obtained with those recommended in the service manual for maximum longitudinal and transverse allowable warpage.

Measure the height of the cylinder head. The service manual will tell you where you should take the measurements. Determine if the cylinder head can be resurfaced by milling or by surface-grinding. Check the cylinder head for cracks, especially between the valves and injector tube. After your inspection is complete, reinstall all the removed plugs and cups. When necessary, install a dummy injector (scrapped injector).

Seal all the coolant openings with suitable steel plates and gaskets. Drill and tap one plate to accommodate either an air- or a water-hose connection.

Air-Testing Cylinder Head   Connect an air hose to the tapped cover plate. Using an air regulator, pressurize the cylinder head to about 40 psi [2.81 kg/cm²], then submerge the cylinder head in water. Check for air leakage, especially around valve seat and injector sleeve locations.

Water-Testing Cylinder Head   One method of water testing is to connect a water hose to the cylinder head and pressurize it to about 40 psi [2.81 kg/cm²]. Steam-clean the cylinder head to raise the temperature of the head and water to approximately 180°F [127°C]. Blow-dry with compressed air. Check carefully around the valve seats and injector sleeve locations for cracks. As a rule, it is not worthwhile to repair a cracked cylinder head.

Injector-Sleeve Service   There are many types of injector sleeves (tubes). Many of them require very little servicing and, if replacement becomes necessary, it is a simple operation unless they are Cummins and Detroit Diesel injector sleeves. To remove the type of injector sleeve shown in Fig. 14–9, tap a thread into the sleeve, screw a suitable eye bolt into the threaded sleeve and pull the sleeve out. Clean the sleeve bore thoroughly. Steam-clean the cylinder head until its temperature is approximately 180°F [127°C]. Coat the new sleeve with an oil and water sealer then, with the appropriate driver, drive the sleeve into place. Some sleeves must be reamed after installation to ensure proper injector cooling and nozzle seating.

Servicing the injector sleeve of a Detroit Diesel or Cummins engine is a little more complex and special tools are required. Although the Detroit Diesel injector sleeve may show no sign of damage or coolant leakage, the injector protrusion and the seating pattern must be checked. If the seating pattern is unsatisfactory, it can be refinished by using the correct

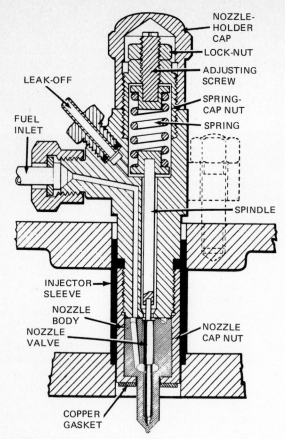

Fig. 14–9 Cross-sectional view of an injector (installed). (*CAV Ltd.*)

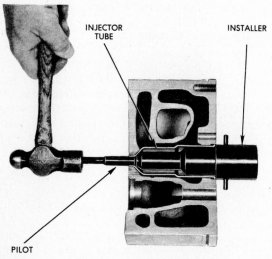

Fig. 14–10 Removing a Detroit Diesel injector tube. (*GMC Detroit Diesel Allison Div.*)

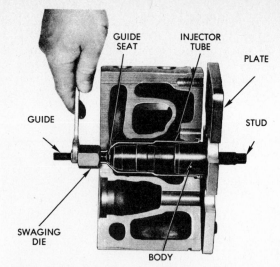

Fig. 14–11 Swaging (flaring) an injector tube. (*GMC Detroit Diesel Allison Div.*)

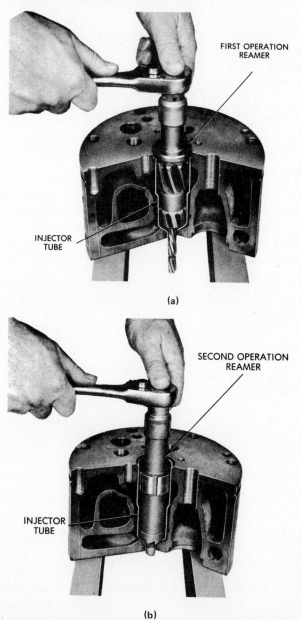

(a)

(b)

Fig. 14–12 (a) Reaming an injector tube for the injector body and spray tip, (b) reaming an injector tube for the injector unit. (*GMC Detroit Diesel Allison Div.*)

reamer, however, the injector protrusion must then be rechecked. If the protrusion is more than specified, the sleeve must be replaced.

To replace a Detroit Diesel engine injector sleeve, place the installing tool (installer) into the sleeve (injector tube), insert the pilot through the small hole of the sleeve and screw it into the installing tool (see Fig. 14–10). Tap the injector sleeve out. The injector-sleeve bore and O-ring counterbore must be cleaned. The O-ring counterbore in the cylinder head must be free of nicks and burrs.

To install the sleeve, slide a new sleeve over the installing tool and screw in the pilot tool. Insert a new O ring into the counterbore and place the sleeve in the injector bore. Drive the sleeve into position, replace the pilot tool with the flaring die (swaging die), and torque the die to 30 lb·ft [4.2 kg·m](see Fig. 14–11). This action causes the injector sleeve to flare out against the tapered shoulder of the cylinder head.

Reaming Injector Tube to Fit Injector The injector sleeve must be hand-reamed. The two reamers shown in Fig. 14–12 are required to ensure proper cooling, protrusion, seating, and sealing of the injector nut. Use the first reamer, as shown in Fig. 14–12a, to ream for correct clearance of the injector body, nut, and nozzle tip. Use the second reamer (Fig. 14–12b) to ream for correct injector-nut-to-sleeve-seat and injector protrusion.

Start each reaming operation by applying cutting oil to the reamer flutes. Applying moderate force, turn the reamer in a clockwise direction to prevent breaking the cutting edges of the reamer and to produce a smooth bore. Throughout the reaming process, repeatedly withdraw the reamer in a clockwise rotation to remove the cutting chips. Be generous with the cutting oil to ensure a smooth surface.

When removing excessive material to achieve specified protrusion, take care not to ream out too much material or to cut a poor seat.

Valve Guides Replaceable valve guides are used on diesel engines. They are made of a cast-iron alloy which has superior wear and is more corrosion-resistant than the alloy used in the cylinder head. These guides are usually half the length of the valves. When inserted into the cylinder head, they are flush with the ports (valve throat) to prevent turbulence of the incoming air or outgoing exhaust gases (see Fig. 14–13). The inner surface is often treated to increase its resistance to chemicals and corrosion. Some valve guides are grooved or knurled to reduce friction and to increase lubrication control. As a result, the service life of the guide, as well as the service life of the valve, is increased. Often the exhaust guides are counterbored to prevent carbon buildup and to reduce heat transfer which could damage the valve stem. The inside diameter is precision honed or reamed to control lubrication and valve-stem movement.

Valve-Guide Service When an engine has been operated for a considerable time its guides will show signs of wear. The guides must therefore be inspected for either replacement or possible reuse. Sometimes the guide can be knurled to restore its inside diameter.

Before you measure the guide with a small-bore gauge, check it visually for chips, burrs, or cracks. When carbon is present, remove it with a power tool connected to a wire brush. To check the guide for wear, adjust the bore gauge diameter to maximum allowable guide wear. Check at several points along the valve guide, especially at the valve head and stem end (Fig. 14–14). If the bore gauge is loose at any point in the guide, the guide must be knurled or replaced.

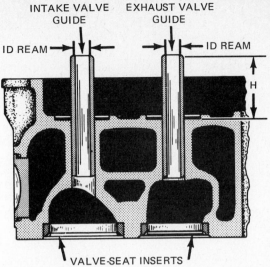

Fig. 14–13 Cross-sectional view of the cylinder head with valve guides. (*Mack Trucks Canada, Ltd.*)

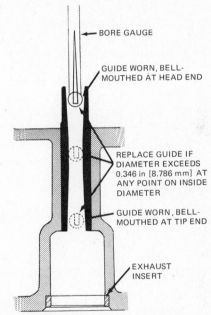

Fig. 14–14 Bell-mouthed valve guide. (*J. I. Case Co., Components Div.*)

Bell mouthing at the valve head end usually is caused by a buildup of carbon on the valve stem. Bell mouthing at the stem end can be caused by a bent valve spring, worn rocker arm, or improper arcing of the rocker-arm valve-stem surface.

Replacing Valve Guides To remove the valve guide, you should press or drive it out from the underside of the cylinder head. After removal of the valve guide, check the valve-guide bore. If the bore is damaged it must be reamed to accommodate a new oversize valve guide. To install a new guide, use a mandrel to press the guide into its bore to its specified height above the valve-spring seat surface. When no height specifications are given in the manual, the usual instructions call for pressing the guide flush with the valve port. Although honing or reaming the valve guide after installation is sometimes recommended, some manufacturers simply suggest checking the inside diameter.

When you hone or ream the valve guide, use lubricant, and ream from the valve head side to achieve a satisfactory surface finish.

Valve Service   Clean the valve with a buffer and polish the stem with crocus cloth.   Check the valve head to determine if it is cupped, cracked, or burned, and whether it has a thick enough margin to allow regrinding (Fig. 14–15).   Magnaflux the valve to reveal hidden flaws, especially if it has been welded. Check the valve stem for nicks, pitting, and scuffing.   Check the keeper (retainer) grooves and valve-stem tip for wear or damage.   Worn keeper grooves will tip the spring retainer as well as the spring, causing the valve seat to leak and the valve to stick. They will also increase valve-stem and guide wear. Beware that refinishing the stem tip may restore its surface but may also remove the case hardening.

Measure with a micrometer the diameter of the valve stem at various points of the guide-bearing surface.   The average allowable maximum wear is 0.001 in [0.025 mm].

Measure the valve-stem straightness with a runout indicator shown in Fig. 14–16.   The valve must be replaced if the runout exceeds 0.002 in [0.050 mm].

Grinding Valves   Make a cursory check of the valve grinder by inspecting the tightness of the drive belts, the lubrication of the bearings, and the coolant level.   Make particularly sure that the chuck is clean.

Dressing Grinding Stone   Select a grinding stone that is correct for the valve being refaced.   Select the correct coolant and install the diamond dressing tool.

CAUTION   Do not use water or a soluble oil when grinding sodium-filled valves.   Use kerosene.   Ker-

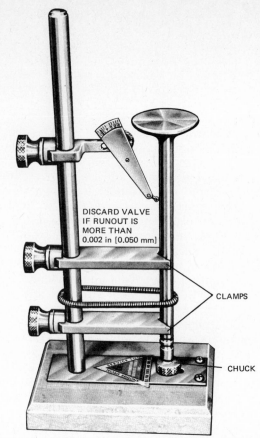

Fig. 14–16   Using a runout indicator to measure valve-stem straightness. (*J. I. Case Co., Components Div.*)

osene will not explode as a result of being mixed with sodium should the valve be damaged.

Start the motor and move the diamond toward its stone, adjusting it so that it will take a fine cut. Turn on the coolant and slowly move the diamond back and forth across the stone.   To prevent roughness, avoid cutting too deep or moving across the stone surface too fast.

When you have finished dressing the stone, remove the diamond and adjust the chuck angle to the desired degree.   Be very precise with your alignment.   Insert the valve stem into the chuck as close as possible to the clamps to avoid valve vibration or bending during the grinding operation (Fig. 14–17).

Start the chuck drive and check the valve runout. If the valve runout is larger than 0.002 in [0.050 mm], relocate the valve in the chuck and try again. If the valve is not centered within the limits of the runout, or if it is bent or worn, or if the valve head is warped, the valve must be discarded because the margin would be uneven after grinding.   This would result in uneven temperature on the valve head and early valve failure.

Lightly cut across the valve face to determine if the valve can be ground.   Grinding may remove too much surface and result in a margin less than half of the original width.   This check will also indicate warpage of the valve head which did not show up previously.

Record the micrometer feed setting at the beginning of the grinding operation.   To avoid valve over-

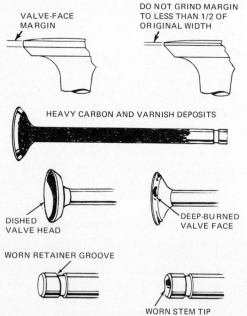

Fig. 14–15   Valve checkpoints. (*J. I. Case Co., Components Div.*)

Fig. 14-17 Grinding a valve face.

Fig. 14-18 Grinding a valve stem.

heating, take only light cuts and move the valve slowly across the full stone face. Move the micrometer feed only 0.001 in [0.025 mm] at a time, and frequently stop the chuck drive to check the grinding progress. When the valve-face surface is smooth and free of pitting, check the micrometer feed dial and record the amount of metal removed. If the amount removed is larger than that specified, the margin will probably be too small and/or the valve head will be lower in the cylinder head than the specified maximum.

CAUTION Do not move the stone beyond the valve face during the grinding operation. This will cause the stone to ridge or groove and the grinding surface to become round. Also be careful not to damage the valve fillet area.

Sometimes it becomes necessary to redress the stone in order to maintain a smooth, even surface and a uniform valve-seat angle during the grinding of one valve. This necessity can arise because of the heat scale on the valve or because the grinding stone was not appropriate for the hardness of the valve.

NOTE Make certain when the old valves are to be reused that each valve is placed in its original guide.

When you have ground all the valves, reservice the stem end using an attachment (shown in Fig. 14-18) to support and hold the valve in position. Turn the micrometer feed until it contacts the stone. Record the micrometer setting. Back off a few thousandths of an inch [hundredths of a millimeter] and start the grinding wheel motor and position the cooling stream to the valve-stem end. Turn the micrometer feed slowly toward the stone to ensure a cleancut surface. Note the amount of metal removed. Do not remove more surface than that specified (about 0.015 in [0.38 mm]) or it will eliminate the surface hardening and accelerate wear. **NOTE** Do not neglect to dress the side of the grinding stone.

When all stem ends are ground, it is sometimes necessary to grind a new chamfer. This is done in the following manner: Adjust the attachment to a 45° angle and lay the valve in place. Adjust the stops so that the valve cannot slide toward the stone. Start the grinding wheel and bring the stem end in contact with the stone and slowly turn the valve in its V block to grind the chamfer. Do not grind too large a chamfer as this will reduce the contact area of the stem end and may cause rapid wear of the rocker-arm surface.

Final Checks of Valves When you have refaced and cleaned all the valves, recheck the margin. Check the refaced valve on a runout indicator. The valve-face runout should not exceed 0.002 in [0.50 mm]. If it does, check the general condition of the valve refacer; it may need to be cleaned or repaired. Check and remeasure the valve stem to determine if it is the cause of the high runout. Do not forget to measure the valve-head depth. This is often neglected, despite the fact that it can create untold trouble.

NOTE New valves must also be checked, measured and, at times, reground (resurfaced) if they have been damaged during shipping or handling.

Valve-Seat Insert A valve-seat insert is a metallic ring which is lodged inside the cylinder head to increase the service life of the valve and valve seat. The cast-iron alloy from which the seat is manufactured will vary to suit various operational conditions. It may be of regular cast iron with the normal amount of iron, carbon, silicon, phosphorus, and sulfur, or a cast-iron alloy with additional elements such as chromium, nickel, manganese, molybdenum, copper, cobalt, and tungsten. It is important for the insert to seat firmly in the counterbore to ensure good heat transfer and to avoid distortion.

Valve-Seat Insert Checks and Service Check the valve-seat insert for cracks or looseness by lightly tapping the cylinder head near the insert. Check the seat area width against that specified. The average valve seat width is between 0.060 to 0.120 in [1.52 to 3.04 mm]. If the width exceeds this specification and cannot be narrowed down during a regrinding operation, the insert must be replaced. If grinding brings the valve head below the size specified, the insert must also be replaced.

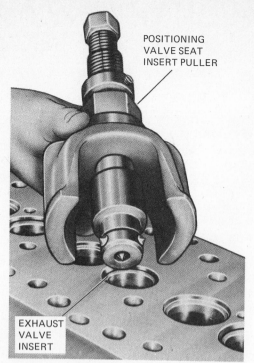

POSITIONING
VALVE SEAT
INSERT PULLER

EXHAUST
VALVE
INSERT

Fig. 14—19 Removing a valve-seat insert. *(Allis-Chalmers Engine Div.)*

Fig. 14—20 Insert boring tool. *(Cummins Engine Co., Inc.)*

Fig. 14—21 Installing a valve-seat insert.

NOTE Carefully check the valve-head height against its specification. When the valve head is too high, it may interfere with the piston and the valve seat may have to be reground.

When the valve head is too low, a loss of compression is unavoidable. The valve and/or the valve insert must then be replaced. If no insert is used and the valve seat is damaged beyond regrinding, a valve seat insert must be installed.

To remove an insert, use a puller similar to that shown in Fig. 14—19. Do *not* use a prybar, punch, or chisel because the hardened material shatters like glass and serious face or eye injury could result. If you do not have a puller, run a couple of welding beads on the inside of the insert. **CAUTION** Protect the valve guide and do not weld the cylinder head.

After the insert has cooled, it can easily be removed by hand or with pliers. Upon removal, check the counterbore for burrs, cracks, or rough edges. If these defects are present, remove them when you clean the counterbore. Some manufacturers deplore replacement of the same size insert, rather they suggest boring the counterbore to accommodate an oversize insert. Other manufacturers recommend that when the counterbore is undamaged, the same size insert be used as that removed.

NOTE When using an insert cutting tool such as that shown in Fig. 14—20, make certain that the valve guide is not worn, that the insert cutter is the correct size, that it is sharp, and that it is installed properly. The insert cutting tool must be fastened tightly to the cylinder head.

When you install a new insert, place it in dry ice. When thoroughly chilled, quickly place the insert over the counterbore with the valve seat up.

CAUTION Take care when handling dry ice. Wear asbestos gloves.

Use a driver as shown in Fig. 14—21 to drive or press the insert down tightly into the counterbore. It is sometimes recommended to peen the insert with a special tool or a round-nose punch to force the metal around the outer insert chamfer edge.

Valve-Seat Grinding Tools Defective valve-seat grinding tools should be repaired and worn out tools replaced. A valve seat, improperly ground, can cause compression losses, combustion changes, increased fuel consumption, burned valves, or breaking off of the valve head. A valve seat must be ground to a precise angle and be concentric with the valve guide. The seat area must be smooth and the width must be within specification.

When valve seats need grinding, they must be ground to a precise measurement. The tools required are: (1) a valve-seat grinder and stone dresser; (2) a pilot, the same size as the valve-stem diameter, to center the grinder; (3) a dial gauge to measure the concentricity of the valve seat; and (4) various grinding stones of different angles to grind the valve seat to a narrower width. The stones must also be of the correct width so that they do not interfere with the walls of the combustion chamber or cut a shoulder in the valve seat. They must be made of a material that will grind either the hardened insert or the cast-iron alloy of the cylinder head. The valve guide must not only be clean, it must conform to specified dimensions, otherwise the pilot will not be square to the valve seat. The valve seat and surrounding area must also be free of carbon and oil to ensure maximum cutting efficiency of the grinding stones.

STONE DRESSING After you select a stone of the correct width and texture, secure it to the drive. Then place the valve-seat grinder on the guide pin of the dressing tool. Back the diamond cutter away from the stone, adjust the diamond holder to the correct angle, and then lock it securely. Start the drive motor and adjust the diamond to the stone. Slide the diamond across the full stone face. Do not take too heavy a cut or slide the diamond too fast over the stone surface. To do so may ruin the diamond or shatter the stone. It would definitely leave the stone face rough.

Interference Angle The valve and the valve seat are ground at slightly different angles to produce a narrow line of contact. Such an angle is known as an *interference angle*. The angle must be positive and must not exceed $1^1/_2°$ (see Fig. 14–22). The purpose of the interference angle is to permit a narrow leakfree valve when the engine is first started. As the valve gets hot, the valve head curls lightly and expands to a full seat contact.

Valve-Seat Grinding (Refacing) Before you grind a valve seat, make sure that the valve guide is clean. Then apply a light coat of engine oil to the pilot, and insert and secure it in the valve guide.

Clean the bore of the valve grinder and place the grinder over the pilot. Adjust the knurl micrometer knob so that the stone just contacts the valve seat. Support the valve-seat grinder and start the drive motor. Feed the stone gently to the seat with the micrometer knob. This will ensure a smooth seat surface. Too much force on the stone will produce a rough seat.

Record the dial number of the micrometer at the start of the grinding operation. Remove only enough material to produce a pit-free continuous seat. Record the amount of material removed by checking the dial number of your micrometer feed. You may have to dress the stone frequently before one valve seat is perfectly ground.

Before you narrow the seat to position the valve-face seat contact, you should measure the valve seat for concentricity. This can be done by using a dial gauge tool as shown in Fig. 14–23. When the gauge is installed and the pointer adjusted, rotate the upper half of the tool to measure the valve-seat concentricity once more. The maximum runout should not exceed 0.002 in [0.050 mm]. If the valve-seat runout exceeds maximum allowable tolerance, recheck valve guide, pilot, and valve-seat grinder for wear. If excessive tolerance is not due to wear of any of these parts, you must regrind the valve seat.

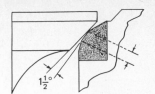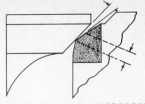

POSITIVE ANGLE – CORRECT   NEGATIVE ANGLE – INCORRECT

Fig. 14–22 Correct and incorrect interference angles. *(Allis-Chalmers Engine Div.)*

Fig. 14–23 Measuring valve-seat runout. *(International Harvester Co.)*

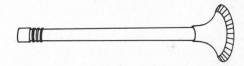

Fig. 14–24 Pencil marks on the valve face. *(Cummins Engine Co., Inc.)*

Locating Valve Face and Seat Contact To check seat contact, wipe a thin film of Prussian blue on the valve seat and rest the valve momentarily on its seat. (Do not rotate the valve.) A thin, continuous line should be evident on the valve face. The contact area must be at the correct height and the valve seat must be the recommended width.

Another method of checking the concentricity and condition of the valve seat is to pencil mark the valve face as shown in Fig. 14–24. Then place the valve against the valve seat and rotate it about 10°. Remove the valve and check your pencil marks. The seat and concentricity are satisfactory when all pencil marks are broken.

To correctly locate the contact seat area on the valve face and at the same time to reduce it to the correct width, grind it with a 30° stone. This will lower the seat and reduce its width. Otherwise use a 60° stone to raise the seat, and reduce it to correct width.

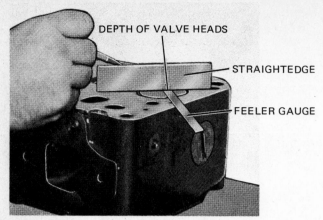

Fig. 14–25 Measuring valve-head height. *(Mack Trucks Canada, Ltd.)*

Measuring Valve-Head Height   The last step to be taken before installing the valve is to measure and compare the height of the valve with the cylinder head surface.   To do this, use a straightedge and feeler gauge as shown in Fig. 14–25.   Check the measurement against the manufacturer's specification.   If the valve head is too high, the seat or the valve must be reground to lower it.   If the valve head is too low, the valve and/or insert must be replaced.

Valve Springs, Spring Retainers, and Keepers   The steel used for a valve spring usually is composed of carbon, manganese, phosphorus, sulfur, and silicon.   It must be resistant to high temperature and corrosion and must maintain its strength throughout the cycle of compression and expansion to control the sealing force.   The coil is designed to ensure valve alignment with the spring retainer and valve keeper.   Valve springs which have evenly spaced coils are based on Hook's law.

To control valve vibration or valve flutter at the time of closure and to reduce valve-seat and valve-seat-insert wear, variably spaced springs are used (Fig. 14–26b).   Springs of this design, unlike those utilizing Hook's law, have variable coil spacing.

NOTE   When variably spaced valve springs are used, the close-wound end (with less resistance) must be installed toward the cylinder head to give the valve spring stabilization.

When the valve lift and/or the engine speed is high, two springs are used to ensure adequate pressure and prevent valve vibration and flutter. The coils usually are wound in the opposite direction to prevent the valve springs from rotating (see Fig. 14–26a).

A spring retainer made of steel or cast-iron alloy stabilizes the valve spring and locks the spring, through a set of keepers, to the valve stem (Fig. 14–27).   Valve keepers have internal grooves or recessed areas which match with the raised area or grooves on the valve stem.   When the two keepers are placed on the valve stem, the outer surface forms a cone and matches with the contour of the valve-spring retainers.   Sometimes a lower valve-spring seat (retainer) is used (see Fig. 14–27).   It acts as a

(a)   (b)   (c)

Fig. 14–26   Valve springs. *(a)* Uniform coil spacing (one coil wound to the left, the other to the right); *(b)* variable coil spacing; *(c)* uniform coil spacing.

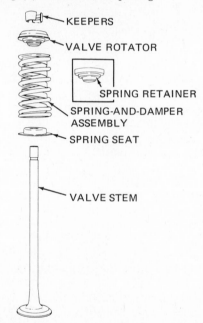

Fig.   14–27 Valve   and   valve   components.   *(Allis-Chalmers Engine Div.)*

heat sink and as a wear washer to prevent the spring from wearing into the cylinder head.

Valve-Spring Service   Clean with solvent any valve springs that have been coated for protection to maintain their corrosion resistance.   After properly cleaning and drying with compressed air, visually check them for rust, pitting, or fractured coils. Check both spring ends; they must be flat and must have a smooth, tapered surface.   Check both ends for squareness by placing the spring on a flat surface. Slide a combination square close to the valve spring and then rotate the spring slowly (Fig. 14–28).   The spring should remain parallel with the edge of the square.   Place the other end of the spring on the flat surface and rotate the spring against the combination square again.   If the rotating spring remains less than specified, the spring must be replaced. An average force variation of 5 percent is allowable. **NOTE** When a total of more than 0.030 in [0.76 mm] is removed from the valve and the valve insert due to refacing, a valve spacer should be used to increase the spring tension.

Valve Rotators   To extend the valve and valve-seat life, the exhaust and/or intake valves of some engines are equipped with positive or nonpositive valve rotators. The nonpositive valve rotator has a Belleville Spring (cone spring) between the two-piece valve-spring retainer, and a coil spring above the cone spring. As the rocker arm forces the valve down, the valve-spring force is momentarily removed from the valve. This gives the valve momentary freedom allowing it to rotate to a different position. A positive rotator is similar in design to the nonpositive except that the cone spring is replaced by a number of small steel balls. They lie in a ramplike groove and are held there by small coil springs (Fig. 14–29). As the rocker arm moves the valve down, the cone spring is compressed. This forces the balls to move in their inclined ramps and the valves to rotate. As the rocker-arm force is removed, the cone-spring force is released. The balls are then forced by their coil springs to return to their starting position.

Valve-Rotator Service   It is not possible to check the function of a valve rotator when it has been removed. At this point your only alternative is to thoroughly clean and visually inspect the rotator for external damage. It can easily be checked when the engine is running by making a chalk mark on the

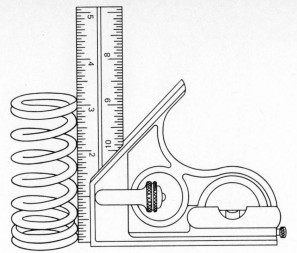

Fig. 14–28 Checking valve-spring alignment. *(Allis Chalmers Engine Div.)*

within 1.55 mm (0.64 in) of the edge of the square as you rotate it, you can assume that the spring is parallel and straight.

Measuring Spring Height and Force   Position the spring between the two anvils of a valve-spring tester and let the movable anvil rest lightly on the spring end.

From the tester scale, read the free length of the spring. If it is below specification, the spring must be replaced.

Compress the spring to a specified length. By doing this, force is put onto the lower anvil and recorded onto a dial calibrated in pounds [kilograms]. If the compressed valve-spring force is

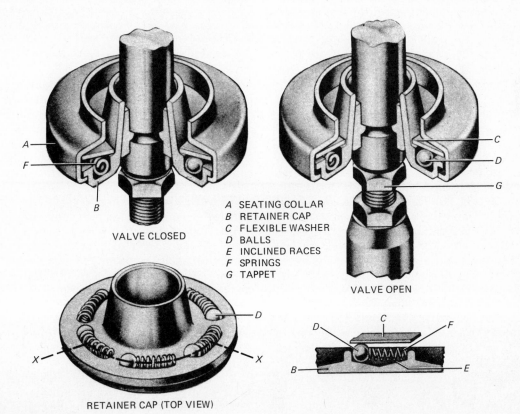

VALVE CLOSED

A SEATING COLLAR
B RETAINER CAP
C FLEXIBLE WASHER
D BALLS
E INCLINED RACES
F SPRINGS
G TAPPET

VALVE OPEN

RETAINER CAP (TOP VIEW)

Fig. 14–29 Positive valve rotator.

rotor retainer and retainer body. When the valve rotator is operating, the retainer mark should wander (rotate) away from the spring retainer mark.

Valve-Bridge Service (Crosshead) Visually check the surface of the rocker-arm lever and the valve-stem contact area for excessive wear or cracks. Check the adjusting screw and bridge threads for wear and distortion. With a small-bore gauge, measure the inside of the bore at several points for out-of-roundness and taper. When the contact surfaces are worn, the valve bridge must be replaced because of resurfacing problems. When the bridge bore is worn beyond specification, it can be reamed to accommodate an oversize valve-bridge guide pin.

Replacing Valve-Bridge Guide Pin With a micrometer, measure the outside diameter of the guide pin and compare the measurement to the manufacturer's specification. Using a square, check the guide pin for straightness. When a replacement is necessary, use a piece of metal pipe to bend the pin back and forth until it breaks. (It should be broken as close as possible to the cylinder head.) Drill and tap a suitable size thread in the guide pin. Use an adapter on a slide hammer to pull the pin from the cylinder head. The guide pin can also be removed with a dowel puller.

To install the guide pin, use a guide mandrel to press the pin into the bore, to the specified protrusion. After installation, recheck the guide pin for straightness.

Valve Seal Some engines utilize valve seals to prevent oil from passing into the combustion chamber through the intake or exhaust guides. These seals, made of Teflon, are placed either over the valve guide or on the valve stem below the valve keeper (see Fig. 14–30). The valve seals effectively control oil losses caused by vacuum, gravity, and inertia at the intake and exhaust valves.

Reassembling Cylinder Head After all the abrasives are removed, the head dried with compressed air, and all plugs reinstalled, swab the valve guides with an oil-saturated brush. Dip the valve stems into clean lubricant (oil) and place them in the valve guides. Make certain after each valve is inserted in the valve guide that the valve-seat and valve-head height are tested. If a valve seal is used, it usually is installed over the guide. Place the lower spring-seat retainer or the valve rotator valve spring, or springs and the upper valve-spring retainer, over the valve stem. **NOTE** Install the valve spring with the closed coil end toward the cylinder head. Compress the assembly with a valve-spring compressor.

Install (if used) the valve seal, and place the two half keepers on the valve stem. Release the spring compressor (Fig. 14–30) or use the tool shown in Fig. 14–31 to install the keepers. Tap on the valve stem to determine if the keepers are in their proper place.

Valve-Seat Testing Before you install the cylinder head, each valve seat should be tested with a vacu-

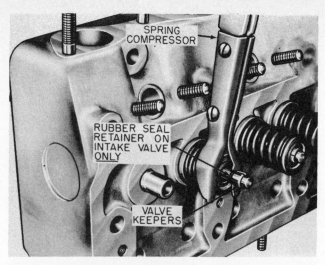

Fig. 14–30 Installing a rubber seal on the intake valve. (*J. I. Case Co., Components Div.*)

Fig. 14–31 Installing valve-stem keepers.

um tester for leakage. A leaking valve can cause a lot of trouble if not detected until after the engine is running. To test for valve seating, place a suction cup over the valve. Insert and squeeze the bulb while positioning the suction cup. If the suction cup stays on the surface of the cylinder head for about 10 seconds after the bulb is released, the valve seat, as well as the insert, is airtight.

A more reliable method of testing is with the use of a tester consisting of a vacuum pump, vacuum gauge, and a vacuum cup (Fig. 14–32). Here, also, place the vacuum cup over the valve and insert. Start the vacuum pump. Observe the vacuum gauge. When the needle of the dial indicates 20 inHg [50.8 cmHg], close off the shutoff valve and stop the motor.

Begin timing as soon as the vacuum gauge needle reaches 18 inHg [45.72 cmHg]. After a time lapse of 10 seconds, the gauge should not read less than 10 inHg [25.4 cmHg].

If the valve seal is unsatisfactory, tap the valve stem gently with a hammer handle or soft-faced hammer and repeat the test. If the seat is still unsatisfactory, check the test cup, the connecting lines

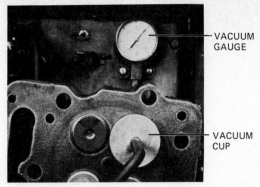

Fig. 14–32 Vacuum-testing a valve for leaks.

and the shutoff valve of the tester. If it checks out satisfactorily, put grease on the insert groove and repeat the test.

If the valve seat is now satisfactory, then the leak was caused by the valve-seat inserts. To seal the insert, stack or peen the insert and repeat the test.

Installing Cylinder Head Make certain that the machined surfaces of the cylinder head and cylinder block are clean. Do not forget to check the cylinder bore. Make sure that there is no oil at the bottom of the cylinder-head threaded bolt holes. This could cause a hydrostatic lock when the head bolts are torqued.

If not previously measured, check the sleeve protrusion. Make certain that the water nozzle or cylinder-head cooling-jet tubes are in place. Install the necessary guide studs to position the cylinder-head gasket and to ensure trouble-free head installation.

Place the head gasket with the side marked "top," facing up, on the cylinder block. If used, install the oil and coolant grommets (seal rings) (Fig. 14–33).

Hoist the cylinder head into position. Make certain that the cylinder head is parallel to the cylinder block when placing it in position and that the head gasket and grommets are located correctly. When multiheads are used, install all cylinder heads, then with a straightedge, align the manifold and/or water manifold mounting surfaces before tightening the head bolts.

Immerse the entire cylinder-head bolt in the recommended lubricant. Remove, and allow the excess oil to drip off. If thrust washers are used,

WATER GROMMET
CAP-SCREW GROMMET
PUSHROD GROMMET

Fig. 14–33 Cylinder-head grommets. (*Cummins Engine Co., Inc.*)

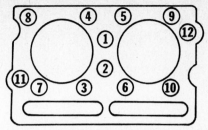

CYLINDER-HEAD CAP SCREWS — 5½-in [139.7-mm] BORE ENGINES

Fig. 14–34 Typical torque sequence. 1. Tighten to 25 lb·ft [3.5 kg·m]. 2. Advance to 80 to 100 lb·ft [11 to 14 kg·m]. 3. Advance to final torque, 280 to 300 lb·ft [38.7 to 41.5 kg·m].

place them on the cylinder-head bolts and insert the bolts, according to their length, into their respective bores. Tighten the head bolts in the recommended sequence and to the recommended torque (see Fig. 14–34).

Installing Injector Refer to the appropriate units in Section 4, "Fuel-Injection Systems," for recommended injector installation procedures.

Questions

1. What is the purpose of a cylinder head?

2. Why should you never remove the cylinder when the engine is hot?

3. List the checks and measurements you must make after the cylinder head has been cleaned.

4. Why should you air-test or water-test a cylinder head?

5. When the injector protrusion is below specification, but has not yet been corrected, what result would this have in regard to combustion?

6. How would you check the serviceability of a valve guide?

7. When pressing a new valve guide into place why is it important to have the protrusion height to precise specification?

8. What is the purpose of valve inserts?

9. What checks and preparations must be made prior to resurfacing the valve seats?

10. Describe how to locate the valve-seat contact?

11. What will result if the valve-head height is 0.020 in [0.508 mm] below specification?

12. Why should you not resurface a valve when the margin is below specification?

13. By what checks and measurements can you determine if a valve is resurfaceable.

14. What result would a weak valve spring have on the valve action?

15. Why are valve rotators used?

16. What is the purpose of a valve bridge?

17. What checks must precede the placing of a cylinder head in position?

18. Why is it important to torque the cylinder head in three separate steps, for example, in the first torque sequence to 6.91 kg·m, then to 20.745 kg·m, then to 34.575 kg·m?

UNIT 15

Valve-Train Operating Mechanism

Servicing Rocker Arm Very little damage or wear will be found on the rocker arm or shaft especially if the engine is not abused or other parts of the valve train have not failed. However, insufficient lubrication, contamination in the oil, or incorrect installation of valve-train components could damage the bushings, shaft, or other components of the valve train (Fig. 15–1).

Before disassembling the rocker-arm assembly, check each rocker arm for identification or punchmark the position of each. Although some rocker arms look alike, their angle may be slightly different to give centralized contact on the valve stem or bridge. It is wise to disassemble one rocker-arm assembly at a time to prevent interchanging of components.

Remove the end-pipe plugs or, when Cup plugs are used, drive them out with a pointed punch and pry the plug from the shaft. In all cases place the shaft in a V block, rather than in a vise, to prevent damage to the shaft.

Clean all the components in solvent and dry with compressed air. Check the rocker-arm shaft and oilholes, and, with a micrometer, measure the shaft for wear. Check the shaft ends of any damaged threads, or the counterbore of the cup seating area.

Check the rocker arms for possible cracks. Check the adjusting screw threads as well as the threads in the rocker arms for distortion. Check the ball end with a radius gauge. When the adjusting screw threads or ball ends are damaged or flattened, replace them.

Check the surface of the valve stem or bridge contact. If it requires resurfacing, be sure to maintain the same contour radius so that the valve is forced straight downward. Do not remove more than 0.010 in [0.25 mm] from the surface, otherwise you may remove the surface hardening. Use the attachment tools of your valve refacer to resurface the rocker arm.

Measure the rocker-arm bushing for wear. When the rocker arm is new, the average running clearance is about 0.001 in [0.025 mm]. When the bushing exceeds specification, press it out, and then press a new one in. Carefully align the bushing oilholes with the rocker arm. Since some rocker-arm bushings have no oilholes, they must be drilled *after* installation.

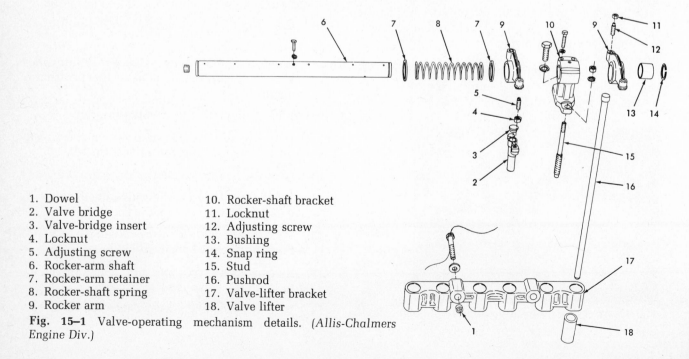

1. Dowel
2. Valve bridge
3. Valve-bridge insert
4. Locknut
5. Adjusting screw
6. Rocker-arm shaft
7. Rocker-arm retainer
8. Rocker-shaft spring
9. Rocker arm
10. Rocker-shaft bracket
11. Locknut
12. Adjusting screw
13. Bushing
14. Snap ring
15. Stud
16. Pushrod
17. Valve-lifter bracket
18. Valve lifter

Fig. 15–1 Valve-operating mechanism details. (*Allis-Chalmers Engine Div.*)

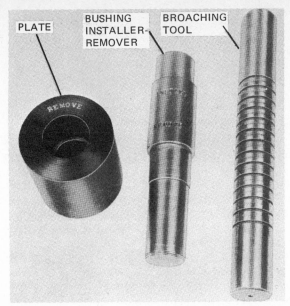

Fig. 15–2 Rocker-arm-bushing burnishing tools. (Allis-Chalmers Engine Div.)

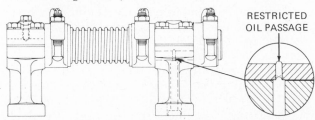

Fig. 15–3 Rocker-arm assembly. (Allis-Chalmers Engine Div.)

To attain alignment and the correct inside diameter, the bushing may be honed or broached with a broaching tool (Fig. 15–2). Check the rocker-arm-bracket mounting surface for flatness and cracks as well as for a smooth wear surface. Make sure that the oil-supply holes are clean. The rocker arms of the Detroit Diesel 53, 71, and 92 series engines have a clevis pinned to the rocker arm into which the pushrod is threaded. Check the threads of the clevis for distortion and the clevis and pin for wear.

Assembling Rocker-Arm Shaft Lubricate the rocker-arm bushing and shaft and assemble them onto the shaft in the sequence in which they were removed. Do not forget to install end plugs or Cup plugs. Position the rocker-arm shaft so that the oilholes in the shaft align with the supply holes (Fig. 15–3). When a special rocker-arm-bracket bolt is used, take care to screw it into its correct location. Be sure that the oilholes of the rocker arm are positioned toward the cylinder head rather than facing upward. Incorrect positioning can allow too much oil to be present above the cylinder head. As a result, the valve guides cannot control the oil and it will find its way into the combustion chamber.

Pushrods or Push Tubes Pushrods or push tubes are made of special steel to withstand the great stress and force to which they are subjected. The advantage of using a push tube is that it has less inertia than the solid pushrod. There is usually a socket on

one end of the pushrod and a ball on the other end to reduce friction and to maintain alignment. The result when motion is transmitted to the rocker arm is that the rocker arm moves in a part circle. The pushrods of the Detroit Diesel series 53, 71, and 92 engines are threaded on one end and screwed into the threads of the clevis. A pushrod spring ensures that the pushrod rests with a predetermined force on the cam roller (cam follower) to maintain it in constant contact with the camshaft lobe.

Servicing Pushrods or Push Tubes Check the straightness of each pushrod by rolling it on a flat surface or by placing the pushrod in V blocks and using a dial indicator to measure the out-of-roundness. Pushrods should not be out-of-round more than 0.020 in [0.508 mm]. Pushrods which are bent beyond specification must be replaced; those within specification should be straightened.

Check the ball ends for looseness, and check them for wear with a radius gauge. To check the socket-end wear, use a new rocker-arm adjusting screw, bluing the ball surface, then rotate the ball in the socket and check the contact surface.

Replace the pushrod when the contact area is less than 80 percent. Cummins push tubes should be checked to see if they have been filled with oil.

Cam Follower (Cam Roller) Cam followers and rollers are made from cast iron or iron alloy which has a high resistance to wear and corrosion (Fig. 15–4). Although the external appearance varies with engine design, all cam followers or rollers reduce friction and evenly distribute onto the camshaft lobe the force placed on them during opening and closing of the valves or injector. The cam follower usually slides up and down in its bore and follows the eccentricity of the camshaft lobe.

Either a socket which accepts the pushrod ball end is machined into the follower or else a replacement socket is pushed into the follower. Some engines, because of the difference in camshaft location and action, have neither a pushrod nor the conventional cam follower. The Detroit Diesel 149 Series engines have the cam roller pinned to the rocker arm which is in direct contact with the camshaft. On some Caterpillar engines, the follower

Fig. 15–4 Typical cam followers.

with the tappet adjustment is placed directly over the valve and no rocker arm or pushrod is used. Some Cummins diesel engines have the cam rollers pinned to a lever which pivots on a shaft located in a separate cam-follower housing. The pushrod rests in a replaceable socket.

Various types of cam followers, also known as *tappets* or *valve lifters*, are used on small diesel engines. The valve lifters do not have cam rollers, but they do have enlarged wear surfaces which are in contact with the camshaft lobe. Since the follower on engines of this type must be installed from the bottom, they must be positioned before the camshaft is installed.

Servicing Cam Follower Regardless of its design, the wear limit of the cam-follower bushing, pin, and bores must be checked very carefully. The cam follower's surface condition must be free of galling, pitting, or scoring. The rollers should be checked for flat spots since any slight defect will affect valve and/or injector timing.

Followers (valve lifters) having a flat surface instead of a roller should be surface-checked. If the damage to the flat surface is not too severe, it can be restored by using a valve refacer. The grinding procedure is the same as that used when refacing the valve-stem end. **NOTE** To retain surface hardness, do not remove more than 0.010 in [0.25 mm].

Valve-Bridge Adjustment To adjust the valve bridge, clamp the bridge in a vise and loosen the adjusting screw locknut. Position the bridge on the valve-bridge guide pin. With light finger pressure on the rocker-arm contact surface, hold the bridge in contact with the valve-stem end opposite to the stem end of the adjusting screw. Using a screwdriver, turn the adjusting screw until it contacts its mating valve-stem end. To compensate for thread looseness, advance the adjusting screw an additional one-eighth of a turn and finger tighten the locknut. Remove the bridge and clamp it in a vise. Hold the screw in position and torque the locknut. To check the adjusted valve bridge, place the bridge in position and check the valve-bridge contact and the clearance between the valve-spring retainer and the bridge.

To check the adjustment (Detroit Diesel), place a 0.001-in [0.025-mm] shim stock between each valve-stem tip and bridge. When forcing the valve bridge down, both shim stocks must be equally tight. If the adjustment is not correct, that is, if there is uneven force on the shim stock, remove the valve bridge and repeat the procedure previously outlined.

Installing Valve and Injector Mechanism Before you install the cam followers (Caterpillar engines), turn the adjusting screws to maximum clearance. This will prevent the valves from being forced against the piston as the camshaft is torqued down. Generously lubricate the bores and cam followers before placing the cam followers into the bores.

When you install the cam follower assembly on a Cummins N, NH, and NT diesel engine, use only one

gasket between the housing and cylinder block. Use a soft-faced hammer to force the assembly into position. Be sure that each housing fits flat against the cylinder block. Tighten bolts in the sequence specified. Install the push tubes. Make certain that each push tube is placed in its correct position and that the ball end fits into the socket.

Timing of the engine is covered in Unit 32, "Cummins Fuel-Injection System."

After the engine (Cummins) is timed, install a new rocker-housing gasket, back off all rocker-arm adjusting screws, and install the assembly. Align the push-tube sockets with mating rocker-arm adjustable ball ends. Install housing hex bolts and torque them in correct sequence to specification.

When assembling the pushrod assemblies of Detroit Diesel 53, 71, and 92, Series engines, place the lower spring seat, pushrod spring, and upper spring seat onto the pushrod. Screw the locknut to the full capacity of the threads onto the pushrod. Place the spring-seat retainer into its seat in the cylinder head, and slide the assembly from the bottom through the follower bore.

Screw the pushrod into the clevis until one thread of the pushrod is visible. Immerse the cam followers in clean Cindol 1705 oil heated to 100°F [73.3°C] for 1 hour to ensure initial lubrication before placing the cam follower into the bore (Fig. 15–5). Bolt the follower guide into place to hold the follower in position. **NOTE** The oilholes in the bottom of the cam follower must point away from the exhaust valve.

Lubricate the rocker-arm bushing and shaft. Slide the rocker-arm shaft through the three rocker-arm bushings. On each end, place a rocker-arm bracket on the shaft with the machined surface facing the rocker arm. If a valve bridge is used, install it on the guide pin making certain that the ends rest squarely in the valve stem. Tilt the rocker-arm assembly into position. Place the hex bolts through the rocker-arm bracket, and then tighten them to the cylinder head.

Valve Adjustment Precise valve clearance, on any type of engine, is essential to compression, combustion, scavenging, horsepower, fuel consumption and running condition (firing) as well as to the service life of the valves and valve seats. A two-cycle diesel engine is more adversely affected by improper valve clearance than a four-cycle engine because its power cycle is completed in one revolution and the cooling of its exhaust valve is about 50 percent less. Remember too, that an increase or decrease in valve clearance will alter the time of the stroke.

CAUTION Whenever some element of servicing disturbs the valve clearance, back off all tappet adjusting screws so that the valves cannot contact the piston when cranked over. The service manual lists valve clearance specifications for a cold engine and a hot engine. Therefore, you must set the valve clearance twice: once before you start the engine and again when it has reached its operating temperature. Readjustment is necessary because the valve

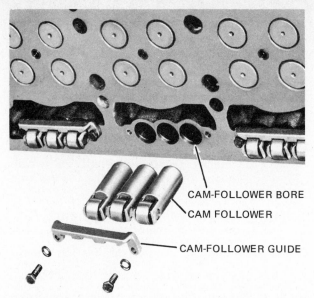

CAM-FOLLOWER BORE

CAM FOLLOWER

CAM-FOLLOWER GUIDE

Fig. 15–5 Installing cam followers. *(GMC Detroit Diesel Allison Div.)*

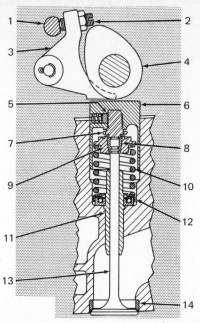

1. Compression-release shaft	7. Gear on adjusting mechanism
2. Adjusting screw	8. Keeper
3. Compression-release rocker arm	9. Spring retainer
4. Camshaft	10. Valve spring
5. Valve-adjusting-mechanism lock	11. Valve guide
6. Cam follower	12. Valve rotator
	13. Valve
	14. Valve-seat insert

Fig. 15–6 Valve mechanism (inlet valve) with compression release. *(Caterpillar Tractor Co.)*

and valve-train components may expand at operating temperature. This can cause a change in the valve clearance.

Let us assume you have repaired a six-cylinder engine with a firing order of 1-5-3-6-2-4. Remember that the valves are closed when the piston is at TDC on compression stroke. When setting all valves with one crankshaft revolution, you should first record the firing order as follows:

$$1 - 5 - 3$$
$$6 - 2 - 4$$

This means that when No. 1 cylinder is on compression stroke at TDC, No. 6 cylinder is also at TDC but has completed its exhaust stroke and the intake and exhaust valves are open.

At the same time, No. 5 cylinder is 120° BTDC and on compression stroke, whereas No. 2 cylinder is 120° BTDC and on the exhaust stroke.

The No. 3 cylinder is 120° ATDC on its intake stroke, and No. 4 cylinder is also 120° ATDC but on its power stroke.

Therefore you can set, at this point, both intake and exhaust valve clearances of No. 1 cylinder, the intake valves of Nos. 2 and 4 cylinders and the exhaust valves of Nos. 3 and 5 cylinders.

To adjust the valve clearance, insert a gauge of the correct thickness between (1) the valve bridge and rocker arm, (2) valve stem and rocker arm, and (3) camshaft and follower. Loosen the locknut and turn the adjusting screw downward (clockwise) to decrease the clearance or upward (counterclockwise) to increase the clearance, between the two contact points of the feeler gauge. The clearance is correctly set when the proper feeler gauge passes with a slight drag between the two surfaces. When you use a no-go feeler gauge, part of the no-go thickness will not pass through the two surfaces.

After determining that the clearance is correctly set, hold the adjusting screw stationary and tighten

the locknut. Recheck the adjustment to ensure that after tightening the locknut, the valve clearance has not changed. Immediately thereafter the adjusted valves should be identified so that none is overlooked.

Rotate the crankshaft 360° so that No. 6 cylinder is on TDC compression stroke. Adjust the valves of No. 6 cylinder, the intake valves of Nos. 3 and 5 cylinders and the exhaust valve of Nos. 2 and 4 cylinders. (See Units 31 and 32 for procedures on adjusting injectors.)

Decompressor (Compression Release) Many engines are equipped with a compression release mechanism which opens either all exhaust valves or all intake valves permitting the engine to be cranked more easily. This mechanism includes the necessary levers and linkage, and a decompression shaft with a pin or with a shaft that is flattened on one side, as shown in Fig. 15–6. The adjusting screw of the release lever is in contact with the flat side of the shaft. In either case, when the decompression shaft is rotated to its engaged lock or stop position, the rocker arms or the followers are forced down to open the intake or exhaust valves.

Servicing and Adjusting the Compression Release The compression linkage, levers, arms, or shafts require very little maintenance or service. However, when servicing, inspect for a possible bent or worn control linkage, a worn setscrew, or worn adjusting screw threads. After replacement of any of these parts, make certain the decompression mechanism cannot contact the rocker arm or follower and that the shaft either rests in its detent (trigger) or stops in the disengage position.

The procedure for adjusting the decompression-release mechanism shown in Fig. 15–6 is as follows:

1. Place the compression release in (disengaged) running position.
2. When you are certain the detent holds the shaft securely, place a 0.30 in [0.76 mm] feeler gauge between the release arm and follower.
3. Loosen the locknut and turn the adjusting screw against the flat side of the release shaft.
4. When the proper clearance is achieved, lock the adjustment.
5. Adjust the remainder of the valves.

Questions

1. What are the major differences between the valve mechanisms shown in Fig. 15–3 and Fig. 15–6?

2. Why is the pivot point of the rocker arm closer to the pushrod than to the valve stem or valve bridge?

3. List the checks and measurements you must make to determine the serviceability of a rocker arm, a rocker-arm shaft, and a pushrod.

4. What is the purpose of a valve bridge?

5. Outline how to adjust a valve bridge.

6. List the checks you should usually make prior to and during installation of the valve-actuating mechanism.

7. When is a valve adjustment mandatory? List the reasons why a valve adjustment is essential.

8. What is the purpose of a compression release?

9. How can you prevent cylinder-head failure?

10. List several reasons (other than those mentioned in this text) why a rocker arm may break.

UNIT 16

Flywheel Housing, Flywheel, and Timing Cover

Crankshaft Rear Oil-Seal Cover Because of the additional weight and/or the engine stand design, many service shops install the rear oil-seal cover, the flywheel housing, and the flywheel after the engine is removed from the engine stand. Rear oil-seal covers differ because of engine design or the purpose for which they are used. Some engines are equipped with a simple cast-iron or aluminium rear cover into which the rear oil seal is pressed (Fig. 16–1). Others use a one-piece flywheel-housing casting, and still others (where the timing gears are located at the rear) have an extended flywheel housing which doubles as a timing-gear housing. The rear oil-seal cover seals the crankshaft and the rear of the cylinder block and is sometimes also part of the oil-pan mounting surface.

Inspecting and Servicing Flywheel Housing To inspect and service the flywheel housing, you should first remove the seals and gaskets. Then clean the cover (or the flywheel housing), and dry it with compressed air. Check it for cracks or damaged mounting surfaces, and check the seal surfaces. Measure the bores, bushings, and shaft for wear and check for pitting and grooving. Where necessary, remove nicks and burrs or replace components

where warranted. Using a straightedge, measure all mounting flanges and sealing surfaces for straightness. Check all threaded holes, especially those of the rear engine mount. Install rear seal and auxiliary shaft seals if used (see Unit 23, "Seals and Gaskets").

Fig. 16–1 Crankshaft rear oil-seal cover. (*Cummins Engine Co., Inc.*)

Installing Rear Cover and Flywheel Housing

Before you install the rear cover and flywheel housing, check all mounting surfaces and dowels, then lubricate the oil seal(s) with clean engine oil. Do not use grease, soap, white lead, etc. Install, when necessary, the O rings and/or gaskets which seal the crankshaft or cylinder block to the flywheel housing. Some manufacturers recommend applying a thin film of nonhardening sealant to both mounting surfaces.

A heavy flywheel housing should be lifted into position with a suitable hoist. To protect the rear seal, push a seal sleeve over the crankshaft before installation. Install pilot stud bolts to the cylinder block to improve alignment and aid installation.

Position the flywheel housing over the aligning bolts, onto the dowel pins, then onto the cylinder block. Some flywheel housings are secured with hex bolts of various sizes and length, others use flat washers, sealing washers, or lock washers. Make certain that the flywheel-housing bolts and washers are placed correctly. Tighten the hex bolts in sequence and to the recommended torque.

When you must install a new rear cover or flywheel housing, do not, at this point, install oil seals, gaskets, or dowels. Place the cover in position and merely hand tighten the bolts because the cover or flywheel housing must first be aligned.

Aligning Rear Cover To align the rear cover, tap it gently until the lower flange is flush with the cylinder block. Then measure the front-cover bore alignment, as shown in Fig. 16–2. When properly aligned, use an accurately ground drill to bore the recommended size holes, then use the correct size reamer to ensure dowel fit. Remove and clean the rear cover. Then install the oil seal and gasket. Position the cover and install all cover bolts. Drive the dowel pin into the reamed holes, and secure the rear oil-seal cover.

Aligning and Measuring Flywheel Housing When you install a new flywheel housing, you must position it to permit the axis of the crankshaft to be concentric with the flywheel-housing bore. To check the position of the flywheel housing, screw an in-

Fig. 16–3 Checking flywheel-housing bore runout. *(Cummins Engine Co., Inc.)*

dicator holder or place a magnetic base onto the crankshaft. Attach a dial gauge to it so that the pointer rests squarely on the surface of the bore (Fig. 16–3). Zero the dial, then turn the crankshaft one complete revolution. Record readings at 90° intervals. The reading at any point must not exceed an average concentricity (runout) tolerance of 0.005 in [0.12 mm]. Tap the housing into alignment and tighten the hex bolts.

Your next check should be the flywheel-housing face runout. To make this check, relocate the dial gauge so that the pointer rests against the flywheel-housing flange. Force the crankshaft forward to remove end play, then zero the dial. Turn the crankshaft one complete revolution, record readings at 90° intervals. **NOTE** Make sure the crankshaft is placed forward when taking the readings. The average allowable maximum face runout is about 0.010 in [0.25 mm].

When the face runout and the flywheel-housing bore concentricity are within specification, drill and ream new dowel-pin holes. Remove and clean the flywheel housing. Install the rear oil seal, O rings (and gasket if used). Install the flywheel housing.

Inspecting and Servicing Flywheel Begin your inspection and service of the flywheel by checking all flywheel mounting surfaces for nicks, burrs, or scoring. Inspect all bores and threaded holes for damaged bores or pulled threads. Check the drive lugs of the intermediate clutch drive plates for wear and, using a square, check their alignment. Use a straightedge to check the clutch contact surface for straightness. Although score marks, grooves, and heat checks are always present to some extent, when the marks are too deep or the clutch surface is tapered beyond specification, the surface must be refaced.

To reface a flywheel, either special grinding machines are needed or a lathe and a special grinder is required to maintain factory standards in regard to surface condition, dimension, and static balance.

Check the ring gear for worn, damaged, or broken teeth caused by abnormal starting motor operation or a damaged cranking motor pinion.

Fig. 16–2 Checking front-cover bore alignment. *(Cummins Engine Co., Inc.)*

Replacing Ring Gear To replace a ring gear, use a blunt chisel to drive it evenly from the flywheel. Alternatively, heat the gear with a torch to expand it before driving it from the flywheel. Before you install a new ring gear, make certain that the gear and the cranking motor pinion correspond. Place the ring gear in an oven or use a heating torch to apply heat to its inner surface.

Check the applied heat with a special crayon (templet stick) that has a rating of 600°F [316°C]. This is done by touching the stick against the ring gear. When the applied heat reaches a temperature of 600°F [316°C], the crayon becomes soft and leaves a mark on the ring gear. At this point use two pairs of pliers and place the ring gear on the flywheel as quickly as possible. If necessary, tap the ring gear against the flywheel shoulder. **NOTE** The chamfered edges should point toward the cranking motor.

Installing Flywheel Make sure that the flywheel and crankshaft flange are clean. Then install new dowels and two guide studs in the crankshaft flange. Lift the flywheel into position. Align the holes and place the flywheel on the guide stud. Place the lock plates (or wear plates, if used) into position and install the flywheel bolts. Tighten them in the correct sequence to the recommended torque. Crimp the lockplates to the bolt head or, when lockwire is used to secure the bolts, be sure to wire the bolts in pairs.

When a new crankshaft or a new flywheel is used, the flywheel pilot-bore runout must be checked and a measurement taken to determine if the flywheel housing and the flywheel face are parallel.

Measuring Flywheel Runout To check the flywheel face runout, place a magnetic base against the flywheel housing and position the dial gauge squarely against the flywheel face (Fig. 16–4). Force the crankshaft forward and zero the dial following the measuring procedure previously outlined. The average maximum runout should not exceed 0.0005 in [0.0012 mm] for each 1-in [25.4-mm] radius. For example, when the pointer of the dial is 10 in [254 mm] from the center of the crankshaft, safe engine

operation dictates that the maximum runout should not exceed more than 0.005 in [0.127 mm].

To check the pilot-bearing runout, relocate the dial gauge so that the pointer rests squarely against the pilot-bearing bore. Check the runout, and if necessary force the flywheel into such a position that it will prevent the bore runout from exceeding 0.005 in [0.127 mm].

After the flywheel is aligned, drill and ream the flywheel and crankshaft flange for proper dowel fit. **NOTE** Do not fully ream the holes in the crankshaft flange. Leave a small shoulder on which the dowels may rest.

Timing-Gear Cover, or Front Cover A timing-gear cover, or front cover, is one-piece cast iron, iron alloy, or aluminium casting, reinforced to withstand the stress placed on it. In addition to sealing the crankshaft, it covers the timing gears. It may become part of the oil-pan sealing and mounting surface, or it may act as the front support or bearing for the engine mount. At other times it may act as the mounting surface for the air compressor, air-conditioning pump, alternator, fuel-injection pump, hydraulic pump, or as the idler pulley for the fan-belt tensioner.

The correct time to install the timing-gear cover is governed by the engine stand or the design of the cover itself. In any event, the cover should be installed as early as possible to prevent dust and dirt from accumulating on the exposed components.

Inspecting and Servicing Timing-Gear Cover Clean the cover, remove the auxiliary seals and scrape off all old gasket material from the sealing surfaces. You can then inspect the cover for cracks, damaged bolt holes, pulled threads, nicks, or burrs. Use a smooth cut file to restore the mounting surfaces and a bearing scraper or emery cloth to restore the bores. See the sections on installation procedure in Unit 23, "Seals and Gaskets."

Some manufacturers recommend installing crankshaft seals after the timing-gear cover has been installed and after the bore runout has been dialed in order to ensure proper bore concentricity.

Installing Timing-Gear Cover From Fig. 16–5 you can see that the gear case is an extension of the cylinder block. It is attached by bolts to the cylinder block. The gear cover must be positioned and aligned before the camshaft, gears, and fuel-injection-pump drive gear are installed.

Before you install the gear cover, lubricate all gears with engine oil. Install two or more guide studs, then place a seal protection sleeve over the crankshaft. Position the oil slinger, if one is used. Apply a nonhardening sealant to the front plate or to the side of the gasket facing the front plate.

When you position the gasket on the front plate, make sure that the oil passages are not covered. Install shims if they are needed to adjust camshaft end clearance.

Lift the gear cover into position, place it on the guide studs and then on the dowel. Tap the cover

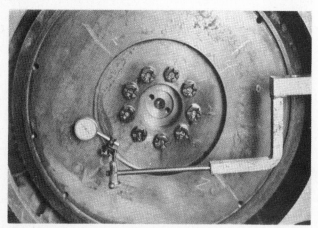

Fig. 16–4 Checking flywheel face runout. (*Cummins Engine Co., Inc.*)

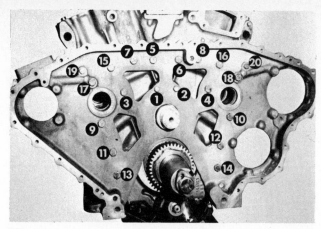

Fig. 16–5 Gear-case torque sequence. (*Cummins Engine Co., Inc.*)

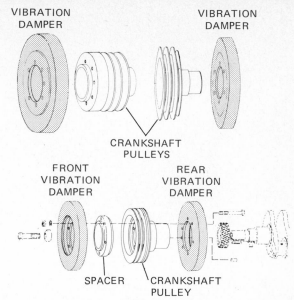

Fig. 16–6 Various crankshaft pulleys and vibration dampers. (*Allis-Chalmers Engine Div.*)

into place, then install and tighten the cover bolts (see Fig. 16–5). If applicable, trim off any excess gasket material at the oil-pan mounting surface. The bottom surface of the cover must be flush with the cylinder block or within the specified limits, that is, within 0.004 in [0.101 mm].

When a new cover is used the dowels must be removed, the front seal bore must be aligned with the crankshaft, and the mounting surface of the oil pan and the cylinder block must be within specification. You may have to compromise alignment to achieve proper bore runout and to achieve tolerance between the cover surface and the cylinder-block surface. After the cover is aligned, the dowel holes must be drilled and reamed to retain cover position.

NOTE Never enlarge dowel bores or file the dowels in order to force the cover to fit. After such modification, the cover bolts will not hold the cover in alignment. Oil will then leak at the joint where the oil pan and cover meet. It may also leak out of the front crankshaft seal.

Upon completion of installation, do not forget to tighten bolts in correct sequence to their specified torque requirements.

Crankshaft Pulley and Hub A crankshaft pulley is one-piece cast iron or cast-iron alloy casting with a machine-tapered or straight bore. Pulleys with a straight bore rest against the shoulder on the crankshaft or against the oil slinger and crankshaft timing gear. A keyway in the pulley bore and crankshaft accommodates a key which positions the pulley or hub to the crankshaft. Other designs have a straight or tapered pulley hub to which the pulley and/or vibration damper is bolted. A machined surface on the crankshaft pulley provides the sealing surface of the front crankshaft oil seal. The vibration damper is bolted against a machined surface at the front or rear of the crankshaft pulley (Fig. 16–6). There are machine grooves (sheaves) on the outer circumference for the drive belts of the coolant pump, air compressor, and/or alternator.

Inspecting and Servicing Crankshaft Pulley It is seldom necessary to replace crankshaft pulleys even though they are often damaged due to improper in-

stallation, removal, or storage, or due to worn sheaves.

Check the pulley bore and keyway for wear and burrs. Check the oil-seal contact area. Check the belt grooves very thoroughly. Be sure the side walls are straight and the surfaces are smooth.

To check the V-belt groove (sheave) for wear, place a new drive belt in the groove. There should be adequate clearance between the belt and the smaller diameter of the V groove (Fig. 16–7). If the new drive belt bottoms out, if the clearance is less than 1/8 in [3.175 mm], or if the sides are rough or dished out, the pulley must be replaced. Any of these defects can cause the drive belt to reduce friction, lose its driving power, and wear excessively.

When the crankshaft pulley has a vibration damper bolted to it, check the flange surface and the threaded bolt holes. If the pulley or hub has a tapered bore, check the taper for sufficient contact, particularly if the pulley, hub, or crankshaft is new.

To check the pulley or hub fit, remove the oil from the crankshaft and the pulley bore. Remove any nicks and burrs and spread a light even coat of Persian blue on the crankshaft taper. Push the pulley or hub straight onto the tapered crankshaft, rotate the pulley 45°, and then pull it straight off. The con-

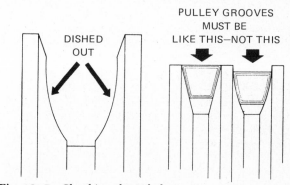

Fig. 16–7 Checking the V-belt groove.

Fig. 16—8 Crankshaft surface finish after lapping. (*Cummins Engine Co., Inc.*)

tact area, defined by a transfer of the Persian blue, must be full 360° and more than three-quarters of the tapered length. Should there be insufficient contact, the pulley or hub must be lapped to the crankshaft. To do this, clean the shaft and pulley bore and apply a 300-grit lapping compound to the crankshaft taper. **CAUTION** Do not allow the compound to come into contact with the lip of the oil seal.

Slide the pulley, or hub, onto the crankshaft taper so that it can be rotated about 90° each way. Although rough at the beginning, the lapping motion will smooth out after a few turns. To check effectiveness of the lapping, wipe off the compound. The surface finish should appear similar to that shown in Fig. 16—8. Even if it looks right, you should recheck the contact area by the Persian blue method.

Installing Crankshaft Pulley Remove evidence of lapping and bluing. Install a new key into the keyway of the crankshaft and slide the pulley onto the taper of the crankshaft. Use a soft-faced hammer to guarantee seating. Place a heavy washer over the hex bolts and lubricate the threads before installing. Then tighten the washer to the recommended torque. **NOTE** Some pulley retainer bolts have a self-locking device.

Pulleys with straight bores must be pressed rather than driven onto the shaft to prevent damage to the pulley and crankshaft thrust bearing.

Inspecting Vibration Damper Before you can inspect any damper, dirt, grease, and residue of corrosion must be removed.

A rubber-element damper should be cleaned with ordinary household detergent and a viscous vibration damper should be cleaned with a suitable solvent cleaner.

When inspecting the rubber-element vibration damper, check the rubber element for deterioration. Check that the index mark on the mounting flange and the inertia weight are in alignment. If they are out of alignment more than 0.064 in [1.62 mm], it is an indication that the rubber has lost its elasticity or has loosened from its bond. The damper must then be replaced.

Check the mounting flanges and the bolt holes for cracks and elongation. Check the surface for straightness.

When you check a viscous damper, inspect it thoroughly for any evidence of external damage. Even small indents make the damper unusable since they prevent the inertia mass from rotating.

Check the mounting flanges as well as the bores. Check for pinholes, broken welds, or cracks by heating the damper in an oven 200°F [93.3°C]. When removed from the oven, no evidence of oil should be visible.

Installing and Testing Vibration Damper Before you mount the vibration damper to the pulley flange, use a dial gauge to check the concentricity and runout (wobble) of the flange. The average allowable runout is about 0.003 in [0.08 mm] and the eccentricity is about 0.004 in [0.10 mm]. The measurement should be taken at the outside radius and the outside diameter.

Before you install the damper, make sure that the mounting surface of the crankshaft and damper are clean. Then position the damper, install the hex bolts, and tighten them in three phases, in sequence, to the recommended torque. Then with a dial gauge, check the concentricity and runout

(a)

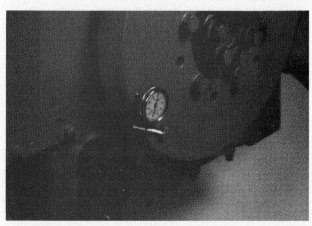

(b)

Fig. 16—9 Vibration-damper checkpoints for measuring (*a*) concentricity and (*b*) wobble. (*Cummins Engine Co., Inc.*)

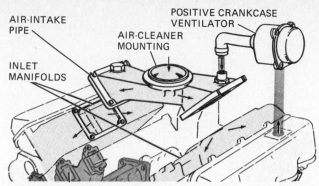

Fig. 16–10 Positive crankcase ventilation. (*Caterpillar Tractor Co.*)

(wobble) as shown in Fig. 16–9. **NOTE** Keep the crankshaft at the front or rear against the crankshaft thrust bearing while making both checks.

Crankcase Ventilation Diesel engines must (breath) circulate air in a precise direction around components which are lubricated in order to remove the harmful vapors. An inadequate ventilation system will reduce the service life of an engine as surely as will an inadequate air-intake system. An inadequate ventilation system will not permit the acid, the condensation, or the pressure to escape. This causes the oil to slush, the oil temperature to rise, and the seals and gaskets to leak.

Most older diesel engines rely on air motion created by the fan blades (and/or the movement of the vehicle) to lower the pressure at the breather pipe. This causes atmospheric pressure to force air through a filter into the engine. The air circulates within the engine and leaves through the breather pipe.

V-type engines may use two ventilation systems, two breather pipes, and two filters. The typical locations for the air breathers (filters) are on the valve cover, on the timing-gear cover, and on the cylinder block.

Positive Crankcase Ventilation Because of the lower efficiency of the older system and the enforce-ment of pollution control, engine manufacturers have been persuaded to change the ventilation system to a positive crankcase system (Fig. 16–10). With this system fumes from the engine are filtered before being drawn into the intake manifold. V-type engines have either two filters, or they use a crossover pipe.

NOTE No control valve of any kind is required on a diesel engine; however, a gas engine requires one or more control valves. This is because the diesel engine inhales air only, whereas a gas engine inhales a mixture of air and fuel.

Questions

1. What is the purpose of the rear oil-seal cover?

2. Why must the flywheel-housing bore and the face runout be within specified tolerance?

3. List the factors which determine the size and weight of a flywheel.

4. Describe how to install a new ring gear.

5. List the checks required to determine the serviceability of a flywheel.

6. What checks must be made after a new crankshaft or flywheel has been installed?

7. Give two reasons why a timing cover or gearcase cover must be aligned with the center of the crankshaft and the cylinder block.

8. List the areas of damage (or wear) on a crankshaft pulley which will necessitate its replacement.

9. List the precautions you must take when installing a vibration damper.

10. What is the difference between a road draft and a positive crankcase ventilation?

UNIT 17

Bearings

Friction *Friction* is defined as the resistance to movement between any two surfaces which are in contact with each other. Friction is classified into three groups: sliding friction, rolling friction, and fluid friction (see Fig. 17–1).

Everything that turns requires a bearing (or bearings) to reduce friction between moving and stationary parts, to reduce power loss, and to decrease wear. Many types of bearings have been designed in an effort to overcome friction and frictional heat.

Bearing Types Bearings may be divided into two main types: friction bearings and antifriction bearings. Both types are used on diesel engines.

Friction bearings can be found in the engine or its components. They are used to support the crank-

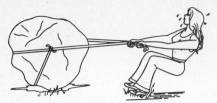

SLIDING FRICTION

ROLLING FRICTION

FLUID FRICTION

Fig. 17–1 Types of friction.

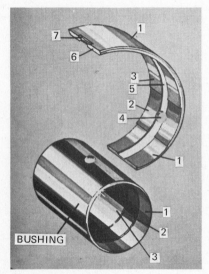

BUSHING

1. Steel back   5. Oil groove
2. Lining   6. Parting face
3. Surface   7. Locating lug
4. Oilhole

Fig. 17–2 Insert bearing. *(Cummins Engine Co., Inc.)*

shaft, the connecting rod, the engine camshaft, or the fuel-injection pump camshaft, the rocker arms, and in some applications also the rocker-arm shaft, the idler gears or pulleys, the turbocharger shaft, the generator rotor shaft, the starter armature shaft, and the oil pump gears.

The use of antifriction bearings is limited. You may find them in such applications as the coolant fan and the belt tightener, in coolant pumps, fuel-injection pumps, governors, the flywheel pilot bearing, and in generators.

FRICTION BEARINGS

The simplest type of friction bearing is a hole drilled into a support plate to guide the shaft. Friction bearings used on diesel engines and for supporting components are called *insert bearings*. They may be designed after bushings or sleeve-type bearings (Fig. 17–2).

Friction Bearing Design and Construction   Bearing construction originates with a strip of steel to supply the strength for the bearing back. A softer lining is bonded to this steel strip. The softer metal is necessary to improve conformability and embedability of the bearing. Figure 17–3 illustrates the manufacturing steps of a typical engine bearing:

1. Blanks cut
2. Facing attached to length (width)
3. Blank pressed into semicircular shape
4. Overplate added
5. Locking lip formed
6. Oilholes punched or drilled
7. Oilholes chamfered
8. Blank grooved and broached to height
9. Inside surface broached and overplated
10. Inside surface overplated

The most popular types of bearings produced by Feder-Mogul are: steel back with a bronze inner layer, and lead- or tin-based Babbitt overlay; steel back with a sintered copper-nickel inner layer and lead-alloy overlay; steel back with a sintered copper-alloy inner layer, a barrier plate, a lead-alloy overplate, and flash tin over the entire bearing; steel back with a pure silver inner layer and lead-alloy overplate; steel back with a cast copper-alloy inner layer, a barrier plate, a lead-alloy overplate and flash lead alloy over the entire bearing; steel back with an aluminum-alloy lining, a lead-alloy overplate and a flash tin plate over the entire bearing; steel back with an aluminum alloy lining and a flash tin plate over the entire bearing.

Function of Friction Bearings   To do its job properly, a friction bearing must be held firmly in place and in full contact with the supporting bore. It must hold and protect the shaft, withstand extreme pressure and heat, and absorb harmful grit. It must also have the ability to maintain a film of lubricant between moving and stationary parts.

The simplest and most common method of holding a bushing in position is to press it into its supporting bore. A locating lug (Fig. 17–4) or dowel is used to hold a half-bearing in place during the assembly. The lug fits tightly into the slot of the support, assuring proper alignment and eliminating the possibility of movement. A half-bearing is manufactured with a spread and is slightly larger than the bearing bore. This is to ensure a precise fit and full contact with the bore surface when the connecting-rod or main-bearing cap screws are torqued. The bearing spread by which the dimension is larger than its bore varies between 0.005 and 0.030 in

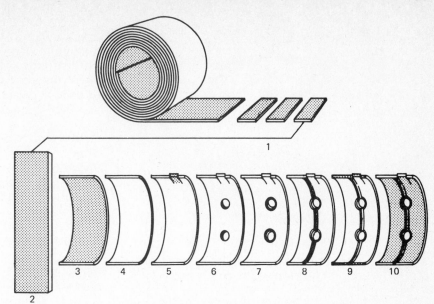

Fig. 17–3 Manufacturing steps of an insert bearing. *(Federal-Mogul, Replacement Sales Div.)*

[0.125 and 0.75 mm] depending on bearing design and size. The bearing crush may be as little as 0.00025 in [0.006 mm].

An insufficient crush or spread allows the bearing to move within its bore, reducing heat dissipation and causing excessive wear.

Bearing Requirements   No singular metal has the strength to withstand all the demands placed on bearings.

A bearing should have a *fatigue strength* (load-carrying capacity) to withstand loadings without the metal flaking away or cracking on the surface or at the bond line. A 100 percent steel-backed bearing or a 100 percent aluminum bearing has high fatigue strength (see Table 17–1).

A bearing should have *conformability* since no journal or bore is perfectly round or straight. The bearing must mold itself to this imperfection so that the load is evenly distributed over the total bearing surface. A soft metal overplate and/or tin plate increases the conformability of the bearing.

A bearing should have *embedability*, that is, the bearing surface must be soft enough to absorb minute particles. This is an important feature since no filter is timelessly effective and no service and maintenance procedure can completely guarantee that no dirt particles can enter the oil to score the journal and bearing.

A bearing should have *resistance to corrosion* because chemicals and water from combustion (or from other sources) enter the crankcase. The chemicals and water in the oil will attack the bearing metal if it is not resistant to them.

A bearing should have *resistance to seizure* because, under certain conditions, there is a metal-to-metal contact between the journal and the bearing. Materials used to prevent this type of failure are shown in Table 17–1.

A bearing should *conduct heat* to a high degree so that most of the heat will be conducted to the connecting rod or cylinder block. (A good bearing fit increases conductivity.)

A bearing should have a relatively *high-temperature strength*, that is, its construction and composition must not weaken from heat.

Table 17-1   BEARING MATERIALS AND PROPERTIES

Property	Back 100% steel	Lining 75% copper 25% lead	Overlay 90% lead 10% tin	Aluminum bearings 100% aluminum
Fatigue strength	High	Medium	Low	High
Conformability	Low	Medium	High	Medium
Embedability	Low	Medium	High	Low
Corrosion resistance	High	Medium	High	High
Friction reduction	Low	Medium	High	Low
Heat resistance	High	Medium	Low	High

SOURCE: Cummins Engine Co., Inc.

Fig. 17–4 Locating lug and slot.

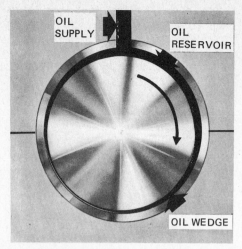

Fig. 17–5 Shaft rotation plus a load forms an oil wedge. *(Federal-Mogul, Replacement Sales Div.)*

Fig. 17–6 A stationary shaft rests on the bearing. *(Federal-Mogul, Replacement Sales Div.)*

A bearing should have *oil clearance* in order to lubricate properly, to cool and to form a wedge to center the journal during rotation (see Fig. 17–5).

Bearing Action When the engine is stopped, the journal rests against the lower bearing shelf (Fig. 17–6). As the crankshaft is rotated (and oil is present in the oil clearance), the journal climbs the bearing and oil slides under the load area. The journal is lifted, or the connecting-rod bearing is moved away from the journal, by the wedge formed by the oil molecules. Some oil molecules tend to stick to the journal and bearings. The oil molecules attached to the journal rotate with it while those on the bearing are somewhat stationary.

The slippage of the oil molecules past each other is known as *fluid friction*. However, the oil film thickness between the journal and the bearing is not always the same. The bearing load during the four strokes, the speed, and the load changes vary the film thickness. Under extreme load conditions only the oil that sticks to the journals and bearing area remains, permitting a metal-to-metal contact which could lead to seizure.

Oil Grooves and Oilholes Oilholes are not required in all bearings. Oil grooves are usually only in the lower half of the bearing. They are used to distribute oil over the total journal and bearing area. In some applications an additional oil groove is used to direct oil to other bearings or to supply oil to other components within the engine.

Main and Connecting-Rod Bearing Evaluation Very reliable evidence can be gained from examining the main bearings, the connecting-rod bearings, and the journals. The extent of their deterioration is evidence of the condition of the engine as a whole.

The cause of excessive bearing wear may be divided into four major groups: dirt failure (45 percent), lubricant and allied failures (25 percent), operation and installation failures (15 percent), surface-reaction failure (15 percent).

Normal Bearing Wear Before you can diagnose abnormal wear, you must first understand what is considered to be normal wear. Most normal bearing wear occurs during the first few hours of operation, thereafter wear is minimal. The bearing shown in Fig. 17–7 was taken from a truck engine which was operated for 4500 hours. It shows normal wear. Under normal usage some of the thin lead-tin surface wears off, exposing the lining (copper or aluminium). The pattern of wear is more concentrated toward the center of the bearing because of its enlarged bearing diameter combined with the reduced diameter of the journal. When truck or crawler engine bearings show this wear within less than 2000 hours or 100,000 miles [160,930 km] of operation, the wear is considered to be abnormal, suggesting that abrasives have entered the oil. Check for the following: poor air filtration, intake manifold leakage, poor lubrication filtration, or overfueling. Fine abrasives may also enter the oil during the engine rebuilding period or through carelessness while making oil and filter changes.

As previously mentioned, most bearing failures are due to foreign matter (plain old dirt) passing between the journals and bearings. This also applies, of course, to other operating components. Depending on the type of foreign matter in the lubricant, the journals, bearings, and components may become scratched, pitted, or discolored, etc.

How to Prevent Dirt from Contaminating Lubricant
1. To begin with, your work area and tools must be clean.
2. Before assembling the engine, make sure that all components and bores are immaculately clean. When the engine is not being worked on, cover it with plastic sheets to keep out any fine dust.
3. Keep all oil storage containers and measuring equipment clean.
4. Follow the manufacturer's recommended procedure when making oil and filter changes.
5. Avoid excessive delay between oil filter changes because this may cause the filter to become plugged and cause the bypass valve to open.

Fig. 17–7 Normal friction-bearing wear after long use. *(J. I. Case Co., Components Div.)*

Fig. 17–8 Damage caused by coarse particles. *(Cummins Engine Co., Inc.)*

6. When adding oil, wipe the area around the dip-stick clean before reinserting.

7. Service the breather regularly.

8. Remember that the entry of even a small amount of dirt into the lubricant will create extensive damage at a later date.

Bearing Failure Due to Coarse Particles in Oil Coarse particles may originate as residue from moving engine components, from improper handling of lubricant or oil filters, or from inadequate removal of honing or boring abrasives.

The bearing shell shown in Fig. 17–8 will fail completely because of the long deep scratches which decrease the efficiency of lubricant and heat dissipation. The visible particles have displaced metal (lead) and have added to the abrasion, causing heat to build up and melt the lead surface.

Bearing Failure Due to Improper Contact between Bearing Shell and Bore Particles, whether large or small, left between the back of the bearing shell and the bore during assembly will prevent the bearing from seating properly (see Fig. 17–9). This causes decreased clearance, localized heating, and excessive wear of those parts of the bearings and journals which are above the imbedded particles. Excessively large imbedded particles will destroy the bearings and journals altogether.

The improperly positioned bearing shell shown in Fig. 17–10 was only in operation for 2 hours. It will fail because of the misaligned oil-supply hole. It may also cause the bearing shells to seize and rotate in the crankpin bore.

Bearing shells which are loose due to improper torque are inadequately crushed and leave the bore of the bearings out-of-round. This increases friction and does not allow sufficient heat dissipation. The affected bearing and journal often are destroyed

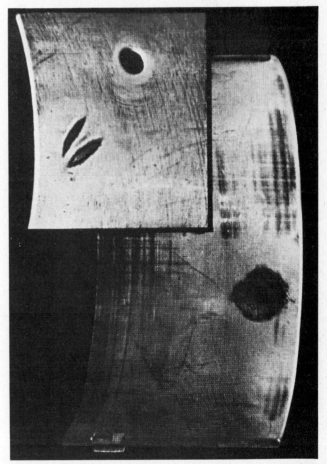

Fig. 17–9 Damage caused by immobile dirt particles. *(Cummins Engine Co., Inc.)*

whereas the other bearings show normal wear (Fig. 17–11).

The back of a bearing which is loose in its bore has a shiny appearance (fretting) and sometimes shows wear at the mating surface of the shells.

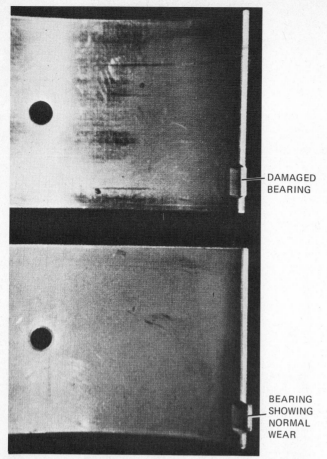

Fig. 17–10 Damage caused by improper installation. (Cummins Engine Co., Inc.)

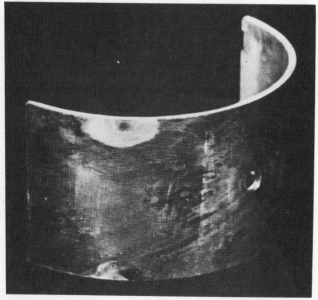

Fig. 17–11 Damage caused by improper torque. (Cummins Engine Co., Inc.)

When the shells are extremely loose, a metal transfer from the shell back to the bore surface may take place. Looseness of bearing shells may also result in an out-of-round or oversized crankpin bore, or distorted bores from engine overloading.

A misassembled connecting rod is shown in Fig. 17–12. Once again this bearing may not have the proper crush and may also misalign. As a result, the bearing could break at the oilhole, damaging the

locating tang area. In extreme cases, the shell assembly spins in the crankpin bore.

Bearing Failure Due to Overspeeding and Long Idle Period The diesel engine mechanic has no control over bearing failure caused by overspeeding and long idle periods. The mechanic does, however, have control over speeding when the governor is misadjusted, or when the fuel is cut off, or when the fuel-injection pump is improperly adjusted.

The bearing shell shown in Fig. 17–13 was damaged by overspeeding. The increased bearing load decreased the oil film and allowed a metal-to-metal contact between the shell and the journal. You will also find, when confronted with an overspeed bearing failure, that the lower half of the connecting rod and the upper half of the main bearing are worn more noticeably.

Bearing failure because of prolonged engine idle period is now uncommon because today's operators are careful to keep the engine rpm above idle speed. Nevertheless, it is possible for the bearings to fail when the engine temperature drops below operating temperature and when unburned fuel has diluted the lubricant, thereby reducing its viscosity.

Bearing Failure Due to Cold Starting This type of bearing failure usually is restricted to areas where the ambient temperature goes below 20°F [6.6°C]. Low ambient temperature reduces the amount of oil supplied to the bearings when the engine is started. The bearing furthest away from the oil supply will deteriorate radically whereas the one closest to the oil supply may have little or no wear. When the ambient temperature is low, the engine should not be run at full speed, or under full load, until the engine temperature has reached operating temperature; if it is, the bearings will be damaged, as shown in Fig. 17–14. During cold weather it is mandatory to follow the manufacturer's recommendations as to the viscosity of the oil and the starting procedure.

Bearing Failure Due to Lack of Lubrication Many conditions can lead to oil starvation of the bearings or running components. The damage caused can vary from light wear to total destruction of all bearings (Fig. 7–15). Some bearing failure which are due to insufficient lubrication can be prevented by a good maintenance program such as: maintaining the oil level, repairing excessive oil leaks, changing oil filters, reducing the period between oil and filter changes, and shutting the engine down only after it has cooled.

Bearing failure due to poor maintenance is rare today because of the extensive training programs attended by both diesel mechanics and operators. Nevertheless, there are some causes of bearing failure which are not easily discernible before the bearings are irreparably damaged. These are: plugged oil pump screens, excessively worn or damaged oil pump or relief valve, and a broken or loose oil-supply line.

If it appears that only one bearing is damaged, check the supply hole or the line; either one can become plugged or restricted. To make a complete

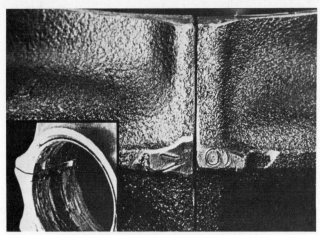

Fig. 17–12 Misassembled connecting rod. *(Cummins Engine Co., Inc.)*

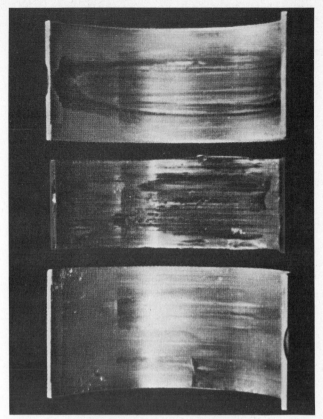

Fig. 17–14 Damage caused by a cold start. *(Cummins Engine Co., Inc.)*

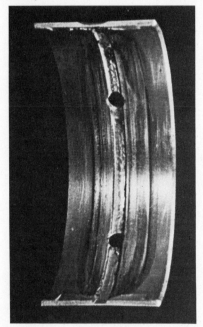

Fig. 17–13 Damage caused by overspeeding. *(Cummins Engine Co., Inc.)*

analysis of the damage, however, you should examine all bearings and journals and compare their wear, and the wear rate of the lower and upper shells.

When starting the engine initially, do not forget to pressurize the lubrication system before testing the engine on the dynamometer. This also applies to factory-new engines. Neglecting to take this step can lead to early bearing failure if the lubricant from the bearing drains into the oil pan.

Bearing Failure Due to Coolant in Lubricant The source of a coolant leak into the oil is sometimes hard to locate since the leakage can come from damaged head gaskets, seal failure at the cylinder sleeves, oil cooler damage, improperly installed injector tubes, or improper torque or damaged gasket and seat of the precombustion chamber. Of a more serious nature is the coolant leakage into the lubrication system through cracked cylinder heads or a cracked cylinder block.

The bearing shown in Fig. 17–16 failed because the coolant entered the lubrication system and

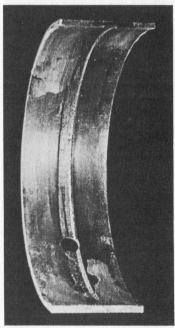

Fig. 17–15 Damage caused by a lack of lubrication. *(Cummins Engine Co., Inc.)*

mixed with the lubricant. As a result, some of the lead surface was destroyed and the remaining lead surface became discolored. Whether the discoloration is black or brown depends on the type of corrosion resistor used. When the lubricant is extremely diluted, the bearing shells become flaked or pitted.

When lubricant becomes grey or milky colored, it could be contaminated with coolant. Ethylene glycol or antifreeze in the lubricant forms a gummy substance. If, after checking the coolant level, you suspect that there is coolant in the oil, do *not* start the engine. Remove the oil plug and check for its presence. Do not judge too hastily! What may first appear to be coolant in the lubricating oil may be water condensation due to temperature changes of the oil and the engine. It may also be the result of an engine operating temperature that is too low.

Bearing Failure Due to Fuel Diluting Lubricant

Some engines because of their fuel-injection systems are more vulnerable to fuel leakage than others. Generally speaking, however, fuel in the lubrication system is not common. It may enter the lubrication system through a worn seal in the fuel-transfer pump or fuel-injection pump, or through a leaking nozzle, or damaged injector O-ring seal (Cummins) or fuel-supply line (GM).

Incomplete combustion caused by worn rings, leaking valves, restricted air intake, worn turbocharger, or worn blower is the most common cause of fuel entering and diluting the lubricating oil. Incomplete combustion (overfueling) can also be due to incorrect adjustment of the injection pump, a damaged or incorrectly adjusted aneroid control or fuel ration control, or to prolonged idling at low ambient and engine temperatures.

A bearing shell damaged because fuel was in the lubrication system is shown in Fig. 17–17. You can see the surface wear in the illustration, and when you find a bearing shell with this type of damage, you will notice a dark green hue where pitting is evident. Pitting becomes even more pronounced when combustion by-products of high-sulfur-content fuel have passed by the piston ring into the lubricant.

Bearing Failure Due to Off-side Wear

When one side of a bearing is excessively worn, it is wise to examine the entire bearing set. The other bearings in the set may be damaged or may have metal embedded in their shells since the abraded material will circulate through the lubrication system.

Figure 17–18 shows a bearing worn on the off-side. This may be the result of a bent connecting rod, an out-of-round journal, a misaligned bearing cap, an improperly installed bearing shell, or a loose bearing cap.

When the whole set of main bearings, as well as the connecting-rod bearings, show off-side wear (Fig. 17–19), the cause is obviously more serious and more expensive to repair. Such wear could be from a sagging (warped) cylinder block, the vibration of engine components, vibration from the power train, a damaged or inadequate viscous damper or rubber vibration damper.

Surface-Reaction Failures

ELECTRIC CURRENT Bearing damage caused by electric current (electrolysis) will show small pits in a defined pattern and/or surface fluting (Fig. 17–20). The pattern or the fluting (when present) may vary from engine to engine because of the differences in the current sources and the attacked metal.

Fig. 17–16 Damage caused by coolant in the lubricant. *(Cummins Engine Co., Inc.)*

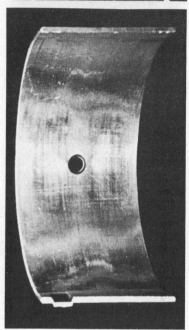

Fig. 17–17 Damage caused by fuel in the lubrication system. *(Cummins Engine Co., Inc.)*

Damage to bearings due to electric current may originate from the electrically actuated components such as motors, switches, relays, etc. Inadequate grounding of the engine to the frame or static current from rotating parts or belts may also damage the bearings.

To prevent bearing damage from electric current, make certain that all the electrical components and the engine are grounded properly to the frame.

Bearing failures have occurred when large areas of the bearing surfaces have disintegrated due to static current because the engine was not used over the cold winter months and was not winterized.

CHEMICAL CORROSION Another type of surface reaction is chemical corrosion (Fig. 17–21). The pitting, discoloration, surface roughness, and fatigue cracks are caused by gaseous chemicals in the air. This

6273

Fig. 17–18 Damage caused by off-side wear. *(Cummins Engine Co., Inc.)*

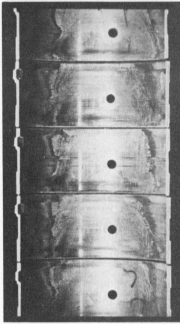

Fig. 17–19 Damage to whole set caused by off-side wear. *(Cummins Engine Co., Inc.)*

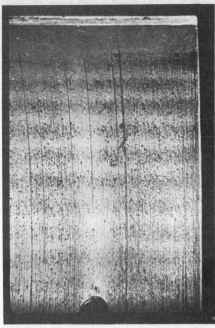

Fig. 17–20 Damage caused by electric current (electrolysis). *(Cummins Engine Co., Inc.)*

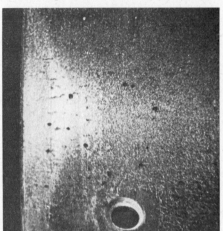

Fig. 17–21 Damage caused by chemical corrosion. *(Cummins Engine Co., Inc.)*

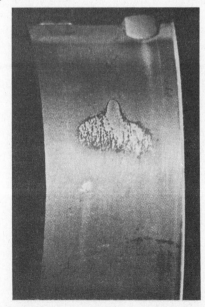

Fig. 17–22 Damage caused by cavitation or corrosion. *(Cummins Engine Co., Inc.)*

problem can be solved by using special air cleaners. Standard air cleaners do not filter out chemicals.

CAVITATION CORROSION Cavitation in the oil film causes localized spalling or erosion of the unloaded bearing shell (the upper main and lower connecting-rod bearing) (Fig. 17–22). Cavitation damage occurs only on the surface layer and it does not affect the bronze or aluminium lining. Since this type of failure is caused by aeration of the lubricating oil, check the inlet side of the lubrication pump for loose connections.

Working on the side of a hill when the oil pan is not specifically designed for this terrain can also cause cavitation corrosion.

Thrust-Bearing Failure As a general rule, thrust bearings, thrust washers, or the thrust surface of the crankshaft are not serious failure areas. Only when the end play is incorrect, when the driver rides the clutch, or the air or hydraulically actuated clutch is not adjusted properly will difficulties arise.

ANTIFRICTION BEARINGS

All antifriction bearings use a rolling element (balls, rollers, or needles) between the inner and outer ring (race) to convert sliding friction into rolling motion. Either the inner or the outer ring remains stationary.

Because of the small contact area between the rolling elements and the inner and outer rings (races) and in order to withstand the high compression stress, the material used for ball bearings is usually of heat-treated chromium-alloy steel and the material for most roller bearings is usually carbonized steel alloy.

The load placed on the bearing can be perpendicular to the shaft and axis of the bearing (radial load), parallel with the axis of the shaft and bearing (thrust load), or a combination of radial and thrust loads (Fig. 17–23).

The radial load is carried by only half of the rolling elements by compressing the inner ring against the rolling element, to the outer ring or vice versa. However, the thrust load is carried equally by all of the rolling elements.

Types of Antifriction Bearings   Antifriction bearings may be classified into six types: (1) ball bearings, (2) cylindrical roller bearings, (3) needle bearings, (4) tapered roller bearings, (5) self-aligning roller bearings, and (6) thrust bearings. However, each type is coded further according to its individual application.

Bearing Identification Code   Ball bearings and roller bearings are identified by a numerical code which indicates the bore in millimeters or in six-

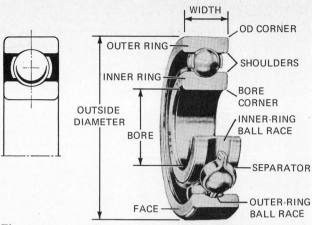

Fig. 17–24   Single-row deep-groove ball bearing. *(International Harvester Co.)*

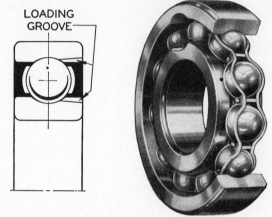

Fig. 17–25   Single-row loading-groove ball bearing. *(International Harvester Co.)*

teenths of an inch. Letter codes indicate the type of bearing, the outside diameter (OD), the width, the cage, the seal or shield, the modification, and the required lubrication. The internal fit, the tolerance, and any special characteristics are coded by number.

Ball Bearings   A single-row deep-groove (no loading) ball bearing is shown in Fig. 17–24. The balls roll in a single deep groove, one in each race. The high supporting grooves and the close fit allow the bearing to take a very high radial load. Because the load is through the axis of the ball, it can also withstand a substantial thrust load.

NOTE   Since little or no movement exists between the inner and outer races, careful installation and alignment between the shaft and the housing is necessary.

A single-row loading-groove ball bearing is shown in Fig. 17–25. It is so named because the balls are inserted through a filling groove or loading groove. This type of bearing has a higher radial load capacity than the bearing shown in Fig. 17–24 because more balls are inserted.

NOTE   Where thrust load may be a factor, this type of bearing should be installed with the loading groove facing toward the thrust side.

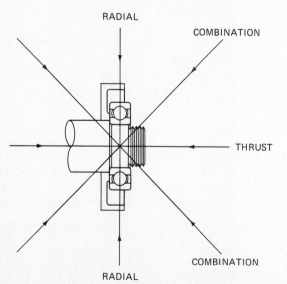

Fig. 17–23   Schematic view of an autifriction bearing, showing types of loads.

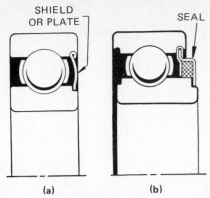

Fig. 17–26 (a) Shield bearing and (b) sealed bearing. (International Harvester Co.)

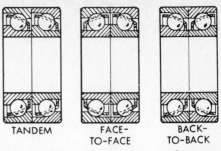

Fig. 17–27 Installing single-row angular-contact bearings. (International Harvester Co.)

SNAP-RING BEARINGS A snap ring placed in the groove of the outer ring is used to provide a shoulder for the axial location of the bearing in the bore.

SHIELD BEARINGS A shield or plate on one or both sides of this type of bearing (Fig. 17–26a) limits the entrance of small particles between races and balls and at the same time reduces the pressure and flow of oil through the balls.

NOTE Extreme care and cleanliness must be exercised during the handling and cleaning of this type of bearing. Do not damage the shield during removal or installation; it may contact the bearing separator and cause wear or increased friction.

SEALED BEARINGS These bearings have a permanent seal either on one side or on both sides (Fig. 17–26b). Bearings having a seal on both sides are prelubricated and cannot be cleaned. When their seals are damaged or the bearing feels rough during rotation, they must be replaced.

NOTE When you install or remove these bearings, be extremely careful not to damage the seals.

Where only one seal is used, take care that no dirt enters the bearings during installation. It is difficult to remove dirt once it has worked its way between the balls, separator, and races and has reached the seal.

SINGLE-ROW ANGULAR-CONTACT BEARINGS Single-row angular-contact bearings are used in injection pumps where radial load and thrust load are combined. However, the thrust can only be applied in one direction. Single-row angular-contact bearings usually are used in pairs, with one bearing on each side of the shaft.

NOTE When you install these bearings in pairs, make certain that the thrust faces point in opposite directions to each other. When they are used for one-sided heavy thrust load, install them as shown in Fig. 17–27.

DOUBLE-ROW DEEP-GROOVE BEARINGS The design of these bearings does not differ from that of the single-row deep-groove ball bearings but, of course, the load capacity in any application is higher because of the two rows of balls.

DOUBLE-ROW ANGULAR-CONTACT BEARINGS These bearings are designed similar to the single-row angular-contact bearings. However, they usually are constructed with a predetermined internal preload. They are used where a shaft must be held tightly, radially as well as axially.

Cylindrical (Straight) Roller Bearings These bearings use straight rollers as their rolling elements and have a very high radial capacity since the roller axis and the inner and outer race (or the contact surface) are parallel (see Fig. 17–28). They are manufactured in single or double rows and vary in their design.

Needle Bearings Initially, needle bearings were considered cageless roller bearings because they use rollers as their rolling element. They are classified as needle bearings when their length is at least six times that of the diameter of the rollers. This type has high radial capacity but no thrust-carrying capacity.

A needle bearing may be any one of the designs shown in Fig. 17–29, that is, with or without a separator, without inner ring or outer ring or cage, or without an outer ring or cage.

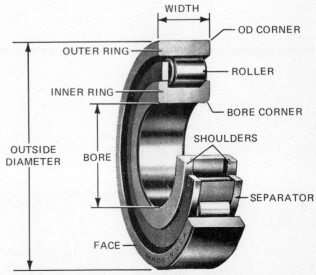

Fig. 17–28 Cylindrical (straight) roller bearing. (International Harvester Co.)

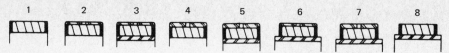

1. Roller only
2. Roller assembly with outer ring
3. Roller assembly with outer and inner rings
4. Roller assembly with outer ring and separator
5. Roller assembly with outer and inner rings and separator

6. Roller assembly with outer ring and wide inner ring
7. Roller assembly with outer ring, wide inner ring, and separator
8. Roller assembly with wide inner ring

Fig. 17–29 Needle-bearing designs. *(International Harvester Co.)*

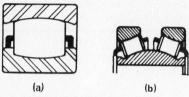

(a)   (b)

Fig. 17–30 Self-aligning roller bearings: *(a)* single-row spherical roller with outer spherical raceway, *(b)* double-row hourglass roller with inner spherical raceway. *(International Harvester Co.)*

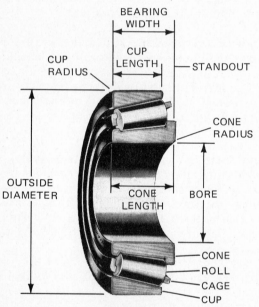

Fig. 17–31 Tapered roller bearing. *(International Harvester Co.)*

NOTE When you install a needle bearing with an outer cage, press it on the side with the bearing identification mark.

Self-aligning Roller Bearings One type of self-aligning roller bearing has single- or double-row spherical rollers and has an *outer* spherical raceway (Fig. 17–30*a*). The contour of the *inner* raceway has the shape of the roller. The second type of self-aligning roller bearing has an *inner* spherical raceway and the rollers have an hourglass shape (Fig. 17–30*b*). The contour of the *outer* raceway has the shape of the rollers.

Tapered Roller Bearings These bearings have tapered rollers as the rolling element between the cone and the cup (Fig. 17–31). The lines of contact between the rollers, cones, and cup, when extended, intersect in the center of the shaft. The cage and the

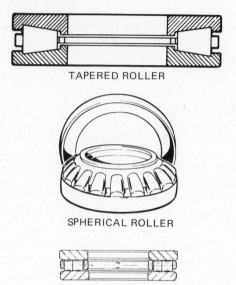

Fig. 17–32 Types of thrust bearings.

high shoulder on the cone keep the roller in alignment. These bearings are designed to carry high combined radial and thrust loads. As a general rule, tapered bearings are used in pairs with one bearing on each side of the shaft.

NOTE Tapered roller bearings need periodic adjustment to take up any looseness which may have developed due to wear or stress.

Thrust Bearings Thrust bearings are designed for axial thrust load application only, their radial load-carrying capacity is only incidental. They use balls, tapered rollers, or cylindrical rollers as their rolling element (see Fig. 17–32). The tapered or cylindrical thrust bearings generally use short rollers to compensate for the variation in speed between the outer and the inner diameters.

Removing Antifriction Bearings The greatest enemies of any bearing are dirt and improper installation. Cleanliness of the tools and the work area in which a bearing is to be removed cannot be overemphasized. Careless handling will most assuredly promote early bearing failure.

The following are a number of rules applicable whether the bearing is to be discarded or reused:

1. Never use a hammer, a hammer and punch, or an impact wrench in conjunction with a puller arrangement when removing bearings. These tools will damage the races and the contact surfaces of the rolling elements and lead to spalling and premature failure.

2. Never heat a bearing with a cutting torch. Use a heating tip and a temperature stick to determine the temperature.

3. When possible, press or pull the bearing or the race from the shaft or housing. Use the correct pullers and attachments, as shown in Fig. 17–33, or drive them off the shaft, as illustrated in Fig. 17–34.

4. When using a press to remove a bearing or race, support the bearing or the housing correctly, as shown in Fig. 17–35.

5. When using a tube-type tool, be careful not to cock the bearing. Drive it alternately on one side and then on the other.

6. When using heat to remove an inner ring which is equal to the diameter of the shaft, use an asbestos glove to prevent your hands from being burned. Alternatively, heat only one part of the ring or cut off part of the ring, and then remove it. You can also use a hammer and cold chisel on the heated area to crack the ring open.

7. After bearings have been removed, handle them with as much care as if they were new. Wrap them in clean paper or promptly place them in a basket and immerse the assembly in clean solvent. **CAUTION** Never "rotate" a removed bearing. It can be damaged by such practice.

8. To prevent shield damage, do not place shielded bearings in the same container.

9. Do not place a bearing with both sides sealed in solvent for cleaning. Wipe the surface clean and wrap it in clean paper for later inspection.

10. Do not place too many bearings in one basket for cleaning. You will reduce the cleaning effectiveness because dirt from one bearing will wash into another.

Cleaning Antifriction Bearings Let the bearings soak in a basket for some time in a recommended cleaning solution, in diesel fuel, or in Varsol.

CAUTION All solvents are highly inflammable. Handle them with care to prevent a fire.

After a period of time, agitate the basket several times. Air can also be used to agitate the cleaning fluid. This will remove the particles (contaminants) from the bearings.

Remove and inspect one bearing at a time. If further cleaning is necessary, use a durable brush (the bristles must not come out or break off during cleaning) to loosen the remaining dirt particles, etc.

After stubborn dirt has been loosened, complete the cleaning by moving the bearings back and forth in the solvent and then repeat this procedure in a clean solvent bath. Use forced air to remove any remaining particles and to remove the solvent, however, hold both races. Never allow a race to spin by the force of the air; the ring may explode. Always use caution when employing forced air.

When you clean bearings which have removable shields, dismantle the shields and then clean the bearings and the shields separately. When you clean bearings with fixed single- or doublesided shields or with a single seal, use only an oil spray cleaner. Otherwise continuously rinse them with clean solvent to wash out the particles. If compressed air is used, take care not to damage the seal or the shields.

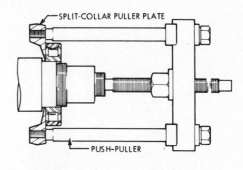

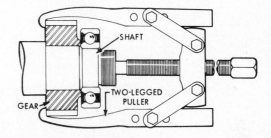

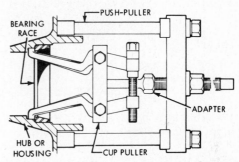

Fig. 17–33 Typical puller arrangements. (*International Harvester Co.*)

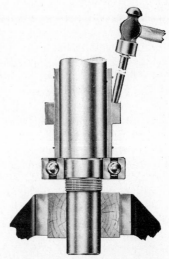

Fig. 17–34 Driving a bearing off the shaft (partial cutaway view). (*International Harvester Co.*)

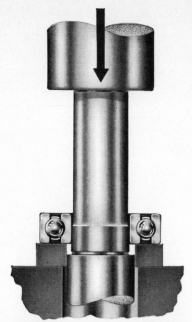

Fig. 17–35 Supporting a bearing when pressing it from the shaft (partial cutaway view). *(International Harvester Co.)*

Fig. 17–36 Holding a bearing for inspection. *(International Harvester Co.)*

Never attempt to clean cage needle bearings. It is much more convenient to replace them.

Inspecting Antifriction Bearings   After the bearings are clean, dip them in light engine oil, gently rotate them a few times, and place them on clean paper to drain off the excess oil.

To determine the serviceability of a bearing, hold it as shown in Fig. 17–36, then check it for discoloration; for cracked or damaged races; damaged shield (or shields) or damaged seals; damaged separator; broken or damaged balls or rollers; brinelled, flaked, or spalled areas on balls, rollers, or races. Indication of any of the above will necessitate bearing replacement.

If, on inspection, the bearing appears satisfactory and during inspection does not feel rough or have a tendency to stick, the bearing should be dipped in clean oil and stored in greaseproof paper until it is to be reinstalled.

If bearings are to be stored for a longer period, coat them thoroughly with grease so that no air can come in contact with their surfaces and then wrap them in greaseproof paper. Identify the wrapping with the bearing part number or indicate the location from which it was removed.

Installing Antifriction Bearings   A bearing is installed by doing precisely the reverse of what you did to remove it. However, before you can press, push, or drive the bearing in place, the shaft and the bore must be cleaned and then checked for nicks and burrs. If any are present, remove them.

Apply a moderate amount of oil to the bearing seat and bore and at this time remove the bearing from its package. Needless to mention, your hands and tools must be spotlessly clean.

Align the bearing on the press bed, and align the shaft with the bore. Then press or pull the bearing until it seats firmly against the shoulder (Fig. 17–37). Make certain there is no dirt on the shaft.

If it is necessary to install the bearing with the shaft in place or if no puller or press is available, tap the bearing lightly with a tool similar to that shown in Fig. 17–38. This will start it square onto the shaft. Be certain it is not cocked; otherwise, the shaft and/or bore will be scraped or burred. To prevent the balls or rollers from being damaged while driving a bearing against the seat, you should place a lintfree cloth around the assembly.

CAUTION   Under no circumstances should you use an impact wrench to turn the puller spindle, the puller nut, or the shaft nut, or to pull the bearing in place. Any of these procedures could damage the rolling elements and the races.

The safest way to install a bearing to a shaft is to heat the bearing to 250°F [121°C]. This may be done in an oven, on a hot plate, or by the method shown in Fig. 17–39. Heating the bearing will expand it so that the inner race slides freely onto the shaft. **NOTE**   Work as quickly as possible when positioning the bearing.

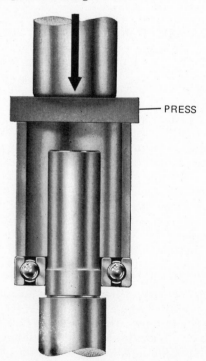

Fig. 17–37   Installing a bearing using a press (partial cutaway view). *(International Harvester Co.)*

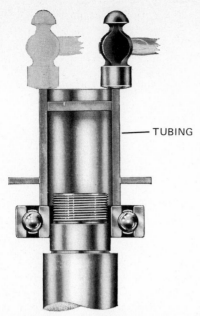

—— TUBING

Fig. 17–38 Installing a bearing using a piece of tubing and a hammer (partial cutaway view). *(International Harvester Co.)*

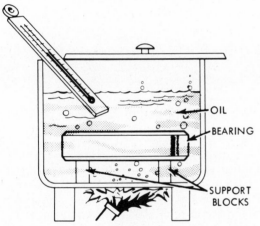

OIL

BEARING

SUPPORT BLOCKS

Fig. 17–39 Heating a bearing for assembly on the shaft. *(International Harvester Co.)*

An alternative method, which is also a safe method, is to cool the shaft or the bearing, depending on bearing installation, using dry ice to shrink the shaft or the bearing.

> **CAUTION** Take care when using dry ice. Always wear asbestos gloves for protection.

Questions

1. Name the two types of friction bearings that are used on diesel engines.

2. What type of materials are used for the bearing backs?

3. Why are half bearings manufactured with a spread? Why do bearings require a crush?

4. Why are oil grooves usually only in the lower half of connecting-rod bearings and in the upper half of main bearings?

5. List the four most common causes of friction bearing failure.

6. What is the main difference between a friction bearing and an antifriction bearing?

7. Why has a roller-type bearing a greater load-carrying capacity than a ball bearing of the same size?

8. For what purposes are single- or double-row angular-contact bearings designed?

9. List the three main causes of antifriction bearing failure.

10. List the checks you must make before installing a bearing.

SECTION 3

Intake, Exhaust, and Cooling Systems

Much of a diesel engine's successful performance depends on its supporting systems, that is, the intake system, exhaust system, cooling system, and their components such as the connecting lines and fittings, filters, and gaskets. Too often, these systems receive insufficient priority in maintenance and service, simply because the problems they create do not immediately affect engine performance. Nevertheless, a reduction in any system's efficiency causes unnecessary wear and strain on the engine components.

UNIT 18

Air-Intake System

Purpose of Air-Intake System　The purpose of the air-intake system is: (1) to supply clean and cool air to each cylinder as required for complete combustion, (2) to supply air for scavenging, (3) to reduce (silence) the airflow noise, and (4) to cool the air going to the cylinders.

Air-Intake-System Components　The air-intake system consists of: an air cleaner, connecting elbows, tubes, hoses, and the intake manifold. When a turbocharger is used, the compressor side of the turbocharger becomes part of the intake system. Sometimes an aftercooler is used to further cool the air and thereby improve engine efficiency (Fig. 18–1).

V-type engines have either two intake systems with two air cleaners and two turbochargers, or else a single intake system with a crossover intake manifold. The required amount of air is directly related to the bhp, atmospheric pressure, and temperature. It ranges from 2.5 to 6 ft³/min (cubic feet per minute) [0.0707 to 0.169 m³/min (cubic meter per minute)]: 2.5 to 2.8 ft³/bhp for naturally aspirated four-cycle engines, 3.0 to 4.0 ft³/bhp for turbocharged four-cycle engines, 3.5 to 6 ft³/bhp for turbocharged two-cycle engines.

Intake Manifold　The intake manifold is a one-piece or multipiece cast-iron alloy or cast-

aluminium alloy casting which connects the intake ports of a row of cylinders and distributes the air equally to each cylinder. The inlet port of the manifold is connected to the air cleaner, the aftercooler, or the compressor side of the turbocharger.

Inspecting Intake Manifold　After you clean the manifold, recheck its internal passages. If any foreign obstructive material still remains, remove

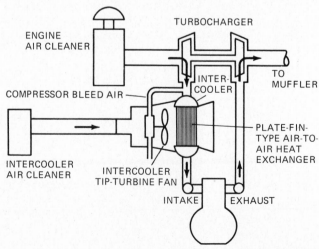

Fig. 18–1　Schematic drawing of an air-intake system using air-to-air intercooling. (*Mack Trucks Canada, Ltd.*)

it. Inspect the manifold for cracks, and the mounting surfaces for corrosion, straightness, or warping. If any of these defects are sufficient to prevent proper sealing, the flanges must be machined or the manifold replaced.

Installing Intake Manifold   Be sure the mounting surface of the manifold and the cylinder head are clean before placing a new manifold gasket between them. Follow the correct sequence and tighten the manifold bolts to their recommended torque. Cover the inlet opening if the turbocharger is not to be immediately installed.

Air Cleaner and Silencer   The efficiency of an engine depends to a large extent on adequate maintenance and servicing of the air cleaner along with the other components of the air-intake system. There is a wide range of air cleaners available to meet any air demand of a given engine and to provide ample clean cool air to the combustion chamber. Insufficient air, because of air-cleaner restriction, will limit the amount of fuel the engine can burn. This will result in a loss in power output as well as excessive exhaust smoke and high fuel consumption. A damaged or leaking air cleaner, flanges, or hoses can lead to excessive engine-component wear, shorter engine life, and higher oil consumption.

Two types of air cleaners are used: the oil-bath cleaner and the dry-type air cleaner.

Oil-Bath Cleaner   Oil-bath cleaners, because of their lower cleaning capacity, are more prevalent where dust conditions are not severe. Their efficiency is about 97 percent compared to a dry-type cleaner which is 99.9 percent.

A typical oil-bath cleaner with a silencer is shown in Fig. 18–2. Air enters through a screen in the hood and passes downward through the stand-up tube toward the oil bath. The stand-up tube ends about ½ in below the oil level. The air enters the oil bath where it is washed, then changes direction. Inertia causes the larger dirt particles to settle at the bottom of the oil cup. As oil and air mix to cause a vapor, the smaller dirt particles are carried with the oil mist (vapor) upward through the wire-type filter element or elements. When the oil mist enters the filter, the heavier particles (oil mixed with dust) cling to the wire mesh and the oil mist is reduced continuously until clean oil-free air leaves the filter.

Servicing Oil-Bath Cleaner   A formidable weakness of an oil-bath cleaner is the impossibility of installing in it a warning device to signal when there is little or no oil in the bowl, when the bowl is loose or has come off, or when the bowl is full of dirt.

There are warning devices or gauges which will indicate excessive airflow restriction [see Air-Intake Restriction and Poor Compression (Unit 51)]. The average permissible restriction for an oil-bath cleaner is about 5 in [12.7 cm] on the water manometer gauge. The only way to be sure if the oil-bath

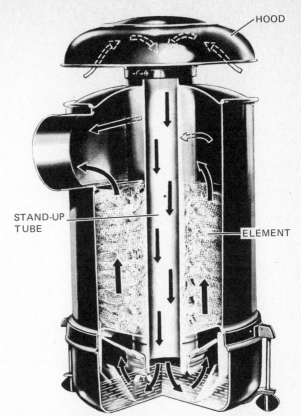

Fig. 18–2   Oil-bath cleaner. (The arrows show the oil flow.) *(GMC, Detroit Diesel Allison Div.)*

cleaner is functioning properly is to check, daily, the general condition of the oil level and the amount of accumulated dirt.

A precleaner cap should be used when operating an engine in an area where sawdust, leaves, and chaff are in the air (Fig. 18–3).

In order to service the oil-bath cleaner, remove the oil bowl, pour out the old oil, and remove the sludge. Wipe the bowl clean and refill it to the "level" mark with the same grade and weight of oil as that used in the engine. Proper oil viscosity is very important to the washing action since oil which is too light (of low viscosity) can cause oil to be drawn out of the bowl. Oil which is too heavy (of high viscosity) can increase restriction to an unacceptable level. In either case, oil may be drawn into the cylinders. Furthermore, the cleaning action may have been lost.

Be sure to check the center tubes, the precleaner cap, and the lower filter element. The filter elements of a light-duty oil-bath cleaner are removable, whereas the filter elements of a heavy-duty oil-bath cleaner are not. Therefore, when the filter elements of a heavy-duty oil-bath cleaner are dirty, they must be cleaned in a cleaning tank similar to that shown in Fig. 18–4.

Do not use an air pressure higher than 5 psi [0.035 kg/cm²]. After washing an air cleaner in solvent, make certain to remove all traces of the solvent either by using a steam cleaner or by blowing it dry with compressed air. If this step is not taken, the engine may speed beyond governor control since it will use the solvent as fuel.

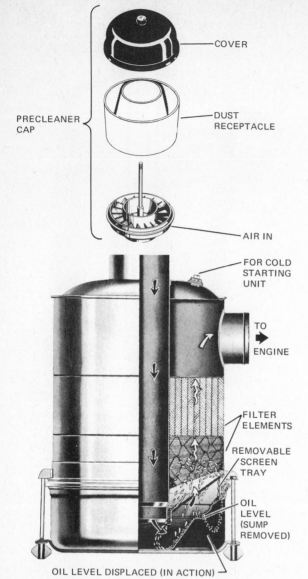

Fig. 18-3 Oil-bath cleaner with precleaner cap. (The arrows show the oil flow.) *(Mack Trucks Canada, Ltd.)*

Fig. 18-4 Cleaning an oil-bath filter element.

Dry-Type Air Cleaner Most (on- and off-highway) engines use dry-type air cleaners since they have an efficiency of 99.8 to 99.9 percent regardless of airflow. Furthermore, the cleaning efficiency is not affected by climatic conditions or the type or size of dust particles. In addition, the operator is instantly alerted to air restriction by means of built-in warning devices. The only drawback to the dry-type air cleaner is that carbon or oil vapor can reduce the service life of the filters.

A great variety of dry-type air cleaners are used. They range from a simple single element (Fig. 18-5)

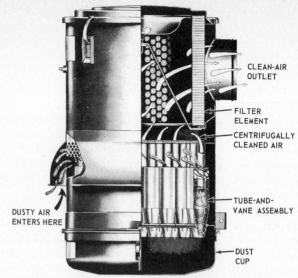

Fig. 18-5 Donaclone single-element air cleaner. (The arrows show the airflow.) *(Mack Trucks Canada, Ltd.)*

to a large two-stage element which may have some type of mechanical cleaning device (Fig. 18-6).

Air-cleaning Action of a Donaclone Air Cleaner The Donaclone is a dry-type cleaner which combines a centrifugal-stage cartridge and a paper-filter cartridge, the latter being protected by a perforated steel shell. The centrifugal-stage cartridge has many tube assemblies. Each tube assembly has a larger outer tube which tapers off at the bottom. The smaller tubes, to which vanes are fastened, protrude into the larger tubes approximately one-quarter of the length of the smaller tubes. The small tubes are fastened to the upper sealing ring and the large outer tubes are fastened to the lower sealing ring.

As the engine is cranked over, a lower pressure is present around the inlet skirt of the air cleaner. Atmospheric pressure forces the air through a large screen into the centrifugal cartridge and around the smaller tubes. The air passes into the larger tubes where the vanes force the air to rotate (cyclonic motion). The heavier foreign particles are thrown outward against the walls of the larger tubes and then drop downward into the dust cup. The partly cleaned air (that is, the center of the rotating air motion) passes upward through the smaller tubes into the paper-element cartridge where particles of the lower micron range are removed.

Servicing Dry-Type Air Cleaner When the dust cup is about two-thirds full of foreign particles, say to within $1/2$ in [12.5 mm] from the bottom of the tapered tubes, the dust cup must be cleaned.

To clean the filter, remove the clamp which holds the cup. Dump out the particles. Clean the bowl and the gasket. Do *not* use a solvent, gasoline, or other fuels. Use only a dry rag or ordinary laundry detergent and warm water.

When the restriction indicator shows that the maximum allowable restriction has been reached, both cartridges must be cleaned or the filter element replaced. To service the filter element, remove both

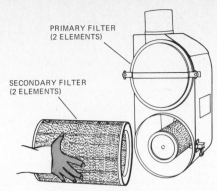

Fig. 18-6 Two-stage air cleaner.

cartridges. Wash the elements or clean them with compressed air. When you use the compressed air method, direct the air through the cartridges opposite to the normal airflow. Hold the air nozzle about 1 in [25.4 mm] from the surface of the paper element and move it up and down while rotating the cartridge (see Fig. 18-7).

When oil or carbon residue is present, the cartridge must be washed. Soak the cartridge about 10 minutes in a solution of warm water and laundry detergent or air-cleaner detergent, then flush it gently with a low-pressure water stream until clean water leaves the filter. Air dry or use a warm airflow to dry the cartridge before reusing it.

Check the paper cartridge to be sure it is clean and free of damage by placing a light on the inside and rotating the element. If an even bright light is visible throughout, the cartridge is clean. If there are holes in the cartridge or if it is ruptured, it must be replaced.

NOTE Do not use compressed air to dry the element. Do not allow the hot air with which you are drying the filter element to exceed 160°F [71°C]. Either will destroy the paper cartridge.

The body assembly can now be cleaned. Wipe off or blow out any loose dirt. Check the sealing flanges and gaskets for damage. Reassemble. Make certain when you reassemble the air cleaner that the cover fits into position smoothly, without being forced, and that the gasket seal is 100 percent effective.

Aftercooler Aftercoolers (also called *heat exchangers*) are small radiators positioned between the compressor housing and the inlet manifold. Sometimes the aftercooler and the inlet manifold are one unit (Fig. 18-8); the aftercooler also can be bolted to the cylinder head.

Coolant enters the aftercooler and passes through the core tubes into the cylinder block or cylinder head. Compressed air from the turbocharger flows around the tubes, is cooled, and then enters the inlet manifold.

Most engines which are turbocharged employ an aftercooler to further improve the brake mean effective pressure (bmep). This increases the power output by about 10 to 20 percent because the incoming air is cooled to within 40°F [22°C] of the engine-coolant temperature and more air enters the cylinder. The result is lower cylinder pressure, more effective cooling of the cylinder components, and a

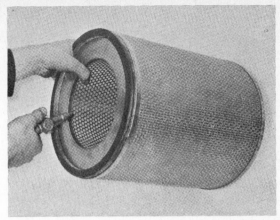

Fig. 18-7 Cleaning a filter element using compressed air. *(Caterpillar Tractor Co.)*

Fig. 18-8 Sectional view of an aftercooler.

lower exhaust-gas temperature, which brings about higher bmep.

Without the aftercooler, atmospheric air at ambient temperature enters the turbocharger. The air temperature then rises because of the heat from compression and from the turbocharger. This results in a loss in air density, less power, and a higher temperature within the cylinder and the exhaust gases.

NOTE For a 1°F [0.56°C] increase in air-intake temperature, the exhaust temperature increases 3°F [1.67°C].

Inspecting and Servicing Aftercooler After you have cleaned and disassembled the aftercooler, make the checks and tests outlined under Inspecting and Testing Oil Cooler (Unit 13). However, when making a leakage test (pressure test) do not pressurize the cooler tubes more than 15 psi [1.05 kg/cm²] of air. When the coolant tubes show a buildup of scale or any other type of residue, use a solution the same as that for cleaning the coolant side of the oil cooler.

Testing Air-Intake System for Leakage When there is evidence of dirt in the compressor housing, on the compressor wheel, or in the connecting link to the intake manifold, you should pressure-test the intake system to locate a leak. This test should also be made after servicing the intake system, or when dirt is found on the inside of the air-filter housing.

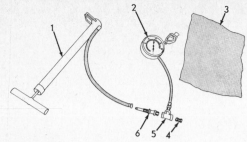

1. Tire pump
2. Pressure gauge
3. Airtight rubberized or plastic material
4. ¼-in [6.35-mm] pipe nipple
5. ¼-in [6.35-mm] pipe tee
6. Valve stem with ¼-in [6.35-mm] pipe plug

Fig. 18–9 Equipment required to pressure-test the air-intake system. *(Allis Chalmers Engine Div.)*

To test for air leaks you need a pressure gauge which registers about 20 psi [1.40 kg/cm²] a shutoff valve, a pressure regulator (tire pump), and either an inner tube or a rubberized plastic material which will withstand a pressure of 5 psi [0.35 kg/cm²] (Fig. 18–9). Cover the air-cleaner intake with the inner tube or rubberized plastic material to make it airtight, and connect the air hose to the intake system. Position one piston so that the valves do not overlap and so that no air can escape through the exhaust valves. The valves do not overlap when the piston is ATDC and the exhaust valve(s) of that cylinder has (have) just closed and the intake valve(s) is (are) open.

Now pressurize the intake system to 5 psi [0.35 kg/cm²], shut off the airflow, and observe the pressure gauge. If the pressure drops back to zero immediately, recheck the piston position. It may be that the exhaust valves are not closed or there may be a serious leak in the air-intake system. Apply the soap solution method to all connections and hoses to locate the leaks. Repair as necessary and then repeat the test. If pressure again drops quickly back to zero, make a thorough check of the air-cleaner housing and the aftercooler. When no leak can be found in the entire intake system, but specified air pressure still cannot be obtained, the cause is due to either a leaking valve or valves, or worn piston rings [see Air-Intake Restriction and Poor Compression (Unit 51, "Troubleshooting and Tuneup")].

Questions

1. List the inspections and checks to be made prior to installing a manifold.

2. What is the major difference between an oil-bath air cleaner and a dry-type air cleaner?

3. What would result (to engine performance) if the oil level in the bowl was, say, 1 in higher than specified?

4. Why does a dry air cleaner have a higher efficiency than an oil-bath cleaner and why is it not affected by climatic conditions?

5. Explain how you would check a dry-type filter element for damage and for cleanness.

6. What is the purpose of an aftercooler?

7. List the areas you should check when an air-intake leak is suspected. How would you locate this suspected air leak?

UNIT 19

Exhaust System

Purpose of Exhaust System The purpose of the exhaust system is to direct the exhaust gases into the atmosphere and to silence the noise by dampening the exhaust pressure waves. When a turbocharger is used, the exhaust gases are directed to the turbine housing, where they drive the turbine, and from there they are vented to the atmosphere.

Exhaust-System Components The exhaust system consists of one or two one-piece exhaust manifolds or of multisectional exhaust manifolds, exhaust pipes, and silencer (muffler) (Fig. 19–1). When a turbocharger is used, the turbine side of the turbocharger becomes part of the exhaust system.

Inspecting and Servicing Exhaust Manifold In most cases the manifold can be cleaned with solvent;

Fig. 19–1 Dual-section exhaust manifold.

however, when carbon deposits or scale are present, the manifold should be cleaned with a sand or glass bead cleaner.

Check the manifold for cracks. Using a straight-edge, check the mounting surface for warpage. If warpage is sufficient to prevent effective sealing, the mounting surfaces must be machined or the manifold replaced. Check the threaded bores for damaged threads or broken studs. If you have not previously checked the stud bolts in the cylinder head for thread damage, do so now. When installing new stud bolts, use an antiseize lubricant to prevent thread corrosion and seizure.

Installing Exhaust Manifold To prevent damage to the engine, make sure that all loose deposits and cleaner dust (residue) are removed from the manifold, particularly when a turbocharger is used.

After you have checked the cylinder-head surface, place new manifold gaskets over the stud bolts or install temporary stud bolts. **NOTE** The bevelled side of the gasket should face you.

When a multisectional manifold is used, install the center section first but do not tighten the bolts. Then slide each end section into place or assemble them on the workbench and install them as a unit. Apply an antiseize lubricant to the threads of the studs or manifold bolts, and tighten the manifold bolts to the correct torque and in the recommended sequence. Check the service manual in the event special washers are required on the manifold bolts.

Install the exhaust elbow or the connecting link to the turbine. If the turbocharger is not to be installed immediately, cover the exhaust opening.

Turbocharger Failure The most prevalent causes of turbocharger failure are: extreme temperature caused by hot shutdown, a restricted air cleaner, air leaks in the intake system, leaks in the exhaust system, overfueling, higher altitude without compensatory fuel-pump adjustment, or a dirty compressor wheel.

Secondary causes of turbocharger failure are the failure to prelubricate the turbocharger after completion of servicing, after an oil filter change, or after a long shutdown period. A malfunction in the lubrication system or the oil supply will also cause the turbocharger to fail.

Improper maintenance is another contributor to turbocharger failure. Dirty air cleaners, oil leakage into the air-intake system, leaking oil lines, air leaks, exhaust leaks, loose or overtorqued mounting bolts or clamps also can reduce the efficiency of the turbocharger.

It is seldom that a compressor or turbine wheel is damaged by foreign material entering the turbocharger or that a turbocharger fails because of improper assembly or component damage during assembly, although both have been know to happen.

Installing Turbocharger Before you install a turbocharger, check the intake and exhaust manifolds for loose foreign materials such as bolts, lock washers, etc. Make sure before placing the turbocharger onto its mountings that all the manifold bolts are torqued to specification, that the mounting flanges are clean, and that the gaskets are in the correct positions.

Install all hex bolts finger-tight using an antiseizing lubricant on the turbocharger mounting bolts. Loosen the V clamps that fasten the turbine housing and compression housing to align the compressor outlet, then tighten the mounting bolts and V clamp to the recommended torque.

Connect the turbocharger to the inlet manifold (or after cooler), the oil inlet, and the oil-return line. When you install the oil-return line, avoid sharp bends and avoid an angle of more than 30° from the vertical. Keep the air-intake cover on to prevent foreign material from entering the turbocharger.

Inspecting and Maintaining Turbocharger Turbochargers (depending on the engine design and torque) are exposed to temperatures of 800 to 1300°F [427 to 701°C] or more, and they may be driven at speeds of 60,000 to 150,000 rpm. Therefore, weekly inspection of a turbocharger is advisable if it is to be kept in good running condition.

Begin your inspection with the air-intake system's air cleaner because a faulty or dirty air cleaner restricts the air flow. A loss in power is then unavoidable since boost pressure is lowered. The restriction also can cause an otherwise serviceable seal to leak because of the vacuum the restriction creates.

Remove the connecting link to the intake manifold and check the compressor housing and connecting link for the presence of oil.

NOTE The compressor housing (and sometimes the connecting link) of all operating turbochargers contains a small but harmless amount of engine oil. This is usually due to the vacuum created behind the compressor under running conditions. However, it can also come from an overfilled oil-bath air cleaner, although if this were the case, the vanes would also show evidence of oil. If you find "heavy" deposits or wet oil in the compressor housing and connecting link, they are an indication of seal leakage. The turbocharger should then be serviced immediately, otherwise extensive damage will result.

Check to be sure that all intake piping and components are aligned, that they fit without stress, that they are properly torqued, and that no evidence of leakage is present.

Check the crankcase breather and the turbocharger oil-return line for restriction. If either is restricted, the oil pressure will build up and cause the turbine end seal to leak.

Check, and when necessary, correct the position of the oil-return line. If there is oil on the external surface of the turbocharger, check the oil-inlet connection and the condition of the oil hose. If neither is defective, it is possible that the oil could have been blown on the turbocharger through air circulation or from a leaking oil cooler or valve cover.

Check the turbine housing for hairpin cracks which occur on its outer surface and near the mounting flange.

Check the exhaust manifold for gas leakage and the exhaust piping for restriction.

Fig. 19-2 Checking turbine-shaft end play with a dial indicator.

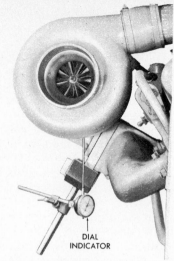

Fig. 19-3 Checking radial clearance. *(GMC Detroit Diesel Allison Div.)*

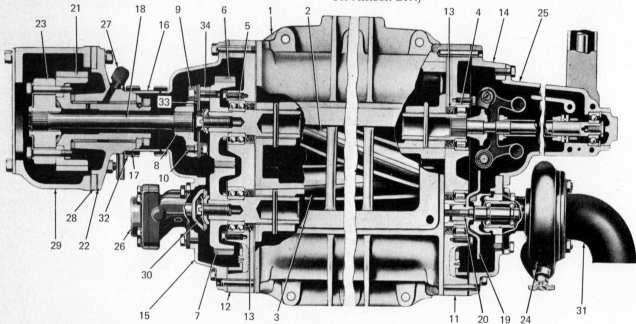

1. Blower housing
2. Blower rotor (upper R.H. helix)
3. Blower rotor (lower L.H. helix)
4. Front roller bearing
5. Rear ball bearing (double-row thrust)
6. Rotor gear (upper R.H. helix)
7. Rotor gear (lower L.H. helix)
8. Rotor drive-gear hub
9. Plate-to-gear bolt
10. Plate-to-hub bolt
11. Front end plate
12. Rear end plate
13. End-plate oil seal
14. Front-end plate cover
15. Rear-end plate cover
16. Blower drive-shaft cover
17. Drive-shaft-cover seal
18. Blower drive shaft
19. Water-pump-drive-coupling assembly
20. Allen-head-bolt coupling
21. Blower drive gear
22. Blower drive-gear-hub support
23. Blower-drive coupling assembly
24. Freshwater pump
25. Governor
26. Fuel pump
27. Oil-line-to-blower-drive elbow (90°)
28. Rear cylinder-block end plate
29. Flywheel housing
30. Fuel-pump-drive fork
31. Water-pump-inlet cover
32. Drive-cover-seal clamp
33. Blower-rotor drive-hub plate
34. Plate-to-gear spacer

Fig. 19-4 Sectional view of the blower, drive assembly, and accessories, including mechanical governor attached to standard blower or small-diameter rotor blower (71 Series engines). *(GMC Detroit Diesel Allison Div.)*

Cleaning Compressor Housing and Compressor Wheel If there is an oil deposit or dirt on the compressor wheel, housing, or connecting link, the components should be cleaned, otherwise maximum performance (boost pressure) cannot be maintained. Some manufacturers suggest cleaning these components after 50,000 mi (miles) [80.450 km (kilometers)] or 1000 hours of operation.

When servicing the components you should, at the same time, check the compressor wheel for nicks and burrs. If the compressor components are only lightly covered with dirt or oil, remove only the compressor housing. Take care not to damage the compressor wheel or the diffuser when removing the housing. Use only a recommended metal-cleaner solvent and a bristle or nylon brush to wash the components. Never use a caustic solution, wire brush, sharp object, or glass bead cleaner because all of these will damage the components. This is particularly true of the compressor wheel which may lose its balance.

CAUTION Never permit cleaning fluid to contact the bearing housing.

Tolerance Checks To check the end play of the turbine shaft, install a dial gauge so that the dial pointer rests on the compressor end of the shaft (see Fig. 19–2). When moving the shaft back and forth against the dial indicator pointer, the total indicated dial movement is the total end play. It should be within 0.004 to 0.006 in [0.101 to 0.152 mm].

If the end play is less than 0.004 in, it is an indication of carbon or oil residue buildup. If the end play is more than 0.006 in, the bearings are worn. In either case the turbocharger should be serviced immediately.

To check bearing and shaft wear (radial clearance), install a dial gauge with an extension through the opening created by the removal of the oil-return line (see Fig. 19–3). The extension must pass through the bearing hole and rest on the turboshaft. Exerting equal force on both ends of the shaft, move it against and away from the dial indicator pointer. When moving the shaft back and forth against the dial indicator pointer, the total indicated dial movement is the total bearing clearance. If the total indicated clearance exceeds 0.003 in [0.076 mm], the turbocharger should be serviced.

Roots-Type Blower The Roots-type blower has two hollow, three-lobe rotors which revolve with very close clearances within the housing (Fig. 19–4). The housing is bolted to the cylinder-block air-box opening. To achieve efficient sealing and a uniform airflow (volume), one rotor lobe is twisted to the right and one is twisted to the left. The two rotors are timed by two drive gears which space the rotor lobes to a close clearance. Since the lobes do not come in actual contact with each other or with the housing, the rotor needs no lubricant. Should the drive gears exceed the backlash clearance, the rotors will then come in contact with each other. The resultant wear will cause a reduction in volume and

pressure. The rotor shafts rotate on roller bearings in the front end plate and on radial thrust bearings in the rear end plate (the drive end). Lip-type seals or ring seals (piston-type) seal the rotor shaft.

The upper rotor is driven by the rotor drive gear. The ratio differs from approximately 1.55:1 to 2.49:1, depending on the engine model. A flexible coupling is used to reduce the transfer of torque fluctuation to the blower and governor drive. The governor drive shaft is splined into the upper rotor. The lower rotor drives the fuel-transfer pump at the front.

Lubricating Blower The timing gears, governor, and fuel-pump drive are pressure-lubricated from the main oil gallery. The main oil gallery leads to an oil passage in each blower end plate. Oil returns to the crankcase via an oil passage in the cylinder block. The rear bearings of the rotor are splash-lubricated from the timing gear, and the front bearings are lubricated through the governor weights (flyweights). The blower-drive support bearings are pressure-lubricated and excessive oil drains over the timing gears into the crankcase.

Inspecting Blower When inspecting the blower on an engine which is in running condition, you should first check the effectiveness of the blower pressure, air-intake restriction, and compression (see Unit 51, "Troubleshooting and Tuneup").

Before you inspect the condition of the rotors, housing, end plates, bearings, and seals, you must remove the air shutdown housing. After it is removed, thoroughly clean each component. Check the screen for damage, and make certain the valve rotates freely and that it is positioned flat on its seat. If it is not flat on its seat, the airflow will not stop when the engine is in shutdown position.

Check the ability of the latch to hold the valve securely open when the engine is running. Check the springs for erosion and broken ends. Examine the rotor and housing for scratches, burrs, and grooves. If these defects are present, it is an indication that foreign matter has entered through the air-intake system. If the damage cannot be corrected with emery cloth, the blower must be serviced.

Check the leading edges of the lower blower and the trailing edges of the upper rotor for evidence of contact or wear. Check the backlash of the timing gears; if it exceeds 0.004 in [0.101 mm], the gears must be replaced. If the rotor crowns and the mating rotor roots show a contact pattern or if there is extensive wear on either the inner housing or end plates, then one or more bearings are worn or else the rotor shaft is loose.

Engine oil on the rotors near the end plates or housing or oil on the end plates indicate leaking oil seals. This leakage could be the result of a worn bearing, a high air-intake restriction, a plugged crankcase breather, or a damaged seal.

To check the flexible drive coupling, force the top rotor to rotate. A springing movement of about ⅜ to ⅝ in [9.525 to 15.87 mm] (measured at the lob crown) should be present. When the rotor is released, it should spring back to within ¼ in [6.35 mm] of its previous position. If the rotor does not spring back, or alternatively, if it springs back further than specified, the flexible drive coupling must be replaced.

Removing Blower When you remove and disassemble a blower, follow the steps outlined in the appropriate service manual. This is essential since the number and location of blowers differ among engine models and according to engine application.

Inspecting the Disassembled Blower Components Refer again to Fig. 19–4. The picture shows the location of components and parts of a six cylinder 71T blower. Begin your inspection of this blower by scrutinizing the inside walls of the blower housing for burrs and scoring. Using a straight-edge, check the side surface and the mounting surface for flatness. Check the bolt holes for elongation and the threaded holes for damaged threads. Remove burrs, nicks, and scratches with emery cloth.

Check the end-plate surface and the bearing bores and seals for grooving or for evidence that the bearings have rotated in their bores. Whenever the engine is placed in the shop for a major overhaul, all seals and bearings should be replaced. Alternatively, if a major overhaul is not being done, the bearings and seal surfaces should be checked for wear (see Unit 17, "Bearings").

Check the splines of the rotor shaft for wear (shaft serration) and, at the same time, check the splines of the timing gears. When the gears are loose on the blower shaft, the gears and/or the rotors must be replaced. Otherwise the backlash will exceed specification. Check the gear teeth for wear or damage.

NOTE Never replace only one gear, always replace both. These replacements reduce noise and also prevent excessive wear of the timing gears. The excessive wear will eventually damage the rotor.

When the seal surface is damaged, the problem can be overcome by using an oversized seal and a seal spacer. When the splines or bearing surfaces are worn, the rotor must be replaced.

Inspect the sealing ribs of the rotors. They must be flawless if they are to maintain sealing and ensure proper air volume and pressure.

Check the spring packs of the blower drive coupling and the cams for wear.

Assembling Blower Begin assembling the blower by installing the seal-ring carriers in the end plates, as shown in Fig. 19–5. Make certain they start straight in their bores and are pressed into place to the specified protrusion. Then press the collar carriers onto the rotor shafts, as shown in Fig. 19–6.

Install oil-seal rings in the carrier and lubricate the seal rings well. Since both end plates are identical, place either one on wooden blocks. The right-hand rotor (right helix lobe) can then be placed in the right opening of the end plate. When placing the rotor in position, take care not to exert too much force as you work the seal ring onto the collar. Align the right-hand rotor to allow the omitted

spline on the rotor shaft to directly face the left-hand rotor.

Carefully install the left-hand rotor. The omitted spline on the shaft and the right-hand rotor, must face in the same direction. Before you can place the blower housing over the rotors and onto the end plate, you must consider the rotation of the rotors. The blower-housing openings (air in and air out) must face in the correct direction. Install the blower housing onto the end plate. If force is required, use a soft-faced hammer.

Lubricate and install the oil-seal rings in the carriers, aligning them so that they protrude an equal distance from the carrier bore. Gently place the front end plate on the blower-housing dowels. Carefully

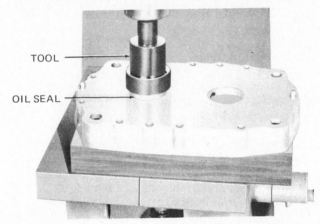

Fig. 19–5 Installing the oil seal (seal-ring carrier) in the blower end plate. *(GMC Detroit Diesel Allison Div.)*

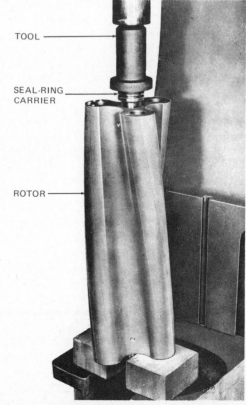

Fig. 19–6 Installing the oil-seal-ring carrier on the blower rotor shaft. *(GMC Detroit Diesel Allison Div.)*

start the seal rings onto each collar. You may have to rotate or move the rotors or the end plate slightly to get the seal rings started. Once they are started, position the end plate over the dowels and force it against the blower housing.

Install, then tighten, the two fillister head screws which hold the front end plate in place.

Remove the two roller bearings from their packing and lubricate them well. With the identification numbers facing you, place the roller bearings over the rotor shaft and onto their bores. Use a bearing installer to drive the bearings into place. (The bearing-retainer plates secure the bearing to the front end plate.)

Position the coolant-pump drive coupling on the recommended rotor shaft, then, with the drive-coupling retainer bolt, draw the oil slinger and coupling against the end of the shaft. Tighten the hex bolt to the recommended torque. Place a new gasket over the dowel pins and install the end-plate cover, then tighten the hex bolts to recommended torque.

Turn the blower 180°. Place it on wooden blocks and install the two fillister head screws, the ball bearings, and the bearing retainers.

NOTE Check your service manual to determine if the blowers require shims between the end of the shaft and the bearings.

At this point, measure the end clearance between each rotor lobe and end plate. This will require ten separate measurements (eight are shown in Fig. 19–7). Should the clearance exceed specification, you must disassemble the blower and replace the two rotors. Check the clearance at E and D (Fig. 19–7). Should the clearances exceed specification, the blower housing and/or rotor must be replaced.

Installing Timing Gears When reusing the rotors, replace the spacers on each rotor shaft. However, when new timing gears and/or new rotors are used, pull the timing gears onto the rotor shaft without using any spacers.

To do this, lay the blower housing on the workbench with the air inlet facing upward and with the timing-gear ends over the edge of your workbench. Align and lubricate the splines of the shaft and the timing gears. Locate the timing gears so that the one with the left-hand helical twist is on the left-hand helix rotor and the one with the right-hand helical twist is on the right-hand helical rotor. Install your pullers. Place a clean rag between the rotors and another one between the rotor and housing. This prevents the rotors from turning as you evenly pull the gears into place (Fig. 19–8).

NOTE Both gears must be pulled onto the shaft at the same time and care must be taken when coming to the end of the gear position not to damage the rotor lobes.

Timing Blower Rotors With the timing gears in position, use a dial gauge to measure the backlash. It should not exceed 0.004 in [0.101 mm]. With a feeler gauge inserted at C and CC (Fig. 19–7), measure the clearance at several intervals along the total

length of the rotors. Compare your measurement with that specified in the service manual. The recommended clearance between the trailing edge of the upper rotor and the leading edge of the lower rotor (CC) is 0.002 in [0.050 mm].

To increase the clearance at CC, pull the bottom timing gear out; to reduce the clearance, pull the gear in. To increase clearance at C, pull the top timing gear out; to reduce clearance, pull the gear in. When you have the clearance properly adjusted, use a feeler gauge to measure between the gear and bearing inner race to determine the required thickness of the shims. Pull both gears evenly from the rotor shafts. Place the selected shims on the shafts against the inner race bearing and install the gears.

NOTE Be sure to recheck the clearance at CC and C.

Installing Fuel-Transfer Pump Place a new gasket on the fuel-transfer-pump mounting surface. Place the pump-drive-coupling fork on the square of the drive shaft and align it with the slot in the drive disk.

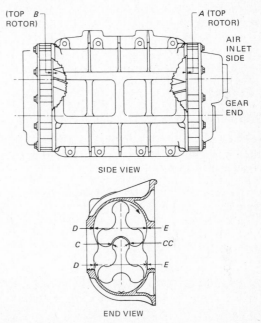

Fig. 19–7 Rotor-to-rotor and rotor-to-housing measuring points. (*GMC Detroit Diesel Allison Div.*)

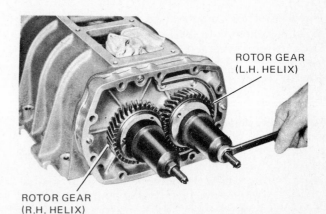

Fig. 19–8 Installing rotor gears on rotor shafts for preliminary clearance check. (*GMC Detroit Diesel Allison Div.*)

Place the pump-drive-coupling fork in position and tighten the transfer pump to the rear-cover end plate.

Installing Governor and Coolant Pump Place a new gasket on the mounting surface of the governor housing and slide the spline of the governor shaft into the spline of the upper rotors. Use new copper gaskets on the hex bolts which secure the governor to the front-cover end plate.

To install the coolant pump, place a new gasket on the mounted surface of the front-cover end plate. Align the locks of the coolant-pump drive coupling with the locks of the intermediate coupling and position the pump on the cover. Secure the pump to the end cover using new seal washers on the three hex bolts.

Installing Blower to Engine Since there are many varied blower locations, only the most common procedures are listed below.

Use nonhardening gasket sealant on all gaskets. Hand-tighten all hex bolts, particularly those of the governor which are screwed into the cylinder head, *before* tightening them to the recommended torque.

Connect and secure the fuel lines, coolant-pump outlet, hoses and clamps to the cylinder block. Connect the oil cooler. Install air shutdown and connect the turbocharger (compressor) with the air inlet.

Questions

1. Which areas of the exhaust manifold and cylinder head must be checked before the exhaust manifold can be installed?

2. What are the main differences between a turbocharger and a Roots blower?

3. List the periodic checks that should be made on a turbocharger.

4. Outline how to check the end play of the turbocharger shaft.

5. How can turbocharger failure be prevented?

6. Why does a Roots-type blower require no lubrication?

7. List the areas or check points which should be inspected to determine the operating condition of the blower.

8. Explain how you would increase, to specification, the running clearance of a blower at CC (Fig.19–7).

9. Why must the timing gear be evenly removed from, or evenly installed to, the rotor shaft? (See Fig. 19–8.)

10. What checks must be made after a turbocharger or blower has been installed to the engine?

11. A turbocharged diesel engine should be idled down before shutdown, to prevent (a) air pressure in the cylinders, (b) oil from going into the engine, (c) turbocharger from running out of oil, (d) oil from boiling in the turbocharger.

12. When hauling a turbocharged diesel engine on a low bed you should (a) fill the turbocharger with oil, (b) plug the exhaust pipe opening, or (c) plug the air-filter opening.

13. An aftercooler unit is used on a turbocharged engine (a) to cool the exhaust gases, (b) to cool the inlet air before it enters the engine, (c) to cool the inlet air before it enters the turbocharger, or (d) to cool the exhaust manifold.

UNIT 20

Cooling System

Purpose of Cooling System The purpose of the cooling system is to dissipate the heat arising from combustion. About 28 percent of the heat must be removed in order to retain the engine's efficiency and to reduce component wear. The required volume that the coolant pump must move through the coolant system, as well as the quantity of coolant needed, are determined by the specific engine horsepower.

Combustion heat is dissipated in three different ways: (1) convection, by means of air currents, (2) radiation, by waves sent out from the vibrating molecules, (3) conduction, by traveling through the metal to the cooling passages (where the coolant picks up the heat and carries it into the radiator). Heat absorbed by the engine oil is partly removed by con-

duction. The remainder is removed in the oil cooler and the oil pan by a combination of the methods just described.

The dissipation of heat, in itself, would be relatively simple if it were not essential that the cooling system maintain an even temperature at any torque range, at any engine-speed range, and at varying ambient temperatures, for example, from -30 to $+110°F$ [-34.4 to $-43.0°C$].

At maximum engine torque and high ambient temperature, the system is forced to dissipate heat at its maximum capacity in order to maintain the top tank temperature around 180°F [82°C]. When the engine torque and the ambient temperature are low, the system must nevertheless maintain the engine at approximately the same temperature.

Cooling-System Components Beginning at the front of the engine, the components which make up an average cooling system are: the radiator, fan, coolant pump, engine oil cooler, cylinder block, cylinder head, water manifold, thermostat, cooling pipes, and hoses (Fig. 20–1).

Some engines have additional components, such as a torque-converter oil cooler, a radiator shutter system, a coolant filter, a surge tank, and a second coolant pump. Engine and equipment manufacturers jointly determine the various interrelated cooling-system components to meet the needs of different on- and off-highway engine applications.

Maintaining Cooling System Quite often, operators and mechanics restrict their checks of the cooling system to the coolant level, coolant leaks, and perhaps to the drive belts. Sometimes the cooling liquid is not inspected or changed, and the coolant filter, thermostat, shutter system, and internal passages also are neglected. Inefficiency in any of these components can cause the coolant temperature to recede to about 150°F [66°C], allowing the oil to become sludge. Coolant deposits then form within the engine where the coolant meets hot metal. These deposits reduce the cooling flow to the cylinder sleeve, piston, and valves and thereby accelerate wear.

Coolant deposits fall into four categories: scale from waterborne minerals, products of corrosion, products of chemical incompatibility, and petroleum contaminants.

Coolant-System Requirements The coolant temperature in the top tank of the radiator should never exceed 200°F [93°C] regardless of ambient temperature or engine torque.

The coolant system must be capable of: (1) raising the coolant temperature quickly to keep engine-component wear to a minimum, (2) providing an outlet for the coolant to escape, (3) maintaining a greater than atmospheric pressure at the inlet side of the coolant pump, and (4) providing a means for venting itself during the filling operation.

Air-cooled Engines The cooling system of an air-cooled engine consists of an engine-driven blower cooling the cylinder fins and metal shields (Fig. 20–2). The cooling fins on the cylinder and cylinder head are precisely calculated and designed according to the required heat dissipation of the area. They are enlarged to increase dissipation of heat and reduced to dissipate less heat. Metal shields direct the air around the fins in a predetermined flow and at a predetermined velocity to help achieve an even temperature.

Advantages of an air-cooled engine are:

1. It is of lighter and simpler construction compared to a liquid-cooled engine of the same horsepower.
2. The cooling system is easily maintained, that is, by simply checking the condition and position of the shields and the fins for breakage, dust or oil ac-

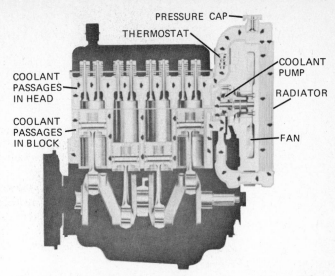

Fig. 20–1 Sectional view of a cooling system. (The arrows show the coolant flow.) (*J. I. Case Co. Components Div.*)

Fig. 20–2 Deutz air-cooled engine.

cumulation, and by checking the blower for bearing wear and general condition.

Radiators Two types of radiator designs are common to diesel engines: where the core, bottom and top tank, and side members are bolted together to form an assembly (Fig. 20–3) or where the core, top and bottom tank, and side members are soldered together to make up an assembly (Fig. 20–4). Core construction may be any one of the three types shown in Fig. 20–5.

Factors which determine the radiator design are airflow, size, and cost.

Radiators should be as nearly square as possible if they are to provide the most effective fan performance, since this shape has the least unswept core area. When operating economy and cost are of equal importance, the radiator fins are spaced with 8 to 10 fins per square inch [1.2 to 1.5 fins per square centimeter]. When heat dissipation is of prime concern, the number of fins per square inch is increased,

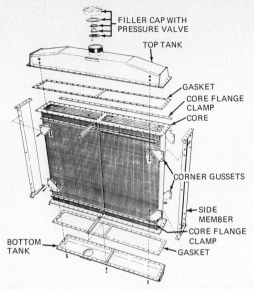

Fig. 20-3 Typical radiator details. (*Allis-Chalmers Engine Div.*)

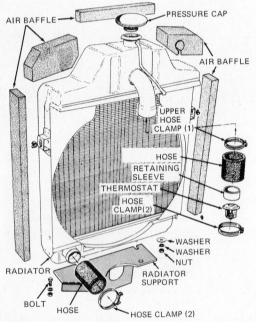

Fig. 20-4 One-unit radiator design. (*J. I. Case Co. Components Div.*)

but this can lead to faster plugging of the core from insects and other alien materials. Since additional fins restrict the airflow, more horsepower is needed to drive the fan.

Engine horsepower determines the size (front square area) of the radiator. It averages 3 to 4 in² [19.35 to 25.80 cm²] for each engine horsepower used and the airflow is between 1000 and 1300 ft/min [304 and 396 m/min] in velocity.

To improve airflow and fan efficiency and to direct the air around the engine, three types of fan shrouds are used. They are bolted to the radiator and allow the fan blades to protrude about two-thirds past the corner of the shroud (Fig. 20-6). Airflow efficiency is greatest when the tip clearance does not exceed 1.5 percent of the fan diameter.

Most radiators have a drain valve located at the

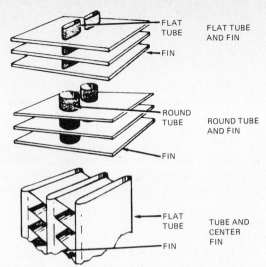

Fig. 20-5 Radiator core designs. (*GMC Detroit Diesel Allison Div.*)

lowest point of the radiator, one outlet connection in the bottom tank, an inlet connection in the top tank, and a radiator filler opening (filler neck) which is sealed by a radiator cap.

Some radiators have a solid baffle which divides the top tank into an upper and lower section to reduce aeration. Coolant enters the lower section from the cylinder head and the upper section from a standpipe. This latter design helps combat the cooling system's biggest enemy—air! Air in the coolant reduces heat dissipation, reduces heat transfer to the coolant, and can cause the coolant pump to lose its prime. It is also the principal cause of corrosive action in the coolant passages.

Coolant-System Piping Two coolant systems are schematically illustrated in Figs. 20-7 and 20-8. The system shown in Fig. 20-7 is designed with a separate top tank and a makeup line (supply line) which connects the top tank and the lower tank. The vent line (deaeration line) connects the cylinder head or cooling manifold to the top tank. Because of these lines, problems associated with filling the radiator and air locks are prevented. This system uses a pressure cap on the radiator.

The bypass line is the link between the thermostat and the cylinder block. Its purpose is to reduce warm-up time and to prevent overheating of the cylinder head. When the engine is shut down, the coolant pump stops but coolant continues to circulate from the hotter cylinder head into the cylinder block via the bypass line.

The system shown in Fig. 20-8 uses an additional reservoir (surge tank) which is mounted higher than the (downflow) radiator. The auxiliary tank may be one as simple as that shown in Fig. 20-8, or it may be a tank similar in design to the top tank of a radiator. In this case the surge tank has a pressure cap.

The main function of the system design shown in Fig. 20-8 is to remove and/or prevent air from entering the system. It also provides additional coolant fluid to the system and reduces the possibility of coolant losses should the system become overpressurized.

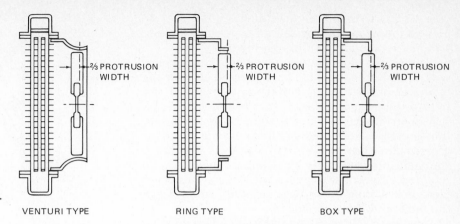

Fig. 20–6 Types of fan shrouds. *(Cummins Engine Co., Inc.)*

VENTURI TYPE RING TYPE BOX TYPE

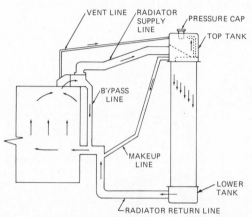

Fig. 20–7 Deaerating-top-tank piping. (The arrows show the coolant flow.) *(Cummins Engine Co., Inc.)*

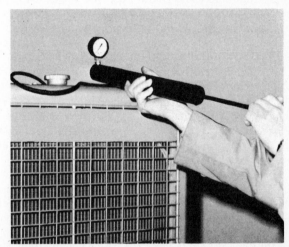

Fig. 20–9 Pressurizing the cooling system. *(Caterpillar Tractor Co.)*

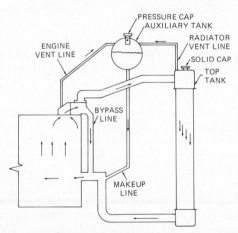

Fig. 20–8 Auxiliary-tank piping. (The arrows show the coolant flow.) *(Cummins Engine Co., Inc.)*

Radiator Testing As soon as it becomes apparent that there is a coolant leak, the faulty component should immediately be repaired to prevent damage to the other components which also rely on the coolant. Radiator damage or leakage may be the result of undue force from external objects, a rough road or engine vibration, a loose radiator mounting, a broken engine mounting, or stiff radiator hoses. Occasionally, corrosion may cause the radiator to leak.

Before you clean the radiator, examine it carefully for a white or rusty colored (leak) stain. Mark the stained area and then thoroughly clean the radiator. Remove all foreign material from and between the cooling cores and fins using plain water or compressed air in reverse airflow.

Before you remove the radiator for repair, it may be expedient to pressure-test the cooling system so as to expose any leaks which were not previously visible.

To pressure-test the coolant system, install a coolant-system pressure tester as shown in Fig. 20–9. Pressurize the system about 10 percent over the normal operating pressure. Check the total coolant system, particularly the coolant manifold and casting-core hole plugs. (Take care not to overpressurize the system.) Excessive pressure can swell the radiator core tubes and tank or damage the coolant-pump seal.

Removing Radiator Some radiators can be removed with relative ease; others require a more time-consuming effort. The simplest procedure is to first drain the coolant from the cylinder block and radiator. Then remove upper and lower radiator hoses. Remove the shroud and lay it over the fan. Finally, remove the mounting bolts. The radiator is now ready to lift up from its mounting.

CAUTION Use a lifting device that you know is safe.

Servicing Radiator Radiators should be inspected daily and cleaned regularly when operating conditions include the possibility of dust, leaves, insects, or other matter being trapped between the cooling fins. If these foreign particles are not removed, they will restrict the airflow through the cores.

To dislodge alien matter, remove the shroud and, with the air hose, direct the compressed air over the total area of the radiator surface in a reverse direction to the normal airflow. It is sometimes necessary to first flush water through the fin area to loosen the foreign matter. When the presence of oil or grease causes dust to adhere to the fins of the core, a spray-cleaning gun carrying cleaning solvent may be required. Never use a wire, screwdriver, or brush to clean debris from the fins because they will damage the tubes and bend the fins.

Any repair to the radiator core should be done by a recognized radiator shop since it has the necessary equipment to make the required tests and repairs. On the other hand, if the leak is caused by a damaged gasket, you should be able to service it.

To replace the leaking gaskets of the radiator shown in Fig. 20-3, hold the radiator by means of its side members. Never lay the core against an object because this will bend the cooling fins. Use an impact wrench to remove the top-tank bolts, but take care that your wrench holds the nut without slipping and damaging the core.

After you have removed the top tank, clean the tank and its flanges. Remove the old gaskets and when necessary straighten the core flanges. Check the core top surface and tubes for corrosion, scale, or sludge. If the radiator tubes are restricted, the bottom tank must be removed and the core cleaned in a hot cleaning tank.

CAUTION Use only the cleaning solution recommended by the manufacturer. Finally flush with clean water to remove any remaining solution.

When reassembling the radiator, apply a nonhardening sealant to both sides of the gasket. Tighten the hex bolts by using an even cross-tightening pattern to ensure a good gasket seal.

Before you install the radiator, straighten all fins. Bent fins restrict the airflow, and result in reduced heat dissipation.

Cleaning Cooling System Do not put off cleaning the cooling system until the continual process of physical and chemical changes in the cooling system has caused insufficient cooling or has restricted the coolant flow. With a chemical cleaner, remove any accumulated deposits since they cause a reduction in heat transfer to the coolant fluid.

Two types of cooling-system cleaners are used: the alkaline cleaner and the inhibitor acid cleaner. The alkaline cleaner is most effective for removing sludge and silicon scale. The inhibitor acid cleaner is most effective for removing rust and carbonate scale.

Your cleaning procedure should include three steps: cleaning with an alkaline cleaner, recleaning with an acid cleaner, and flushing the system with a neutralizing fluid. Follow the cleaner manufacturers' instructions regarding the use of their prod-

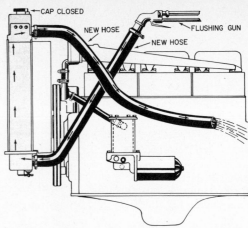

Fig. 20-10 Reverse-flow flushing the radiator. (The arrows show the water flow.) *(J. I. Case Co. Components Div.)*

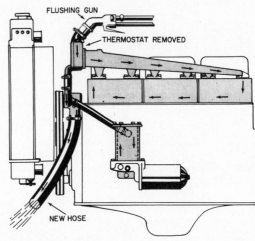

Fig. 20-11 Reverse-flow flushing the engine block. (The arrows show the water flow.) *(J. I. Case Co. Components Div.)*

ucts. Do not hesitate to seek advice on cleaning problems from your local supplier.

Simply stated, the cooling-system cleaning procedure is as follows:

1. Drain and, if necessary, flush the system with water to remove as much contamination as possible.
2. Remove the thermostat.
3. When the system has a bypass line (Figs. 20-7 and 20-8), this line must be plugged in order to allow concurrent cleaning of the radiator and to prevent overheating of the cylinder block.
4. Fill the cooling system with your ready-mixed alkaline solution and run the engine for the recommended length of time. (You may have to use the shutters or cover the radiator to raise the coolant temperature to that recommended by the supplier.)
5. After the recommended running time, cool the engine down by reducing the speed and removing the cover from the radiator.
6. Drain the system, flush it with clean water, and refill it with the acid solution. **CAUTION** Do not fill or flush the system with cold water when the

engine is hot because rapid cooling distorts the engine casting.

7.   When the acid solution is drained from the cooling system, neutralize the system by flushing it with water and refilling it with a neutralizing solution. Another method is to reverse-flow flush first the radiator and then the cylinder block, in separate steps.

REVERSE-FLOW FLUSHING RADIATOR   For reverse-flow flushing you need a flushing gun with an air and water connection and an air and pressure regulator. Connect a return hose to the radiator top tank so that the water does not go over the engine.   Connect the flushing gun to the lower tank, as shown in Fig. 20–10.   Open the water supply to fill the radiator, then regulate the air pressure to that recommended by the engine manufacturer.   With the water running out of the top tank, intermittently apply air pressure for about 2 to 3 seconds at a time until clean water runs out of the top-tank hose.

REVERSE-FLOW FLUSHING CYLINDER BLOCK   It is time-consuming to reverse-flow flush the cylinder block because the coolant pump must be removed.   (You must also remove the thermostat and plug the bypass if this has not yet been done.)

Your first step is to install the flushing gun to the top coolant inlet, as shown in Fig. 20–11.   Open the water valve to let water flow through the cylinder block.   Adjust the air-pressure regulator to the recommended pressure and blow the water out of the cylinder block.   Repeat this procedure until clean water runs out of the opening.

Install the coolant pump and thermostat.   Connect the bypass line.   Fill the coolant system with an antifreeze concentration which is not less than 30 percent, otherwise the cooling system will not be protected sufficiently from corrosion.

Radiator Filler (Pressure) Cap   The radiator or the surge tank (expansion tank) is equipped with a pressurized filler cap (Fig. 20–12).   Its purpose is to maintain a coolant-system pressure during engine operation and to equalize pressure when the engine is cooled off.

When the pressure in the system shown in Fig 20–7 rises higher than the valve-spring tension, coolant leaves through the overflow, thereby reducing the coolant-system pressure.   However, in the system shown in Fig. 20–8, when the pressure rises above the spring tension, air alone, and not the coolant fluid, leaves the system thus reducing pressure.

It is very important to the entire cooling system that the pressure valve and vacuum valve seal properly and that the valve springs have the proper tension.   When the pressure valve has a weak spring or a faulty seal, the boiling point of the cooling system will be lowered because the faulty valve lowers the system pressure.

For each 1 psi [0.07 kg/cm²] above atmospheric pressure within the cooling system, the boiling point is raised by about 3°F [1.66°C].   The average cooling-system operating pressure is about 15 psi [1.05 kg/cm²].   This allows the engine to operate at a higher

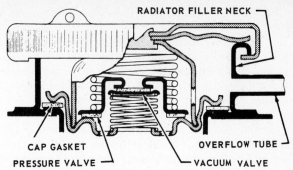

Fig. 20–12   Radiator filler cap. (*Mack Trucks Canada, Ltd.*)

temperature since the water then boils at 257°F [125°C].   The boiling point is higher when antifreeze is added.   For example, when the antifreeze concentration is about 40 percent at atmospheric pressure, this mixture will boil at about 223°F [109°C].   With a 7-psi [0.49-kg/cm²] pressure cap, the boiling temperature will be higher: 243°F [117°C].

Testing the Radiator Filler Cap and Filler Neck   Check the top and lower seat and the expansion tank filler neck.   A filler cap cannot hold the cooling-system pressure if the seat is damaged.   The seat must be smooth; when it becomes rough, the filler neck must be replaced.

Check the general condition of the filler cap, the synthetic rubber seals, the gaskets, the pressure, and the vacuum springs.   If any of these components are damaged, the cap must be replaced.   When the cap is in good condition, use a filler-cap tester to check the effectiveness of the seals and the release pressure of the valve.   The pressure should be within 1 psi [0.07 kg/cm²] of specification.   When checking the sealing effectiveness of a nonpressure cap (solid cap), pressurize the tester about 10 percent above normal system pressure.   The pressure on the gauge should not vary at all.

Cooling Fans   There are many pusher (blower) or suction fans in use, with varying airflow capacity and of contrasting design.   On smaller diesel engines the fan is bolted to the coolant pump whereas on larger engines, the fan is bolted to a fan-hub assembly (Fig. 20–13).   The coolant pump and

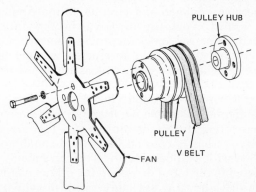

Fig. 20–13   Typical fan and fan-hub assembly. (*Allis-Chalmers Engine Div.*)

the fan-hub assembly are driven by one or more V belts.

It is the engine manufacturers who select the cooling-system components to be used on a given engine. They choose the radiator size, the shroud and fan size, the fan design and its rotating speed. Together, these components ensure that (1) the cooling temperature in the top tank of the radiator does not exceed maximum prescribed temperature; (2) the horsepower required to drive the fan does not exceed 6 percent of the engine horsepower; (3) the speed of the tip of the fan is not greater than 18,000 ft/min [6000 m/min] and therefore keep the fan noise to an acceptable level; (4) the airflow does not exceed 1300 ft/min [396.24 m/min]; and (5) there is no dead area on the surface of the radiator.

The manufacturers also consider the weight of the fan as it affects the horsepower, the life of the drive belts, and the bearings. Weight, however, is only one of the factors that have influenced many companies to use fiber-glass fans instead of steel fans. The flex of certain fiber-glass fan blades provides maximum cooling at any speed. Their fan-blade pitch changes automatically reducing or increasing with the change in engine speed. Furthermore, fiberglass fans require less power than is needed to drive a fixed steel-bladed fan running at the same speed. Bearing and drive-belt life is increased and noise reduced.

Whether a pusher or suction fan will be used depends on engine application. Loaders, and wheel or track machines commonly use a pusher fan since it is less likely to draw as much dirt, sand, and small stones into the radiator as the suction fan. Fastmoving vehicles, however, use suction fans since the airflow through the radiator at their normal vehicle speed acts against the pusher fan.

Fan Inspection Check the fan for loose rivets, cracks, or bent blades. Remove any oil or dirt from the fan blades since contamination causes an unbalanced condition which can lead to blade breakage and bearing wear. Do not attempt to weld a cracked fan. Welding will destroy the fan's balance as well as the tensile strength of the steel. Slight bends may be straightened but if the blade is sharply bent, the fan must be replaced.

Thermomodulated and Viscous Fans Inherent in the thermomodulated and viscous designed fans is their control capability to drive the fan only as required. For instance, to provide a fast engine warm-up and to maintain an efficient coolant temperature. They also require less horsepower to drive than the standard-design fan, resulting in increased drive-belt life and reduced noise.

The thermomodulated fan is an integral unit with no controlling device or piping (see Fig. 20-14). The drive assembly consists of a sealed clutch housing filled with a silicone fluid. The drive plates are connected to the input drive and the driven plates to the fan-drive-housing assembly. The silicone-fluid film between the drive and driven plates transmits the torque from one member to another, depending

on engine temperature. Changes in air temperature, to which the temperature-control unit is exposed, can vary the viscosity of the silicone fluid. Variations in viscosity reduce or increase friction and cause an alteration in fan speed.

Maintenance and service of the thermomodulated and viscous fans is restricted to a periodic check of the fluid level, the condition of the bearings, and the cleanness of the temperature-control assembly.

Viscous Fan Drive The viscous fan drive is a sealed unit. It operates by changing the amount of fluid between the input and output member, which results in a controlled speed drive. A leaf-type bimetal sensing unit is located on the front cover. It senses the temperature coming through the radiator core. The sensing unit is coupled to an internal valve. As the bimetal sensing unit reacts to temper-

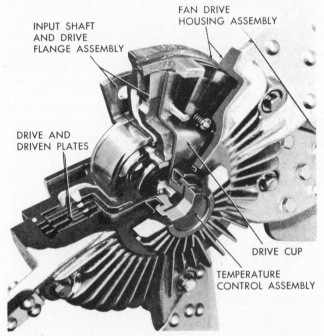

Fig. 20-14 Sectional view of a thermomodulated fan. (GMC Detroit Diesel Allison Div.)

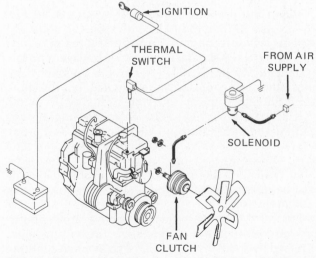

Fig. 20-15 Components of a Horton fan clutch. (Horton Industries, Inc.)

ature, it opens or closes the valve and regulates the flow of viscous fluid from a reservoir area into the drive area.

Thermatic Fan Drive and Horton Fan Clutch The thermatic fan drive and the Horton fan clutch (Fig. 20–15) use a single-plate clutch which is disengaged by spring force and engaged by air pressure. As the coolant-system temperature approaches 184°F [84.60°C], the sensing switch closes the electric circuit and the solenoid opens the air valve. Reservoir pressure of 90 to 110 psi [6.33 to 7.74 kg/cm²] is directed to the fan-clutch piston which then forces the clutch plate against the friction lining and connects the fan with the pulley. As the coolant-system temperature drops below the sensing switch temperature, the switch opens the electric circuit. The solenoid then closes off the air from the reservoir and opens the exhaust valve. Air is released from the piston and the spring force disengages the clutch, causing the fan to stop.

Servicing Fan Clutch, Sensing Switch, and Solenoid To check the operation and adjustment of the sensing switch, immerse the lower part of the switch in water. Connect an ohmmeter to the leads (see Unit 37, "Electric Circuits and Test Instruments"). Immerse a thermostat in the water and heat the water. When the ohmmeter indicates circuit continuity, take the thermometer reading. If the thermometer reading does not correspond with the recommended opening temperature, adjust the switch to allow it to open and close at the specific temperature.

NOTE If a shutter system is used, the sensing switch of the fan must be adjusted about 5 to 10°F [2.8 to 5°C] higher than the sensing switch of the shutterstat.

To check the solenoid for an open, short, or grounded circuit, connect an ohmmeter to the electrical leads. Measure the resistance of the solenoid coil. Compare this measurement with your specification.

Check and inspect the air valves and seat. Make certain that the rubber is not damaged, that the seat is not rough and that it has no grooves.

To service the fan-clutch assembly (bearings, piston seals, O rings, friction clutch or pulley), clamp the assembly in a vise and remove the eight, socket-head cap screws which hold the air chamber to the bearing cage (see Fig. 20–16). Remove the air chamber, slide the key out of the adjusting nut, and remove the nut. The spindle, spacer, and piston assembly can now be removed from the spindle (journal).

Replace all seals and gaskets. Repack bearings with the recommended grease. Check the pulley grooves, the friction disks, and the spline hub for wear. Replace the air-seal cartridge. Reassemble the components.

NOTE When you lubricate the seals and O rings, check that they are not twisted. Tighten the adjusting nut, the stud bolts, and the cap screws to their recommended torque.

Fan Hub and Fan-Belt Tighteners Although many types of fan hubs, idlers, or fan-belt tighteners are used, they are generally of the same construction (Fig. 20-17). Each has external V grooves and internal machine bores to accommodate its bearings and seals. The shaft (spindle) is part of the adjustable or fixed mounting bracket or is bolted to the bracket. Tapered roller bearings, ball bearings, or bushings support the pulley.

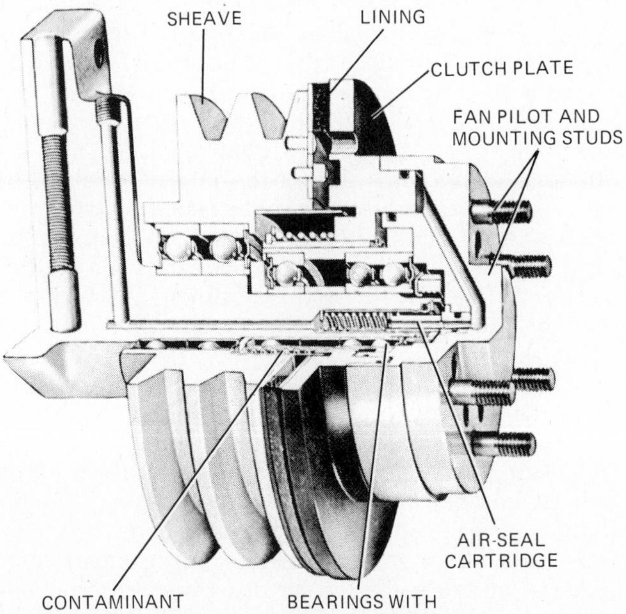

Fig. 20–16 Cutaway view of a Horton fan clutch. (*Horton Industries, Inc.*)

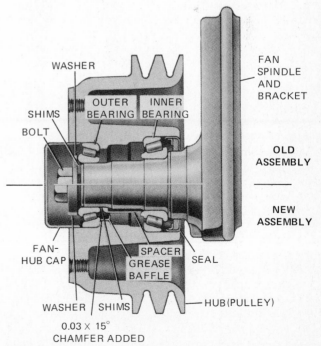

Fig. 20–17 Typical fan-hub assembly. (*GMC Detroit Diesel Allison Div.*)

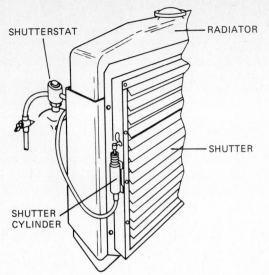

Fig. 20–18 Radiator shutter system.

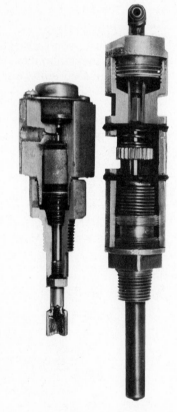

Fig. 20–19 Cutaway view of a shutterstat.

NEW STYLE   OLD STYLE

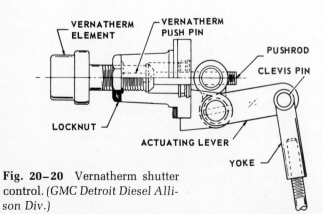

Fig. 20–20 Vernatherm shutter control. *(GMC Detroit Diesel Allison Div.)*

Inspecting and Servicing Fan Hub and Fan-Belt Tighteners Begin servicing by cleaning all parts thoroughly and then drying them with compressed air. Check the pulley grooves for wear and remove any corrosion or rough spots. If the grooves are damaged, or if the sides are rough, the pulley must be replaced. Check the bearings and seal surface on the spindle. The latter is seldom found to be defective, still it must be checked. When new bearings are required, press the old cups out and the new ones in, using the correct bearing pusher. Support the pulley only by its bore surface and not by the groove surface. Do not hammer the seals into place; use a press to place them in position.

When you press the bearing cones onto the spindle, rest the cones squarely on the bore and then press them tightly against the bore shoulder. Do not forget to pack the bearings or lubricate the seal(s) before placing the pulley on the spindle.

Support the spindle in a press or in a vise and press or drive on the outer bearing cone. Install the washer and nut. Tighten the nut to seat the bearings, then back off and tighten the nut until a slight drag is felt when the pulley is rotated. (Back off only enough to allow the cotter pin to fit.)

When your service manual recommends an end play, then you must install a dial gauge on the spindle so that the pointer of the gauge rests against the pulleys. Tighten the nut to accomplish the required end play. (The preload or end play of some pulleys is adjusted by shims.)

Radiator Shutter System The radiator shutter system helps to maintain engine temperature by closing off the airflow to the radiator (Fig. 20-18). This system consists of the shutter, shutter operating bar, shutter cylinder, shutterstat, and in some cases an air filter.

The shutter is bolted to the front of the radiator and each individual shutter blade is connected to the shutter bar. The shutter return spring on the shutter bar keeps the shutter open, whereas the shutter cylinder allows it to close. The shutterstat, located in the radiator inlet or radiator outlet connection, is a temperature-sensing air valve (Fig. 20-19).

SHUTTER OPERATION When the engine is stopped and there is no air pressure in the reservoir, the shutter is opened by the shutter spring. As soon as the air pressure is high enough to overcome the shutter return spring, air enters the shutterstat and passes through the filter, around the needle valve, and out of the air outlet into the shutter cylinder. Air pressure forces the piston to move to the right, causing the linkage to rotate. This rotation pulls the shutter bar and brings the shutter blades into the closed position. The coolant temperature then rises quickly because air cannot be drawn through the radiator. As the coolant temperature reaches the sensing-switch adjustment setting, the thermostatic material has expanded so far that it moves the pushpin against the needle valve, closing the air-inlet valve and opening the exhaust valve. The air exhaust from the shutter cylinder and the shutter spring opens the shutter blades.

An air filter should be used when there is no other valve in the air system to provide for the removal of the accumulated water and sludge.

Other shutter systems use a Vernatherm element (actuator) which serves as the sensing unit and the actuator (Fig. 20-20). It is screwed into the actuating mechanism. This element is a sealed unit, made of a thermostatic material which is a special mixture of hydrocarbon that expands as the desired operating temperature is reached. The expanding hydrocarbon forces the diaphragm upward to move the piston. This causes the actuating lever to rotate and the shutter operating bar to open the shutter. As the coolant temperature is reduced, the hydrocarbon retracts into the cup and the shutter-return spring closes the shutter.

To have a precise coolant temperature, all three sensing units (thermostat, shutter system, and thermostatic-fan sensing unit) must be adjusted, first to allow the thermostat to open, then to allow the shutter to open and, finally, to start the fan rotating. Service manuals show the precise degree at which each sensing unit (element) should start to open, actuate, or close the electric circuit.

Inspecting, Testing, and Adjusting Shutter System First check all linkage for wear. With solvent, remove accumulated dirt from the assembly. When possible, use a graphite powder on all movable joints, rather than a 10-weight engine oil, since oil increases the amount of dirt accumulation.

When necessary, adjust the yoke so that the cylinder stop nut rests against its stop when the shutter is closed.

To check the fit of the shutter blades, start the engine and run it about 800 rpm. Hold a large piece of paper against the shutter. If the paper slides off, the shutter blades are closed properly. If the paper stays in place, any one of the following malfunctions may be present: the shutter blades may not be sealing properly; the relation between the shutter bar and cylinder may be out of adjustment; the shutter cylinder may be leaking; the linkage may be binding; or the shutter frame may be bent. Whatever the problem, it should be corrected promptly.

Adjusting Shutterstat To check and adjust the shutterstat, immerse the bulb of an accurate thermometer into the top tank of the radiator. Start the engine. The thermostat should start to open at 170°F [77°C]. When the temperature reaches about 185°F [85°C], the shutter should open. Reduce the engine speed and within 7°F [3.90°C], that is, at a temperature of 177°F [80.9°C], the shutter should close.

To delay shutter opening (and increase coolant temperature), increase the spring tension against the bellows by turning the adjusting wheel in a right-hand direction. To open the shutter earlier, turn the wheel in left rotation. Five clicks on the wheel will alter the opening of the shutter by about 2°F [1.1°C]. If a Vernatherm element is used, the shutter opening can be altered by replacing the element with one that has a 5°F [2.8°C] higher or lower setting.

To check the fan-clutch sensing unit, disconnect the air line from the shutter cylinder so that the coolant temperature can be raised. Run the engine at high idle. As the coolant temperature approaches 190°F [88°C], the fan clutch should be fully engaged. At this point, open the shutter by hand and observe the coolant temperature. At about 182°F [83.5°C] the air should exhaust from the fan-clutch piston to release the fan clutch.

Thermostat The thermostat is a valve located between the cylinder head and the top tank of the radiator which reacts to temperature changes. Its exact location varies with engine application. It is usually in the end of the cooling manifold or in a separate thermostat housing bolted to the cylinder head or cooling manifold.

Because of their higher gallons (U.S.) per minute (gal U.S./min) [liters per minute (l/min)], larger engines utilize two thermostats to reduce restriction of the coolant flow.

The thermostat valve is connected to a heat-sensing element which may be either a bellows, a wax pellet, or a hydrocarbon-mixture pellet (see Figs. 20-21 and 20-22). A pressure spring, placed between the heat-sensing element and the thermostat frame, opposes the expanding force of the element.

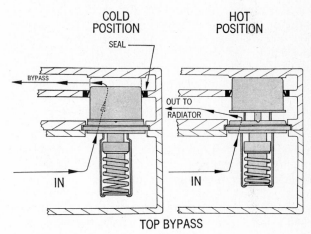

Fig. 20–21 Bellows thermostat.

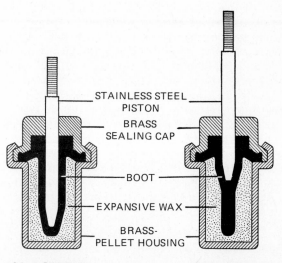

Fig. 20–22 Wax-pellet thermostat.

The thermostat, depending on its design and the coolant temperature, directs the flow of coolant within the cooling system. Three types of thermostats are used: the full-blocking, the partially blocking, and the nonblocking thermostat. Small diesel engines do not have a bypass line or a vent line. The thermostat blocks the coolant flow to the radiator when the valve is closed and coolant flows through the cylinder block, the cylinder head, and back into the inlet side of the coolant pump. A small vent hole in the valve of the thermostat allows the coolant to flow to the radiator and allows air to escape. As the coolant approaches thermostat temperature, the valve gradually opens and directs coolant to the radiator.

The system shown in Fig. 20-23 has a full-blocking thermostat. At a coolant temperature below 170°F [77°C], the thermostat valve closes off the flow to the radiator. Only a small amount of coolant passes through a hole in the valve to the radiator. During this period most of the coolant is directed, by the thermostat, through the bypass line to the cooling passages of the engine and back into the inlet side of the coolant pump. The system vents itself through the orifice in the thermostat valve. At a thermostat temperature of about 170°F [77°C], the heat-sensing element expands gradually, forcing the valve to open. The coolant flow is diverted to the radiator and to the engine. As the coolant temperature increases, the flow to the engine (bypass) line is restricted. At about 185°F [85°C] it is completely stopped, and all coolant flows into the radiator.

The cooling flow of a nonblocking thermostat system, after the thermostat is open, is shown in Fig. 20-24. A partially blocking thermostat system is illustrated in Fig. 20-25.

NOTE The radiator shutter system controls the coolant temperature after the thermostat has brought the coolant to operating temperature.

Inspecting and Testing Thermostat When you have removed the thermostat, check the valve seat. Check the heat-sensing element for external damage or rupture. Check the bleed (vent) hole to be sure it is open. If the valve is bent, distorted, or easily pushed off its seat, it must be replaced.

It is imperative to test the thermostat to determine whether its valve opens and closes at the proper temperature and whether it fully opens or fully closes when immersed in heated water. The thermostat is manufactured to allow the valve to open at a predetermined temperature in the range of 150 to 180°F [66 to 82°C]. The valve should reach the fully open position when the temperature increases by approximately 15°F [8.33°C] above opening temperature.

Let us now assume that you have just removed a thermostat. It has the part number stamped on its flange or it is stamped to identify its opening temperature at 170 to 175°F [77 to 79°C].

NOTE Do not clean the thermostat before you test it if you suspect the coolant-system trouble was caused by the thermostat valve or thermostat action.

Immerse the thermostat in a container of water

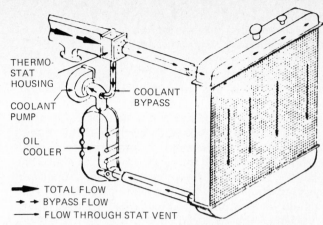

Fig. 20-23 Coolant flow of a full-blocking thermostat when the thermostat is closed. *(GMC Detroit Diesel Allison Div.)*

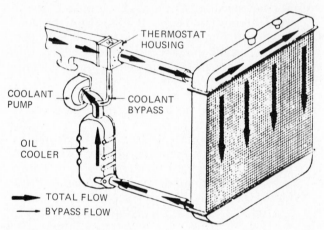

Fig. 20-24 Coolant flow of a nonblocking thermostat when the thermostat is open. *(GMC Detroit Diesel Allison Div.)*

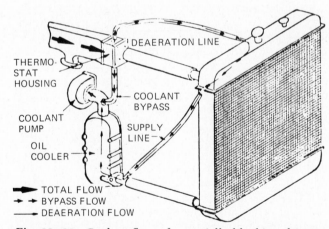

Fig. 20-25 Coolant flow of a partially blocking thermostat when the thermostat is closed. *(GMC Detroit Diesel Allison Div.)*

and heat the water. As the valves open, remove the thermostat and insert a long strip of paper between the valve and the seat. Let the water cool (or add cold water to reduce the temperature) to below that indicated as the thermostat opening temperature. Again, immerse the thermostat in the water, holding it by the paper strip. Gradually heat and stir the water. The paper will be released when the thermostat valve opens. At this point in the test, the temperature of the water should be within specification, in this instance, 170 to 175°F [77 to 79.5°C]. If the valve does not open until a higher temperature is reached, the accumulated sludge or corrosion must be removed and the test must be repeated.

Should the thermostat continue to open earlier or later than specified, it must be replaced. When the valve opens at its specified temperature, heat the water to a higher temperature to determine at which temperature the valve is fully open. *Never assume the valve is fully open. Always check it!*

If a thermostat fails to open to its maximum or if the bypass valve (door) does not close, the thermostat must be replaced.

NOTE When making this test do not let the thermostat or the thermometer rest against the container walls. This will give you a false reading.

Testing for Combustion Leakage into Cooling System To determine if air or combustion gases are leaking into the cooling system, run the engine until it reaches normal temperature 180°F [82°C]. Then drain out sufficient coolant to allow removal of the upper radiator hose and thermostat. Remove the thermostat, upper radiator hose, and drive belt(s).

Supply the system with coolant until it reaches the level of the thermostat-housing neck. Start the engine and accelerate five to six times while watching the outlet opening for bubbles or a sudden rise of liquid. Appearance of bubbles or a rise of liquid indicates that combustion gases are entering the cooling system.

CAUTION Perform the test as quickly as possible; otherwise the coolant will boil and steam and bubbles will rise from the thermostat neck resulting in misleading test results.

Testing for Air Leaks in Cooling System Two methods are used to determine if air is circulating within the cooling system. One method is to pressure-test the cooling system as previously outlined, the other is as follows:

1. Drain as much coolant from the system as is necessary to place a short visible plastic tube between the thermostat housing and radiator top tank.
2. Refill the engine and run it until it reaches normal temperature.
3. Observe the coolant flow. Air in the coolant will be visible as white round spots passing out of the cylinder head into the radiator through the plastic hose.

Coolant Pump Coolant pumps are either driven directly or indirectly by V belts from the crankshaft pulley or by a gear from the timing gears. A typical coolant pump is shown schematically in Fig. 20-26.

The shaft rotates on single- or double-row ball bearings. The bearings may be sealed, lubricated by engine oil, or packed with grease.

The size and design of the impeller, and the rotating speed at which the impeller is driven, depends

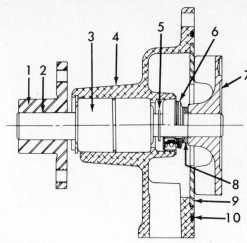

1. Pulley hub
2. Shaft
3. Bearing
4. Water-pump body
5. Slinger
6. Seal assembly
7. Impeller
8. Ceramic seal
9. Body plate
10. O ring

Fig. 20–26 Schematic view of a typical coolant pump.

on the amount of coolant flow required to cool the engine components. The coolant flow can vary between 10 and 180 gal (U.S.)/min [37.85 and 681.3 l/min]. As the pump capacity increases, so does the size of its internal components, that is, the shaft, bearings, housing, seals, and impeller.

Coolant-Pump Failures A coolant pump is said to have failed when it has lost its pumping capacity. Loss of the pumping capacity can be the result of bearing failure since bearing failure increases clearance between the impeller and the housing and causes increased internal slippage within the pump.

The cause of bearing failure can sometimes be traced to a damaged seal assembly which has allowed coolant to pass into the bearings. It also may be the result of any one or more of the following: overtightened drive belts, misalignment, vibration of the pump shaft, or overheating of the coolant (hot shutdown).

A damaged seal assembly can be the result of bearing failure, overheating, contaminated coolant corrosion, scale buildup, excessive wear of carbon face or ceramic face, excessive wear of seat or damaged bellows.

Loose bearings in the housing or on the shaft can also cause early pump failure because they allow the impeller to come in contact with the housing. Also, when there is scale buildup on the internal housing and on the impeller or when they have become corroded, the resultant rough surfaces will reduce coolant flow.

Servicing Coolant Pump Let us assume you have to service a coolant pump similar to the one shown in Fig. 20-27. Clean the coolant pump externally and remove the fan-hub-retainer nut and impeller-retainer nut. With a suitable puller, pull the impeller and fan hub off the shaft and remove the keys. You may have to tap the impeller holes to install the puller bolts.

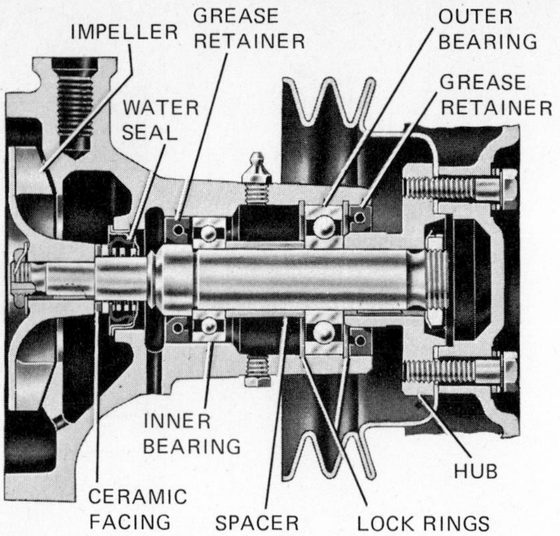

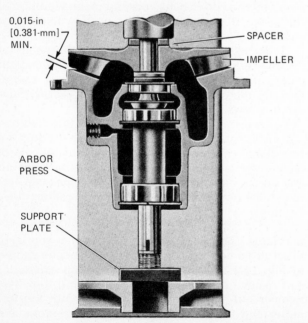

Fig. 20–27 Cutaway view of a typical coolant pump.

NOTE Use a shaft protector to protect the pump shaft when pulling the hub and impeller off the shaft. If a ceramic seal is bonded to the impeller, take care not to damage it.

Use a hammer puller to remove the front lip-type seal, then remove the bearing retainer (snap ring). Place the coolant pump on a press, supported by the bearing bore, and press out the shaft and bearings from the impeller side.

Remove the other front bearing retainer. Remove the rear lip-type seal and press the coolant seal out. If the pump shaft is reusable, press the bearings from the shaft.

Inspecting Coolant-Pump Components Clean all components thoroughly and dry them with compressed air. Check the impeller ceramic-seal face. If it is scored or damaged, you must replace the impeller. If the impeller is damaged externally, or if

Fig. 20–28 Pressing the impeller onto the pump shaft (sectional view). *(Mack Trucks Canada, Ltd.)*

Table 20–1 SUITABLE WATER

	Parts per million
Total hardness (max.)	170
Chlorides (max.)	40
Sulfates (max.)	100
Total dissolved solids (max.)	340

the vanes are worn, damaged, or cracked, the impeller must be replaced.

Check the pump shaft for wear, where it contacts the lip-type seal, the bearings, and the coolant seal. Replace the shaft when necessary.

Check the pump housing for cracks or other damage caused by worn bearings. Replace all seals and bearings to reduce the possibility of early bearing failure or coolant leakage.

Reassembling Coolant Pump Reassemble the coolant-pump components in precisely the reverse order in which they were disassembled. When pressing the bearings and seals into place, use the correct adapters and sleeves. If specified in your service manual, pack between the bearings and the space between the bearings with applicable grease before you press the assembly into the pump housing. Apply a thin coat of water-sealing compound to the outside diameter of the bearings before installation.

NOTE New impellers sometimes have a waxlike coating over the ceramic seal. Remove this coating before you install the impeller, but take care not to damage the impeller.

When pressing the impeller onto the shaft, make certain that the coolant seal and the impeller surface are clean. Do not apply any kind of liquid to their surfaces. Support the pump shaft, then press on the impeller until the specified clearance between the pump housing and impeller is achieved (Fig. 20-28).

Coolant Liquid Only water with an acceptable mineral content should be used in the cooling system of an engine. Water that is within the limits specified in Table 20-1 is satisfactory, nevertheless, proper inhibitors must be added to protect the cooling system against corrosion and sludge.

The inhibitors used to minimize corrosion or to protect iron surfaces exposed to the coolant are soluble chemicals such as chromates, borates, nitrates, and nitrites. Borates, nitrates, and nitrites are often chosen because they can be used with plain water and an ethylene glycol solution. Inhibitors which maintain a soft, acid-free solution and prevent the system from forming a mineral deposit are of many aromatic derivations. A coolant-inhibitor chart recommended for Detroit Diesel is shown in Table 20-2.

NOTE When servicing a coolant-filter element seek the *latest* information about that filter.

Most diesel engines use an ethylene glycol antifreeze solution since it requires no additional inhibitors. However, this solution must not decrease below 30 percent in volume, otherwise the inhibitors are no longer strong enough to protect the system against corrosion and sludge. Many engine compa-

Table 20-2 COOLANT INHIBITOR COMPATIBILITY CHART

Inhibitor or inhibitor system	Corrosion inhibitor type	Complete inhibitor system	Coolant	
			Water	Ethylene glycol solution
Perry filter element 5030 (Type O and OS)	Nonchromate	Yes	Yes	Yes
Lenroc filter	Nonchromate	Yes	Yes	Yes
Nalcool 2000	Nonchromate	Yes	Yes	Yes
Sodium chromate	Chromate	No	Yes	No
Potassium dichromate	Chromate	No	Yes	No
Perry filter element 5020 (Type O and OS)	Chromate	Yes	Yes	No

SOURCE: Mack Trucks Canada, Ltd.

nies recommend the use of a test kit to check the inhibitor level of the coolant system.

CAUTION Do not add chromate inhibitors to the ethylene glycol antifreeze. Do not use an antifreeze containing sealer additives. Either can create plugging problems.

Drive Belts Standard drive belts are divided into five groups, each identified by a code letter. Each V belt has a code number which varies among manufacturers. For example, C-200-101: C represents the cross section of the belt, 200 represents the length, and 101 identifies the tolerance in length. In this case, it is 0.1 in [2.54 mm] per belt length.

Although construction of V belts varies, they are all intended to transfer rotary motion (without stretching), to expand sideways in the grooves, to increase friction and, to some extent, to resist abrasive wear.

Drive-Belt Failure and Preventive Maintenance Drive-belt failure caused through improper alignment, improper tightening, or some other inadequacy has been the cause of many expensive repairs to the cylinder sleeve, piston, and cylinder head. Sometimes these malfunctions exist because the belt was difficult to tighten, inaccessible, or simply not maintained because of forgetfulness.

When you replace an inadequate drive belt, replace it with a belt from the manufacturer who supplied the original. Replace dual belts with a matched belt set; *never* replace only one belt.

Do not pry off the belt from the sheaves. This will damage the pulley as well as the side wall (fabric) of the belt. Slack off the idler or tightener so that the belt can be put onto the sheave(s) by hand. Make certain the shafts are parallel so that the sheaves are aligned.

Before you replace the drive belts, check the sheaves for oil, grease, corrosion, rough spots, or rough sides. Thoroughly clean the sheaves and smooth out any rough spots. If this is not possible, the pulley must be replaced. Be sure to correct the cause of oil or grease leaks.

Follow the engine manufacturer's recommendation about belt tension and the use of a drive-belt tension gauge. A common rule of thumb is to measure the belt length from the center of each pulley and measure the belt width. When pressing down on the belt centrally, between pulleys, the belt should flex one width at each foot length.

When new drive belts are installed, run the engine at least 20 minutes and then retighten all belts. This is essential because new belts stretch and seat themselves in the sheaves.

Indications of V-Belt Malfunction When a drive belt squeals, it is loose, worn, or has lost its friction. Belts will also squeal when the sheaves are worn or the belt is contaminated. When the drive belt has been operated beyond its capacity, it will stretch. It should be replaced with a stronger belt.

If a drive belt runs crooked, it has been stretched on one side as it was forced over the pulley, or else it has run misaligned. When a drive belt whips and flaps, it may be because it is too loose, because one or both pulleys are loose on the shaft, because the pulley shaft is bent, or because the idler tightener, fan hub, or generator, etc., is not rigidly supported.

Cracks on the outside of the belt indicate that the pulley is too small or that there is severe reverse bend from the idler pulley. However, it may have been damaged from an external force.

When the drive belt, the side fabric, or the rubber is cracked, either the belt is running too loose or an oil or grease condition has caused it to lose its friction and it has burned.

A drive belt which has turned over completely either has broken cords, worn sheave grooves, excessive belt vibration or the pulleys are misaligned.

Questions

1. Why does an air-cooled engine maintain a uniform temperature throughout the cylinder block and cylinder head more so than a liquid-cooled engine?

2. What maintenance checks of the cooling system should be made daily?

3. List the ways in which the radiator can dissipate heat.

4. List each step to be taken when testing a radiator for leaks.

5. Why must a radiator or cylinder block be flushed in reverse flow?

6. What would happen to the radiator and coolant hoses if the vacuum valve seized in the open position? What would happen if the vacuum valve seized in the closed position?

7. During the cooling system checkup, what checks would you make to the cooling fan?

8. List the main reasons why a thermomodulated drive or fan-clutch drive is used.

9. Why is it so important that the opening of the thermostat precede the opening of the shutter and the engagement of the fan drive follow the opening of the shutter?

10. How would you test thermostat operation?

11. By what test would you determine whether the cooling system has an air leak or combustion leak?

12. Give several reasons why coolant pumps are of the nonpositive displacement design rather than the positive displacement design.

13. Why do all coolant pumps use a mechanical seal instead of a lip-type seal or an O ring?

14. Why is ethylene glycol antifreeze used to protect the coolant from freezing, rather than alcohol or some other synthetic substance?

15. What is the purpose of a coolant filter?

16. Correct procedure is essential when installing a drive belt if failure is to be prevented. List several don'ts when installing the drive belt.

UNIT 21

Hydraulic Lines and Fittings

Need for Hydraulic Lines and Fittings The hydraulic lines and fittings are as important as any component in the hydraulic system. In transporting the fluid they become an integral part of the system. Poor or inadequate piping, regardless of the type of hydraulic line used, will not only reduce the efficiency of the system but will also reduce the service life. A variety of hydraulic lines and fittings are available. It is essential for the mechanic to know about the various attributes and shortcomings of each type.

Types of Hydraulic Lines The three most common types of hydraulic lines used are: pipes, tubings, and hoses. When selecting the appropriate hydraulic line, a mechanic should know where and how the line is located, at what temperature and pressure the line will operate, what volume of liquid the line will carry, and what type of liquid will be used.

The location, temperature, and pressure will determine the type of material used for the hydraulic line. The volume of the fluid flow will determine the size of the line. Note that manufacturers recommend a maximum velocity of 15 ft/s (feet per second) [4.572 m/s (meters per second)] on pressure lines and 5 ft/s [1.524 m/s] on inlet lines. Supply charts and data tables, obtainable from most manufacturers, can be used as a further aid in selecting the most suitable line.

The number of bends must be kept to a minimum regardless of the type of hydraulic line used, the length of the line, or the radii of the bends. Every section of the line must be fastened securely to keep vibration to a minimum and to remove the weight from the fittings or joints.

When making your selection, remember that bends are preferable to elbows because they reduce friction and turbulence. The bend radius of the pipe, tube, or hose should not be less than 2½ to 3 times the inside diameter. When hoses are used instead of pipes or tubing, you must consider the applied pressure and the movement of the hose caused by the actuator as well as the reducing of bends. **NOTE** Pressure can cause the hose to expand and contract. You should also avoid a bend radius which is too sharp. It causes friction and this will increase turbulence.

Pipes Many manufacturers classify piping and tubing indiscriminately. But actually there are a number of differences between them. When ordering or selecting supplies, you should first check the individual manufacturer's terminology in order to avoid misinterpretation.

To the mechanic, the words *pipe* and *tubing* have many meanings. Distinctions lie not in the material or the design but in the dimensions, that is, the outside diameter (OD), the inside diameter (ID), and the wall thickness. Nominal dimensions are used to simplify the standardization of pipe fittings. The wall thickness is identified by a schedule number.

Although manufacturer classifications of pipe and tubing overlap, you should know when to use one as opposed to the other. Tubing is used much more frequently for hydraulic lines because of the difficulty in bending and sealing pipes. The sealing problem, however, has been improved as a result of the cooperation between industry and the Society of Automotive Engineers in setting up improved standards for pipe thread. The result has been the dry-seal pipe thread. Dry-seal threads form a tighter and safer seal because they can be tightened to make a leakproof joint without excessive torque.

Threading Pipe   The actual threading of pipe is not a common occurrence for a diesel mechanic, but if it should become necessary the following procedure should be followed:

1. Hold the pipe in a chain vise in order to prevent internal and external damage.
2. Remove any burrs from the pipe since they reduce flow and increase turbulence.
3. Use a sharp die and cutting oil to achieve a clean thread. Damage to the threads can result in leakage.
4. Cut the pipe to the appropriate length and, whenever possible, use a pipe cutter instead of a hacksaw. The pipe cutter will leave no residue which may enter the pipe and damage the hydraulic system.
5. Use pipe dope or pipe tape to improve the sealing and prevent corrosion and seizure. This is placed on the male fitting only, leaving the first two or three threads completely clean. If the sealing compound enters the system, it may cause serious damage.

Pipe Fittings   A great variety of pipe fittings (connectors) are manufactured. Most common are the threaded pipe fittings of the tapered and straight (dry-seal) designs (Figs. 21–1 and 21–2). (Flange-type and welded-type fittings are less common.) Standard fittings are adequate to accommodate an operating pressure up to 3000 psi [210.9 kg/cm²] and have a proof-test rating of 15,000 psi [1050.0 kg/cm²]. This makes the threaded design most suitable for the diesel engine.

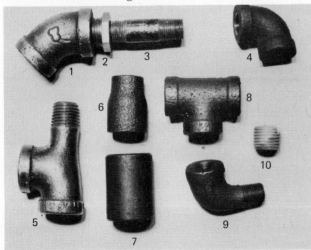

1. 45° elbow	5. Street tee	9. Street elbow
2. Bushing	6. Reducer	10. Slotted plug
3. Nipple	7. Coupling	
4. 90° elbow	8. Tee	

Fig. 21–1   Tapered pipe fittings.

All tapered pipe threads should be tightened to a minimum of 2½ full turns past hand-tight. The straight pipe threads, on the other hand, can be positioned and then sealed with a seal nut and O ring.

Tubing   Tubing made from steel, copper, nylon, polyethelene, or other synthetic materials are the most common hydraulic lines used in diesel engine application.

A tubular product is called a *tube* if its nominal size is the actual measurement of the outside dimension. Standard tubings vary in size from ⅛ to 3 in [3.18 to 76.2 mm]. When examining the tube-size designation chart in Table 21–1, note that the tube sizes increase in 1/16-in increments to ⅜ in and then in ⅛-in increments to ¾ in. Within each tube size there is a selection of wall thicknesses to fill the various pressure demands.

To identify a standard rigid tube size, a dash number system is used. This dash number system relates to the outside diameter (OD) of the tube size in 1/16 in.

Table 21–2 lists standard tube size increments and corresponding dash numbers.

Cutting Tubes   Tube cutting should never be considered so routine that it is done carelessly. Avoid, under all conditions bending a tube more than is necessary.

1. Hold the tube roll in an upright position on a clean and straight surface. The tube end is held firmly against the surface while the tube coil is rolled out to the desired length. Never pull the tube end from the tube roll. Remember to keep the open end closed off to avoid the entry of foreign material.
2. In cutting the tube, use a tube cutter to assure a clean, square cut on the edges. Do not force the cutting wheel into the tube. Instead, let the wheel cut into the tube by rotating the tube cutter 360° and gently increasing wheel pressure after each revolution.
3. While cutting, support the waste section of the tube so that its weight does not interfere with the rest of the cutting.

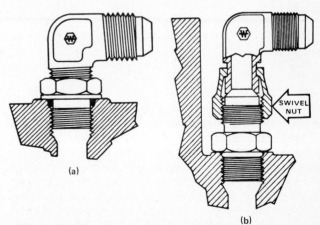

Fig. 21–2   Straight pipe fittings: (a) straight thread O-ring fitting, (b) swivel-end fitting used with connector. (*The Weatherhead Co. of Canada, Ltd.*)

Table 21-1 STANDARD SIZE HYDRAULIC TUBING

Tube OD, in*	Tube ID, in	Wall, in
1/8 0.125	0.055	0.035
	0.061	0.032
	0.065	0.030
	0.069	0.028
3/16 0.1875	0.117	0.035
	0.123	0.032
	0.127	0.030
1/4 0.250	0.120	0.065
	0.134	0.058
	0.152	0.049
	0.166	0.042
	0.180	0.035
	0.190	0.030
5/16 0.3125	0.182	0.065
	0.196	0.058
	0.214	0.049
	0.228	0.042
	0.242	0.035
	0.248	0.032
3/8 0.375	0.245	0.065
	0.259	0.058
	0.277	0.049
	0.291	0.042
	0.305	0.035
	0.311	0.032
1/2 0.500	0.310	0.095
	0.334	0.083
	0.358	0.072
	0.370	0.065
	0.384	0.058
	0.402	0.049
	0.416	0.042
	0.430	0.035
	0.436	0.032
5/8 0.625	0.435	0.095
	0.459	0.083
	0.481	0.072
	0.495	0.065
	0.509	0.058
	0.527	0.049
	0.541	0.042
	0.555	0.035
3/4 0.750	0.532	0.109
	0.560	0.095

* Refer to Conversion Tables for metric equivalents of English tube OD sizes.

4. Do not let your cut become too harsh because it will burr both the inside and outside of the tube. Should this occur, however, use a reamer blade to remove the inside burrs. (The reamer blade is mounted on the tube cutter and can be folded away when not in use.) The outside burrs can be removed with a file or an outside-burr remover. In all cases, try to achieve smooth, square edges. Above all, make sure that all debris from the cutting operation is removed from the tube.

Bending Tubes Steel or copper tubing which is small in size and has thin walls is easy to bend if a spring tube bender is used. After you have selected the correct spring size, slide the tubing over the bend tube and gently shape it to the desired bend. A slight overbending will facilitate removal.

Larger tube sizes and thicker walls are bent much more easily with a mechanical tube bender. Here also, select the proper size mechanical tube bender to avoid impairment of the tube. It is recommended that the bend be made prior to the flaring or cutting of the final length.

Pointers for Bending Tubes **1.** Mark the start of the bend on the tube. (For example, you might be bending a 1/2-in [12.7-mm] steel tube to a 90° angle. Note that if the bend must be located immediately after the flare or compression nut, the bend should be started a distance from the nut of two to three times the length of the nut. This ensures easy installation. **2.** Position the tube so that your mark coincides with the centerline mark of the tube bender. Then, hook the hold-down foot over the tube. **3.** Gently move the handles together until the desired angle is reached (for example, 90°). Always overbend a little. This is to compensate for the tube's tendency to straighten out slightly after removal.

Tube Fittings and Adapters A tube may be connected to a component, hose, pipe, or to another tube by one of several kinds of tube connectors. Quite often, a welding-type tube connector is used. The most common tube connectors, however, are those of the threaded-type design and these, therefore, are the ones you should become most familiar with.

In general, threaded tube fittings and adapters are divided into two types: the flare-type fitting and the compression-type fitting.

With the flare-type fittings and adapters, connection and sealing is achieved by spreading (flaring) the ends of the tube to an angle of 37 or 45°. The fitting nut and adapter, when properly torqued, will make a secure, leakproof seal.

Connection and sealing of the compression-type fittings and adapters is achieved by screwing the fitting nut to the adapter, thus compressing the sleeve or sleeve nut onto the tube.

The six threaded-type connectors are:

- SAE 37° JIC flare twin
- SAE 45° flare
- 45° inverted flare
- Compression fitting
- Self-aligning
- Threaded sleeve

Because of the many types of fittings and adapters, various thread sizes are used. A tube fitting thread-size comparison chart of the six threaded fittings (in each classification of tube size) is given in Table 21-3.

By examining Table 21-3 you will recognize the ambiguity arising from the countless number of fittings, adapters, and thread designs. For example, good sealing, usually thought of as the result of the tube being properly flared, installed, and torqued,

Table 21-2 TUBE SIZES AND DASH NUMBERS

Actual OD, in*	3/16	1/4	5/16	3/8	13/32	1/2	5/8	3/4	7/8	1	1 1/8	1 1/4
Standard dash nos.	-3	-4	-5	-6	-7	-8	-10	-12	-14	-16	-18	-20

* Refer to Conversion Tables for metric equivalents of English tube OD sizes.

Table 21-3 THREAD-SIZE COMPARISON CHART

Tube		SAE 37° JIC flare twin		SAE 45° flare		45° inverted flare		Compression self-align		Threaded sleeve	
Size, in*	Dash no.	Size, in	Threads per inch	Size, in	Thread per inch	Size, in	Threads per inch	Size, in	Threads per inch	Size, in	Threads per inch
1/8	-2	5/16	24	5/16	24	5/16	28	5/16	24	5/16	24
3/16	-3	3/8	24	3/8	24	3/8	24	3/8	24	3/8	24
1/4	-4	7/16	20	7/16	20	7/16	24	7/16	24	7/16	24
5/16	-5	1/2	20	1/2	20	1/2	20	1/2	24	1/2	20
3/8	-6	9/16	18	5/8	18	5/8	18	9/16	24	9/16	20
7/16	-7	...	...	11/16	16	11/16	18	5/8	24	5/8	18
1/2	-8	3/4	16	3/4	16	3/4	18	11/16	20	11/16	16
5/8	-10	7/18	14	7/8	14	7/8	18	13/16	18		
3/4	-12	1 1/16	12	1/16	14	1 1/16	16	1	18		
7/8	-14	1 3/16	12								
1	-16	1 5/16	12	...	...	...	...	1 1/4	18		

* Refer to Conversion Tables for metric equivalents of English tube OD sizes.

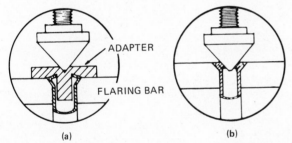

Fig. 21-3 (a) First step in making a double flare; (b) finishing the flare. (*The Weatherhead Co. of Canada, Ltd.*)

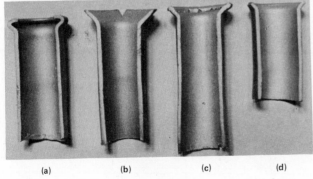

Fig. 21-4 Improper flares and their causes: (a) tubing cut at an angle, (b) improper flare length, (c) burrs not removed, (d) improper flare length.

actually depends on correct selection of components (thread design).

Inverted Flare Inverted flare fittings, although found on fuel and lubrication lines, are uncommon on diesel engines because the short nut is unable to support the hydraulic line. More suitable to diesel application are the SAE 45° and the SAE 37° JIC flare-twin fittings, which will be discussed later.

Forming Double Flare To form a good double flare, you must keep the tube straight. Also, the pointers for cutting tubes, outlined previously, should be strictly· adhered to. Further rules for forming a good flare are:

1. Select the proper adapter and hole size in your tube-holding fixture.
2. Slide the nut on the tube.
3. Insert the tube. Clamp it securely when the tube extends to the height of the adapter.
4. Lubricate the tube slightly and insert the adapter.
5. Locate the cone of the yoke over the center of the adapter and run it down until it bottoms against the tube-holding fixture. This action causes the tube to bell out or flare. It is the first step in forming a double flare (Fig. 21-3a).
6. Remove the adapter and run the cone tightly against the belled tube to form the finished flare (Fig. 21-3b).
7. Remove the tube and check the flare diameter. Also, examine the flare for excessive thinning out of material (see Fig. 21-4).

Correct Method of Installing Fittings Before you begin the installation, be sure that all foreign materials have been removed from the nut, the adapter, and the inside of the tube. Lubricate the threads before aligning the tube with the adapter. Excessive force should not be used in alignment; this can cause a cross-threaded fitting and stress on the tubing.

Hand-tighten the tube nut. Then using a flare wrench until a solid feeling is encountered, apply a sixth of a turn.

SAE 45° Flare A single SAE 45° flare (using a long nut) is shown in Fig. 21–5. Various adapters are shown in Fig. 21–6. For the diesel engine component it is advisable to use only long nuts and a double flare in order to decrease connection failure.

Flaring and installation procedures are the same as those suggested for a double flare fitting.

SAE 37° JIC Flare Twin Two- and three-piece SAE 37° JIC flare-twin fitting assemblies are shown in Fig. 21–7. Various adapters are shown in Fig. 21–8. From the illustrations you will recognize how these types of fittings give greater tube support and why they are recommended for diesel applications.

Flaring and installation procedures for the SAE 37° JIC flare twin are the same as those suggested for the inverted flare and the SAE 45° flare. The flaring angle for the JIC flare twin is 37° instead of 45°.

Flareless Connectors Threaded sleeve, self-aligning, and compression fittings are not recommended for diesel engines because the connectors cannot withstand vibration. By using a long nut on self-aligning and compression fittings, improved vibration resistance is achieved. A correctly assembled threaded sleeve is shown in Fig. 21–9.

NOTE When you install a tube with this type of fitting, the tube must rest on the adapter shoulder not over the tight nut. Also, after the nut is hand-tightened, an additional 1½ turns with a wrench is recommended.

A self-aligning fitting is shown schematically in Fig. 21–10. A compression fitting is shown in Fig. 21–11. The assembly and torque procedure is identical to that outlined above for a threaded sleeve.

Flexible Tubing Tubes made from nylon, neoprene, or other synthetic materials are used in industry today to connect fuel, oil, and lubrication lines to instrumentation panels. This type of tubing has gained wide acceptance because of its adaptability, durability, and high vibration resistance. Nylon tubing is preferred in diesel engines because of its resistance to deterioration by petroleum products.

Table 21–4 outlines the required data for proper selection of tube size, wall thickness, bend radius, working pressure, and fitting size.

The compression-type, self-aligning fitting is the least used. However, the synthetic grommet-type fitting shown in Fig. 21–12 is quite common. This fitting with flexible connections can be used with steel, copper, or synthetic tubing. The nut is bottomed against the adapter, thus making overtorquing of the nut impossible. Consequently, there is no distortion of the tube or restriction to the fluid flow. For these reasons the synthetic grommet-type fitting is highly recommended for diesel applications.

However, when using nylon or synthetic tubings it is advisable to insert a sleeve in order to maintain a good seal and avoid tube distortion (see Fig. 21–13).

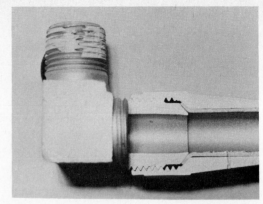
Fig. 21–5 Cutaway view of an SAE 45° flare fitting.

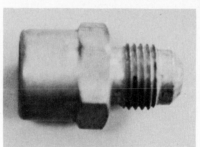

FEMALE CONNECTOR MALE CONNECTOR

MALE ELBOW UNION

Fig. 21–6 SAE 45° flare fitting adapters.

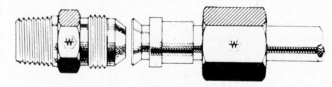
THREE-PIECE TYPE

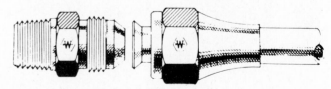
TWO-PIECE TYPE

Fig. 21–7 SAE 37° JIC flare-twin fitting assemblies. (The Weatherhead Co. of Canada, Ltd.)

This should be done regardless of the type of compression fitting used.

Cooling-System and Heater Hoses In this textbook, when reference is made to hydraulic hoses,

Table 21-4 FLEXIBLE NYLON TUBING

For fittings dash no.	Size	OD, in*	ID, in	Wall thickness, in	Max. working pressure, psi at 70°F	Min. burst pressure, psi at 70°F	Min. bend radius, in
-2	1/8	1/8	0.096	0.015	333	1000	3/4
-3	3/16	3/16	0.138	0.025	333	1000	1 1/2
-4	1/4	1/4	0.190	0.030	333	1000	2 1/2
-5	5/16	5/16	0.242	0.035	333	1000	3
-6	3/8	3/8	0.295	0.040	333	1000	3 1/2
-8	1/2	1/2	0.375	0.062	333	1000	4

* Refer to Conversion Tables for metric conversions.

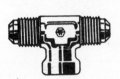
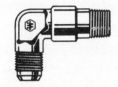

FEMALE BRANCH TEE   MALE ELBOW (LONG)

90° FEMALE ELBOW   BULKHEAD UNION

Fig. 21-8 Hydraulic tube fitting adapters. (The Weatherhead Co. of Canada, Ltd.)

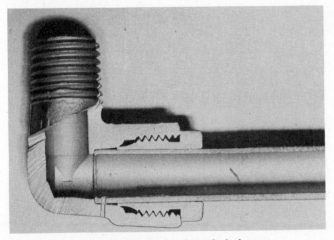

Fig. 21-9 Cutaway view of a threaded sleeve.

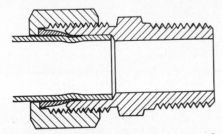

Fig. 21-10 Schematic view of a self-aligning fitting. (The Weatherhead Co. of Canada, Ltd.)

they are limited to those used in cooling systems or to heater hose, low-pressure and medium-pressure hoses only.

Cooling-system hoses and heater hoses differ in design, construction, and composition. Cooling-

Fig. 21-11 Cutaway view of a compression fitting.

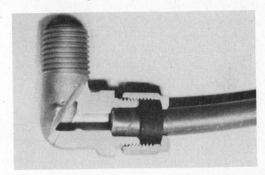

Fig. 21-12 Cutaway view of a synthetic grommet-type compression fitting.

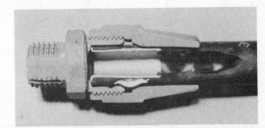

Fig. 21-13 Cutaway view of a compression fitting using nylon tubing with an insert.

system hoses, heater hoses, and all standard hose styles use a dash number system to identify the inside (ID) dimension.

Today's diesel engine cooling systems operate at a maximum pressure of 26 psi [1.82 kg/cm²]. The burst pressure of a 2-in [50.8-mm] cooling-system hose is about 70 psi [4.92 kg/cm²]. The hose must be flexible to a temperature of −40°F [−44°C] and must not expand more than 2 percent at 212°F [100°C].

The inner hose is usually made from buna, Teflon, or silicon because these materials are oil-resistant and do not swell or peel. They are reinforced with a single or double fiber or coarse steel braid. The outer cover is usually neoprene, silicon, or synthetic rubber. Some radiator hoses, especially on the inlet

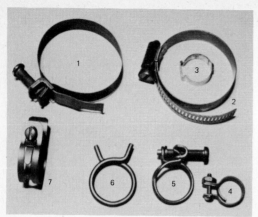

Fig. 21-14 Hose clamps.

1. Type D   3. Crimp type   5. Type A   7. Type A
2. Type F   4. Type B   6. Type E

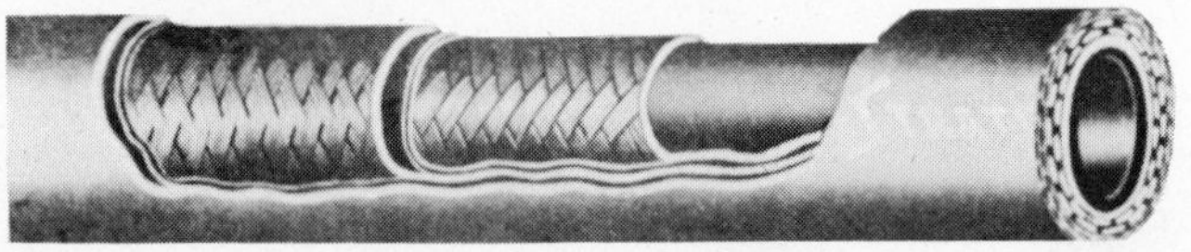

Fig. 21-15 Stratoflex hose construction. This hose is made up of a seamless synthetic rubber inner tube reinforced with one braid of fabric cord and one braid of stainless steel wire, and covered with a synthetic rubber cover. (*Stratoflex of Canada, Inc.*)

Part no.	Hose size*	Hose ID, in	Hose OD, in	Min. burst pressure, psi	Max. working pressure, psi	Min. bend radius, in
205-4-L	−4	3/16	33/64	8000	2000	2
205-5-L	−5	1/4	37/64	8000	2000	2 1/4
205-6-L	−6	5/16	43/64	6000	1500	2 3/4
205-8-L	−8	13/32	49/64	6000	1500	4 5/8
205-10-L	−10	1/2	59/64	6000	1500	5 1/2
205-12-L	−12	5/8	1 5/64	5000	1250	6 1/2
205-16-L	−16	7/8	1 15/64	2500	600	7 3/8

*Dash number.

Fig. 21-16 Weatherhead hose construction. This hose is made up of a Buna-N inner tube reinforced with a two-fiber braid and covered with neoprene. (*The Weatherhead Co. of Canada, Ltd.*)

Catalog no.	Hose ID, in	Hose OD, in	Max. working pressure, psi	Min. burst pressure, psi	Min. bend radius, in
H-17-1/4	1/4	9/16	1250	5000	3
H-17-3/8	3/8	3/4	1125	4500	4
H-17-1/2	1/2	15/16	1000	4000	5
H-17-3/4	3/4	1 1/4	750	3000	6
H-17-1	1	1 1/2	560	2250	8

Fig. 21-17 Aeroquip hose construction. This hose is made up of a synthetic rubber inner tube, cotton inner braid, single-wire braid reinforcement, and abrasion-resistant rubber cover.

Part no.	Hose ID, in	Hose OD, in	Recom. working pressure, psi	Min. burst pressure, psi	Min. bend radius, in	Vacuum, inHg
2651-4	0.188	0.516	3000	12,000	3.00	28
2651-5	0.250	0.578	3000	10,000	3.38	28
2651-6	0.313	0.672	2250	9000	4.00	28
2651-8	0.406	0.766	2000	8000	4.62	28
2651-10	0.500	0.922	1750	7000	5.50	28
2651-12	0.625	1.078	1500	6000	6.50	28
2651-16	0.875	1.234	800	3200	7.38	20

side of the water pump, are susceptible to collapse. To compensate for this, spiral steel wires are molded into the hose or inserted over the inner cover, giving the hose more uniform strength. Radiator hoses for diesel uses are usually either molded or flexible.

Hose Clamps Six types of hose clamps types are shown in Fig. 21–14. The clamp size is indicated by a numbering system which can range from size 6 to 195. This means that there is a clamping diameter range from 0.038 to 6 in [0.95 to 152.4 mm].

The type E clamp is more suitable for small hoses, whereas the type F clamp is suitable for all hoses.

Low- and Medium-Pressure Hoses Hose styles and sizes are as numerous and varied as those of pipes and tubings. To distinguish a hydraulic hose by quality (style) and size, a lettering and numbering system is used. No doubt you are already familiar with the names Weatherhead, Aeroquip, and Stratoflex. All three companies use a numbering system. The letter stamped on the hose identifies the manufacturer and the first digit describes the basic hose style. Naturally, the numbers vary from company to company.

To accommodate the many hydraulic applications, manufacturers have included in their product (hose style) four or five groups classified according to working pressure. Each individual group includes different hose compositions to suit different types of liquids. The hose group styles are:

- *Group 1* low pressure with a maximum working pressure up to 400 psi [28.16 kg/cm²]
- *Group 2* medium working pressure up to 2000 psi [140.8 kg/cm²]
- *Group 3* high working pressure up to 3000 psi [211.2 kg/cm²]
- *Group 4* very high working pressure up to 5000 psi [352.0 kg/cm²]

Some companies have special groups for special products.

The next number stencilled on the side of the hose is the dash number. This dash number system is used on both tubing and hoses. However, wherever the dash number indicates the inside diameter of a hose, the dash number of the hose is matched with the number of the tubing in order to determine the proper sizes when connecting a hose with a tube. Figures 21–15 to 21–17 show three hoses from Group 2 hose style. When examining the hoses and charts you will notice that the following four items differ with each manufacturer: (1) the material and construction, (2) the outside diameter of the hose size, (3) the working and burst pressure, and (4) the bend radius. (This change correlates with the variation in pressure.) Note that the above variations are also encountered in the other three hose styles.

Hose Ends Because of the great variety in hose styles and connections, a large assortment of hose ends are needed. Six examples of hose ends are shown in Fig. 21–18. The hose end has only one purpose—to connect a hose (troublefree) to an adapter or component. Each hose style uses the same

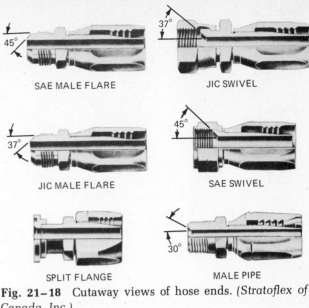

SAE MALE FLARE JIC SWIVEL JIC MALE FLARE SAE SWIVEL SPLIT FLANGE MALE PIPE

Fig. 21–18 Cutaway views of hose ends. (*Stratoflex of Canada, Inc.*)

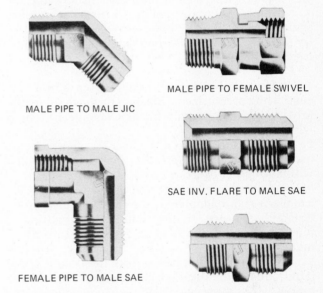

MALE PIPE TO MALE JIC MALE PIPE TO FEMALE SWIVEL SAE INV. FLARE TO MALE SAE FEMALE PIPE TO MALE SAE MALE SAE TO MALE SAE

Fig. 21–19 Cutaway views of hose-end adapters. (*Stratoflex of Canada, Inc.*)

socket; the nipple is the only variable factor in connecting with the different types of adapters.

To classify hose ends, all manufacturers use a numbering system. As with most reference systems, the numbering will vary from manufacturer to manufacturer.

Adapters With the great assortment of hose ends an even greater number of adapters are needed (see Fig. 21–19). Here also, to facilitate identification, a numbering system is used. For the adapters illustrated the manufacturer is Stratoflex (SF). The "male pipe to male JIC" in Fig. 21–19 is a 45° elbow. The thread and flare design is JIC on one side and pipe thread on the other side. The thread size is ½ in [12.7 mm] and the hose size is ½ in [12.7 mm].

Hose Service When hoses are installed and assembled correctly, they seldom break down or require servicing. Any problems which do arise in

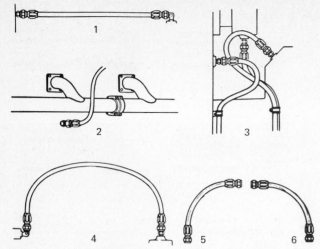

Fig. 21–20 Improper hose installations. (*Sperry Rand Canada, Ltd., Vickers Div.*)

this area will be due to improper assembly of hose ends, incorrect length, or improper installation.

A number of improper hose installations are shown in Fig. 21–20. Before you read any further, try to figure out what's wrong with each installation. Here are the answers:

1. A hose will change in length (from 2 to 4 percent) when pressurized. Slack in the hose should be provided to compensate for any change in length which may occur.
2. Hose or tubing passing close to a hot exhaust manifold will deteriorate. It should be protected with a fireproof baffle.
3. For neater appearance and easier maintenance, use elbows and adapters if the flow volume is low. Remember, however, that where flow volume is high, the use of elbows should be kept to a minimum.
4. Any vibration or external force here could cause failure due to the insufficient support of the hose or tubing. If rubber-coated clamps had been used, external damage and excessive vibration would have been prevented.
5. The bend radius on thise hose does not allow sufficient flow volume, so pressure is reduced. Any high pressure here might cause line failure. To assure maximum connection security, the bend radius should be increased.
6. A twisted hose will reduce flow volume. Also the socket will loosen under high pressure.

Servicing Hydraulic Hoses The fitting of a hose to a hose end need not be difficult if the correct assembly procedure is used. However, because of the many varieties of hose sizes, fittings, and styles, small but sometimes very costly mistakes can be made. Another problem that creates hose failure is the lack of uniformity among manufacturers in the numbering system for the outside dimension. Therefore, when assembling hoses and hose ends, it is essential to make all purchases from the same manufacturer.

Perhaps the next greatest factor contributing hose

failure is the mechanic's disregard for the recommended assembly procedure. The following points may seem obvious, but they are the key if accuracy and perfection are desired.

Assembly Procedure The correct assembly procedures for the four most popular hose styles follow:

SOCKETLESS
1. Cut the hose to the required length.
2. Oil the inside of the hose and the outside of the nipple.
3. Push the hose on the fitting underneath the protective cap. (If available, the Aeroquip socketless fitting assembly machine may be used.)

MEDIUM PRESSURE
1. Cut the hose square to the length required with a fine-tooth hacksaw or a cutoff wheel.
2. Put the socket in the vise and screw the hose counterclockwise into the socket until it bottoms, then back off one-quarter turn. **NOTE** An assembly tool is required for all single wire-braid hoses in sizes of ¼ to ¾ in [6.35 to 19.05 mm].
3. Liberally oil the nipple threads, the assembly tool mandrel, and the inside of the hose.
4. *Male ends:* Push the assembly tool into the nipple. Oil the nipple, and then screw it clockwise into the socket and hose. Tighten the nipple until it is snug against the socket.
5. *Swivel ends:* Tighten the nipple and the nut on the assembly tool. Oil the nipple and then screw it clockwise into the socket and hose. Leave ¹/₃₂ to ¹/₁₆ in [0.79 to 1.59 mm] clearance between the nut and the socket so that the nut will swivel.

HIGH PRESSURE
1. Cut the hose to the length required.
2. A double wire-braid hose must be stripped of its rubber cover before inserting it in the socket. To locate the stripping point, lay the hose end next to the high-pressure fitting. The hose end should align with the notch on the socket. To strip the hose, cut around down to the wire braid, slit it lengthwise, raise the flap and pull it off with pliers. Then clean the wire braid with a wire brush or soft wire wheel, being careful not to fray or flare it.
3. Put the socket in the vise and screw the hose into the socket counterclockwise until it bottoms.
4. Oil the nipple threads and the inside of the hose liberally. (No assembly mandrel is needed for a double wire-braid hose.) Use grease instead of oil for larger sizes.
5. Screw the nipple into the socket and hose. Leave ¹/₃₂ to ¹/₁₆ in [0.79 to 1.59 mm] clearance for take-up.

TEFLON
1. Cut the hose to the correct length. Trim any loose wires flush with the tube stock. Remove any burrs on the tube end with a knife.
2. Slip the two sockets back to back over the neckdown end of the hose, positioning them 3 in [76.2 mm] from each end.

3. Mount the nipple hex in a vise. Work the hose bore over the nipple to size the tube and aid in separating the braid prior to fitting the sleeve. Then remove the hose from the nipple.

4. Start the sleeve over the tube and under the wire braid by hand. Complete the positioning of the sleeve by pushing the hose squarely against a hard surface. Make sure that the tube butts against the shoulder of the sleeve.

5. Lubricate the nipple and socket threads. For stainless steel fittings, use carbon tetrachloride; for other material combinations use standard petroleum lubricants.

6. Push the hose over the nipple with a twisting motion until it is seated against the nipple chamber. Push the socket forward, and hand-start the threading of the socket to the nipple. Wrench-tighten the nipple hex until the clearance with the socket hex is $1/32$ in [0.79 mm] or less. Then tighten a little further in order to align the corners of the nipple and socket hexes.

7. To disassemble, unscrew and remove the nipple. Slide the socket back on the hose by tapping against a flat surface, then remove the sleeve with pliers.

Questions

1. List the materials of which pipes and tubes may be made.

2. List three different ways in which pipes may be constructed.

3. Explain how to cut a ¾-in pipe to a given length and cut a thread onto that pipe.

4. When pipes are assembled, why is a compound or a pipe tape used?

5. When is a double-flare tube required?

6. When double or single flaring a ½-in tube (either SAE 45° or JIC 37°), why is it important that the tube be cut square and the inside reamed?

7. Why is it important not to overtorque a fitting?

8. If a hose replacement has a smaller inside diameter than that of the original it would (a) increase velocity, (b) increase the volume, or (c) decrease the volume?

9. What is the difference between the identification on a hose fitting to be used (a) on a medium-pressure hose, or (b) on a high-pressure hose?

10. Why is it necessary to cut a hose very close to the vise and use a hacksaw blade which has 32 teeth?

11. Why is it advisable during assembly to hold the fitting socket, rather than the hose, in the vise?

12. Why is it important to have the correct hose length between fittings?

UNIT 22

Filters

Need for Filters The most common cause of trouble in the engine is contamination due to the presence of foreign matter. Therefore, the most sensible way to maintain and preserve a high-performance hydraulic system is simply to keep the system clean. Many different types of filters are made in order to accommodate the various types of hydraulic systems. Filters can be a life line because they remove contamination and thus protect the system. The manufacturers originally install the filters, strainers, and breathers. They also provide service manuals with precise instructions in order to ensure troublefree operation of the components. Foreign matter usually enters the system through careless or inadequate maintenance or through the normal wear of the components in the system itself.

Filter-Element Materials There are three classes of filter materials: mechanical, inactive absorbent, and active absorbent. The mechanical filter consists of closely woven metal screens or metal disks. It generally removes only fairly coarse insoluble particles.

Inactive absorbent filters are composed of materials such as cotton, yarn, cloth, impregnated cellulose paper, or porous metal. They will remove quite small particles and some types even remove water and water-soluble contaminants. The elements often are treated to give them an affinity to the contamination found in the system.

Active absorbent filter materials, such as charcoal or Fuller's earth, remove particles by absorption as well as by filtering. They are not used as filter material for the lubrication or fuel-injection systems. They are, however, used as a filter material for the exhaust system.

Micron The unit of measurement for determining the effectiveness of a filter is the micron (μm). One micron is equal to 0.000039 inch (1 μm = 0.000039 in), or in metric units of measurement, 0.000001 meter (1 μm = 0.000001 m). When new and clean, a filter will prevent a specific percentage of particles measuring a specific minimum size from entering the fluid.

Filter Types Four basic types of filters are used: screen and strainer, surface type, deep type, and edge type. The type, size, and micron rating is dictated by the system itself.

SCREENS AND STRAINERS Screens or strainers are surface-type filters and sometimes are referred to as mechanical filters (Fig. 22–1). These inlet screens prevent large foreign particles from entering the system. They are classified according to a sieve number which relates to the micron rating.

SURFACE-TYPE FILTER The surface-type filter element shown in Fig. 22–2a is composed of a specially treated micronic cellulose paper. The paper is formed in vertical convolutions (wrinkles) and is made in a cylindrical pattern. It is reinforced on the inside and outside and is equipped with a seal on the top and bottom. Sometimes multifolded paper elements are used.

DEEP-TYPE FILTER The design of the deep-type filter is quite different from that of the surface filter (see Fig. 22-2b). The deep-type filter is more efficient and has a longer service life. Figure 22-3 shows two types of filter materials and illustrates the substantial depth of the filter material.

Deep-type filters of porous materials consist of fine woven copper or cinder bronze elements pressed to fit the filter housing (Fig. 22-4). They can also be made of minute stainless steel balls joined as one inflexible piece.

EDGE-TYPE FILTER Edge-type filters are used as primary filters for the lubrication and fuel-injection systems. In this design, many copper, bronze, paper, or steel disks are positioned over a tube. The tube, acting as an oil line, directs oil to the outlet port. Some edge-type filters have the added convenience of an automatic scraper or a hand-operated scraper for cleaning the outside of the disk (Fig. 22–5). This, of course, helps to extend the life of the filter.

Liquid Flow through Filter Regardless of the design or the type of a filter element, a filter element is sealed in a housing. O rings or gaskets separate the filtered from the unfiltered liquid. Liquid enters near the top of the filter housing or, with a screw-on-type oil filter, it enters via the adapter plate, and flows into the outer area of the filter element (Fig. 22–6). The system pressure forces the liquid through the filter element into the center area. Filtered liquid then passes through the center and on to the outlet port. Some filter designs incorporate an antidrain check valve to prevent fluid from draining from the filter bowl when the engine is stopped.

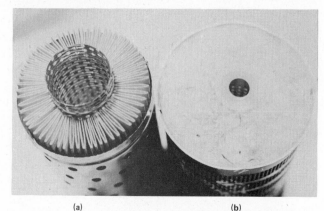

Fig. 22–2 *(a)* Surface-type filter; *(b)* deep-type filter (cotton).

V-TYPE PAPER ELEMENT   WOVEN COTTON ELEMENT

Fig. 22–3 Deep-type filter materials.

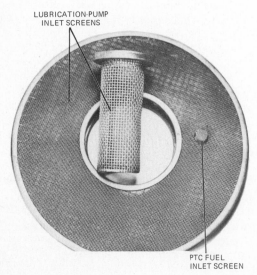

LUBRICATION-PUMP INLET SCREENS

PTC FUEL INLET SCREEN

Fig. 22–1 Inlet screens.

WOVEN STEEL ELEMENT   CINDER BRONZE ELEMENT

Fig. 22–4 Deep-type filter materials (porous).

Fig. 22–5 Edge-type filter with hand-operated scraper.

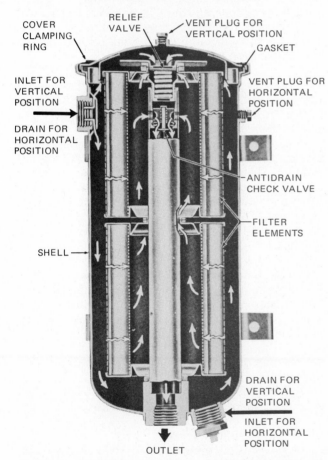

Fig. 22–6 Liquid flow through an oil filter. *(Mack Trucks Canada, Ltd.)*

This ensures instant oil or fuel pressure when restarting.

Full-flow oil filters have either a built-in bypass valve or a bypass valve which is mounted separately. The purpose of the bypass valve is to bypass oil in the event that the filter element becomes plugged, ensuring lubricant to the system. Relief valves are also used in oil coolers for the same purpose.

Full-Flow and Bypass Filters Both full-flow and bypass filter arrangements are used on today's diesel engines. Full-flow filters are connected directly to the pump outlet. All fluid must pass through the filter element before entering the system (Fig. 22–7). In the bypass filter arrangement, the filter is connected to the oil galley or cooling system and only a portion of the liquid flow passes through the filter element. This filtered liquid then returns to the crankcase or radiator and is recycled. Eventually, all the fluid finds its way through the filter.

Servicing Oil Filter
1. In order to replace a filter element, remove the drain plug and drain the filter housing (see Fig. 22–6).
2. Remove the cover and filter element.
3. Clean and inspect the bowl and cover.
4. Clean and examine the relief valve. Make sure that the return outlet holes and orifices are free of any foreign materials.
5. Replace the drain plug and install a new filter element.
6. Fill the filter. Use the correct amount, type, and grade of lube oil, fuel oil, or coolant.
7. Place a new gasket on the cover. Make sure that the cover is positioned correctly.
8. Check the oil level. Start the engine and run it at a low idle.
9. Loosen the vent plug as soon as the oil appears.
10. Run the engine at least 4 to 5 minutes to verify that there are no oil leaks.
11. Stop the engine and check the oil level. Add more oil if it is required.

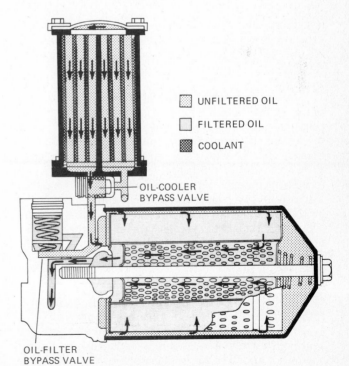

Fig. 22–7 Full-flow filter with oil cooler. Note the bypass valves for the oil cooler and filter. *(J. I. Case Co. Components Div.)*

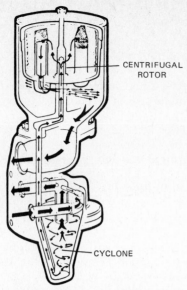

Fig. 22-8 Centrifugal bypass oil filter. (The arrows show the oil flow.) *(Mack Trucks Canada, Ltd.)*

Servicing Coolant Filter The recommended procedure for changing a coolant-filter element is similar to that for the lube oil and fuel filters. In addition you must:

1. Close off the inlet or outlet valves to prevent coolant losses. If there is no shutoff valve the radiator must be drained.
2. Clean or replace the corrosion-resistant plate if it has become thin or pliable.
3. With the engine running, check (by means of the condition indicator) the coolant flow and color. Also check for the presence of air locks.

Servicing Fuel-Filter Elements The recommended procedure for changing a fuel-filter element is similar to that for changing a lube-oil filter. The exceptions are:

1. Any air present in the filter housing must be removed by using the hand primer, gravity, or a low-pressure source to create a fuel flow.
2. When an edge-type element is used, the filter element must be washed in solvent and blown with shop air from the inside to the outside to remove any foreign particles. The procedure must be repeated, if necessary, to ensure that the element is clean.
3. Never blow air from the outside to the inside. This may damage the element by forcing foreign material into the metal shell or disk.

Centrifugal Bypass Filters A centrifugal bypass oil cleaner is shown in Fig. 22-8. The cleaner consists of a cyclone and centrifugal cleaner housing and rotor. The oil is forced under pressure into the cyclone where heavier impurities are thrown out against the walls and then driven downward by the current of oil. The oil in the center of the cyclone is then allowed to bypass the rotor. The rest of the oil is then forced through the passage into the centrifugal rotor.

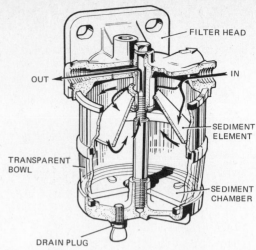

Fig. 22-9 Diffuser-type fuel filter. (The arrows show the oil flow.) *(CAV Ltd.)*

The reaction of oil squirting out of the two nozzles causes the rotor to revolve at high speed. The centrifugal force separates foreign particles from the oil and deposits them on the wall of the rotor. The clean oil flows through an internal oil line back to the crankcase.

Diffuser-Type Filter and Its Operation The diffuser-type fuel filter is a full-flow filter (Fig. 22-9). It has no moving parts; however, this simplicity does not hinder its effectiveness.

Fuel entering through the inlet connections flows over and around the conical section, which acts as a diffuser. The fuel flow passes through a narrow gap around the periphery. It then flows radially toward the center and out via the outlet connection. During this period of radial flow, water and the heavier abrasive particles separate out by gravity and settle into the collecting bowl.

Servicing Centrifugal and Diffuser-Type Filters It is simple to service the centrifugal or the diffuser-type filter. A thorough cleaning and washing will restore its serviceability. Some horse-sense rules for working with these filters are as follows:

1. Change or clean strainers, filters, and breathers regularly.
2. When you change filters or strainers, also clean the filter housing and bowl thoroughly.
3. Check for corrosion.
4. Always use new seals.
5. Repair faulty components as soon as possible.
6. Make certain that repair or exchange components are clean before installation.
7. Maintain proper oil and coolant levels.
8. Drain fuel tank and fuel-filter traps weekly, or daily if necessary.

Questions

1. How and when can the mechanic prevent foreign matter from entering the system?

2. Why have strainers or screens been installed in the pump-inlet side?

3. Explain how a full-flow filter operates.

4. Why is a bypass valve necessary with a full-flow filter?

5. Why is the oil flow in all filters from the outside to the inside?

6. There are four basic types of filters. Briefly outline the difference in the construction of these three elements.

UNIT 23

Seals and Gaskets

Purpose of Seals and Gaskets Many different types of seals and gaskets are used in hydraulic systems for confining water, lube oil or fuel oil. Various seals have also been designed to prevent the entry of unwanted air, dust, dirt, foreign material, and the buildup of a vacuum. Manufacturers have made great efforts to combat such sealing problems as high and low temperatures, expansion and contraction, vibration, pressure or vacuum, and corrosion and oxidation. However, even today with the great variety of seals available to us, it is difficult to find a seal which will be 100 percent effective under all operating conditions.

Engine manufacturers specify the type of seal to be used on their equipment and their instructions must be followed wherever possible. If the specified seals or gaskets are not available, careful consideration should be given to the selection of a suitable substitute. Remember that inadequate sealing reduces the life and efficiency of the components and results in higher costs due to the downtime.

Although seals and gaskets do sometimes cease to function properly, they should not automatically be considered the source of the trouble. More often the problem has been caused by the mechanic's inadequate preparation, improper alignment installation, or incorrect assembly and torque.

Classification of Seals There are two seal classifications: static seals and dynamic seals.

A seal which is used between two stationary components is called a *static seal*. If the seal is also capable of sealing two components which move in relation to each other, it is called a *dynamic seal*.

Mechanical Seals Mechanical seals are found in fuel-injection systems, hydraulic systems, and pumps. The most efficient seals are metal to metal, metal to carbon, and metal to synthetic material. The two lapped mating surfaces are forced together by a hold-down device or by a spring force to form the sealing surface (Fig. 23–1).

O-Ring Seals The construction material of today's O-ring seals is such that they can be used as either static or dynamic seals. O rings of a shape similar to that shown in Fig. 23–2 are made of synthetic rub-

Fig. 23–1 Mechanical seal in a water pump (partial cutaway view).

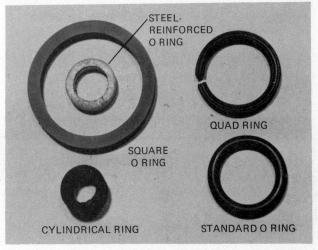

Fig. 23–2 Rings made of synthetic rubber.

ber. They are molded to close tolerances in the cross-sectional areas and to the inside and outside diameters.

To identify the various synthetic materials from which O-ring seals are made, a lettering system is used. The color has no bearing on the proper selection. When a replacement is necessary, use the manufacturer's replacement seals. If these are not

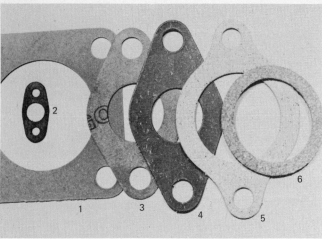

1. Paper
2. Synthetic rubber
3. Asbestos and paper combination
4. Chloroprene and asbestos
5. Asbestos
6. Cork
7. Multilayer of asbestos and aluminum
8. Steel
9. Fiber

Fig. 23–3 Gaskets of different materials.

available, review the requirements carefully before selecting from another source.

When an O ring is applied statically or dynamically, it is fitted into a right angular or curved groove which has been machined into the components to be sealed. The mating surface compresses the O ring about 10 percent. When it is necessary to improve sealing because of higher pressure, extrusion rings (backup washers) are used. The extrusion ring acts as a dirt wiper, permits a wider clearance between the sealed parts, and forms a seal across the clearance gap.

Removing and Installing O Rings When you remove an O ring, you must be careful not to damage the grooves or corners. A suitable O-ring removing tool should be used. Check the O-ring groove and mating surface for nicks and burrs and make certain that the replacement seal is of the recommended type and size. Check the O ring by rolling it inside-out. Stretching it lightly will help to reveal any small cracks or other irregularities. If it is damaged in any way at all, the O ring must be discarded.

During installation, make certain that the O ring is well lubricated and that it is not twisted. Make sure that it is properly inserted into the groove. When you install a backup washer, remember to insert it toward the lower pressure side and lubricate it well.

Gaskets The most common static seal is the gasket. Gaskets are designed to suit particular needs and, therefore, are manufactured in many different materials. When you select a gasket material, you must consider pressure, temperature, hole size, flexibility, smoothness, and the material of the mating surface. You also must consider the function of the gasket, such as whether or not it must seal off oil, water, dirt, and/or air, or whether it is needed to support a vacuum.

Gasket Material Copper, aluminum, steel, cork, fiber, asbestos, synthetic rubber, paper, and various combinations of these materials are used to make gaskets (Fig. 23–3). Copper, aluminum, steel, or fibers frequently are used as gasket material under bolts and adjusting caps or between components. Copper gaskets can be reused in an emergency by annealing them. (*Annealing* means softening.) They can be heated until the material changes color slightly and then cooled in water. This process makes copper more pliable.

CORK The physical properties of cork make it the most versatile of the gasket materials (see 6 in Fig. 23–3). It is a suitable seal where vibration, expansion, or uneven surfaces occur. The cork gasket is improved by coating its surface with synthetic rubber or with a combination of synthetic rubber and cork. When using cork gaskets, however, frequent retorquing is necessary in order to maintain a leak-free seal.

SYNTHETIC RUBBER Although wide ranges of synthetic rubber gaskets are available, they are seldom used on diesel engines (2 in Fig. 23–3).

PAPER (CHLOROPRENE) Chloroprene is widely used as a material for making gaskets. It is water- and oil-resistant, varies in strength and softness, and is adaptable to low and high temperatures and to high pressure. To increase chloroprene's ability to withstand temperature changes and high pressure, rubber, asbestos, or other fibers are added to the original chloroprene composition (1, 3, and 4 in Fig. 23–3).

Gasket Construction On the whole, the majority of gaskets have a simple construction. This means that a single layer of gasket material is cut to fit the opening and surface contour to be sealed, without reinforcement.

Where heavy stress may be encountered the common gasket cannot be used. Under these circumstances, the gaskets must be of a more sophisticated composition.

1. Synthetic O rings to seal oil and coolant passages
2. Detroit Diesel
3. Reinforced multilayer: aluminum and asbestos
4. Reinforced multilayer: copper, asbestos, and steel
5. Reinforced multilayer: copper and asbestos

Fig. 23–4 Head gaskets.

Head Gaskets Cylinder-head gaskets are the most complicated in design and construction because they must withstand extreme pressure, high temperature, and expansion changes. They must seal against compression and vibration, against gases, oil, and coolant. They must resist extrusion, elongation, and oxidation chemicals. A number of head-gasket designs are shown in Fig. 23–4.

A head gasket consists of a multilayer of materials. The two outer gasket layers may be of similar materials or of different types of alloys. The material used for the center layer is asbestos or an asbestos composite. Some head gaskets include two outer layers of asbestos with a center layer of alloy. Some are of the single- or double-sheet design. Regardless of which design is used, the openings for the cooling passages, oil passages, and the piston sleeves are reinforced. Grommets are used to increase sealing, to assure an unrestricted flow of oil or coolant fluid, and to strengthen the head gasket. Some head gaskets include a copper or a special rubber O ring around the lube oil and coolant openings. A special alloy ring may be located above the cylinder sleeve to improve sealing.

Since meticulous care must be taken when installing a head gasket, this subject was covered separately in Unit 14, "Cylinder Head and Valves."

Rules to Prevent Sealing Failure
1. Whenever possible avoid reusing a seal or gasket. Their elasticity deteriorates and small nicks and cracks which are invisible to the naked eye reduce the sealing capacity.
2. Make certain that the mating surfaces are straight and that the grooves or shafts are free of any foreign material before reassembling. Inspect for cracks, nicks, burrs, or dents.
3. Always check the seal or gasket for proper fit. The opening which holds the seal should not be enlarged and the openings of the gasket must not restrict the flow.

4. Before you install cork or paper gaskets, which may have shrunk due to long or improper storage, soak them for a few minutes in lukewarm water. Since wet cork or paper gaskets have a tendency to enlarge, put the wet gasket in a warm place before use. However, remember that temperatures over 100°F [37.8°C] will cause the gasket to curl.
5. Do not remove seals or gaskets from their packing kits until you are ready to use them. Inspect fittings for possible entry of foreign material before installation.
6. Follow installation instructions carefully.
7. Torque the bolts or nuts in the proper sequence, taking care not to overtorque.

Sealing Compound At one time or another you may have used a black sealing-liquid compound to hold a gasket in place or to repair an old gasket. Alternatively, you may have used this compound to seal a hose connection or to improve a gasket seal which had an oxidized surface or an uneven surface. You might recall that this sealing compound was not always satisfactory. Its ineffectiveness may have been due to your selection of sealing compound. Applying the correct sealing compound is a very important factor. It is obvious then that you should be thoroughly familiar with all the different types of sealers, their application, and their capacity to withstand temperature, or pressure, or to resist liquids. The chart in Fig. 23–5 analyzes some common sealants.

Liquid sealants should be used sparingly. (Too thick a coat will plug small oil, water, or air passages.) Always follow the manufacturer's directions to assure a good seal.

Antiseizure Compounds As its name indicates, the function of an antiseizure compound is to prevent seizure of bolts, nuts, or shafts. It also prevents corrosion, oxidation, galling, and chemical reaction between two elements and reduces friction. This compound should be applied to exhaust-manifold bolts and to pipes where a corrosive condition or chemical reaction is likely to occur. Where precise torquing is important, a light coat should be applied on the bolts and nuts to reduce friction.

Lock-Type Compounds A lock-type liquid compound is used to prevent bolts, nuts, seals, bearings, cones, and other components from becoming loose or moving out of position. This compound fills the space between the components and, in principle, makes them inflexible. It should, of course, only be used in cases of emergency. Once it has been applied and allowed to set, this compound often makes it difficult to remove the components without damage. If it does become necessary to loosen the grip of the compound, use moderate heat. Fuel or lubricant will also help dissolve this compound.

Making a Gasket Sometimes it becomes necessary to make your own gasket. If a gasket made from cork, polyacrylic, or asbestos is required, the procedure is fairly simple. Two different methods of making gaskets are available to you.

Product	Method of application	Temperature range and pressure	Uses	Resists	Drys/sets/ solvent
Permatex High Tack Spray-A-Gasket^R Adhesive-Sealant No. 99	Aerosol spray can	−65 to +500°F [−53.8 to +260°C] 5000 psi [351.5 kg/cm²]	All engine gaskets, transmission and rear-end housing gaskets. All thread-ed connections, radiator and heater hose connections, antislip agent for fan belts, adhesive for general use. Tubeless-tire beads, battery terminal protector and wire waterproofer. Can be used on any type of gasket, felt, cork, metal, paper, asbestos, etc.	Gasoline, oil, kerosene, lube oils, water, steam, antifreeze solutions	Sets up fast . . . super tacky . . . yet allows gasket movement. Under engine operating temperatures, it converts to super sealant Lacquer thinner
Permatex High Tack Brushable Adhesive-Sealant No. 98	Brush in can	−65 to +500°F 5000 psi	All engine gaskets, transmission and rear-end housing gaskets. All thread-ed connections, radiator and heater hose connections, antislip agent for fan belts, adhesive for general use. Tubeless-tire beads, battery terminal protector and wire waterproofer. Can be used on any type of gasket, felt, cork, metal, paper, asbestos, etc.	Gasoline, oil, kerosene, lube oils, propane, butane gases, water, steam, antifreeze solutions	Sets up fast . . . super tacky . . . yet allows gasket movement. Under engine operating temperatures, it converts to super sealant Lacquer thinner
Permatex High Tack Super Adhesive No. 97	Tube has applicator tip or use spatula or putty knife	−60 to +450°F [−51.1 to +232.2°C] 5000 psi	All-purpose adhesive-sealant for bonding weatherstrip, trim, feathering sanding disks, leather, vinyl, cloth, rubber, glass, metal, wood, silencer pads, mats, cork, asbestos, paper, insulation, repairing arm rests, holding engine gaskets, fastening headers and padded dash. Sealant for drip rail, car seams, convertible tops, water-proofing cracks, etc.	Gasoline, oil, kerosene, glycol, trans-mission fluid, brake fluid, antifreeze solutions, grease, glycerin, lube oil, propane or butane gases, or water	Fast Firm, nonbrittle stage Lacquer thinner
Form-A-Gasket^R No. 1 Mil Spec. Type I MIL-S-45180B(Ord.)	Spatula, putty knife, etc.	−65 to +400°F [−53.8 to +204.4°C] 5000 psi	Permanent assemblies, repair gaskets, fittings, uneven surfaces, thread connections, cracked batteries	Water, steam, kerosene, gasoline, oil grease, mild acid, alkali and salt solutions, aliphatic hydrocarbons, antifreeze mixtures	Fast Hard Alcohol
Form-A-Gasket^R No. 2 Mil Spec. Type II MIL-S-45180B(Ord.)	Spatula, putty knife, etc.	−65 to +400°F 5000 psi	Semipermanent reassembly work. Cover plates, threaded and hose connections	Water, steam, kerosene, gasoline, oil, grease, mild acid, alkali and salt solutions, aliphatic hydrocarbons, antifreeze mixtures	Slow Flexible Alcohol
Aviation Form-A-Gasket^R No. 3 Mil Spec. Type III MIL-S-45180B(Ord.)	Brush	−65 to +400°F 5000 psi	Sealing of close-fitting parts. Easy to apply on irregular surfaces	Water, steam, kerosene, gasoline, oil, grease, mild acid, alkali and salt solutions, aliphatic hydrocarbons, antifreeze mixtures	Slow Flexible Alcohol
All-Purpose Cement No. 50	Spatula, putty knife, etc.	−40 to 225°F [−40 to +100°C]	Glass to glass, glass to metal, glass to rubber	Water, polishes, and cleaners	Fast Hard Toluene
Pipe Joint Compound No. 51 New, improved formula. U.L. approved	Brushable, viscous liquid	−65 to +400°F 5000 psi	Threaded fittings, flanges. Can be applied over oil and grease film	Hot and cold water, steam, natural gas, propane, butane, fuel oils, ker-osene, lubracating oils, petroleum-base hydraulic fluids, antifreeze mixtures	Slow Flexible Alcohol
Super "300" Form-A-Gasket^R No. 83	Brush	−65 to +425°F [−53.8 to +218.3°C] 5000 psi	Assembly work on high-compression engines, diesel heads, cover plates, high-speed turbine superchargers, automatic transmissions, gaskets	High-detergent oils and lubricants, jet fuels, heat-transfer oils, glycols 100%, mild salt solutions, water, steam, aliphatic hydrocarbons, diester, lubricants, antifreeze mix-tures, petroleum-base hydraulic fluids, aviation fuels	Slow Flexible Alcohol
Indian Head Gasket Shellac No. 5	Brush	−65 to +350°F [−53.8 to +176.4°C] Variable	General assembly work and on gaskets of paper, felt, cardboard, rubber and metal	Gasoline, kerosene, greases, oils, water, antifreeze mixtures	Slow Hard Alcohol

Fig. 23–5 Permatex sealant chart.

METHOD 1

1. Lay the gasket material over the components to be sealed.
2. Secure the material so that when you are tracing the outline of the component the gasket material will remain in place.

3. Cut the inner and outer circumferences with tin snips or scissors, and use a gasket punch to cut any openings.

METHOD 2 This method is recommended only when the edges of the component are sharp. If the edges are not sharp, the material will not cut properly and damage will result.

1. Lay the gasket over the component and use a ball peen hammer to gently tap out one bolt hole. **CAUTION** If the hole is threaded, take care not to damage the threads.
2. Insert a bolt in the tapped out hole.
3. Begin by tapping out only a minimum number of holes to secure the gasket material to the component.
4. Finally, tap out the remaining bolt holes and openings.
5. Tap gently around the edges of the material so that it is cut to the contour of the component.

Lip-Type Seals Lip-type seals are found in large numbers on diesel engines and their accessories. The designs vary depending on whether they are to be used for sealing off oil, water, dirt, or low pressure or for sealing in two directions.

TYPES AND DESIGNS OF LIP SEALS The most common lip-type seal designs are illustrated in Fig. 23–6. All these designs have two common factors: a sealing element is bonded to the seal metal case, and the opening side faces the higher pressure side.

There are several variable factors in the design of lip-type seals. Some seals utilize a garter spring to increase lip contact. A rubber coat over the outer circumference of the seal case can be utilized to improve housing sealing. The sealing itself can be either internal or external.

The composition of the lip seal is very important and depends on its intended use and location. Silicones, polyacrylics, and Viton are heat-resistant and the least vulnerable to chemicals. However, these products are not as effective against abrasives as nitrile rubber compounds; certain chemicals used in lube oil can cause them to deteriorate when exposed to high temperature.

It follows, of course, that the best replacement possible is the one specified by the manufacturer. Replacement should be done as early as possible to safeguard the engine or component against costly failure. Needless to say, proper storage, handling, and installation procedures are also important.

Removing and Installing Seals
1. Before you clean the components, inspect them to determine the cause of seal leakage. After locating the leakage area, check for accumulation of oil, carbon, or rust.
2. Check the alignment of the seal on the shaft and in the bore.
3. Look for the presence of tool marks.
4. Check the shaft and the seal cage housing as a possible origin of leakage.

Almost any method is acceptable for the removal of a seal providing no damage occurs to the shaft or the housing bore. The three most common ways to remove a seal are shown in Figs. 23–7 to 23–9.

After you have removed the seal, examine the components to determine the cause of the oil leak.

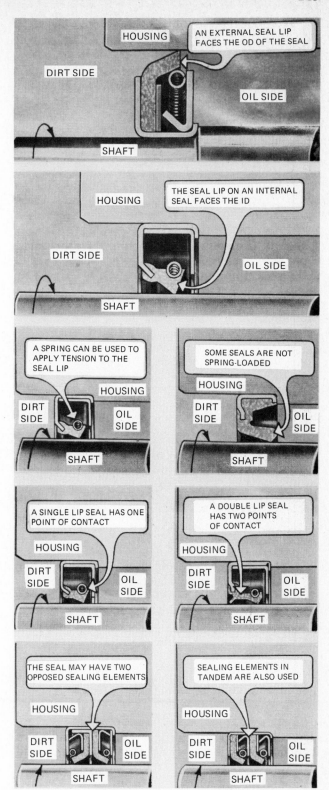

Fig. 23–6 Various lip-type seals. (*Caterpillar Tractor Co.*)

1. Measure the shaft and housing for roundness.
2. Check the shaft for roughness, misalignment, or grooves.
3. Check the shaft for runout or bearing wear.
4. Check the seal-wear pattern with respect to problems of misalignment.
5. Check for evidence of abrasive wear. Note the type of abrasive.

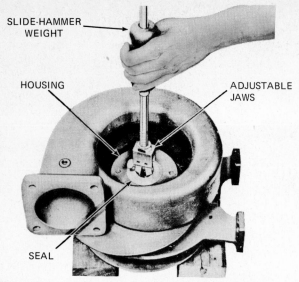

Fig. 23–7 Removing an oil seal from the turbine housing with a slide-hammer puller. *(Allis-Chalmers Engine Div.)*

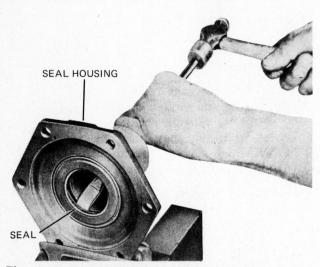

Fig. 23–8 Removing an oil seal with a punch and hammer. *(General Motors Corp., GMC Truck and Coach Div.)*

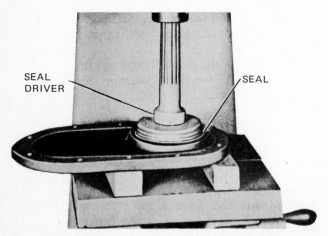

Fig. 23–9 Using a driver to remove an oil seal from the bearing cover. *(General Motors Corp., GMC Truck and Coach Div.)*

6. Check for previous installation damage.
7. Check for evidence of foreign material.

Pointers for Installing Lip-Type Seals   Before you begin installation, recheck to be certain that you have selected a seal with the correct inside and outside diameters.   Make sure that the seal case is also of the correct dimension.

Clean the bore opening or recess and remove any nicks or burrs which may be present.

Polish the shaft with 600-grit oil paper or crocus cloth.

When the shaft is grooved, use either a spacer or a sleeve to make a new lip contact (see Fig. 23–10). When using this method however, you must first check that the lip has a sealing surface and that the seal case is not rubbing on the rotating surface.

ALLOW FOR VARIATION IN OLD AND NEW LIP HEIGHTS WHEN DETERMINING POSITION FOR NEW SEAL

LIP HEIGHT

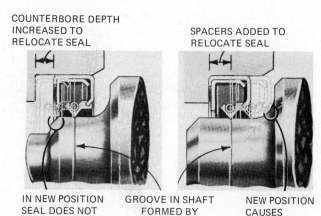

COUNTERBORE DEPTH INCREASED TO RELOCATE SEAL

SPACERS ADDED TO RELOCATE SEAL

IN NEW POSITION SEAL DOES NOT CONTACT SHAFT

GROOVE IN SHAFT FORMED BY OLD SEAL

NEW POSITION CAUSES INTERFERENCE

Fig. 23–10   Making a new lip contact. *(Caterpillar Tractor Co.)*

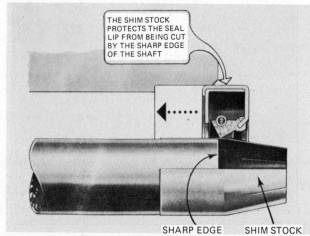

THE SHIM STOCK PROTECTS THE SEAL LIP FROM BEING CUT BY THE SHARP EDGE OF THE SHAFT

SHARP EDGE   SHIM STOCK

Fig. 23–11   Using a shim-stock sleeve when sliding the seal over the shaft. *(Caterpillar Tractor Co.)*

If you are using a seal case without rubber coating, first apply a thin coat of nonhardening sealing compound to the counterbore in order to assure easy installation. This will improve the sealing as well.

Having taken the above precautions, you may then place the open side of the seal with its face toward the oil side into the bore. Since fluid pressure increases lip force when the component is moving, a reversed seal may cause the oil pressure to lift the seal lip off its shaft.

After the shaft is in place, take care not to damage the lip or the seal element when you slide the seal over the shaft. When the shaft end and counterbore are aligned, or when the shaft is splined or has a keyway, use the sleeve guide provided by the manufacturer. If this is not available, use a protective shim-stock sleeve, as shown in Fig. 23–11. If it is at all possible, you should solder the seams and then smooth the surface. This will help prevent lip damage during installation.

If the seal must be driven below the face, select a seal driver which fits loosely in the counterbore (Fig. 23–12). Under all circumstances the force on the seal case must be exerted on its strongest point, that is, the outer circumference. In the event that a seal driver is not available, use a socket or pipe of a size close to the outside diameter of the seal. Avoid driving the seal in place with a hammer because this will deform the seal housing as well as the sealing element. Never use a punch to install a seal.

Questions

1. When is a seal called a *static seal?*

2. When is a seal called a *dynamic seal?*

3. List the precautionary measures which must be taken when installing a new O ring.

4. Explain how to cut a new gasket for an inspection cover.

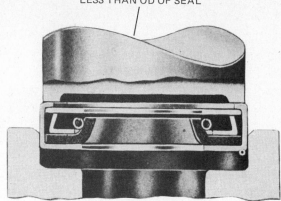

PRESS-FITTING TOOL 0.01 in [0.25 mm] LESS THAN OD OF SEAL

PRESS-FITTING TOOL SEATED AGAINST FRONT OF SEAL

Fig. 23–12 Installing an oil seal in the housing. (*Mack Trucks Canada, Ltd.*)

5. When is a cork gasket used?

6. Why are cylinder-head gaskets commonly made of multilayer materials?

7. Why should you avoid reusing a seal or gasket?

8. Why should gaskets and/or seals remain in their packing kits until you are ready to use them?

9. Why is it essential to use specified torque and specified sequence when installing components which have a gasket between them?

10. List six reasons or circumstances under which a gasket could fail.

11. List all the checks to be made before installing a new seal.

12. Under what circumstances is it essential to use a sleeve guide?

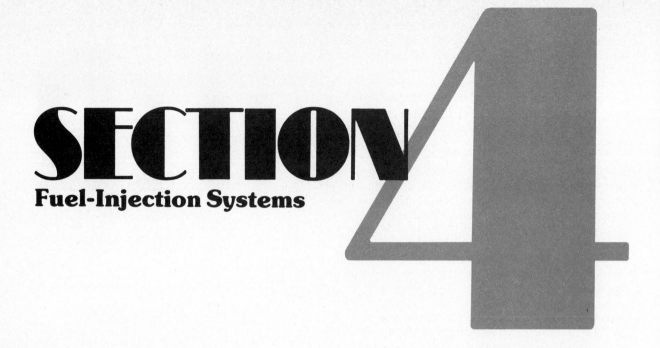

SECTION 4
Fuel-Injection Systems

UNIT 24

Diesel Fuel

Fuel Classification (Grades)   Diesel engine fuel ranges from a highly refined distillate fuel obtained from fractional distillation of crude oils, to a product of residual materials from crude-oil distillation, cracking processes, and fractions removed during the production of crankcase oil.   However, the fuels with which the diesel mechanic is most concerned are those known as middle distillates, that is, grades 1D, 2D, and occasionally 4D.

To meet the needs of the various engine designs and sizes, and to accommodate frequent speed and load changes, permissible limits of fuel properties have been suggested by the American Society of Testing and Materials (ASTM) and by the United States Bureau of Mines.   Choice of the correct fuel allows easier starting and decreases wear of fuel-injection equipment, piston rings, valves, and cylinder sleeves.

The ASTM defines three grades, 1D, 2D, and 4D, which are equivalent to types C-B, TT, and RR of the United States Bureau of Mines (see Table 24-1). Recommended fuel specifications are listed in Table 24-2.

Cetane Number (Ignition Quality)   The cetane number of a diesel fuel expresses its ignition quality. The higher the cetane number, the shorter the ignition delay period.

The cetane number may be defined as the percent-age by volume of the normal cetane in a blend with alpha-methylnapthalene to match the ignition quality of the test fuel.   The cetane number affects cold starting, combustion roughness, acceleration, and exhaust smoke density.   However, the ignition requirement of a diesel engine changes with the design of the combustion chamber, engine speed, operating condition as well as ambient temperature and altitude.

High-speed engines require a higher cetane number to prevent engine roughness and knock, but the shorter delay period may lead to a darker exhaust smoke under maximum torque.   This is because the raw fuel which is sprayed into the combustion chamber at this point ignites due to the shorter delay period.

Distillation Range   The distillation range influences the Btu, and viscosity of the fuel.   It affects power output, starting, smoke, and exhaust odor.

Btu (Heating Value) and API° Gravity   Although heating value and API° gravity are not specified in the ASTM classification, they are very important. They affect the power output, combustion, starting, and exhaust emission.

The heat value of a fuel is expressed in Btu per pound or per gallon.   A fuel with a higher heat value (more Btu/gal) will produce more power than a fuel

Table 24–1 PROPERTIES OF ASTM GRADES OF DIESEL FUELS

Grade[a]	Flash point, °F [°C] Min.	Pour point, °F [°C] Max.	Water and sediment, vol % Max.	Carbon residue on 10% residuum, % Max.	Ash, wt % Max.	Distillation temperatures, °F [°C] −90% point Min.	Max.	Viscosity at 100°F [37.8°C] kinematic, centistokes, (or Saybolt Universal, sec) Min.	Max.	Sulfur, wt % Max.	Copper strip corrosion Max.	Cetane number[e] Min.
1D	100 or Legal [37.8]	b	Trace	0.15	0.01		550 [287.8]	1.4	2.5 [34.4]	0.50	No. 3	40[f]
2D	125 or Legal [51.7]	b	0.10	0.35	0.02	540[c] [282.2]	640 [338]	2.0[c] [32.6]	4.3 [40.1]	0.7[d]	No. 3	40[f]
4D	130 or Legal [54.4]	b	0.50		0.10			5.8 [45]	26.4 [125]	2.0		30[f]

[a] To meet special operating conditions, modifications of individual limiting requirements may be agreed upon between purchaser, seller, and supplier.

[b] For cold weather operation, the pour point should be specified 10°F [5.6°C] below the ambient temperature at which the engine is to be operated except where fuel oil heating facilities are provided.

[c] When pour point less than 0°F [−17.8°C] is specified, the minimum viscosity shall be 1.8 cSt (32.0 SSU) and the minimum 90 percent point shall be waived.

[d] For all products outside the United States, the maximum sulfur limit shall be 1.0 wt.%.

[e] Where cetane number by Method D 613 (Test for Ignition Quality of Diesel Fuels by the Cetane Method) is not available, ASTM Method D976 (Calculated Cetane Index of Distillate Fuels) may be used as an approximation. Where there is disagreement, Method D 613 shall be the referee method.

[f] Low-atmospheric temperatures as well as engine operation at high altitudes may require use of fuels with high cetane ratings.

SOURCE: Chevron Research Company

Table 24–2 RECOMMENDED FUEL SPECIFICATIONS

Characteristics	Fuels 1D	2D
Cetane number	45	40
Gravity	35–40	26–34
Pounds per gallon [kg/gal]	6.95 [3.16]	7.31 [3.32]
Btu per gallon average	137,000	141,800
Sulfur, percent by weight max.	0.5%	0.1%
Viscosity 100°F [37.8°C]	45 seconds	32 seconds
Flash point	100°F [37.8°C]	125°F [51.7°C]
Pour point	10°F [5.6°C] below ambient temperature	
Water sediment	No more than a trace should be present	
Carbon residue	0.15	0.35
Ash	0.01	0.02
Corrosion	Pass	Pass
Volatility:		
10% point	350–475°F [175–250°C]	400–500°F [209–260°C]
90% point	450–600°F [230–315°C]	550–650°F [280–340°C]
End point	500–625°F [260–320°C]	625–700°F [320–360°C]

of lower heating value. However, the difference of power output between 1D and 2D is between 4 and 11 percent depending on combustion chamber and fuel-injection pump design.

Btu and API° gravity are directly related. The higher the API° gravity (or the lower the specific gravity), the higher the heat value (see Fig. 24–1). Fuel may range from 26 to 40 API° gravity which corresponds with a range in specific gravity of 0.8963 to 0.8155. This is equivalent to 142,000 to 135,500 Btu/gal (U.S.), or 170,400 to 138,210 Btu/gal (Imp.).

This is about 7.53 to 6.79 lb/gal (U.S.) at 60°F [15.6°C], or 9.03 to 8.14 lb/gal (Imp.).

Heat Value For the moment let us approximate the heat value of a fuel that has 19,000 Btu/lb and consists of 15 percent hydrogen and 85 percent carbon. To burn the fuel, air is needed. Atmospheric air at sea level is composed of 20.95 percent oxygen (O), 78.9 percent nitrogen (N), and miscellaneous gases at 0.06 percent. 1 ft³ [28.32 l] of air at 32°F [0°C] sea level registers 29.9 inHg [75.96 cmHg] and weighs

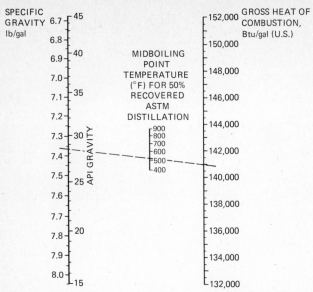

SPECIFIC GRAVITY lb/gal
6.7
6.8
6.9
7.0
7.1
7.2
7.3
7.4
7.5
7.6
7.7
7.8
7.9
8.0
45
40
35
30
25
20
15
API GRAVITY
MIDBOILING POINT TEMPERATURE (°F) FOR 50% RECOVERED ASTM DISTILLATION
900
800
700
600
500
400
GROSS HEAT OF COMBUSTION, Btu/gal (U.S.)
152,000
150,000
148,000
146,000
144,000
142,000
140,000
138,000
136,000
134,000
132,000

Fig. 24–1 Relation between API gravity and Btu. *(Chevron Research Co.)*

0.0807 lb [0.367 kg]. Under the ideal conditions outlined above, 1 lb of air equals 12.4 ft^3 [351.1 l]. This figure should be adjusted to actual temperature and atmospheric pressure and air density. In this respect SAE standard test code for a diesel engine is:

Barometric pressure = 29.38 inHg [74.4 cmHg]
Temperature = 85°F [29.4°C]

This calculates out to approximately 1 lb equaling 14 ft^3 [0.396 m^3] of air.

For complete combustion of 1 lb of fuel, about 260 ft^3 [7.36 m^3] of air is required. It is essential to remember these facts when servicing the air-intake system.

Sulfur The presence of excessive sulfur content in the fuel can cause deposits and corrosion. These deposits will increase wear of cylinder, piston, and piston ring. The corrosion is caused when sulfur produced during combustion combines with the water of combustion to form strong acids which attack the engine components. Diesel engines using direct injection are more affected than engines using a precombustion chamber, a turbulent chamber, or power cells.

Viscosity A viscosity lower than specified reduces the effectiveness of the fuel as a lubricant and accelerates wear of injection components. Component wear can cause loss of the amount of metered fuel due to leakage in the injection pump or injector.

High viscosity can also reduce the amount of fuel metering within the system due to higher flow resistance. However, a lower viscosity changes the spray pattern droplet size which causes a change in the time of ignition (combustion) (Fig. 24-2).

Flash Point The temperature at which a substance will give off a vapor that will flash or burn momentarily when ignited is called the *flash point.* The flash point has no effect on engine performance. It is specified simply as a guide for handling and storage safety.

Pour Point The *pour point* is the minimum temperature at which the fuel can flow through the filters or be transferred by the transfer pump to the fuel-injection pump. The fuel flow is affected only when the ambient temperature is below the pour point temperature.

Water and Sediments If the amount of water exceeds specification, it can cause corrosion and ad-

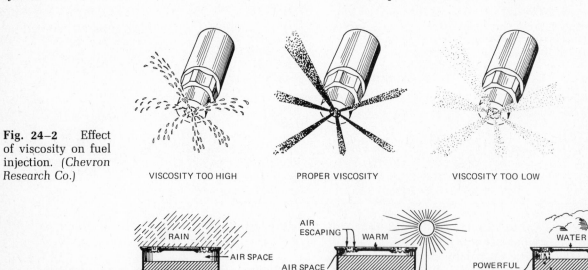

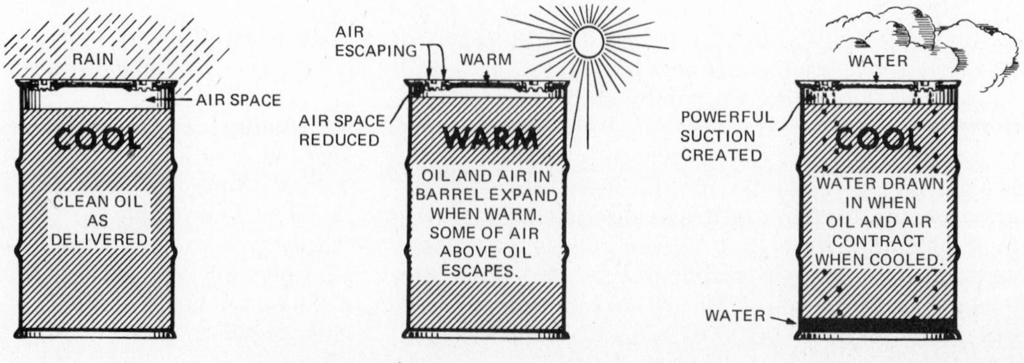

VISCOSITY TOO HIGH
PROPER VISCOSITY
VISCOSITY TOO LOW

Fig. 24–2 Effect of viscosity on fuel injection. *(Chevron Research Co.)*

RAIN
AIR SPACE
COOL
CLEAN OIL AS DELIVERED
AIR ESCAPING
WARM
AIR SPACE REDUCED
WARM
OIL AND AIR IN BARREL EXPAND WHEN WARM. SOME OF AIR ABOVE OIL ESCAPES.
WATER
POWERFUL SUCTION CREATED
COOL
WATER DRAWN IN WHEN OIL AND AIR CONTRACT WHEN COOLED.
WATER

Fig. 24–3 Effects of weather on drums stored outside. *(J. I. Case Co., Components Div.)*

versely affect all moving parts of the fuel-injection system, especially the injector nozzle. Water droplets will not pass through the multihole nozzle, therefore, it can be damaged. Sediments cause sludge buildup within the system. This causes wear of injection components and plugs the filters.

Carbon Residue Carbon residue is the deposit left in the combustion chambers and fuel-injection equipment as a result of the use of fuel made from residual blends or from inefficient combustion.

Ash Ash originates from additives or from the crude oil itself. It causes wear of fuel-injection components, pistons, and piston rings.

Volatility Any substance that may be readily vaporized at relatively low temperatures is said to have volatility. Volatility affects power output, performance, starting, warm-up, and exhaust smoke.

Fuel Additives Additives are sometimes used to improve certain characteristics of diesel fuel. Inhibitors and preventitives may be used to improve ignition quality, to reduce oxidation and corrosion, and to keep the fuel-injection system clean. Sometimes additives are used to reduce smoke and odor.

Fuel Storage and Handling When it is not practical to use a large storage tank and storage of fuel in drums is the only alternative, take all the precautions outlined in the section Storing and Handling Crankcase Oil (Unit 12). As far as possible, avoid storing the drums outside. The effects of the weather on drums stored outside is illustrated in Fig. 24-3.

If fuel is being stored in drums, do not use them as a convenient method for transporting the fuel to the tractor or truck because this further handling will stir up sediment. Drive the equipment to the fueling location. Use several drums so that the sediment is allowed to settle.

When a hand dispensing pump is used to pump fuel out of the drum, make certain the standpipe is about 2 in [50.8 mm] shorter than the drum and make certain the drum is tilted slightly.

Questions

1. What is the difference between fuel acquired from fractional distillation and the product of residual materials?

2. List the major differences between 1D and 2D fuel.

3. What advantage (additional to improving starting) has fuel with a high cetane content over regular fuel?

4. When the viscosity of the fuel is too high or too low, what effect does it have on ignition, combustion, and exhaust smoke?

5. Why does ASTM specify maximums on certain properties of diesel fuel, that is, water and sediments, carbon residue, and ash?

6. What would result if the ash content in a fuel were higher than specified?

7. Grey exhaust smoke can be the result of (a) a fuel of high cetane rating, (b) a fuel of high Btu rating, (c) a 1 percent sulfur content, or (d) a fuel of high ash content.

8. Fuel of high sulfur content will cause (a) corroded injector tips and corroded spray orifices, (b) low engine power output, (c) pitting of the main bearings, or (d) clogged fuel filters.

9. An engine will operate more efficiently in cold weather by (a) lowering the sulfur content of the fuel, (b) increasing the Btu rating, (c) increasing the cetane rating, or (d) lowering the cetane rating.

UNIT 25

Governors

Function of Governor Only a few engine applications do not require a governor. An outstanding example of engine application without a governor is in the conventional passenger car. Here the engine speed and engine load changes are sensed by the driver. The driver's movement of the throttle is an extension of conditioned reflex; the driver acts as a governor (the driver controls the airflow).

However, human beings are not capable of reacting to load and speed changes quickly enough in-sofar as the diesel engine is concerned, since it can accelerate at a rate of more than 2000 rps (revolutions per second). The driver must, therefore, rely on modern mechanics, hydraulics, or the electronic robot (governor).

Types of Governors Numerous types of governors are manufactured, each for a specific engine application and each capable of readjusting automatically to its load and speed. The five different types of governors are: (1) mechanical, (2) pneumatic, (3) servo, (4) hydraulic, and (5) electronic governors.

The name of the governor indicates its design and its ability to control speed and/or load. However, regardless of the type of governor required to do a particular job, each must perform two major tasks. The governor must measure engine rpm and it must have the capability (power) of moving the fuel rack or controlling the fuel flow or the pressure. Furthermore, it must (1) limit maximum speed and prevent the engine from overspeeding; (2) it must control idle speed to prevent the engine from stalling; (3) it must regulate engine speed between low and high idle; (4) it must be capable of providing for more fuel when the engine lugs down to maximum torque.

Governor Terminology Before studying the basic fundamentals of the different types of governors, you should first understand certain terms related to the subject:

- *Low-idle speed* is the lowest speed (no load) at which a governor (maintains) controls the engine speed. Under the *no-load* condition, the engine is operating without producing work. Under the *load* condition, the engine is operating and also producing some work.
- *High-idle speed* is the maximum speed at which a governor allows the engine to run.
- *Droop speed* is the reduced speed from high idle after load is placed on the engine. The average speed droop of a mechanical governor is between 5 and 10 percent.
- *Maximum torque speed* (sometimes also called *rated speed* or *stalled speed*) is the speed at which the engine develops its maximum torque. At maximum torque, the engine applies maximum force on the camshaft, which thereby exerts its maximum twisting effort.
- *Overspeed* is a speed above high idle.
- *Governor cutoff speed* is the speed at which the governor moves the control rack to the no-fuel position or cuts off the fuel flow.
- *Sensitive* is the response of a governor to speed change expressed in percentage.
- *Momentary speed change* is the change of the engine speed immediately after a sudden change in engine load or throttle setting.
- *Speed drift* is the gradual variation in the engine speed above or below set governor speed.
- *Hunting* is a constant or repeated rhythm change in the engine speed.
- *Stability* is the ability of a governor to maintain engine speed without hunting.
- *Promptness* is the ability of a governor to respond to engine speed changes.
- *Speed regulation* is the change of the engine speed within high and low idle through the action of the governor lever.

The most common governors used on diesel engines for trucks, tractors, marine and industrial applications are of the limiting-speed or of the variable-speed, mechanical, servo, or hydraulic type.

Less common, and used for special engine applications, are the constant-speed, load-limiting, torque-converter, and the load-control (mechanical, hydraulic, or electronic) governors. Some small diesel engines use a variable-speed pneumatic governor.

A *limiting-speed* governor maintains the engine speed at the set idle speed and limits high-idle engine speed. A manually operated throttle controls the speed between idle and maximum speeds.

A *variable-speed* governor controls the engine speed over the entire speed range and load. A throttle control sets the governor to the desired engine speed.

A *constant-speed* governor has the ability to hold the engine at constant speed, regardless of the load, providing the load is within the power of the engine.

A *load-limiting* governor limits a load applied to the engine at any given speed. Its purpose is to prevent overloading of the engine.

A *load-control* governor controls and adjusts the amount of load applied to the engine to save output capacity of the engine.

A *torque-convertor* governor senses the speed of the torque-convertor output shaft and thereby limits the set engine speed.

MECHANICAL GOVERNOR

All mechanical and most hydraulic governors have a speed-measuring mechanism and a fuel-changing mechanism. The speed-measuring mechanism consists of the components shown in Fig. 25-1.

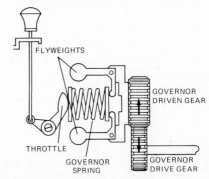

Fig. 25-1 Simplified view of a governor speed-measuring mechanism. (*Caterpillar Tractor Co.*)

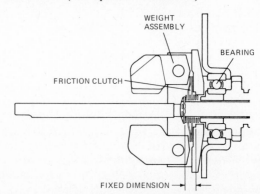

Fig. 25-2 Cutaway view of a friction-clutch arrangement. (*American Bosch-Ambac Industries Inc.*)

The two L-shaped arms with weights are pivot-mounted to the governor drive gear or to the governor drive shaft. A friction clutch (Fig. 25-2) or a spring arrangement is placed between the drive and the weight assembly to reduce drive pulsation. To increase the centrifugal force, most weight assemblies are driven at higher speed than the fuel-pump camshaft speed. To oppose the weight force, the governor spring (speeder spring) is placed against the weight (flyweight) fingers. The spring force is increased or decreased by a control mechanism placed above the governor spring. The high-idle and low-idle adjusting screw is used to limit the expansion and control compression of the spring. A thrust bearing is placed between the stationary spring and the rotating flyweights to reduce friction and smooth motion.

The fuel-changing mechanism of all mechanical governors consists of mechanical levers and linkages. It connects the fuel control rack with the component that is placed between the governor weight fingers and the governor spring. One such component is the right governor spring retainer, as shown in Fig. 25-3; another is the sliding sleeve assembly, as shown in Fig. 25-4. In the governor shown in Fig. 25-4, the throttle lever is connected to the fulcrum lever. The fulcrum lever arm is connected through a linkage with the fuel control rack and slide-fitted through the pivot pin with the sliding sleeve assembly. Limited-speed governors will be discussed in detail later in this unit.

Mechanical Governor Operation When the engine is not operating, the governor flyweights are forced toward each other by the governor spring, and the

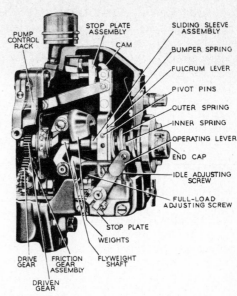

Fig. 25-4 Sectional view of a limiting-speed governor. *(American Bosch-Ambac Industries Inc.)*

fuel control rack is moved toward the increased fuel position by the force of the governor spring (see Fig. 25-5 and Table 25-1).

If the operator moves the throttle, say to half engine speed, its movement through linkage increases the force on the governor spring. When the engine is running, the centrifugal force of the rotating flyweights forces the weights outward and moves the weight finger against the spring force until both forces are equal. At the same time the fuel control rack is moved toward the low-fuel position and the governor maintains a constant engine speed until either force is changed. When the operator increases the spring force, by moving the throttle to the high-idle position, the spring force collapses the weights and allows the fuel control rack to move and thereby increase the fuel injected per cycle (Fig. 25-6). As the engine speed increases, the weight force increases. This moves the fuel control rack toward the low-fuel position until an equilibrium exists between both forces.

When the engine is operating at high idle and its speed drops, due to added load, the weights move inward and the governor spring expands and moves the fuel control rack toward the full-fuel position. In response to the fuel increase, the engine's maximum torque speed is maintained until the load is reduced.

When the throttle is moved to the low-idle position, the force on the governor spring is reduced and the flyweights move outward. This causes the fuel control rack to move in the direction of the no-fuel position, resulting in an engine speed drop and a reduction in the weight force. The spring force now moves the control rack toward the full-fuel position. When both forces are equal, the engine operates at a constant-idle speed.

Torque-limiting Devices On some governors a stop collar and bar, or stop plate, limits the fuel metered per cycle (see Figs. 25-5 and 25-6). On other

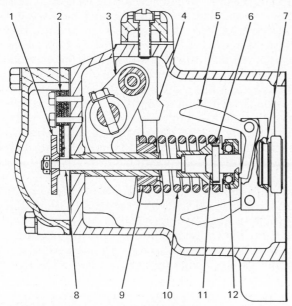

1. Collar
2. Torque spring
3. Lever assembly
4. Shoulder low idle
5. Weight
6. Spring seat
7. Bearing
8. Bolt
9. Spring seat
10. Governor spring
11. Fuel rack
12. Thrust bearing

Fig. 25-3 Caterpillar mechanical governor. *(Caterpillar Tractor Co.)*

Table 25-1 ENGINE SPEEDS AND GOVERNOR POSITIONS

Speed	Flyweights	Governor spring	Fuel control rack
High idle, no load	Maximally outward	Maximally compressed	High-idle fuel position
Droop speed	Slightly inward (from maximally outward)	Slightly expanded (from maximally compressed)	Increased fuel position (from high idle)
Maximum torque	Slightly inward (from droop speed)	Slightly expanded (from droop speed)	Maximum fuel position
Low idle, no load	Slightly outward (from low idle, load)	Slightly compressed (from low idle, load)	Low idle, no-load fuel position
Low idle, load	Slightly outward (from maximally inward)	Slightly compressed (from maximally expanded)	Low idle, load fuel position
Not operating	Maximally inward	Maximally expanded	Starting fuel position

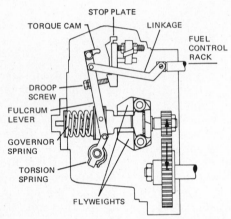

Fig. 25-5 Position of governor components with engine not operating. (*American Bosch-Ambac Industries Inc.*)

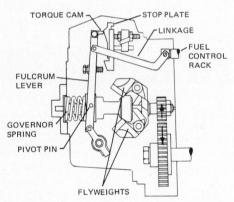

Fig. 25-6 Position of governor components at high speed and load. (*American Bosch-Ambac Industries Inc.*)

governors, the fuel stop is modified so that additional rack travel can occur when the engine lugs down to torque speed. On these governors the stop bar is replaced with a torque spring (Fig. 25-3). As the engine picks up a load, the engine speed and the flyweight force are reduced and the governor spring forces the fuel control rack toward the full-fuel position to maintain high idle.

When the droop speed (high-idle, load) is attained, the stop collar contacts the torque spring. If the load on the engine increases further, the engine

speed and weight force decrease, allowing the governor spring to expand. This action bends the torque spring against the governor housing causing the fuel control rack to move a fixed distance in the full-fuel direction. With decreased engine speed, more air can enter the cylinders, and more fuel is metered per cycle, giving the engine maximum torque.

A differentially designed governor and its action are shown in Figs. 25-7 and 25-8. When the throttle is in the high-idle position, the governor spring exerts the maximum pull on the tension lever and moves the guide bushing assembly with maximum force against the heavy weight force. The guide bushing assembly then moves the guide and fulcrum lever with the fuel control rack to a position slightly before full load. As the engine picks up a load and reaches droop speed, the weight force decreases due to decreased engine speed, allowing the guide bushing assembly and floating lever to move to the left in Fig. 25-7, thereby moving the fuel control rack to the full-load position. The movement of the tension lever is stopped because the torque-capsule housing butts against the full-load stop screw. If the engine speed is reduced even more, due to additional load, the weight force decreases also and the heavy torque spring moves the guide bushing assembly, the floating lever, and the fuel control rack to the maximum-fuel position.

Detroit Diesel Governor A limited-speed mechanical governor (used with a V71 Detroit Diesel engine) is shown in Figs. 25-9 and 25-10. When the engine has ceased to operate, the force of the high-speed spring holds the high-speed plunger against its stop; the low-speed spring holds the low-speed spring cap against the gap adjusting screw (which is located on the operating shaft lever) (see Fig. 25-9). The expanded springs have pivoted the operating shaft lever and the operating fork now rests against the thrust bearing and the riser rests against the flyweight finger of the low- and high-speed weights (see Fig. 25-10).

The semirotation of the operating shaft lever has caused the differential lever to pivot on the fulcrum pin (which is located on the speed control shaft).

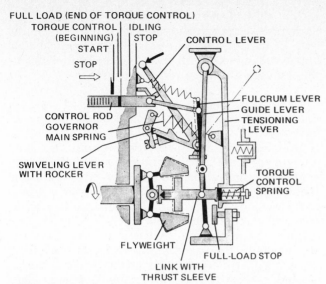

Fig. 25-7 Position of governor components at high-idle speed and load (beginning of torque control). (*International Harvester Co.*)

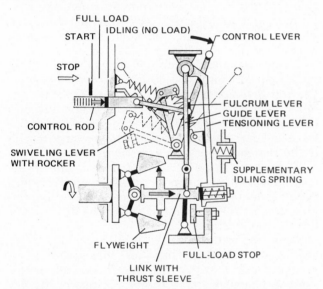

Fig. 25-8 Position of governor components at low-idle speed. (*International Harvester Co.*)

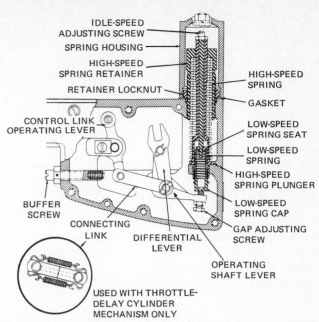

Fig. 25-9 V71 limiting-speed governor (top view). (*GMC Detroit Diesel Allison Div.*)

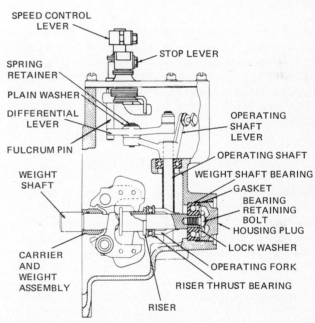

Fig. 25-10 V71 limiting-speed governor (side view). (*GMC Detroit Diesel Allison Div.*)

This moves the connecting link so that it pivots the control-link operating lever on its shaft. The fuel rod (which is connected to the control-link operating lever, shown in Fig. 25-9) moves to the maximum-fuel position.

When the operator moves the speed control lever so that the engine operates at 1000 rpm, the fulcrum pin then moves the differential lever to the left, pivoting the operating shaft lever and adding force to the low spring without dislodging the fuel rod from the full-fuel position. Once the engine begins to operate, the flyweight force of both the low- and high-speed governor weights moves the riser, the thrust bearing, and the fork to the right, causing the operating shaft to pivot and the operating lever to place a greater force on the lower spring cap. The low-speed spring is compressed and closes the gap 0.0015 in [0.038 mm] between it and the high-speed plunger.

During this time the differential lever was forced (by the operating-shaft lever) to pivot on its fulcrum pin. The pivoting operating-shaft lever moves the connecting link and pivots the operating lever which in turn moves the fuel rod to decrease fuel. When the flyweight force is equal to the low-speed spring force, the engine operates at the set speed.

When the operator moves the speed control lever to the full-fuel position, the differential lever is instantly moved through the fulcrum pin which causes the differential lever to pivot on the pin of the operating-shaft lever. This action moves the connecting link and the control-link operating lever, and as a result moves the fuel rack to the full-fuel position.

The engine responds to the increased fuel and the flyweight force increases. This action moves the low-spring cap against the high-speed spring plunger. From approximately 1200 rpm on, only the high-speed weight force acts on the riser. The low-speed governor weights meanwhile have contacted their limiting stops. When the flyweight force is in balance with the high-speed spring force, the engine operates at high-idle speed.

MAXIMUM-FUEL POSITION   When the speed control lever is in the full-fuel position and the engine is operating at high idle, the flyweight force and the high-speed spring force are equal and the fuel rod is near the maximum-fuel position. When the engine speed is reduced, due to a load increase on the engine, the flyweight force reduces and the high-speed spring slightly expands. This causes the operating shaft with its lever to pivot. The pivoting motion of the operating shaft moves the differential lever to the right (in Fig. 25-10) because the differential lever slides on the fulcrum pin. The connecting link is then moved to the right, pivoting the control-link operating lever and causing the fuel rod to go into the maximum-fuel position.

HIGH- AND LOW-IDLE ADJUSTMENT   The idle speed is adjusted by increasing or decreasing the low-speed spring energy with an idle-speed adjusting screw (Fig. 25-9). The high-idle (maximum no load) speed is adjusted by increasing or decreasing the energy on the high-speed spring by turning the high-speed spring retainer.

PNEUMATIC GOVERNOR

Pneumatic governors, as mentioned previously, are used on smaller diesel engines and effectively control engine speed from idle to maximum speed. The movement of the fuel control rack is a joint effort of the atmospheric pressure, the governor spring, and the low pressure. As illustrated in Fig. 25-11, the pneumatic governor consists of two separate main units: the venturi unit and the diaphragm unit.

The venturi unit is positioned between the air cleaner and the air-intake manifold. The venturi opening (venturi throat) limits the maximum airflow and, therefore, controls the maximum engine speed. The butterfly valve is connected through linkage to the accelerator pedal and controls the airflow and, therefore, the engine speed between low idle and high idle. The auxiliary venturi, butterfly valve, and the vacuum connections are positioned at the narrowest point of the venturi.

The diaphragm unit is mounted to one end of the fuel-injection-pump housing and a leather or rubber diaphragm divides the unit in two to form the atmospheric and the vacuum chambers.

The left side is always under atmospheric pressure either directly through a filter or through a pipe connection to the crankcase. The diaphragm is connected by a pin to the fuel control rack.

The vacuum chamber is connected by a vacuum pipe to the auxiliary venturi. The governor spring in the vacuum chamber maintains the diaphragm under pressure and, therefore, holds the fuel control rack in the full-fuel position. A mechanical link connected to the shutoff lever, moves the fuel rack to the no-fuel position to stop the engine.

Pneumatic Governor Operation   When the engine is stopped, the governor spring has moved the fuel control rack to the full-fuel position and the diaphragm rests against the full-fuel stop. When the engine is running and the butterfly valve is resting against the idle stop screw, the valve is nearly closed and a low pressure (vacuum) is created at the engine side of the butterfly valve because the airflow to the cylinders is restricted. A vacuum (low pressure) also exists in the vacuum chamber and, therefore, the atmospheric pressure forces the diaphragm and the fuel control rack to the no-fuel position. The engine speed reduces in response to reduced fuel, and the vacuum at the engine side of the butterfly decreases. The governor spring and the remaining pressure in the vacuum chamber move the diaphragm and fuel control rack against the atmospheric pressure to increase fuel, and the engine operates at idle speed (Fig. 25-11).

When the butterfly valve is opened, say to half engine speed, more air is allowed to enter the cylinder and the vacuum at the engine side of the butterfly decreases. Instantly the diaphragm and fuel control rod move to the left until the governor spring force and the remaining pressure in the vacuum side of the chamber equal atmospheric pressure.

When the engine speed is reduced, due to increased load, the vacuum decreases and the pressure in the vacuum chamber increases, moving the diaphragm and the fuel control rod in the direction of the full-fuel position.

When the butterfly valve is fully open, it rests against the adjustable full-load stop screw. The pressure (vacuum) in the manifold and in the vacuum chamber are very low because of the decreased air velocity in the venturi. The diaphragm is momentarily moved against the extended piston of the torque control spring (Fig. 25-12a). The fuel control rod is moved to the maximum-fuel position. As the engine speed increases (to high idle), the air velocity and the vacuum increase in the auxiliary venturi and in the vacuum chamber. Atmospheric pressure moves the diaphragm and fuel control rack away from the full-load stop in the direction of the no-fuel position (Fig. 25-12b). When the engine speed decreases (due to added load) to its maximum torque speed, the air velocity decreases. The pressure in the vacuum chamber approaches atmospheric pressure. This causes the governor spring to force the diaphragm against the torque control piston, compress the torque spring, and cause the fuel rack to move to its maximum position.

Maximum fuel is controlled by the full-load stop or the movement of the torque control piston. The maximum airflow is controlled by the venturi. To have a smoke- and sootfree engine over a wide range of engine loads, the governor spring and the torque control spring must control required fuel delivery.

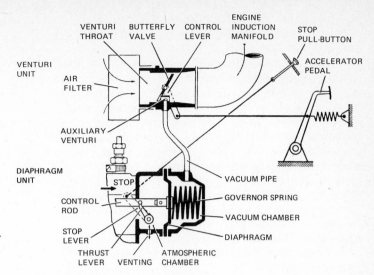

Fig. 25–11 Position of pneumatic-governor components at high idle. *(Robert Bosch GmbH, Stuttgart, West Germany.)*

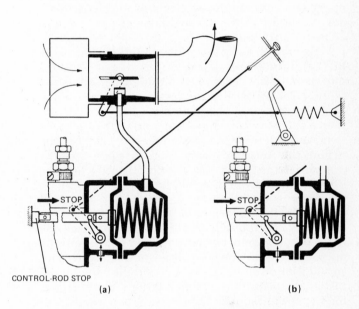

Fig. 25–12 *(a)* Momentary governor position at maximum fuel; *(b)* governor position at high idle, no load. *(Robert Bosch GmbH, Stuttgart, West Germany.)*

SERVO–TYPE GOVERNOR

Where the engine speed changes constantly due to operating conditions, servo or hydraulic governors are used to compress the governor spring or to move the fuel rack through hydraulic action. Both types of governors reduce the effort required to move the throttle since only a small force is necessary to move the governor control mechanism.

Servo Construction Servo-type governors are mechanical governors to which a servo mechanism has been added. The servo consists basically of a hydraulic piston and an oil-control valve (the cylinder) (Fig. 25-13). The piston which forces the governor spring against the sleeve assembly is loosely fitted inside the end cover and the cylinder is fitted over the outside of the end cover. The sleeve assembly is connected through levers to the fuel control rack and the cylinder is linked, through levers, to the throttle. The spindle is center- and cross-drilled to supply engine oil to the piston cavity and the sleeve assembly. One end of the spindle is supported in the end cap and the other end is fastened to the drive gear and the flyweight carrier.

Servo-Type Governor Operation Study Fig. 25-13 as you read the following section. "Left" and "right" refer to directions in the illustration.

When the engine is not operating, the piston rests against the cylinder and the sleeve assembly is

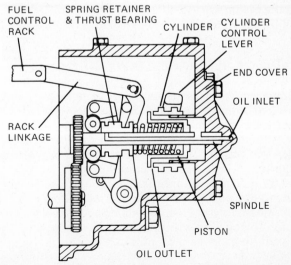

Fig. 25–13 Servo-type governor. *(Caterpillar Tractor Co.)*

forced to the left by the governor spring. This causes the flyweight to move inward and causes the fuel rack to move to the full-fuel position.

Assume that the operator positions the throttle so that the engine will operate at 1000 rpm. Under this condition, when the engine is running, oil instantly enters the end cover, lubricates the sleeve assembly, and pressurizes the area between the end cover and the piston. The oil pressure forces the piston to the left and increases the force on the governor spring. This holds the sleeve assembly in a position to increase fuel. During this time the return oil passage opens through the movement of the piston, allowing oil to drain and the pressure to be reduced between the piston and the end cover. The flyweight force, meanwhile, has become greater and, therefore, increases the force on the governor spring. This moves the piston to the right and reduces the drain oil passage. When the flyweight force and the force created on the piston are equal, the engine operates at a constant speed.

When the operator moves the throttle to high idle, the cylinder closes the drain oil passage but its movement is limited through the high-idle adjusting screw. The oil pressure increases. The force on the piston increases. The governor spring moves the sleeve assembly to the left, and allows the fuel rack to move to increase fuel. With the movement of the piston, the drain oil passage opens again and the pressure in the piston area is reduced until a balance between the governor weight force and the force on the piston is established.

When the operator moves the throttle to the low-idle position, the cylinder instantly increases the drain oil passage between it and the piston and the oil pressure then drops. Subsequently the governor spring is allowed to expand to the right, moving the piston to the right and reducing the drain opening. Engine speed reduces and the governor weight collapses, allowing the spring retainer and thrust bearing to move to the left. The governor weight force and hydraulic force are now in balance, and the engine operates at idle speed. However, the reduced opening results in an increase in oil pressure and the piston stops (or moves slightly to the left). Once again the governor weight force and the force on the piston are in balance, and the engine operates at idle speed.

HYDRAULIC GOVERNOR

Numerous types of hydraulic governors are manufactured to meet the requirements of today's diesel engines. The simplest of these is the governor used on DPA distributor-type injection pumps (Fig. 25–14). (See also Unit 33, "DPA and Roosa Master Distributor-Type Fuel-Injection Pumps").

The governor shown in Fig. 25–14 has neither flyweights nor a fuel-changing mechanism. The fuel-measuring device, the metering valve, the governor spring, the idling spring, the rack, and the throttle-shaft lever with pinion make up the governor.

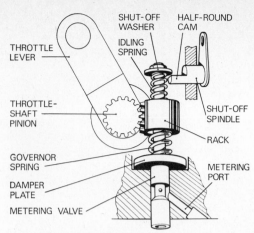

Fig. 25–14 DPA hydraulic governor. *(CAV Ltd.)*

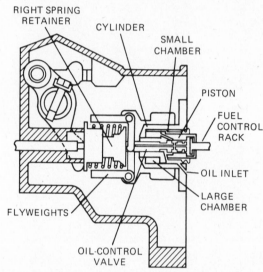

Fig. 25–15 Direct-acting hydraulic governor. *(Caterpillar Tractor Co.)*

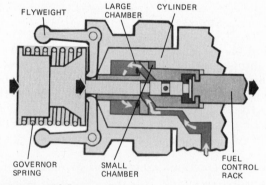

Fig. 25–16 Oil flow as speed increases. *(Caterpillar Tractor Co.)*

Not all hydraulic governors are so simply designed; some are very complex. However, nearly all hydraulic governors use a mechanical speed-sensing mechanism known as flyweights. A single-acting piston, a double-acting piston, or a servo piston acts as the fuel-changing mechanism.

Direct-acting Hydraulic Governor Figure 25–15 illustrates a direct-acting hydraulic governor with a single-acting piston. The flyweights and the carrier are fastened to the governor drive gear and the assembly rotates on the stationary cylinder. The piston is hooked to the fuel control rack and is lap-fitted within the cylinder. The oil-control valve is center- and cross-drilled and fastened to the right spring retainer. Oil passages within the cylinder lead into the small chamber, to the piston, to the oil-control valve, and into the large chamber (between the piston and the cylinder).

Hydraulic Governor Operation When the engine is not operating, the governor spring forces the right spring retainer, the oil control valve, the fuel control rack, and the piston to the right in Fig. 25–15. This places the control rack in the full-fuel position because the oil control valve then butts against the fuel control rack. When the operator moves the throttle to half engine speed, the force on the governor spring is immediately increased but no other action takes place. However, as soon as the engine begins to operate, engine oil enters the cylinder, flows into the small chamber, and passes by the open land on the oil control valve into the larger chamber. It pressurizes both chambers, but the fuel control rack remains in the full-fuel position. At the same time the engine responds to the injected fuel and the engine speed and the flyweight force increase. The flyweight finger moves the oil control valve to the left which aligns the cross-drilled land of the valve with the piston passage and the large chamber. Oil drains from the large chamber, through the center of the valve into the governor housing (Fig. 25–16). The pressure in the large chamber then drops. The pressure in the small chamber, however, remains the same, forcing the piston and the fuel control rack to the left in Fig. 25–16. When the piston catches up to the valve land, it closes the drain passage. The engine then operates at constant speed because the piston is hydraulically locked, holding the fuel rack in a fixed position.

When the engine speed increases due to downhill operation, the flyweight force increases. The flyweight force instantly moves the control valve, opening the drain passage and reducing the oil pressure in the large chamber. The pressure in the small chamber now moves the piston and the fuel control rack toward the no-fuel position, causing the engine speed and the flyweight force to reduce. The governor spring moves the oil control valve to close the drain passage. The engine then operates at the set speed. When the engine speed reduces to less than the set speed, due to increased load, the governor spring moves the oil control valve slightly, and opens the oil-supply passage to the large chamber. This moves the piston toward the increased-fuel position until the governor spring force and the flyweight force are equal. The piston is then again hydraulically locked and the engine operates at the set speed.

High- and low-idle speed are adjusted by limiting the throttle movement and thereby the energy exerted on the governor spring. To stop the engine, the oil control valve is moved, through linkage, to

the left and the enlarged end of the valve moves the piston so that it places the fuel control rack in the no-fuel position.

Indirect-acting Governor A schematic drawing of an indirect-acting hydraulic governor, with a single-acting piston is shown in Fig. 25–17. This governor is of the variable-speed design with an adjustable droop-speed mechanism.

The governor flyweights are pinned to the ball-head assembly which is driven through a set of gears by the camshaft or the blower. The ballhead shaft at the lower end, drives the drive gear of the auxiliary governor pump (28 in Fig. 25–17). This pump has a pressure relief valve to limit the maximum operating pressure. The engine lubrication system supplies oil to the auxiliary pump.

The lower governor spring seat is part of the lap-fitted pilot-valve plunger and fits into the upper part of the ballhead shaft. The governor spring (speeder spring) is located between the lower spring seat and

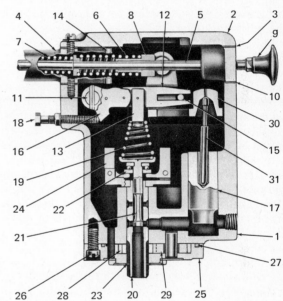

1. Governor housing	18. Maximum-speed adjusting screw
2. Governor cover	
3. Gasket cover	19. Speeder spring
4. Subcap	20. Ballhead assembly
5. Fuel rod	21. Pilot-valve plunger
6. Shutdown spring	22. Plunger bearing
7. Fuel-rod spring	23. Lock ring
8. Fuel-rod collar	24. Flyweight
9. Fuel-rod knob	25. Governor base
10. Housing-to-subcap gasket	26. Base-to-housing screw
11. Speed-adjusting shaft	27. Housing-to-base seal ring
12. Floating lever	28. Oil-pump drive gear
13. Spring fork	29. Oil-pump driven gear
14. Stop pin	30. Terminal lever
15. Droop-speed adjusting bracket	31. Terminal-lever-to piston pin
16. Speed adjusting lever	
17. Power piston	

Fig. 25–17 Sectional view of an SG hydraulic governor. (GMC Detroit Diesel Allison Div.)

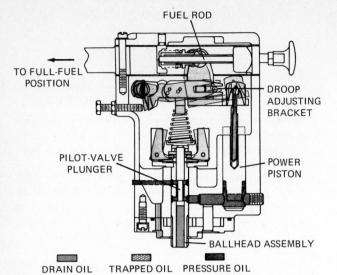

Fig. 25–18 Governor position and oil flow when spring force increases (or as load increases and speed tends to decrease). *(GMC Detroit Diesel Allison Div.)*

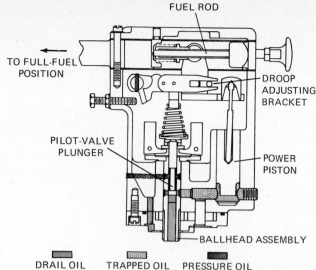

Fig. 25–19 Governor position when engine operates at constant speed. *(GMC Detroit Diesel Allison Div.)*

the spring fork. The spring fork is pinned to the floating lever and the floating lever is pinned to the speed adjusting lever. The speed adjusting lever is pinned to the speed adjusting shaft.

The maximum-speed adjusting screw (18 in Fig. 25–17) limits the travel of the speed adjusting lever, thereby limiting the energy on the governor spring. The power piston is slide-fitted into its cylinder (which is part of the governor housing) and is linked through the floating pin to the terminal lever. The terminal lever is connected to the fuel rod and pivots on the terminal lever shafts. The load-limiting screw limits the pivoting of the terminal lever and therefore limits the travel of the fuel rod. When the engine is not operating, the fuel rod is in the no-fuel position, having been forced there by the fuel-rod spring. The power piston has been forced downward by the terminal lever.

To overcome prolonged cranking of an engine using this type of governor, the terminal lever shaft must be rotated to the fuel-on position, as this moves the fuel rod into the full-fuel position. (A long cranking period is otherwise required to create enough pressure to move the power piston upward and to move the terminal lever and the fuel rod to the fuel-on position.)

An advantage to this type of governor arrangement is that the governor acts as an automatic shutdown device. In the event that the oil supply to the governor is lost, the auxiliary pump cannot supply oil to the power piston. The power piston is then forced downward, causing the fuel rod to move to the no-fuel position.

When the engine is operating and the operator has moved the speed adjusting shaft to half engine speed, the speed adjusting lever rotates, the floating lever pivots on the droop-speed adjusting pin, causing the spring fork to move downward and compress the governor spring. This action moves the pilot-valve plunger against the flyweight force. The land on the pilot-valve plunger opens the oil passage to the power piston. (During this time, engine oil from

the engine lubrication system is supplied to the auxiliary pump and the auxiliary pump raises the pressure to the set value of the relief valve.)

The oil (under pressure) then enters the area below the power piston (see Fig. 25–18). The power piston and the floating pin rise, pivoting the terminal lever and causing the fuel rod to move toward the full-fuel position. However, when the terminal lever pivots, it immediately reduces the spring force on the governor spring because the droop-speed adjusting bracket is bolted to the terminal lever. The pin of the droop-speed adjusting bracket lifts the floating lever and reduces the spring force on the governor spring. The engine responds to the increased injected fuel, the flyweight force increases, and the flyweight finger causes the thrust bearing to move the plunger upward. The upward movement of the pilot-valve plunger opens the oil passage in the ballhead, reducing the oil pressure below the power piston. The fuel-rod spring now moves the terminal lever and the droop-speed pin. The fulcrum lever then moves slightly, increasing the force on the governor spring. At the same time the floating pin moves the piston downward. Upon equalization of the force on the governor spring and the force of the flyweights, the pilot-valve plunger closes the oil port in the ballhead. The oil pressure remains fixed in the passage and below the power piston (Fig. 25–19).

NOTE To decrease the droop speed, move the droop-speed adjusting bracket to the right. (This will increase the fulcrum.) To increase the droop speed, move the droop-speed adjusting bracket to the left. (This will reduce the fulcrum.)

When one force is changed (say the operator moves the throttle to the high-idle position), the force on the governor spring is increased and the pilot-valve plunger opens the oil passage to the ballhead and the power piston. When the engine speed drops due to increased load, the flyweight force decreases and the governor spring force moves the

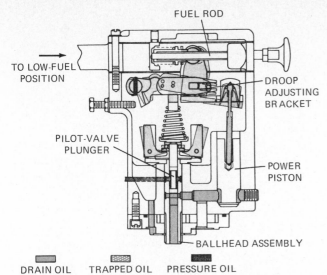

FUEL ROD

TO LOW-FUEL POSITION

DROOP ADJUSTING BRACKET

PILOT-VALVE PLUNGER

POWER PISTON

BALLHEAD ASSEMBLY

DRAIN OIL   TRAPPED OIL   PRESSURE OIL

Fig. 25–20 Governor position when spring force is decreased or when the load decreases and engine speed tends to increase. *(GMC Detroit Diesel Allison Div.)*

pilot-valve plunger downward, opening the oil passage. In either instance, the power piston is forced upward, resulting in an increase in fuel.

When the operator moves the throttle to reduce speed, the force on the governor spring is reduced and the flyweight force moves the pilot-valve plunger upward. The passage in the ballhead opens, and the pressure below the power piston reduces. The power piston then moves downward resulting in reduced fuel and reduced engine speed (Fig. 25–20).

When the engine is operating at a fixed speed and suddenly some of the load is taken from it, the flyweight force increases because of the increased engine speed. The governor finger moves the lower spring seat and the pilot-valve plunger, thus opening the oil passage in the ballhead. Oil drains from below the power piston and the fuel rod moves toward the low-fuel position.

Questions

1. Why do diesel engines require a governor?

2. What are the two functions of a governor?

3. List the major differences between (a) a mechanical governor and a servo governor, (b) a mechanical governor and a hydraulic governor, and (c) a mechanical governor and a pneumatic governor.

4. Why do most governors use some kind of a pulsation device?

5. List the basic components of most governors.

6. What would happen to speed and governor action if the governor spring were broken?

7. Name the components which limit the maximum engine speed on an engine having a pneumatic governor.

8. If the diaphragm of a pneumatic governor were punctured, what result would it have on rated speed and on idle speed.

9. List the main differences between the hydraulic governor shown in Fig. 25–15 and the governor shown in Fig. 25–17.

10. Name the components which limit the maximum speed of the governor shown in Fig. 25–17.

11. In order to accomplish maximum engine torque, which of the following is necessary: (a) the plunger must turn more and the governor weight must move further outward; (b) the fuel rack must move inward and the speeder spring must expand; (c) the engine speed must drop from high idle to torque speed and the fuel rack must move to the more-fuel position.

12. The narrow band of speed variation through which the governor makes no corrective movement is called (a) speed group, (b) dead band, or (c) speed droop.

13. Which of the following factors are involved in Robert Bosch torque backup arrangement: (a) torque-cam adjustable stop plate, (b) control rack and bumper spring, or (c) torque-cam stop plate and torsional spring.

UNIT 26

Antipollution Control Devices

Function of Control Devices A diesel engine which emits excessive smoke usually requires a tuneup or servicing because, evidently, one or more components are not operating effectively. However, a diesel engine shows excessive smoke during acceleration because the governor momentarily overacts, causing too much fuel to be metered and injected.

Antipollution regulations are becoming increasingly strict and engine manufacturers are therefore working extensively to minimize engine noise and to reduce exhaust smoke. As a solution to the problem of dense exhaust smoke emission during acceleration, control devices have been designed for turbocharged engines to limit the fuel-rack travel or the metering valve rotation, to reduce the manifold pressure, or to slow down the motion of the injector control lever. Each of these devices reduces or delays the quantity of fuel injected during engine acceleration to counteract the quick governor action. It is necessary to counteract the quick governor action because of the pressure lag in the intake manifold which remains until the exhaust gases force the turbocharger to turn at a higher speed (creating a higher manifold pressure). The end result is a more proportionate fuel/air ratio.

Aneroid Control The control devices referred to above are known as *aneroids*, *fuel-ratio control valves*, or *throttle-delay mechanisms*. The word *aneroid* is derived from the Greek language and means "without liquid." The aneroid control senses changes in (manifold) air pressure without the use of liquids such as mercury or water.

Cummins' turbocharged engine was one of the first to use an aneroid control valve. This valve bypasses fuel thereby reducing the fuel pressure until there is sufficient turbocharged pressure in the intake manifold.

The aneroid is connected with hoses to the fuel-injection system and to the intake manifold (Fig. 26–1). The upper hose connects the bellows chamber with the intake manifold; the center hose connects the inlet of the aneroid with the throttle fuel passage; and the lower hose connects the bypass outlet with the inlet side of the gear pump.

The bellows divides the aneroid into two parts. The upper bellows chamber is airtight and the bellows are forced upward by the bellows spring. An air breather in the lower part of the housing allows the bellows to displace air as it oscillates during engine operation (Fig. 26–2). The bellows are fastened to a shaft which is linked with the lever of the valve throttling rod. The rotation of this valve throttling rod opens or closes the fuel bypass port leading to the return fitting (Fig. 26–3). The spring-loaded

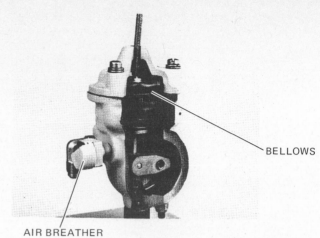

AIR BREATHER   BELLOWS

Fig. 26–2 Cutaway view of the aneroid control. (*Cummins Engine Co., Inc.*)

STARTING PLUNGER

VALVE THROTTLING ROD   FUEL BYPASS PORT

Fig. 26–3 Cutaway view of the valve throttling rod with open port (no air pressure in bellows). (*Cummins Engine Co., Inc.*)

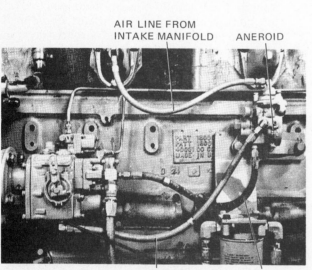

AIR LINE FROM INTAKE MANIFOLD   ANEROID

FUEL PRESSURE LINE   FUEL RETURN LINE

Fig. 26–1 Aneroid-control connections. (*Cummins Engine Co., Inc.*)

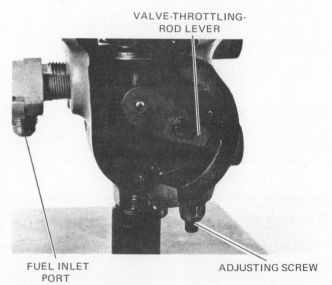

VALVE-THROTTLING-ROD LEVER

FUEL INLET PORT   ADJUSTING SCREW

Fig. 26–4 Sectional view showing valve-throttling-rod lever against the adjusting screw. (*Cummins Engine Co., Inc.*)

plunger (starting plunger) above the valve throttling rod blocks the fuel flow to the valve so that the engine can be started.

Aneroid Operation When the engine is not operating, the bellows are held in an upward position by the force of the bellows spring. The adjusting screw has limited the rotation of the valve-throttling-rod lever (Fig. 26–4), thereby limiting upward travel of the bellows and maximizing fuel bypass opening.

During cranking, the gear pump supplies fuel to the injectors and to the aneroid inlet. The starting plunger blocks the fuel passage and this increases the fuel pressure in the fuel manifold (injectors). When the fuel pressure has increased to say, 12 to 20 psi [0.84 to 1.4 kg/cm²], the starting plunger is forced against its spring force and the aneroid is then connected to the fuel-injection system.

After the engine has started and its speed has increased, the turbocharger pressure also increases. This pressure acts on the bellows. The bellows and its rod are forced downward, rotating the valve throttling rod (in proportion to the intake manifold pressure). The rotation of the valve reduces the fuel bypass opening and the fuel pressure in the fuel manifold increases. When the intake-manifold pressure reaches, say, 30 inHg [76.2 cmHg], the bypass valve is fully closed and the PT fuel pump takes over the control of the fuel-manifold pressure. If the intake-manifold pressure drops below 30 inHg [76.2 cmHg], the bellows spring forces the bellows upward, and the valve throttling rod starts to rotate. The bypass then opens, causing a reduction in the fuel-manifold pressure and resulting in a more proportionate fuel/air ratio.

The correct selection of the bellows spring as well as the correct shim adjustment (when testing on the test stand) can determine when and how much fuel will be bypassed. However, the maximum fuel bypass opening is set by the adjusting screw.

Aneroid or Fuel-Ratio Control Used with Helix-metering Injection Pumps A simplified drawing of an aneroid used with helix-metering injection pumps is shown in Fig. 26–5. This type of aneroid differs from that used with the PT fuel-injection system. The diaphragm (bellows) also separates the aneroid into two parts and (the diaphragm) is bolted to the operating rod. The rod is directly linked to the fuel rack. The left side of the aneroid (in Fig. 26–5) is the sealed chamber and is connected to the intake manifold. The right side of the aneroid is vented to the atmosphere. The operating spring forces the diaphragm to the left (which moves the fuel control rack) to reduce fuel.

An indirect-acting aneroid is shown in Fig. 26–6. From this illustration you can see that the operating shaft is indirectly linked to the fuel rack. However, both units operate on the same principle. (See Unit 34, "American Bosch Distributor-Type Fuel-Injection Pump" and Unit 30, "Caterpillar Fuel-Injection System.")

Indirect-acting Aneroid Operation When the engine is not operating and the operator moves the

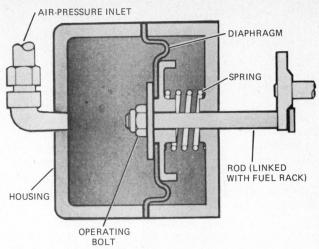

Fig. 26–5 Simplified view of the aneroid used with a helix-metering injection pump.

Fig. 26–6 Cutaway view of an indirect-acting aneroid used with a helix-metering injection pump.

throttle to half engine speed, the aneroid spring limits the travel of the fuel rack. When the engine is running at half speed the intake-manifold pressure is low and the operating spring, assisted by the flyweight force, forces the diaphragm to the left (Fig. 26–5). The operating bolt then moves the fuel rack to a position where it meters less fuel so that the fuel/air ratio is correct and the engine operates with a near clear exhaust smoke.

When the throttle is moved to the full-fuel position, the governor instantly responds. At this time the governor cannot move the fuel rack to the full-fuel position because the operating spring has a greater force on the diaphragm than the turbocharger pressure has. As the intake-manifold pressure increases, the force on the diaphragm increases and so it moves to the right (in Fig. 26–5), allowing the fuel rack to move to the increased-fuel position. When the maximum turbocharger boost pressure is reached, the fuel rack is free to move by the action of the governor. When the aneroid is adjusted properly, the correct amount of fuel is injected as air is available for combustion. Maximum engine acceleration and load acceptance is assured with the minimum of exhaust smoke.

Throttle-Delay Mechanism The emission device used on some Detroit Diesel engines is a throttle-delay mechanism (Fig. 26–7). (See also the sections on the throttle-delay mechanism in Unit 31.)

This device also reduces the exhaust smoke during acceleration. It delays the action of the governor by holding the control tube hydraulically until the spring force moves the throttle-delay piston. The movement of the piston is restricted by the oil it must force out of the cylinder through a small orifice. This action slows the rotation of the control tube and thereby the movement of the fuel rack. The engine speed increases, gradually reducing the density of the exhaust smoke.

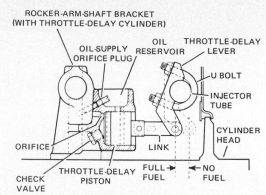

Fig. 26–7 Throttle-delay mechanism.

Questions

1. What is the purpose of antipollution control devices?

2. What would occur if the spring-loaded plunger (Cummins) seized in the open position?

3. In what way would it be noticeable if the bellows were ruptured?

4. If the bellows spring were broken, would the exhaust smoke density increase or decrease?

5. How would engine performance be affected if the atmospheric side of the aneroid or fuel-ratio control were plugged?

UNIT 27

Fuel-Injection Nozzles and Holders

Injector Assembly To a large degree the successful operation of any diesel engine depends on the functional efficiency of its injector assembly.

The selection of the injection holder and nozzle depends on the construction of the combustion chamber, the location of the holder and nozzle assembly and the type of fuel-injection system. This unit, however, will be restricted to the injector assemblies used with Robert Bosch, American Bosch, CAV, Sims, and Roosa Master fuel-injection pumps.

Fuel-Injection Holder The main purpose of the fuel-injection-holder assembly is to position and to hold the fuel-injection nozzle in the cylinder head. The assembly consists of a steel body and the components shown in Fig. 27–1. Machined passages within the body connect the fuel inlet with the injection nozzle. The leak-off fuel (used to lubricate the nozzle valve) fills the spindle and adjusting device area and returns via the leak-off connection to the fuel tank. The lower surface of the nozzle holder is lapped and forms a fuel-tight joint with the fuel-injection-nozzle body.

Fuel-Injection Nozzle The main purpose of the fuel-injection nozzle is to direct and atomize the metered fuel into the combustion chamber. The combustion chamber design dictates the type of nozzle,

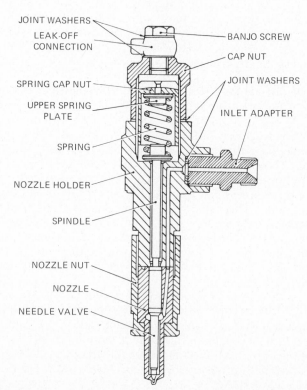

Fig. 27–1 Sectional view of a multihole injector assembly. *(CAV Ltd.)*

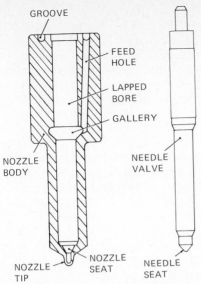

Fig. 27–2 Sectional view of a typical nozzle. *(CAV Ltd.)*

the droplet size, and the spray required to achieve complete combustion within a given time and space [see Combustion Chamber Design (Unit 4)].

Differential-type injection nozzles which open inwardly are the most common. Their designs vary and include the short- or long-reach multihole type, single hole (straight or offset) type, standard pintle type, throttling or pentaux-type fuel-injection nozzles. Multihole nozzles are used with direct-injection combustion chamber designs and the pintle or throttling types are used in engines with precombustion chambers, turbulence chambers, air or energy cells.

Each fuel-injection nozzle has a nozzle body and a needle valve (Fig. 27–2). The lapped joint face of the body has an angular groove and one-fuel or three-fuel passage leading to the pressure chamber

(gallery). The nozzle-body bore and the needle valve are lapped to a close tolerance and are matched sets.

The nozzle-body seat is commonly 1½° smaller than the angle of the needle-valve seat. When the axis of the spray cone is not in line with the axis of the nozzle, two dowels are used to hold the nozzle body in proper position with the fuel-injection holder.

The pressure pin of the needle valve protrudes into the holder and rests in the reset of the spring-loaded spindle (see Fig. 27–1). The upper shoulder of the needle valve protrudes inward. The distance from the shoulder to the face of the nozzle body is called the *needle lift*. It varies from 0.012 to 0.027 in [0.30 to 0.68 mm] depending on nozzle type, fuel flow, and spray pattern. The predetermined pressure taper just above the needle-valve seat on which the fuel pressure acts, lifts the needle off its seat against the spring force.

Injector Action Fuel from the injection pump enters and pressurizes the fuel in the supply passages and pressure chamber. When the force on the lift area is greater than the set spring force on the spindle, the needle valve lifts off its seat and comes to rest with its upper shoulder against the face of the holder. Fuel is forced out into the combustion chamber in a spray pattern, which depends on the type of nozzle used (see Fig. 27–3). The opening pressure of pintle-type nozzles varies from 1000 to 2200 psi [70.308 to 154.66 kg/cm²] whereas multihole-type nozzles vary in opening pressure from 2000 to 3200 psi [140.600 to 224.960 kg/cm²].

As fuel is sprayed into the combustion chamber, a pressure drop occurs within the pressure chamber. However, you will notice when testing an injector of this type on an injector tester, that the needle valve will seat momentarily and you will hear a chatter.

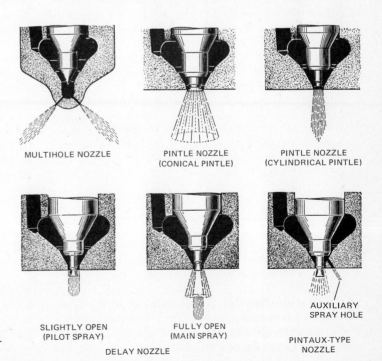

Fig. 27–3 Spray patterns of various nozzle types. *(Robert Bosch GmbH, Stuttgart, West Germany.)*

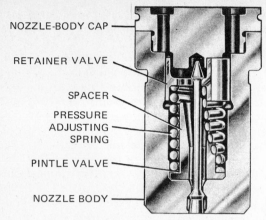

NOZZLE-BODY CAP

RETAINER VALVE

SPACER

PRESSURE ADJUSTING SPRING

PINTLE VALVE

NOZZLE BODY

Fig. 27–4 Sectional view of an ADE pintle-type nozzle. *(American Bosch-Ambac Industries Inc.)*

When the fuel-injection pump spills back, delivery from the pump ceases, the pressure in the pressure chamber drops instantly, and the pressure spring snaps the needle valve onto its seat, preventing raw fuel from leaving the nozzle. The pintle-type injection nozzle (capsule type) shown in Fig. 27–4 requires a different type of injection holder. The holder has a single fuel passage and the nozzle consists of the illustrated components. This nozzle design is used on diesel engines manufactured in Europe and is of the outward-opening pintle-valve design.

Operation of Outward Pintle Nozzle When the pintle valve is at rest position the pressure spring forces the retainer valve and the pintle valve upward because the valve is connected to the retainer valve, forcing the pintle valve against the nozzle-body seat (Fig. 27–4). As fuel from the injection pump increases the pressure within the assembly, the forces on the retainer and pintle valve are increased. When these forces exceed the spring pressure, the retainer valve and the pintle valve are forced outward. The spring is compressed until the retainer valve bottoms against the stop shoulder of the nozzle, thereby limiting the outward movement of the pintle. The opening pressure is governed by the spring force and is not adjustable.

The valve remains in an open position until the fuel-injection pump spills back. This causes a rapid drop in pressure. The pressure spring force then moves the retainer valve and the pintle valve upward, forcing the pintle valve to seat against the nozzle-body seat.

FUEL–INJECTION NOZZLE AND HOLDER SERVICE

The main factors which reduce the service life of an injector are abrasives (dirt), heat, water, and improper installation.

Abrasives in the fuel result in wear to the fuel-carrying components, enlarge the close tolerances of the needle valve and bore, and sometimes damage the injector seat and the spray hole.

Excessive heat changes the expansion rate of the nozzle resulting in a higher wear rate.

Water causes corrosion and weakens the lubricant. Furthermore, since water will not pass through a multihole-type nozzle tip, it may break the tip off.

Improper installation can result in inadequate cooling, seizure of the needle valve, or poor valve seating.

Locating Faulty Injector An injector should only be removed when there is trouble in the assembly or when the engine is in the shop for a major overhaul. However, some manufacturers recommend a readjustment of the opening pressure after 2000 hours of operation to maintain a good spray pattern. This readjustment should preferably be made with the injector installed. If this is not possible, an injector tester should be used.

To locate a faulty injector which may have caused incomplete combustion, engine misfiring, rough running, or excessive smoke emission, start the engine and check for leaking high-pressure connections. Crack open the individual fuel-injection lines one at a time, either at the injector or at the fuel-injection pump. With one fuel-injection line cracked open, listen for a change in engine sound. A faulty injector is present when the sound of the engine has *not* changed.

Removing Injector The injector position and its hold-down device may differ from engine to engine, but the removal procedure is generally the same.

It is essential, first, to clean the surrounding area thoroughly (cleanliness cannot be overemphasized) and, when possible, dry it with compressed air. Remove the high-pressure fuel lines from the injector inlet and from the leak-off fuel lines. Instantly cap all openings to prevent dirt from entering. Use two wrenches if necessary, when loosening the connections, in order to prevent damage to the fittings or twisting of the fuel lines. Remove the hold-down device which secures the injector to the cylinder head. Use two injector prybars to pry the injector up parallel from its bore. When the injector is removed, cap the nozzle tip to protect it from damage and cover the opening in the cylinder head.

Testing Injector Prior to Disassembly An important step in the procedure for servicing an injector is to test it before disassembling it. In addition to preventing unnecessary service, a superficial check will often help you to establish the cause of the failure and/or determine if the assembly needs to be repaired or merely cleaned and readjusted.

Before you install the injector to the injector tester, visually check the spray tip for discoloration, damage, and corrosion.

CAUTION When installing the injector to the tester, install it in such a way that the spray cannot injure you or a fellow worker. The skin can easily be punctured by the high pressure of the spray and such punctures may result in blood poisoning.

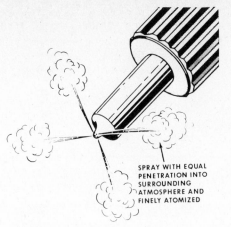

SPRAY WITH EQUAL PENETRATION INTO SURROUNDING ATMOSPHERE AND FINELY ATOMIZED

Fig. 27–5 Checking the spray pattern. *(J. I. Case Co., Components Div.)*

After the air is removed, tighten the connection and switch on the air blower, so that the fuel spray will be drawn back and will drain into the reservoir.

Testing Opening and Closing Pressure   To test the opening and closing pressure, first close the pressure-gauge valve and slightly open the fuel-supply valve. Operate the hand pump rapidly for a few strokes to clear the air from the injector. With the air removed from the assembly, open the pressure-gauge valve about one-eighth of a turn. Slowly operate the hand pump to raise the pressure until the needle valve lifts off its seat and fuel sprays from the nozzle.

Observe and record the opening and closing pressure indicated on the pressure gauge. The valve closing pressure should not be lower than 300 psi [21.092 kg/cm²] of the opening pressure. If it is lower, there is a sticking needle valve.

Valve-Seat Test   The valve-seat test should be made next. Wipe off the nozzle tip and bring the pressure to about 100 psi [7.031 kg/cm²] below the opening pressure. Check the spray tip. The tip must stay dry. If there is any evidence of fuel at the tip, the needle-valve seat is defective.

Back Leakage Test   Next, check for leak-off (back leakage). This test is used to check the fit of the needle valve and nozzle bore and the lapped surface of the nozzle body and nozzle holder. Raise the pressure within 100 psi [7.031 kg/cm²] of the opening pressure and time the pressure drop. An average pressure drop of not more than 880 psi [618.728 kg/cm²] in 6 seconds is an indication that the mating surfaces are not leaking and the needle valve is adequately lubricated. If the pressure drop is higher than specified, then the needle valve fits too loosely or the lapped surfaces are not sealing. If the back leakage time is high, it is an indication of a tight needle valve.

Spray Pattern Test   Finally, check the spray pattern. Close the pressure-gauge valve and operate the hand pump with a sharp stroke about twice every second. The spray should be without distortion and no unatomized fuel should be visible. Each spray (in a multihole nozzle) should have the same pattern and be spaced evenly from the others (Fig. 27–5). You should hear a chatter. When the spray pattern is distorted, there may be a damaged or choked hole.

When an injector tester is not available, remove one injector at a time and reconnect the injector to the fuel line so that the spray pattern and nozzle-seat leakage can be observed when cranking the engine.

NOTE   After the injector has passed all its tests, use a fine brass wire brush to clean the external deposits from the nozzle and nozzle holder. Retest and adjust the opening pressure before installing or storing the injector.

Disassembling Injector   If the injector has failed any of the tests, it should be disassembled, cleaned, examined, reassembled, and adjusted.

Injector servicing is simple, providing you follow the correct procedure and have the necessary equipment and facilities to do the job. A clean workshop is the first essential. Necessary equipment for this job includes a holding rack for the injectors, trays to hold the dismantled components, wash containers, a holding vise, and tools for cleaning the components (Fig. 27–6). An injector tester is essential for readjusting the units after reassembly.

NOTE   Disassemble one injector at a time and place the components in a container filled with fuel. Nozzle bodies and needle valves are never interchangeable. They are lapped together as a unit.

First wash and brush off all external dirt and carbon from the injector. Place the injector in a holding fixture and dismantle the injector in the correct sequence. Remove the cap nut and washer first to avoid damaging the locating dowels, needle valve, and body. Remove the locknut, spring adjusting screw, spring cap, spring seat, spring and spindle, and place them all in clean fuel.

Turn the injector upside down in the holding fixture and remove the nozzle nut and the nozzle assembly. Be careful not to let the needle valve fall from the assembly because it is damaged easily.

Fig. 27–6 Tools and equipment needed to service an injector.

Cleaning and Inspecting Injector

CAUTION Never use hard or sharp tools, emery cloth, crocus cloth, jeweller's rouge, grinding compound, or other abrasive material to clean the nozzle-valve body or any part of the assembly.

Some service shops clean the disassembled injector in a 15 percent caustic soda solution, an air-agitated solvent tank, or in an ultrasonic cleaner. When the nozzle valve cannot easily be removed even after soaking in a carbon solvent, use a hydraulic nozzle-valve extracter. Apply hydraulic pressure from the nozzle test stand to force the nozzle valve out of the nozzle body.

First clean the needle valve with a brass wire brush and then use a soft brass tool to remove any remaining carbon from the needle tip or tip area. Next, clean the needle valve with mutton tallow and a soft cloth or felt pad. Check for pitting and discoloration. Nicks and scratches on the nozzle valve usually are attributable to poor fuel storage and handling practices and/or poor fuel-filter maintenance.

The nozzle body should first be cleaned externally with a soft brass brush and visually checked for discoloration and damage to the pressure face and spray holes. The lapped pressure surface of the body that contacts the holder must be clean. If it is not, the surface must be lapped on a lapping plate, as shown in Fig. 27–7. Check and clean the lapping plate first. Apply tallow or jeweller's rouge evenly on the plate so that there is no danger of a dry surface contacting the nozzle body. Hold the body so that an even force will be exerted on the entire surface while moving the nozzle body smoothly and steadily in the form of a figure 8. **CAUTION** Do not rock the nozzle body during lapping.

Rinse the nozzle body thoroughly in clean diesel fuel. Inspect the lapped surface to ensure that it is completely smooth. Then use a drill or a wire to clean the internal fuel duct (Fig. 27–8). Use the special scraper tool provided to clean the fuel pressure chamber (cavity). Press the hook end of the tool hard against the side of the cavity and rotate it to free any carbon (Fig. 27–9). After selecting a suitable tip size, clean the dome cavity as shown in Fig. 27–10. A bronze scraper should be used to remove any remaining dirt and carbon from the nozzle seat (Fig. 27–11). To clean carbon from the spray holes, use a hand (pin) vise and the correct size wire (supplied with the cleaning kit). Before you use the wire, hone the end until it is smooth and free of burrs. Sometimes it is necessary to taper the end with the oil stone. Insert the wire in the hand vise allowing it to extend about ⅛ in [3.175 mm]. Take extreme care when cleaning the holes to gently rotate rather than push the wire through the hole (Fig. 27–12). Thoroughly rinse the valve and nozzle body in clean diesel fuel before you reassemble them.

A better cleaning method, however, is to utilize a nozzle flushing device. Insert the nozzle and secure it in the flushing device and then mount it to the injector tester. Start pumping the oil. The oil will

Fig. 27–7 Lapping the nozzle body.

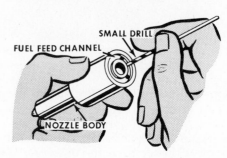

Fig. 27–8 Cleaning the fuel feed channel. (*J. I. Case Co., Components Div.*)

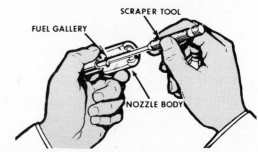

Fig. 27–9 Cleaning the fuel pressure chamber. (*J. I. Case Co., Components Div.*)

flow from outside the nozzle through the spray holes to the inside, flushing out the main carbon particles.

To check freedom of the valve within the body, lift the needle valve about one-third of its length out of its body and hold the body at a 45° angle. The valve should slide onto its seat, as shown in Fig. 27–13. If the valve is sticky, clean it with mutton tallow to remove any varnish or carbon from the surfaces. If the valve continues to stick, the nozzle assembly must be replaced.

Checking Needle-Valve Lift It is important that you measure the lift height of the needle. To do this, place the spindle of the depth dial indicator against the upper shoulder of the needle valve. Zero the dial. Then place the spindle against the face of the nozzle body. The difference on the dial reading indicates the lift height of the needle. The lift height must be within specification to ensure the

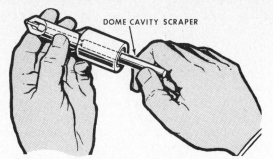

Fig. 27–10 Cleaning the dome cavity. (J. I. Case Co., Components Div.)

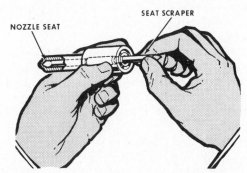

Fig. 27–11 Cleaning the nozzle seat. (J. I. Case Co., Components Div.)

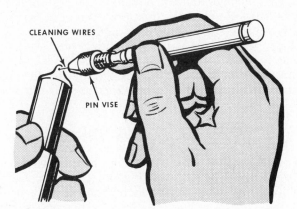

Fig. 27–12 Cleaning the spray hole. (J. I. Case Co., Components Div.)

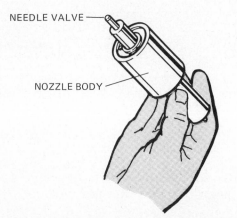

Fig. 27–13 Checking valve freeness. (J. I. Case Co., Components Div.)

proper spray pattern and atomization. If it is greater than specified, you may lap the face of the nozzle body to bring the lift height to specification.

Holder Service   Next, clean the external and internal surfaces of the remaining parts of the holder assembly. After cleaning, carefully inspect the holder. The pressure face must show no sign of scoring or corrosion and the dowel pins, if used, must be in good condition.

Remove the dowel pins and lap the mating surface of the nozzle body to a mirror surface finish.

NOTE   After using the lapping plates, wipe them clean and apply a protective coating of petroleum jelly to protect them against corrosion.

Reassembling Injector   After the final rinse, position the nozzle holder in the holding fixture and place the nozzle assembly over the dowels (if dowels are used). Install the nut and, using a torque wrench, tighten the nut to the recommended torque. Remember that an over-torqued nut can cause distortion and needle-valve seizure. If it is too loose, metered fuel may be lost.

Turn the injector over. Install the spindle, spring, spring retainer, spring cap nut, adjusting screw, and locknut.

When positioning the new gasket and cap nut, remember that injectors which use adjusting shims rather than an adjusting screw require the same number of shims as were used before the injector was dismantled (Fig. 27–14).

NOTE   With American Bosch and CAV pintle-type nozzles, use a centering sleeve before you tighten the nozzle nut to ensure proper alignment.

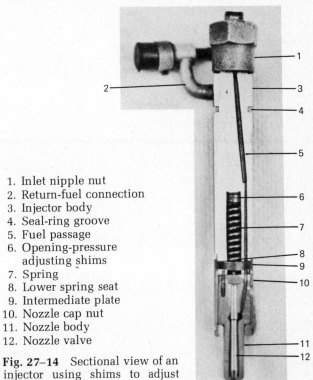

1. Inlet nipple nut
2. Return-fuel connection
3. Injector body
4. Seal-ring groove
5. Fuel passage
6. Opening-pressure adjusting shims
7. Spring
8. Lower spring seat
9. Intermediate plate
10. Nozzle cap nut
11. Nozzle body
12. Nozzle valve

Fig. 27–14 Sectional view of an injector using shims to adjust opening pressure.

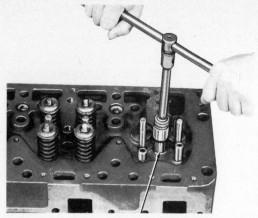

CARBON REMOVING TOOL

Fig. 27–15 Cleaning the injector bore. *(Allis-Chalmers Engine Div.)*

Testing and Adjusting Injector   The reassembled injector is now ready to be adjusted and tested. First, adjust the opening pressure to specification by using the adjusting screw or by removing or installing shims.

As a general rule when new injector springs are used, the opening pressure should be set about 50 psi [3.515 kg/cm²] higher than the specified opening pressure to allow for initial settling of the spring. Next, check or test in the following order: spray pattern and atomization, back leakage, and needle-valve-seat (forward) leakage. When the injector performs satisfactorily, remove it from the test stand and replace all protection caps.

Reconditioned Needle Valve and Nozzle Seat   If the needle valve and the nozzle bore are not damaged but the seats do leak, or if the seats are corroded, the seats should be reconditioned (lapped). However, such reconditioning should be done by someone who is not only particularly skilled in this field but who also has the required special equipment necessary. It is vital that the components be lapped to precision.

Installing Injector   Regardless of the type of injector you are installing, the injector bore (recessed in the cylinder head) must first be cleaned thoroughly and the seating surface of the nozzle or nozzle nut made smooth. Use only the recommended reamer to clean the bore and the seating surface (Fig. 27–15).

Turn the reamer in a clockwise direction only; otherwise the reamer can be damaged. Apply a light coat of grease to the cutting edge of the reamer; the scraped off carbon will cling to this. The coating will prevent the carbon from entering the combustion area.

Clean all the bores. Then crank the engine over to blow out the remaining loose deposits. Remove the nozzle protection cap. Install a new gasket onto the nozzle or into the bore.   **NOTE**   Be sure to remove the old gasket.

Install the injector into the bore and take care not to damage the nozzle tip. If force is required to place the injector into the bore, it is an indication that some of the carbon has not been removed. The injector should slide into position without being forced.

Tighten the hold-down device according to the manufacturer's specification. When two hold-down nuts are employed, use a torque wrench and tighten alternately one and then the other to specification.

Remove the dust-protection cap from the high-pressure fuel lines. Pull the decompressor, crank the engine over to flush, and bleed the high-pressure fuel line. Remove the dust-protection cap from the injector and connect the high- and low-pressure fuel line. Make certain that the connections are tightened correctly. Start the engine and check all the fuel lines for leakage. Check the injector for compression blow-by past the nozzle or nozzle nut seat.

ROOSA MASTER FUEL INJECTOR

Roosa Master fuel injectors have multihole differential-type nozzles (Fig. 27–16). The nozzle body is permanently fastened to the injector body.

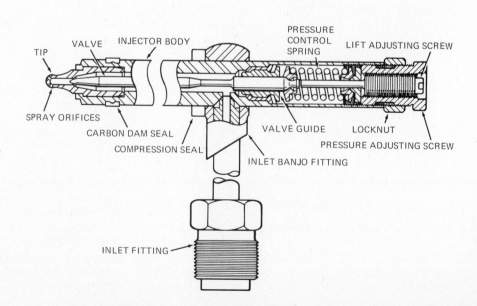

Fig. 27–16 Sectional view of a Roosa Master injector. *(J. I. Case Co., Components Div.)*

The needle valve and the spindle are one unit. The valve is guided by the nozzle and by the closely fitted valve guide. The lower spring seat rests against the needle valve whereas the upper spring seat rests against the pressure adjusting screw. The lift adjusting screw is threaded into the pressure adjusting screw.

The inlet connections and fittings vary between different injectors. Two seals are used, a Teflon seal to prevent combustion accumulation in the cylinderhead bore, and a nylon seal to prevent loss of cylinder compression or engine oil from entering the injector bore. The injector body is Teflon-coated from the carbon seal to the compression seal.

Although the operating principle of the Roosa Master injector is the same as the injectors previously discussed, the Roosa Master needle valve opens and closes more rapidly and when being tested, a distinct chatter is noticeable. The leak-off fuel that passes through the closely controlled clearance of the valve and valve guide lubricates the upper half of the injector and returns through the leak-off connection to the tank. The lift of the needle valve is limited by the extension of the lift adjusting screw.

Removing Roosa Master Injector   It is essential when loosening the high-pressure fitting to use one hand and two wrenches to prevent damaging the inlet connections. After removing the clamp assembly, turn, and at the same time, pull the injector from its bore. If you cannot remove the injector easily by hand, use an injector puller attached to a slide hammer.   **NOTE**   Never use any other device because the injector body may be distorted and it will then have to be replaced.

Testing Prior to Servicing   A Roosa Master injector must also be tested prior to servicing. Use the same test procedure as previously outlined. However, when you make the seat-leakage test, a slight dampness after 10 seconds is permissible with used injectors. When you make the leak-off test, install the injector as shown in Fig. 27–17, then raise the pressure to within 200 psi [14.062 kg/cm²] of the opening pressure. As soon as leak-off drops appear from the return end of the injector, start counting them. When using 2D fuel at a temperature of 70°F [21.1°C], the leak-off should be 3 to 10 drops in 30 seconds. If the leak-off is higher, the injector must be replaced because of the high clearance between the valve guide and the needle valve, or an oversized needle valve must be lapped to the worn valve guide.

Servicing Roosa Master Injector   Remove both seals. Take care not to damage the carbon dam groove or the Teflon coating on the injector body. Clean the nozzle tip and inlet connection with a brass wire brush.

Mount the injector in the holding tool and loosen the pressure adjusting locknut. Then hold the injector in one hand, tilt it backward and remove the pressure adjusting screw, ball washer, upper seat, spring

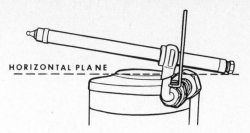

Fig. 27–17   Injector leak-off test. *(J. I. Case Co., Components Div.)*

lower seat, and needle valve. If the valve does not slide out, push the needle valve retracter onto the valve, secure it to the valve by turning the knurled nut and pull the valve from the body.

The cleaning procedure and methods are the same as previously outlined under Disassembling Injector and Cleaning and Inspection. If cleaning does not restore the needle-valve seat, the valve may have to be serviced by lapping it to the tip seat (nozzle seat). Excessive lapping, however, will destroy the interference angle, causing a loss of chatter and poor atomization.

To lap the seat, install the retracting tool to the needle valve and apply a moderate amount of the recommended lapping compound to the seat. Insert the valve and rotate it five times, clockwise, and then five times, counterclockwise, applying gentle pressure on the valve. Do not use power apparatus to lap the seat. Always handle the valve with care because it is easily bent.

If testing indicates that the leak-off is too low, clean the valve guide thoroughly and, if necessary, also lap the guide and needle to increase the leak-off.

Reassembling Roosa Master Injector   Before being reassembled, the components must be thoroughly flushed clean. Hold the injector in one hand while reassembling it in precisely the reverse order to which it was disassembled.

CAUTION   Take care not to dislodge the lower spring seat during reassembly. Turn the adjusting screw sufficiently inward to hold the components in place (Fig. 27–18).

Testing and Adjusting Roosa Master Opening Pressure   After the air is removed from the injector, adjust the opening pressure. This adjustment is made by turning the lift adjusting screw inward until it bottoms, then giving it a three-quarter turn counterclockwise. Using the test-stand pump, raise the pressure and check the opening pressure. If an adjustment is necessary to correct the opening pressure, hold the lift adjusting screw with a screwdriver while turning the pressure adjusting screw.

Setting Roosa Master Valve Lift   To set the valve lift, turn the lift adjusting screw slowly inward but hold the pressure adjusting screw. Then pump fuel through the injector until the valve ceases to open. Raise the pressure to about 400 psi [28.123 kg/cm²] above the opening pressure to check valve seating.

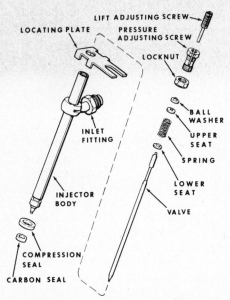

Fig. 27-18 Components of a Roosa Master injector. *(J. I. Case Co., Components Div.)*

(The presence of dampness on the fuel nozzle is acceptable but the formation of droplets or dribbles is not acceptable.)

Next, turn the lift screw the specified fraction of a turn, counterclockwise, to obtain the specified lift. Always recheck the opening pressure. After the opening pressure and lift height have been adjusted, check the spray pattern, leak-off, and chatter.

When the injector has passed these tests, remove it from the test stand and place a protection cap over the nozzle tip. Insert it in the injector holding tool and torque the locknut to the specified torque, as shown in Fig. 27–19.

Installing Roosa Master Injector Clean the cylinder bore with the bore-cleaning tool. Clean and check the sealing surface of the cylinder head. Install the compression seal and with a carbon seal tool, install the carbon seal.

When you install the injector, use a turning or twisting motion to place the injector in position but do not use any lubricant. Install the spacer and clamp assembly, and torque the hex bolt to the recommended torque. Use one hand and two wrenches to tighten the high-pressure connection.

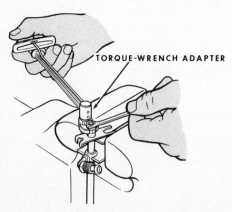

Fig. 27–19 Tightening the locknut. *(J. I. Case Co., Components Div.)*

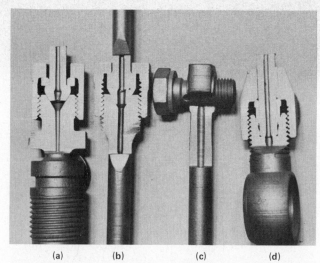

Fig. 27–20 Cutaway views of fuel-injection-line connections. *(a)* Swaged-seat with collar connector; *(b)* swaged-seat connector; *(c)* ferrule-type connector; *(d)* flared-type connector.

FUEL-INJECTION TUBING AND CONNECTIONS

The injection tubings used to connect the fuel-injection pump with the injectors are cold-drawn from low-carbon steel billets, then pierced to remove all internal surface defects. They are afterwards annealed to prevent formation of scale and to produce an accurate size having a smooth inside surface with no inside or outside surface imperfection. Other important characteristics necessary to fuel-injection tubings are:

1. They must have a tensile strength of 45,000 to 55,000 psi [31,500 to 38,500 kg/cm^2] to assure that the tubing will not expand under the high injection pressure.
2. They must be capable of being easily shaped to small bend radii without creating restriction to flow.

As indicated in Table 27–1, the basic outside diameter is 1/4 in [6.35 mm]. However, 5/16 in [7.938 mm], 3/8 in [9.525 mm], 7/16 in [11.113 mm], and 1/2 in [12.7 mm] are also used.

The inside diameter varies and is identifiable by a color code to suit the various injection pressures and fuel flow.

NOTE Use extreme caution when bending the injection tubing. Use only a mechanical tube bender to prevent flat spots or kinks. Do not make a bend radius of less than 1 in [25.4 mm].

Fuel-Injection-Line Connections The four most common fuel tubing connections are shown in Fig. 27–20. (a) and (b) show the swaged-tapered-seat type. This connection consists of an upset tapered male seat formed on the end of the tubing itself, or the swaged tapered seat is formed and then brazed to the end of the tubing. The nut compresses the seat against the mating seat in the fuel-pump-discharge fitting or nozzle-inlet fitting. The style shown in Fig. 27–20a has a collar and a short tapered seat.

Table 27–1 FUEL-INJECTION TUBING IDENTIFICATION

Nominal tubing OD, in	Outside diameter, in					Inside diameter, in			Color code for basic ID
	Basic	Tolerance				Basic	Tolerance		
		Standard		Optional			Standard	Optional	
		Plus	Minus	Plus	Minus		±	±	
¼	0.250	0.005	0.000	0.002	0.000	0.063	0.0025	0.001	Red
	0.250	0.005	0.000	0.002	0.000	0.067	0.0025	0.001	Black
	0.250	0.005	0.000	0.002	0.000	0.078	0.0025	0.001	Yellow
	0.250	0.005	0.000	0.002	0.000	0.084	0.0025	0.001	Blue
	0.250	0.005	0.000	0.002	0.000	0.093	0.0025	0.001	White

NOTE: In this chart, the color code identifies the basic tubing ID. The color code shall be applied to the OD of the tubing in the form of an intermittent stripe (gaps not longer than 12 in), using Dykem dye or other suitable marking material. See Conversion Tables for metric equivalents of English tube sizes.

Some have an extended collar and nut to support the tubing beyond the end of the nut.

The ferrule-type connector shown in Fig. 27–20c consists of a loose ferrule which is slipped over the tubing. It is compressed by the nut into the seat formed by the fuel-pump-discharge fitting or nozzle-inlet fitting and the tubing, causing the ends of the ferrule to bite into the tube wall.

NOTE When installing the ferrule type of connection, make sure that the tubing butts against the fitting counterbore and that it is held there until the nut is drawn up tight.

The tube connection shown in Fig. 27–20d is of the flared type. This style consists of a flared union formed on the end of the tubing, an inlet union screw, and two copper or special-alloy gaskets. The inlet union screw threads into the threads of the fuel-pump-discharge fitting or the nozzle-inlet fitting. On one side it compresses the gasket against its shoulder and against the surface of the inlet union. On the other side it compresses the gasket against the surface of the inlet union and the fuel-pump-discharge fitting or the nozzle-inlet fitting.

The same type styled with the inlet union brazed to the supply tubing is used as a fuel-supply connection.

NOTE When you use the flared style, make certain that you use a new gasket of the specified material. Use a torque wrench to draw the union screw to specification.

Questions

1. What is it that determines the type of injector holder and injector nozzle to be used?

2. Name the four most common fuel nozzles.

3. Why does the needle lift differ between various types of nozzles?

4. Why do all needle-valve nozzles chatter to some extent?

5. What changes would occur in the spray pattern if the needle lift were greater than specified?

6. What are the main differences between an inward-opening and an outward-opening pintle nozzle?

7. Outline the steps to be followed to remove an injector.

8. Give the reason why an injector should be tested before it is disassembled.

9. In what way does a needle valve indicate that one or more orifices are clogged?

10. How would spray pattern and combustion be affected if the lapped surface of the nozzle and holder were nicked or burred?

11. List the main differences between the injector shown in Fig. 27–2 and a Roosa Master injector.

12. Why should only a hammer puller be used to remove a Roosa Master injector?

13. When servicing a Roosa Master injector, what tests and adjustments must be made in addition to those required when servicing a CAV or Bosch injector?

14. When removing or installing a high-pressure injection line from a Roosa Master injector, why should you use the "one-hand, two-wrench" method to tighten and loosen the fittings?

UNIT 28

Fuel-Injection Systems

Importance of Fuel-Injection System   The fuel-injection system (of any design) is the heart of a diesel engine.   In many ways it controls the performance of the diesel engine; therefore, its constant maintenance cannot be overemphasized.   If a cooling or lubrication system is poorly maintained, evidence of neglect only shows up at a later date. However, poor engine performance is immediately apparent when the fuel-injection system is improperly adjusted, damaged, worn, or inadequately serviced.

Many fuel-injection systems are available.   Many of their components (that is, the fuel tank, fuel filters, transfer pump, and connecting lines) are similar in design (see Fig. 28–1).   Differences between systems tend to occur in the fuel-injection pump, in the metering principle, in the injectors, and in the injection principle.

When malfunctions develop they usually originate in the low-pressure side of the fuel-injection system.   The malfunction is often caused by incorrect handling or improper storage of the fuel or by improper maintenance of the fuel tank and fuel filters.   Nevertheless, all fuel-injection systems (except the Cummins PT injection system which varies slightly) have the common purpose of storing, cleaning, and transferring fuel at low pressure to the injection pump.

Objectives of Fuel-Injection System   In addition to the storing, cleaning, and transferring of fuel, other functions of the fuel-injection system are:

1.   To meter the quantity of fuel required by the engine at all loads and speeds (fuel quantity must be equal to ensure the same power from each cylinder)
2.   To start injection at the correct time within the cycle of the engine with regard to load and speed
3.   To ensure the prompt beginning and ending of injection so that the injected fuel is evenly atomized
4.   To inject the fuel at the rate necessary to control combustion and to control the pressure during combustion
5.   To direct, distribute, and atomize the fuel uniformly, as required by combustion chamber design

To determine just how precise a fuel-injection pump must be, let us assume you have a modern four-cylinder four-cycle engine running at 2500 rpm and using 2½ gal (U.S.)/hr [9.45 l/hr] of fuel.   It would require, therefore, 5000 fuel injections per minute (300,000 fuel injections per hour).   The quantity of 2½ gal would have to be divided (metered) into 300,000 proportions and then sprayed at the correct time into each cylinder against the high pressure existing within the cylinder.   The minute quantity, that is, in this case 0.031 cm^3 must be me-

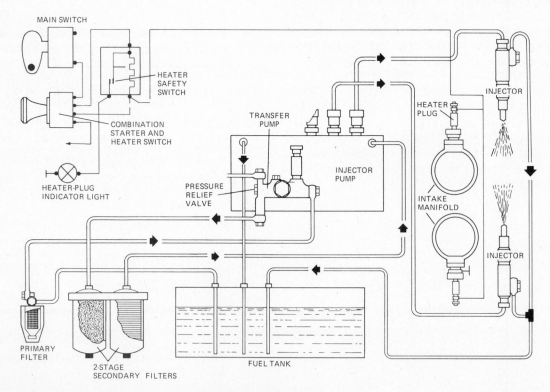

Fig. 28–1   Schematic view of a typical fuel-injection system. (The arrows show the fuel flow.) *(Robert Bosch GmbH, Stuttgart, West Germany.)*

tered within the time in which the crankshaft rotates approximately 600°. At this engine speed that would be about 0.04 second. Furthermore, this quantity must be injected within a certain time, which, on an average diesel engine, is about 15° of crankshaft rotation, or in other words, about 0.001 second. Of course, at low idle the quantity would be much smaller, about 0.009 cm³, but the time factor for metering and injection is increased.

Types of Injection Systems The following units will be devoted to the basic operating principles, service and adjustment, and troubleshooting of (1) fuel-injection systems which use port and helix metering, (American Bosch, Robert Bosch, Caterpillar, CAV, Sims, and GM); (2) controlled bypass, (PT) system; (3) opposed plunger inlet-metering distributor types (Roosa Master and DPA); and (4) sleeve-metering pumps (American Bosch and Caterpillar).

UNIT 29

American Bosch, Robert Bosch, and CAV Fuel-Injection Systems

PORT AND HELIX METERING FUEL–INJECTION PUMPS

The fuel-injection pumps which have been in use the longest and which are the most popular today use the port and helix metering principle. They are of the constant stroke, lapped plunger, port and helix design. Although they are manufactured universally, they vary to only a minor degree in their construction and method of adjustment. The designs available range from a single plunger element (plunger and barrel), with or without a camshaft, to 12 pumping elements in line in one housing or in a V shape with 4 pumping elements on each side (see Fig. 29–1).

Fig. 29–1 APE 8V BB injection pump.

Single-Cylinder APF Pump The nameplate of a fuel-injection pump identifies the manufacturer, the design, and the construction model. The first letter identifies the manufacturer, the next letter (in the caption) stands for injection pump, and the last letter identifies the pump model. Accordingly, APF is the identification for an American Bosch pump (flange-mounted) without a camshaft. (The camshaft makes up part of the engine.) However, regardless of the number of pumping elements used, or whether the pump has a built-in camshaft or not, the components which make up a fuel-injection pump are the same (see Fig. 29–2).

Components of an injection pump must include the following (the numbers refer to those listed in Fig. 29–2):

- A *housing* (1) to hold and support the components
- A *fuel connector* (2) to connect the transfer pump to the injection pump
- A *fuel manifold* (3) to connect the barrel ports to the transfer pump
- A *bleeder screw* (4) to remove air from the inlet manifold
- A *pumping and metering element* (plunger-and-barrel assembly) (5) to meter and deliver the fuel
- A *barrel-locating screw* or *dowel pin* (6) to locate

the barrel in the bore of the pump housing and to act as a fuel deflector for the high return fuel pressure
- A *delivery-valve assembly* (7) to assist in the beginning and ending of injection
- A *delivery-valve spring* (8) to hold the valve on its seat and to assist in maintaining residual pressure in the injection line
- A *seal ring* (9) to seal the high-injection pressure
- A *delivery-valve holder* (10) to connect the high-pressure fuel line to the pump and to hold the pump element and delivery valve in the pump housing
- A *union washer* (11) and *nut* (12) to connect the fuel line
- A *gear segment* (13) bolted to the control sleeve
- A *control sleeve* (14) to transmit control-rack motion to the plunger
- A *control rack* (fuel rack) (15) to transmit governor motion to the gear segment
- A *control-rack bushing* (15)
- A *plunger spring* (16) positioned between the two spring seats to maintain the tappet roller on the camshaft
- Two *spring seats* (17, 18) to maintain spring-end position
- A *tappet roller* (follower) (19) to transmit rotary motion from the camshaft and reciprocating mo-

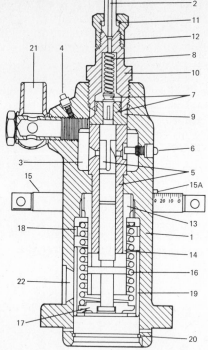

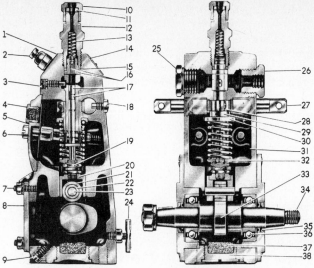

1. Housing
2. Fuel connector
3. Fuel manifold
4. Bleeder screw
5. Plunger-and-barrel assembly
5A. Control-rack bushing
6. Barrel locating screw
7. Delivery-valve assembly
8. Delivery-valve spring
9. Seal ring
10. Delivery-valve holder
11. Union washer
12. Union nut
13. Gear segment
14. Control sleeve
15. Control rack
16. Plunger spring
17. Lower spring seat
18. Upper spring seat
19. Tappet shell
20. Snap ring
21. Fuel-inlet connection
22. Timing window

Fig. 29–2 APF fuel-injection pump. *(American Bosch-Ambac Industries Inc.)*

tion to the plunger. (The lower spring seat is hooked to the plunger.)

- A *snap ring* (20) to hold the tappet roller in the pump housing
- A *fuel-inlet connection* (21)
- A *timing window* (22) to time the injection pump to the engine
- A *screw* to limit the control-rack travel (not visible in Fig. 29-2)

Single-Cylinder APE Pump   Refer to Fig. 29–3, which shows an APE fuel-injection pump. Note that the components in the upper part of the APE pump are similar to those in the APF pump. However, the design has now changed to include the camshaft as part of the pump; therefore, the last identification letter has been changed from F to E. The components added in the upper half of the injection pump are a *bleeder petcock* (2), to bleed air from the manifold, and an *inspection cover* and *fastening screw* (5, 6). With the enclosed camshaft, the *plunger-and-barrel assembly* (17) now includes the

1. Housing
2. Bleeder petcock
3. Barrel locating screw
4. Breather plug
5. Inspection cover
6. Fastening screw
7. Fuel-supply-pump stud
8. Fuel-supply-pump pad cover
9. Drain plug
10. Union nut
11. Washer
12. Delivery-valve holder
13. Delivery-valve spring
14. Nameplate
15. Delivery-valve gasket
16. Delivery-valve assembly
17. Plunger-and-barrel assembly
18. Control-rack screw
19. Tappet assembly
20. Tappet shell
21. Tappet roller
22. Bushing
23. Pin
24. Overflow petcock
25. Closing plug
26. Fuel-inlet plug
27. Control rack
28. Gear segment
29. Control sleeve
30. Upper plunger-spring seat
31. Plunger spring
32. Lower plunger-spring seat
33. Cam
34. Camshaft
35. Oil seal
36. Ball bearing
37. Felt pad
38. Plug

Fig. 29–3 APE fuel-injection pump. *(American Bosch-Ambac Industries Inc.)*

following (the numbers refer to those listed in Fig. 29–3):

- A *tappet adjusting screw* and *lockwasher* (19) to adjust the plunger height
- An *inner roller* (bushing) (22) to reduce rolling friction
- An *outer roller* (tappet roller) (21) held in by the pin, to reduce rolling friction
- The *pin* (23) to hold the tappet roller to the tappet body
- A *cam* (33) and *camshaft* (34) to transmit rotary motion to the tappet
- *Ball bearings* (36) to support the camshaft
- Two *seals* (35) to prevent dirt from entering and to confine oil in the base of the pump
- A *felt pad* (37) to clean and lubricate the lobe
- A *plug* (38)
- An *overflow petcock* (24) to check the oil level
- A *drain plug* (9) to drain the lubrication oil
- A *pad cover* (8) to cover the transfer-pump

1. Delivery-valve holder
2. Barrel
3. Volume reducer
4. Delivery valve
5. Plunger
6. Maximum-fuel-stop screw
7. Excess-fuel device
8. Bridge link
9. Trip lever
10. Telescopic link
11. Speed control lever
12. Damper
13. Governor idling spring
14. Governor main spring
15. Crank lever
16. Speed lever shaft
17. Governor sleeve
18. Governor flyweight
19. Stop control lever
20. Camshaft
21. Tappet assembly
22. Control rod
23. Control fork

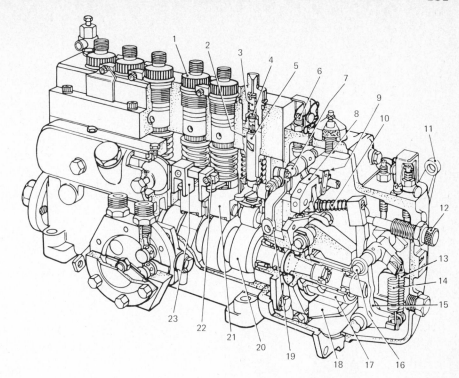

Fig. 29–4 SPE fuel-injection pump (Majormec). *(CAV Ltd.)*

mounting surface when the pump is not being used

Multicylinder SPE and PEP Pumps The differences between the SPE (Fig. 29–4) or PEP (Fig. 29–5) and APE pumps are mainly in the method by which the control rod is linked to the plungers and the method by which the plunger height is adjusted. The SPE and PEP are multicylinder pumps, and their governors are directly bolted to the injection pump housing.

Pumping Element Design The function of a pumping element is to meter and deliver the metered fuel. The plunger has two motions within the barrel to make this possible: a rotary motion which regulates the quantity of fuel (metering) and an upward motion which delivers the metered quantity of fuel (pumping). (These two motions are described in detail in the pumping- and metering-principle sections of this unit.)

Each pumping element consists of a barrel and a plunger (see Fig. 29–6). The plunger fits in the barrel with very close tolerance (a few thousandths of a millimeter), so that the fuel, at high or low pressure, lubricates the element but prevents leakage to the lower side of the barrel. The shoulder of the barrel rests against the counterbore in the pump housing. The barrel usually has two round openings, the inlet port and the spill port, which (when the barrel is mounted in the pump housing), are open to the fuel manifold. The spill port generally is recessed to form a slot which serves to locate the barrel in the housing by means of a deflector screw. On some barrels the two holes are offset at different heights, as shown in Fig. 29–6. Other barrels have only one opening to the fuel manifold. Still other barrels

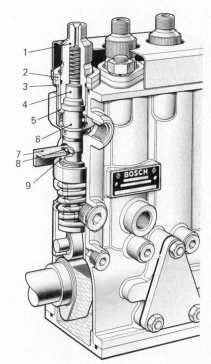

1. Delivery-valve holder
2. Thrust plate
3. Gasket
4. Delivery valve
5. Flanged bushing
6. Pumping element
7. Control rod
8. Ball
9. Control sleeve

Fig. 29–5 PEGP fuel-injection pump. *(International Harvester Co.)*

have an internal passage from the lower side to the inlet port to allow the fuel used for lubrication to return to the fuel manifold.

The upper part of the plungers have a vertical and a horizontal machined-out groove and, depending

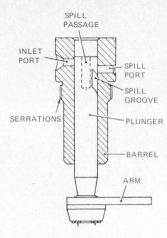

Fig. 29–6 Pumping element of a Majormec fuel-injection pump. *(CAV Ltd.)*

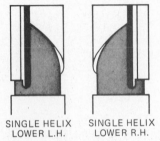

SINGLE HELIX LOWER L.H.   SINGLE HELIX LOWER R.H.

Fig. 29–7 Plungers of the lower-helix design. *(American Bosch-Ambac Industries Inc.)*

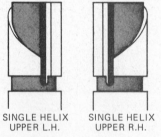

SINGLE HELIX UPPER L.H.   SINGLE HELIX UPPER R.H.

Fig. 29–8 Plungers of the upper-helix design. *(American Bosch-Ambac Industries Inc.)*

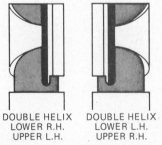

DOUBLE HELIX LOWER R.H. UPPER L.H.   DOUBLE HELIX LOWER L.H. UPPER R.H.

Fig. 29–9 Plungers of the upper-and-lower-helix design. *(American Bosch-Ambac Industries Inc.)*

on the purpose for which it was designed, has one or two spiral machined-out grooves. The spiral machined-out groove is called the *helix*. The shape and the position of the helix govern the fuel delivery curve and either the beginning or the ending of injection.

The plungers shown in Fig. 29–7 are of the lower-helix design. Their beginning of delivery is con-

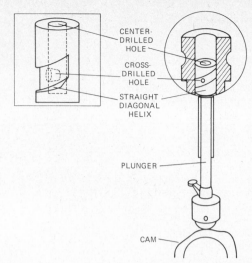

Fig. 29–10 Lower-helix design used on smaller plungers. *(CAV Ltd.)*

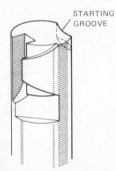

Fig. 29–11 Pump plunger with upper and lower helices as well as a starting groove. *(Robert Bosch GmbH, Stuttgart, West Germany.)*

stant and their ending of delivery is variable. The right-hand helix is used when the governor is at the right side. The left-hand helix is used when the governor is on the left side or when the fuel rack is in front of the plunger.

The plungers shown in Fig. 29–8 are of the upper-helix design. These plungers have a variable beginning and a constant ending of delivery. The plungers shown in Fig. 29–9 have upper and lower helices. Both the beginning and the ending of delivery are variable with these plungers.

Smaller plungers, also of the constant-beginning and variable-ending (lower-helix) design, are manufactured with a center- and cross-drilled hole and have a diagonally machined groove (see Fig. 29–10).

Some plungers have a small starting groove cut above the helix or as shown in Fig. 29–11 (double-helix design). This assists the engine in starting by supplying more fuel and retarding delivery.

On the lower part of some plungers are two flanges which slide in the control sleeve. Others have either a lever arm or a gear segment bolted to the plunger to connect the control rack (fuel rack) with the plunger.

Pumping Principle Using Lower Helix When the tappet roller (follower) is on the base circle of the camshaft, the upper surface of the plunger is below the inlet port and spill port. Fuel under transfer-pump pressure enters the ports and fills the area above the plunger and the helix (Fig. 29–12a). As the camshaft moves the tappet and plunger upward, both ports are closed (Fig. 29–12b). Fuel pressure builds up rapidly, causing the delivery valve to lift off its seat and allow fuel to enter the high-pressure injection line (Fig. 29–12c).

As the lower helix opens the spill port, the fuel pressure quickly decreases while passing through the vertical groove into the helix area and through the spill port to be deflected by the barrel-locating screw (Fig. 29–12d). As soon as the pressure drops, the delivery valve seats, preventing fuel from leaving the injection line. Once the spill port opens, the plunger can no longer pressurize the fuel as it continues to move upward. The upward movement of the plunger stops when the follower rides on the cam nose.

As the plunger moves downward, fuel at transfer-pump pressure enters the spill port and fills the area vacated by the plunger. When the spill port closes momentarily, no fuel can enter the area until the plunger uncovers the inlet port.

Pumping Principle Using Upper Helix During the upward and downward movement of the plunger, the mechanical action is the same, whether using a plunger with an upper or a lower helix. However, when using a plunger with an upper helix, the inlet port does not close at the same plunger height; thus, we say that the beginning of delivery is variable.

Closure of the inlet port depends on the position of the plunger helix as it is placed by the governor. When the governor is in the full-fuel position, the beginning of delivery is earlier (advanced) (Fig. 29–13a). When the governor is in the idle-fuel position, delivery begins later (retarded) (Fig. 29–13b). Regardless of the plunger position, the ending of delivery always occurs when the edge of the horizontal groove on the plunger opens the spill port.

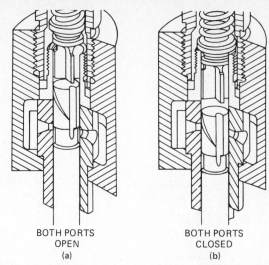

BOTH PORTS OPEN
(a)

BOTH PORTS CLOSED
(b)

Fig. 29–13 Pumping principle with the upper-helix design. (a) Advanced delivery—governor in full-fuel position (ending injection); (b) retarded delivery—governor in half-fuel position (starting injection). (Robert Bosch GmbH, Stuttgart, West Germany.)

When using a variable-beginning and variable-ending plunger design, the beginning of delivery occurs as if using an upper-helix-design plunger, and the ending of delivery occurs as if using a variable-ending plunger.

Metering Principle The governor, which measures the engine speed and senses engine speed changes, transmits its motion to the fuel-changing mechanism to control the engine speed. The fuel-changing mechanism consists of various lever arrangements, the fuel rack, the connecting link to the plunger, and the pumping element.

NO-FUEL POSITION When the governor moves the fuel rack to the no-fuel position, the plunger rotates

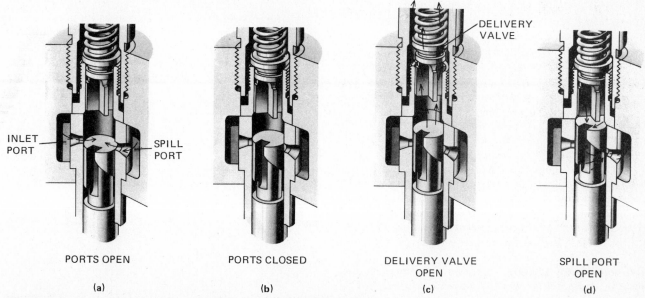

INLET PORT   SPILL PORT

PORTS OPEN
(a)

PORTS CLOSED
(b)

DELIVERY VALVE

DELIVERY VALVE OPEN
(c)

SPILL PORT OPEN
(d)

Fig. 29–12 Pumping principle with the lower-helix design. (Arrows show the fuel flow.) (a) Fuel enters barrel; (b) start of delivery; (c) delivery; (d) end of delivery. (Robert Bosch GmbH, Stuttgart, West Germany.)

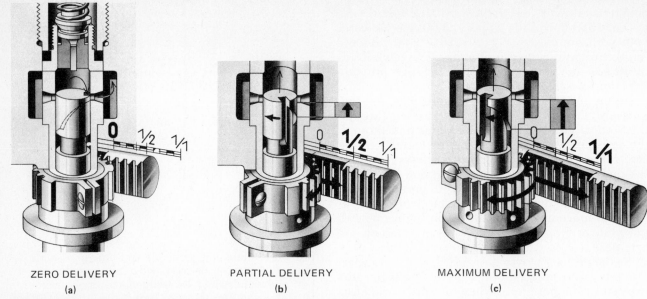

ZERO DELIVERY
(a)

PARTIAL DELIVERY
(b)

MAXIMUM DELIVERY
(c)

Fig. 29–14 Metering principle. *(a)* Governor in no-fuel position; *(b)* governor in half-fuel position; *(c)* governor in full-fuel position. *(Robert Bosch, GmbH, Stuttgart, West Germany.)*

within the barrel to place the vertical groove in line with the spill port (Fig. 29–14a). The port never closes since the vertical groove of the plunger rides over the spill port on its total upward and downward stroke. Fuel displaced during the plunger movement runs up or down the vertical groove to and from the base of the helix and runs out or in, through the spill port.

HALF- AND FULL-FUEL POSITIONS When the governor moves the fuel rack to the half-fuel position, the plunger rotates from the no-fuel to the half-fuel position. This rotation is caused by the horizontal movement of the fuel rack which rotates the gear segment, the control sleeve, and the plunger (Fig. 29–14b). This changes the position of the helix relative to the spill port and the effective stroke. (When the governor moves the fuel rack to the full-fuel position, the helix is as shown in Fig. 29–14c). The longer the spill port is closed, the longer the pumping period (effective stroke) and the shorter the spilling period. Less fuel is delivered during the upward stroke when the spill port opens earlier.

Delivery Valves All port and helix-metering fuel-injection pumps (with the exception of Detroit Diesel unit injectors) use delivery valves, though some vary slightly in design. Delivery valves are essential components of a fuel-injection pump. They reduce the fuel pressure in the injection line, maintain a fixed pressure in the line to prevent fuel from draining into the pumping element, and assist in the quick starting and ending of injection.

Delivery-Valve Design Although delivery valves vary in design (Fig. 29–15), they usually include a valve body with a valve seat and a delivery valve with a displacement plunger, seat, relief piston (unloading piston), and a fluted stem. Most delivery

valves are accurately lap-fitted in the valve body and are, therefore, not interchangeable.

Delivery-Valve Operation As illustrated in Fig. 29–15, the delivery-valve body is seated on the barrel. The valve is held on its seat by the valve spring.

As the plunger rises, it closes the inlet and spill ports. Fuel becomes trapped above the plunger and in the base area of the helix, causing pressure to rise rapidly. The force of this pressure is exerted on the fluted stem and the relief piston. When the force increases beyond the combined force of the spring and the fuel pressure in the injection line, the valve moves upward. With the upward movement of the valve, the displacement piston is forced into the spring area. This reduces the fuel volume in the spring area and increases the pressure in the injection line. As the relief piston leaves its bore, fuel is forced past the relief piston into the fuel line, and injection takes place.

A valve stop located above the delivery valve limits the upward movement of the delivery valve and, at the same time, acts as a volume reducer. The open position is maintained until the helix on

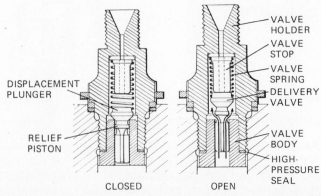

VALVE HOLDER
VALVE STOP
VALVE SPRING
DELIVERY VALVE
DISPLACEMENT PLUNGER
RELIEF PISTON
VALVE BODY
HIGH-PRESSURE SEAL
CLOSED
OPEN

Fig. 29–15 Diagrams of a delivery valve. *(CAV Ltd.)*

the plunger uncovers the spill port. As this happens, the sudden drop in pressure in the base area of the helix and above the plunger, plus the higher fuel and spring pressure on the displacement piston, cause the delivery valve to retract quickly and snap onto its seat. Simultaneous with delivery-valve retraction, the relief piston closes the fuel flow to the pumping element and then the valve seat. The retraction of the valve into the bore increases the spring area by the volume of the withdrawn displacement piston. The effect of this is that the volume in the injection line, before the valve rests on its seat, is suddenly decreased and the pressure reduced, causing the valve in the injector to snap onto its seat. This prevents dribbling or secondary injection due to pressure wave reflection.

FUEL–INJECTION–PUMP SERVICE

Examination Before Removal   Although most fuel-injection-system trouble is the direct result of dirty fuel, you should not assume it is the only problem source. Because damaged or defective components are also a contributing factor to trouble, a number of checks should be made if the problem is to be corrected, (see also Unit 51, "Troubleshooting and Tuneup").

Begin with a test run of the engine to analyze its general condition. Look for fuel leakage, exhaust smoke, acceleration, governor action, governor response, and note the engine temperature.

Check throttle freeness and the high- and low-idle speeds. Check the fuel-transfer-pump pressure. When necessary, change the fuel filters, and then repeat the fuel-pump-pressure test.

If hard starting is noticed, check the compression and the air-intake systems after making certain that the cranking speed is adequate. It is sometimes necessary to check the timing (by the spill method) to ensure that the injection pump is properly timed to the engine.

To determine if all injectors are operating properly, use a screwdriver to pry each pump element upward. The nozzle should tend to boss. Another method of testing an injector is to install a pressure gauge in the high-pressure fuel line to check the opening pressure of the injector. Do not remove the injection pump until it's proven faulty.

Injection-Pump Removal   Although the removal procedures are similar for all injection pumps, individual service manuals should always be referred to before you remove a pump. Before you dismantle an injection pump, obtain the necessary data regarding the pump's construction, its performance, and the engine manufacturer's specifications.

Generally, the removal procedure is as follows:

1.   Clean the surrounding injection-pump area thoroughly to prevent dirt from entering the system.
2.   Before you remove the injection pump, turn the crankshaft to bring the No. 1 piston on compression stroke so that the timing marks are aligned.

3.   Close off the fuel supply. Remove or disconnect the throttle and/or shutoff lever.
4.   Disconnect and cap the inlet fuel connection (and the return line connection, if used).
5.   Disconnect the high-pressure lines from the injectors or from the injection pump, whichever is easier. Cap the openings as the fuel lines are being removed.
6.   Remove the mounting bolts and, when necessary, remove the drive-coupling bolts. Finally, remove the injection pump.

Injection-Pump Nameplates   The construction and design of injection pumps can be ascertained from the nameplate. The anticipated pump performance is obtainable from the test specification and/or the engine manufacturer's service manual.

Assume for example, that the nameplate of an injection pump is: SPE6BADES319A. This number reveals the following:

S	=	Sims Manufacturing Company
PE	=	Pumps with enclosed camshaft
6	=	number of pumping elements
B	=	Size and stroke of the plunger (45-mm camshaft center height and a 7.5-mm stroke)
80	=	Plunger dimension in millimeters
E	=	Internal design changes
S319	=	Special installation features (Engine). In this case the camshaft notch is at No. 1 end (left side) and the governor is at the right side.
A	=	Internal design change for a particular application; only added when applicable. For example, the injection-pump performance when used with a certain engine is high-idle=2300 rpm, low-idle=525 rpm, and no-fuel delivery=1500 rpm. Maximum fuel delivery should be 52.6 to 51.6 cm³ in 200 shots at 600 rpms. Idle fuel delivery should be 12 cm³ in 200 shots at 250 rpm.

Disassembly Procedure   The disassembly procedure for injection pumps varies to some extent among different manufacturers and even among different models within the same company. As a result, it is impractical to include here the many types of injection pumps; therefore only the major differences are discussed in this section.

Regardless of the type of pump to be disassembled, the exterior of the injection pump must first be cleaned. The work area and tools must also be clean. It is also essential to keep all the components of the same element in a separate container (Fig. 29–16). Never interchange the barrels and the plungers or the delivery valves and bodies. Keep the tappet spacer or shims with the pump elements to speed up reassembly.

Begin the disassembly by mounting the injection pump in a swivel vise or in a holding fixture. Remove the transfer pump, inspection cover, and gov-

Fig. 29–16 Typical work area, showing storage of components. *(Robert Bosch, GmbH, Stuttgart, West Germany.)*

ernor cover. Remove the necessary pins and springs which connect the governor with the control rod (fuel rack).

Using a puller, withdraw the weight assembly from the camshaft. Turn the camshaft to bring the No. 1 tappet roller to the TDC position. Then install the tappet holder to keep the rollers away from the cam nose. (Some tappet rollers have a precut hole into which a wire is inserted to hold the roller away from the cam nose.)

When disassembling a PEP injection pump, insert the special tappet lifter into the hole provided, and then turn the tappet lifter 180° (Fig. 29–17). The eccentricity of the special tappet lifter then raises the tappet roller from the cam nose.

To disassemble a CAV Minimec or Majormec (see Fig. 29–18), remove the screws (7), lock washer (8), and clamps (9, 10). Then loosen the delivery-valve holders (93). Holding the pump body (6), loosen all the pump-body screws (11, 12) and sealing washers. When lifting the pump body from the housing, take care that the plungers do not fall from the barrel (90 in Fig. 29–19). The tappet assembly and the tappet locater can now be removed.

Loosen the control-fork screws. Remove the control-rod cover. While sliding out the control rod, remove the control forks. The camshaft can now be removed.

Turn the pump over and remove the inspection plate. If an intermediate camshaft bearing is used, remove the lower bearing holder. With a puller, remove the drive coupling or mechanical timing device. After removing the screws from both end plates, use a puller to withdraw the end plates from the housing. Remove the camshaft.

When you remove the plungers from the type of pump shown in Fig. 29–20, use a removal tool or a tool with a plastic handle to push the tappet roller upward, and then remove the tappet holder. Next, withdraw the plunger with the lower spring seat, the spring, the control sleeve, and the upper spring seat.

To remove the tappet roller and the plunger from a PEP injection pump, install the removal tool as shown in Fig. 29–21. Use the pushrod to force the tappet roller down to allow removal of the special tappet lifter. After the tappet lifter has been re-

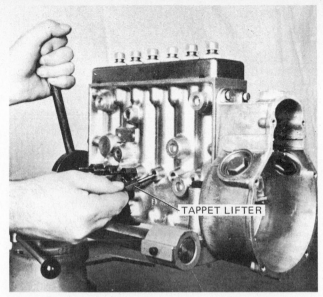

Fig. 29–17 Inserting a tappet lifter. *(International Harvester Co.)*

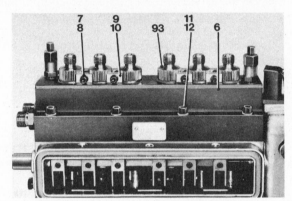

6. Pump body	11, 12. Body screws
7. Screw	93. Delivery-valve holder
8. Lock washer	
9, 10. Clamps	

Fig. 29–18 Preparation to remove the pump body. *(CAV Ltd.)*

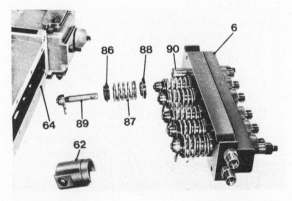

6. Pump body	86. Lower spring seat
62. Tappet assembly	87. Spring
64. Lower injection-pump housing	88. Upper spring seat
	89. Plunger

Fig. 29–19 Body with one plunger removed. *(CAV Ltd.)*

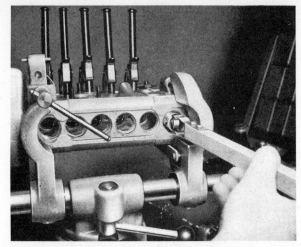

Fig. 29–20 Removing a tappet roller. *(Robert Bosch GmbH, Stuttgart, West Germany.)*

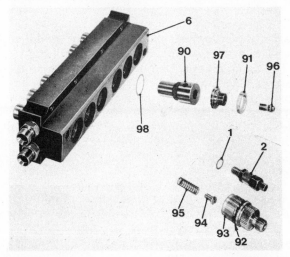

Fig. 29–21 Tensioning tool installed to remove the tappet roller. *(Robert Bosch GmbH, Stuttgart, West Germany.)*

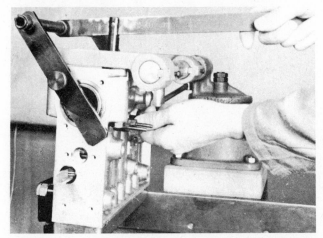

1. Joint washer
2. Air vent
6. Pump body
90. Barrel
91. Joint ring
92. O seal
93. Delivery-valve holder
94. Volume reducer
95. Delivery-valve spring
96. Delivery valve
97. Delivery-valve seat
98. Barrel-seat washer

Fig. 29–22 Pump body with one delivery valve removed. *(CAV Ltd.)*

moved, take out the tappet roller. Bend the end of a small piece of copper wire into a hook to withdraw the plunger and lower spring seat. Remove the spring, the control sleeve, and the upper spring seat.

To remove the delivery valve from a Majormec injection pump (Fig. 29–22), first remove the two air-vent screws, then the delivery-valve holder (93), the valve stop [volume reducer (94)], the delivery valve (96), and the spring (95). Withdraw the plunger, the lower spring seat, the spring, and the upper spring seat (see Fig. 29–19 for these components). Using a soft-faced hammer, tap on the bottom of the barrel (90 in Fig. 29–22) to dislodge the barrel and to force the sealing ring [joint ring (91)] and the delivery-valve body from its bore. Remove the barrel-seat washer (98).

To remove the delivery valve and the barrel from a PEP injection pump, first remove the cover from the pump housing. Remove the nuts from the stud bolts with a special puller, and pull the pump element out of the pump housing. From the pump housing remove the shims (7 in Fig. 29–23) which adjust the port closure. From the pump element (or the housing bore) remove the O ring (1) and the spring ring (2). Slide off the baffle sleeve [impact cap (3)] and the buffer washer (6). Remove the two O rings (4, 5) and the flange bushing (9) from the barrel. Place the unit in the holding fixture and then unscrew the delivery-valve holder (8 in Fig. 29–24) and remove

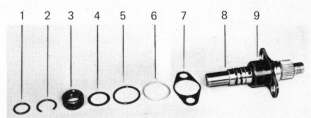

1. Viton ring
2. Retainer
3. Baffle sleeve
4. O ring
5. O ring
6. Buffer washer
7. Shims
8. Pump barrel
9. Flange bushing

Fig. 29–23 Disassembled components from the barrel lower end. *(Robert Bosch GmbH, Stuttgart, West Germany.)*

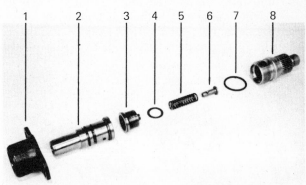

1. Flange bushing
2. Pump barrel
3. Delivery-valve assembly
4. Seal ring
5. Valve spring
6. Volume reducer
7. O ring
8. Delivery-valve holder

Fig. 29–24 Disassembled components from the barrel upper end. *(Robert Bosch, GmbH, Stuttgart, West Germany.)*

its O ring (7). Lift off the valve spring (5), the valve stop [volume reducer (6)], the seal ring (4), and the delivery valve (3). Push the barrel from the flange bushing (1).

Cleaning and Inspecting Pump Housing Remove all gasket materials from the pump housing and end plates. With a puller, withdraw the bearings or the bearing races. Remove all the O rings and the sealing washers. Clean the components thoroughly and dry them with compressed air.

Inspect the pump housing for external damage to sealing, bearing, and mounting surfaces. Smooth out any slightly damaged spots.

Check the stud bolts and the threaded bores for damaged threads.

Check the tappet-roller bores. If they are rough or badly grooved, the housing must be replaced. Otherwise, smooth out the bore surface with a piece of fine emery cloth. If the barrel seats are damaged, use the recommended reseating cutter to reseat their surfaces.

Control Rod and Bushings Examine the control rod to determine if the seal seats are damaged or if the bearing surface is worn. Check for wear in the control-rod notches in which the balls of the control sleeve ride (PEP), or the gear teeth on the control rack. When the control rod (rack) is in the housing, it should slide smoothly, but not too loosely.

Check the control-rod bushings for wear and, if necessary, replace them. It is essential to ream the newly installed bushings with a special reamer to ensure free control-rod movement.

Camshaft Bearings If there is any damage to the camshaft lobes, bearing surfaces, or sealing surfaces, the camshaft must be replaced. Large camshafts should be checked for runout. If there is any doubt about the serviceability of the bearings, they must be replaced.

Tappet Rollers Press out the tappet pin from the tappet body (see Fig. 29–25). Then examine the inner and outer rollers (81 and 82) for damage to their bearing surfaces. If there is any evidence of wear, the defective part should be replaced. If the tappet body is rough or grooved, or if the guide

80. Tappet pin
81. Inner roller
82. Outer roller
83. Adjusting spacer
84. Retainer
85. Tappet body

Fig. 29–25 Tappet-roller assembly (Majormec). *(CAV Ltd.)*

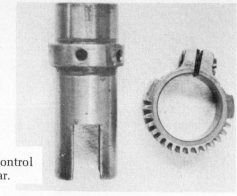

Fig. 29–26 Control sleeve and gear.

Fig. 29–27 Majormec fuel-injection pump with control fork and plunger arm. *(CAV Ltd.)*

grooves are worn, replace the tappet body also. However, slight wear marks or longitudinal marks may be removed with a piece of fine emery cloth.

If the tappet screws are used, check them for damage to the screw threads and to the threads of the threaded bore. Make certain that the plunger contact surface is smooth. Slightly damaged surfaces can be restored by regrinding the surface on a valve refacer.

Barrel and Plunger If there is any evidence of damage to the helix or top area of the plunger, the entire assembly must be replaced. Do *not* replace only one element.

Longitudinal grooves on a plunger indicate that dirty fuel has been used. Horizontal pressure marks indicate that the pump element has been forced to operate against excessive pressure, usually caused by one or more choked or plugged orifices in the nozzle.

Before you clean the barrel and the plunger, apply a moderate amount of mutton tallow to the plunger with a felt pad. Then clean the plunger in diesel fuel, and again apply tallow. Insert the plunger into the barrel, then rotate it while moving it back and forth within the barrel.

Before you clean the component parts, lap the surface of the barrel body to ensure a leakfree seal between it and the delivery valve. Place the plunger and the barrel in clean diesel fuel. While still in the diesel fuel, insert the plunger into the barrel and move it back and forth a few times. This will clean both components.

To check the plunger for freeness, pull the plunger out of the barrel about one-quarter of its length vertically. It should slide slowly back into the barrel by its own weight. If it slides back too quickly, the tolerance between the plunger and the barrel is too great and the element must be replaced.

Plunger Spring   Check the plunger spring for corrosion and straightness. Measure its free length and energy with a spring compressor. Never use a spring which is corroded because it may break or, if its energy is too low, it may not return the plunger quickly enough.

Control Sleeve and Gear   If there is any damage to the teeth of the gear segments or to the slots of the sleeve in which the plunger flange rides, the damaged component must be replaced. If a replacement is not necessary, slightly loosen the screws which clamp the gear to the control sleeve to assure trouble-free calibration (Fig. 29–26).

If the injection pump has a control fork and a plunger arm (Fig. 29–27), they should be checked for wear because they must fit snugly with only slight clearance. When you service a pump which has a control sleeve with a ball socket (Fig. 29–28), make certain the socket has no flat spots.

Delivery Valve   Check the relief valve and the valve seat. The seats must show no sign of wear and no uneven seating area. If the relief piston shows any sign of wear, the assembly must be replaced. If the valve is acceptable, lap the body surface to assure a leak-free seal between it and the barrel.

Reassembling Fuel-Injection Pump   Some manufacturers recommend installing the camshaft first, then positioning it, and adjusting the end play

before installing the pumping element. Other manufacturers recommend installing the pumping element first and then the camshaft. The applicable procedure depends on the construction of the pump.

When you service a Majormec fuel-injection pump, install the camshaft first. If the camshaft is supported by an intermediate bearing, first install the bearing shell into the upper and lower housing. Then install the upper housing into the pump housing before you install the camshaft.

Replace the original set of shims (Fig. 29–29) on each end of the camshaft after making certain that the sets are equal in thickness. Lubricate the bearings and press them into the camshaft. Press the oil seal (70 in Fig. 29–30), into the bearing housing (67). Install the washer (60), and press the outer bearing race (66) against the shoulder of the bearing housing.

Install the camshaft into the housing with the identification mark to the side indicated by the nameplate number. Install a jointing (gasket) and tap the bearing housings into place. Secure them to the pump housing.

CAUTION   While tightening the screws, rotate the camshaft to prevent it from seizing.

With the screws torqued to specification, measure the end play by mounting a dial indicator to the housing so that the pointer rests against the end of

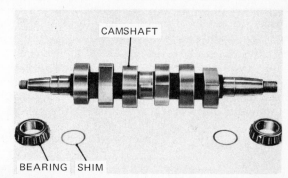

Fig. 29–29   Injection-pump camshaft with bearings and shims. *(CAV Ltd.)*

55.	Key	68, 69.	Screws and
60.	Washer		spring washers
64.	Cambox	70.	Oil seal
66.	Outer bearing race	71.	Camshaft
67.	Bearing housing	72.	Larger key

Fig. 29–30   Bearing-housing components. *(CAV Ltd.)*

Fig. 29–28   Cutaway view of the control rack (PEP).

the camshaft. End-play movement will show on the scale when prying the camshaft to the left and to the right. Average camshaft end play is 0.002 to 0.005 in [0.058 to 0.127 mm].

To adjust the end play, the bearing housing and the bearing must be removed, and shims added or removed as required.

Note that some camshafts (for example, those of PEP injection pump) require a given protrusion on the drive side (see Fig. 29–31).

Assembling and Installing PEP Pumping Elements Place the flange bushing in a vise and assemble the element (Fig. 29–23). Lubricate the O rings with the recommended grease. (The barrel should slide without force into the locking device of the flange bushing.)

When assembled, tighten the connector to the specified torque. Always insert the special O ring (Viton ring) into the housing to prevent damaging it when placing the pumping element into the pump housing.

Return the shim pack, previously removed, to the position on top of the housing. Place the large O ring into the groove of the housing. Next, install the pumping element with the flange bushing pin toward the control-rod side, and then tighten the nuts to the specified torque.

When you assemble a Majormec pumping element to the body (Fig. 29–22), first place the barrel-seat washer in the body. Then insert the lubricated barrel into the bore so that the missing serration locates with the corresponding missing serration in the body. This aligns the ports with the bores in the body. Install the delivery-valve-seat joint ring, delivery valve, spring, valve stop (volume reducer), and a new O ring onto the delivery-valve holder. Tighten the delivery-valve holder to the specified torque.

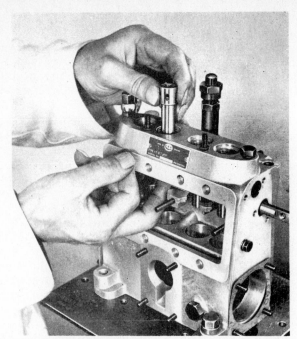

Fig. 29–32 Installing the barrel. (*CAV Ltd.*)

When you assemble the pumping element of the injection pump shown in Fig. 29–32, insert the lubricated barrel into the barrel bore with the spill port toward the deflector screw bore. Align the spill port with the bore. Secure the barrel with the deflector screw, and then install the delivery-valve components in the order shown in Fig. 29–33. Position the control rack so that the center punch marks protrude evenly from the housing. Slide the control sleeve with the gear segments over the barrel so that the index mark on the sleeve faces toward you (see Fig. 29–34). **NOTE** The gear segment clamp opening must be in line with the index mark and the gears snug with the control sleeves.

Install the other control sleeves. When the control rack is in the center position, the control sleeves must all face toward you. Next, install the plunger spring and the plunger with the lower spring seat. Use a tool similar to that shown in Fig. 29–35 to install the pumping element in the housing. The flange mark on the plunger should face upward. Test to make sure that the plunger turns smoothly.

With a tool similar to that shown in Fig. 29–36, install the tappet roller into its bore.

Testing Barrel-Seat and Fuel-Gallery Leakage Regardless of the type of port and helix fuel-injection pump you are servicing, always test the partly assembled pump for barrel-seat leakage and fuel-gallery leakage (Fig. 29–37). To do this, install all plungers (with or without their springs, depending on assembly sequence) into their barrels and hold them in place with the tappet holder.

Connect an air line with a pressure regulator to the inlet manifold and plug all the other openings. Immerse the pump in clean Diesel fuel and pressurize the manifold to about 36.7 psi [2.58 kg/cm²]. Check for air bubbles at the barrel seat, but disregard

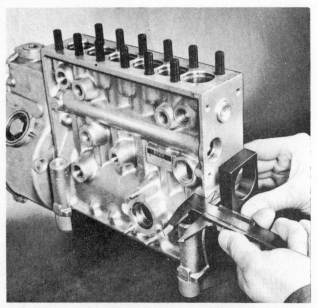

Fig. 29–31 Measuring camshaft protrusion. (*Robert Bosch, GmbH, Stuttgart, West Germany.*)

1. Delivery-valve body
2. O-ring gasket
3. Delivery valve
4. Delivery-valve spring
5. Delivery-valve holder

Fig. 29–33 Delivery-valve components. (Numbers indicate assembly order.)

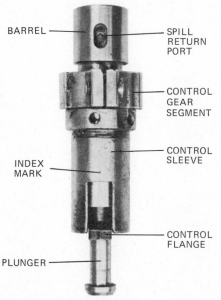

BARREL

SPILL RETURN PORT

CONTROL GEAR SEGMENT

CONTROL SLEEVE

INDEX MARK

CONTROL FLANGE

PLUNGER

Fig. 29–34 Correct assembly of the sleeve, gear, and plunger.

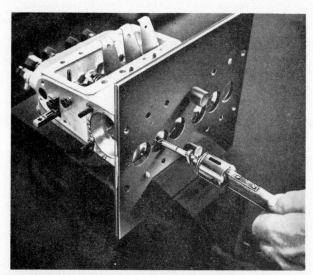

Fig. 29–35 Installing a plunger. *(CAV Ltd.)*

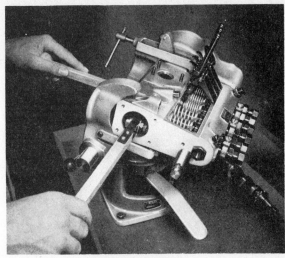

Fig. 29–36 Installing a tappet roller. *(Robert Bosch GmbH, Stuttgart, West Germany.)*

Fig. 29–37 Testing the barrel-seat and fuel-gallery leakage. *(Robert Bosch GmbH, Stuttgart, West Germany.)*

those which may appear at the pump plungers and at the delivery-valve holders.

Injection-Pump Timing Preparation Diesel engines are very sensitive to injection-pump timing and to the (phase) angle between pump elements. The simplest and most accurate method of timing or retiming a pump is by the "spill" method. This method is used to determine the beginning or ending of injection for all pump elements. However, the procedure differs to some extent according to the model.

When timing or phasing the injection pump, mount the pump to a baseplate to prevent damage to the pump body, and then secure it in a vise. Remove the inspection cover and the inlet manifold. Connect a standard fuel filter with a filter element and a shutoff valve. Fill the filter with clean diesel fuel. Install the degree wheel to the drive coupling and a pointer to the housing (Fig. 29–38).

POINTER —

Fig. 29-38 Fuel-injection pump prepared for phasing.

Fig. 29-39 Measuring head clearance or plunger lift.

Turn the camshaft until the (No. 1) cam is in the TDC position, that is, when the line on the camshaft coincides with the line on the bearing housing.

Before you can time the injection pump, you must first measure the head clearance (to prevent the head of the plunger from protruding through the barrel and hitting against the delivery-valve body). Head clearance can be adjusted by means of a spacer, a shim, or a tappet screw.

Measuring and Adjusting Head Clearance (1-mm Specification) To measure the head clearance, remove the delivery-valve holder, valve, and spring. Replace the delivery-valve holder, and tighten it to the specified torque. Place the extension of a depth dial indicator through the delivery-valve holder as shown in Fig. 29-39). With the plunger at TDC, the extension resting on the plunger head, and the base of the dial gauge resting firmly on the holder, zero the dial. Gently lift the tappet roller until the plunger contacts the valve body. Since the lift is transmitted to the dial gauge, the dial will indicate

the distance the plunger has moved. With the tappet screw, adjust the head clearance until the plunger travel is 1 mm.

When a tappet spacer or a shim is used instead of a tappet screw, record the reading, and then bring the tappet roller to the BDC position. Lift out the snap ring, spacer, or shim. Measure its thickness with a micrometer. Select another shim or spacer to give a head clearance of 1 mm. Install the new shim or spacer. Recheck the head clearance.

Another method of adjusting the head clearance is to bring the plunger to the TDC position. Turn the tappet adjusting screw until the plunger lightly contacts the delivery-valve body. Then make one complete turn, counterclockwise, and tighten the locknut. This procedure also gives a 1-mm head clearance since the tappet screw has a 1-mm pitch.

Spill-timing for Port Closure (Lower Helix) After the head clearance is adjusted to specification, install a swan-neck pipe onto the delivery-valve holder in order to see the spill cut off.

Set the control rod to the full-fuel position. Then in the "stop" position the plunger will not close the inlet port. Bring the No. 1 tappet roller to the BDC position and open the fuel shutoff valve to allow fuel, under gravity, to enter the manifold. The fuel will displace the air from the manifold into the inlet port of No. 1 element and force it out of the swan-neck pipe. As airfree fuel leaves the pipe, turn the camshaft slowly (in the direction of rotation) in order to raise the plunger. As the plunger gradually closes the inlet port, the flow from the pipe diminishes. From this point on, turn the camshaft very slowly so that the flow decreases to a dribble and finally to one drop in 5 seconds. If you overshoot closure, the procedure must be repeated from the beginning.

When you have established port closure, adjust the degree wheel to zero. If necessary, mark a new timing line on the drive coupling to coincide with the line on the bearing housing. Turn the fuel supply off and remove the swan-neck pipe from the delivery-valve holder. Wash the delivery-valve spring and valve stop in clean diesel fuel. Reassemble the components.

Spill-timing for Port Opening (Upper Helix) The preparation when spill-timing an injection pump for port opening (upper helix) is the same as when spill-timing a pump for port closing. However, the timing on the injection pump is marked for spill port opening because the beginning of delivery is variable whereas the ending is constant.

When the head clearance is adjusted and the No. 1 plunger is at the BDC position, turn on the fuel and allow it to flow out of the swan-neck pipe. Slowly turn the camshaft in the rotation of the engine firing order. This will cause the upper helix on the plunger to gradually close the inlet port, decreasing the flow of fuel. From this point on, gently turn the camshaft. As the helix closes the inlet port, the fuel flow will be reduced to a dribble because now the fuel is pumped out of the pipe through the action

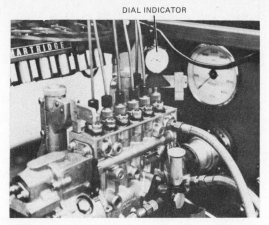

DIAL INDICATOR

Fig. 29–40 Measuring plunger travel (PEP injection pump) using a depth dial indicator.

of the pump plunger. As soon as the upper edge of the vertical groove opens the spill port and/or inlet port, fuel under gravity pressure again starts to run out of the swan-neck pipe. To be positive that the point of port opening is correct, this procedure should be repeated several times before placing the new timing mark on the drive coupling.

Phase-Angle Adjustment To adjust the phase angle of the remaining pump elements, remove the delivery valve, delivery-valve stop and spring from the element next in the firing order. Then replace the valve holder and retorque it to specification. Install the swan-neck pipe to this holder. The point of port opening or closure for the No. 5 element is determined by the method outlined for the No. 1 element.

When the tappet screw, shim, or spacer is correctly adjusted, the port closure or port opening should be exactly 45° from the adjacent pump element for an eight-cylinder pump, 60° for a six-cylinder pump, or 90° for a four-cylinder pump. If the phase angle is greater, the plunger must be raised. When the phase angle is smaller, the plunger must be lowered. This can be done with the tappet adjusting screw, the spacer, or the shim. Phase the remaining elements to obtain the correct degree between them.

NOTE After the port closure or port opening of each element is adjusted, turn the camshaft to the TDC position, and check the head clearance. A clearance is essential, but it need not be measured.

Recheck port closure of the No. 1 element to be sure that the pointer or degree wheel has not moved during the phasing operation.

Timing and Phasing Majormec and PEP Injection Pumps Timing and phasing can be done by the vise-mounted method previously outlined, that is, with a degree wheel, pointer, filter, and fuel line connector, or the injection pump can be installed to a test bench.

When you use a test bench, mount the injection pump and connect the drive coupling to the drive of the test stand. With a fuel line, connect the test-stand transfer pump to the inlet manifold of the in-

jection pump. Remove the delivery-valve holder, delivery-valve stop, delivery valve, and spring. Then reinstall the delivery-valve holder and tighten the holder to specified torque.

Insert a 1/16 6-in [1.55 mm, 127 mm] drill wire through the delivery-valve holder onto the top of the plunger to find the TDC position of the No. 1 pump element. Make sure that the wire is clean.

Up to this point, Majormec and PEP procedures are similar, but from this point on the timing and phasing recommendations tend to vary. Nevertheless, the final results are the same.

When working with a PEP pump, connect the injection line to the delivery-valve holder. Switch on the transfer-pump motor and adjust the transfer-pump pressure to about 10 psi [0.703 kg/cm²]. Open the drip-tube valve, turn the test-stand drive in the rotation specified for the engine's firing order, and spill-time the pump element for port closure.

Set the movable degree wheel of the test stand to zero but do not turn the camshaft. Remove the injection line. Install the dial indicator so that the preload reading of 2.5 mm appears on the dial (Fig. 29–40). Slowly turn the camshaft until the No. 1 tappet roller is on the base circle of the cam. If the plunger traveled the specified distance, the dial indicator should show 2.0 to 2.1 mm.

If the plunger travel is incorrect, remove the pump element, and correct the shim thickness to arrive at the specified plunger travel.

Indexing Timing Pin When you have established the plunger travel and port closure, check that the timing pin, when pushed inward, slides unrestricted into the camshaft notch. If it does not slide in easily, remove the timing pin assembly, remove the dowel pin from the timing-pin cover, reinstall the cover, and loosely secure the assembly to the pump. Next, index the timing pin and tighten the cover bolts to specified torque. Then drill a hole for a new dowel pin.

The next step is to adjust the phase angle of the remaining pump elements.

NOTE Most injection pump specifications allow a tolerance of ±½° in the phase angle.

Timing and Measuring Sims and Majormec Plunger Travel The Sims timing and measuring procedure is to install the dial indicator and zero the dial when the tappet roller of the No. 1 element is on the base circle of the camshaft. Then you should spill-time the No. 1 pump element to establish its port closure (see Fig. 29–41).

When you have determined port closure, check the reading on the dial indicator for plunger travel. If the plunger travel (distance) is correct, check the timing mark on the drive coupling. It should be in line with the mark on the timing indicator. If it is not, scribe a new port closure line on the drive coupling.

If the travel distance and the timing mark are correct, set the degree wheel to zero, and phase the remaining pump elements.

Fig. 29–41 Measuring plunger travel using a dial indicator. *(CAV Ltd.)*

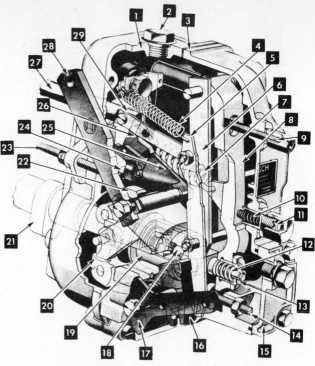

1.	Governor weight housing	16.	Shutoff shaft-spring-housing arm
2.	Plug	17.	Shutoff lever
3.	Linkage housing	18.	Guide lever link
4.	Starting spring	19.	Thrust sleeve
5.	Floating lever	20.	Flyweight
6.	Guide lever	21.	Injection-pump camshaft
7.	Governor spring		
8.	Tension lever	22.	Governor weight carrier
9.	Shutoff or low-idle adjusting screw	23.	Shutoff adjusting stop screw
10.	Bumper spring	24.	Rate adjusting screw
11.	Bumper-spring adjusting screw	25.	Control shaft
		26.	Rocker
12.	Torque-capsule spring	27.	Injection-pump-plunger control rack
13.	Torque capsule	28.	Control lever
14.	Full-load stop screw	29.	Floating lever arm
15.	Rear plate		

Fig. 29–42 Cutaway view of a PEP governor. *(International Harvester Co.)*

To change the phase angle, the tappet spacer must be changed to lower or raise the plunger. Tappet spacers are available in various thicknesses.

Reassembly Continued   After the injection pump is timed and phased, install the rebuilt governor, the transfer pump, and the drive or automatic timing device.

Servicing Governor   To service the PEP governor shown in Fig. 29–42, first remove the components in the order suggested by the service manual. Wash all disassembled parts in clean diesel fuel, and dry them with compressed air.

Inspect the governor housing for cracks or warpage. Check the fit between the control shaft and the bushings. Replace the bushings and/or the shaft if they are worn.

Inspect the bumper spring for evidence of wear and for spring weakness. Inspect the tension, guide and floating lever for excessive wear.

Remove and replace the thrust bearing to the guide lever by using a punch in the holes provided to drive it from the guide lever.   **CAUTION**   Do not lose the adjusting shims.

Check the flyweights for wear. If the fingers have flat spots or if the pins are excessively worn, the weight assembly must be replaced. Use a puller especially designed to pull the assembly from the camshaft.

If the serviceability of the starting and governor springs is doubtful, replace them. Check all pins and bores which connect the various levers. There must be no evidence of looseness between them.

When you reassemble the governor, install new O rings and gaskets. Lubricate all the components during assembly. Governor adjustment is discussed later in this unit, under Adjusting, Calibrating, and Testing Fuel-Injection Pump on Test Stand. (See also Unit 26, "Antipollution Control Devices.")

Aneroid   The aneroid and the connecting lever assembly must be checked and tested. To service the aneroid, remove the components in the order shown in Fig. 29–43. After you have thoroughly cleaned

the components (except for the diaphragm), inspect the housing for cracks and warpage. Check the mounting surfaces for nicks and burrs and the threaded holes for damaged threads. (Nicks and burrs can be removed with a scraper or 600-grit paper.) If the diaphragm is damaged or if oil was noticeable in the lower chamber at the time of disassembly, then the diaphragm must be replaced.

If the operating shaft, cam-activating shaft, activating pin, or aneroid cam show any signs of wear, they should be replaced. If there is any evidence of rust or corrosion on the operating spring, it should also be replaced.

Reassemble all the components in the reverse order to disassembly, leaving the cap nut off the adjusting screw, and the side cover off the housing.

TESTING AND ADJUSTING ANEROID   The aneroid must be tested and, if necessary, adjusted to hold the ex-

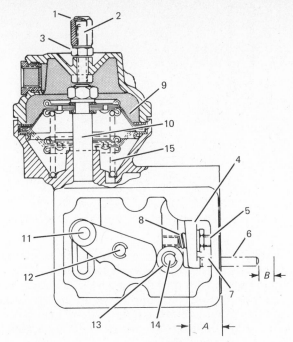

1. Cap nut
2. Adjusting screw
3. Locknut
4. Actuating lever
5. Actuating-lever adjusting screw
6. Actuating pin
7. Nut
8. Spring
9. Diaphragm
10. Operating shaft
11. Cam activating shaft
12. Cam activating pin
13. Aneroid cam
14. Roller
15. Operating spring

Fig. 29–43 Sectional diagram of an aneroid and connecting linkage. (Numbers also indicate order of removal of components.) (*International Harvester Co.*)

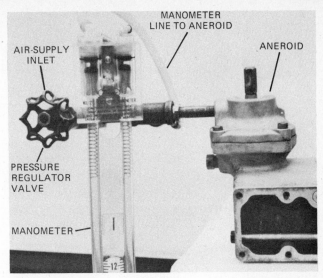

Fig. 29–44 Manometer connection to the aneroid.

haust smoke during acceleration to a minimum without limiting engine performance. To check the aneroid, connect an air supply and a mercury manometer to the inlet port of the aneroid through a T fitting. The air supply must have a pressure regulator (Fig. 29–44).

Loosen the locknut and turn the adjusting screw until it stops. Now apply a pressure of 2 to 3 inHg [50.8 to 76.2 mmHg] to the chamber and observe the cam. It should have barely moved. Then increase the pressure to 12 inHg [304.8 mmHg]. The cam should have moved out to its fullest extent. If not, remove one of the shims which are between the diaphragm and housing to extend the travel. Increase the pressure to 20 inHg [504.0 mmHg]. Shut off the air supply and check the pressure drop. It should not exceed 2 inHg [50.4 mmHg] in 10 seconds. If it does, perhaps there is an air leak in the diaphragm. If there is no leak, then the mating surfaces should be checked.

Open the air supply and maintain a pressure of 20 inHg [504.0 mmHg] in the chamber and measure the distance A in Fig. 29–43. If necessary, use the actuating-lever adjusting screw until the specified distance is achieved. This adjustment ensures a clearance of 0.010 to 0.030 in [0.254 to 0.761 mm] between the actuating pin and the activating lever. The adjusting screw limits the retraction of the diaphragm which, in turn, limits the maximum travel of the actuating pin.

To adjust the travel distance, again apply a pressure of 20 inHg [504.0 mmHg] to the chamber, and then measure the distance B. Then reduce the chamber pressure to zero. Measure the distance again. It should have decreased by [0.165 in ± 0.010 in (4.191 mm ± 0.25 mm].

Turn the adjusting screw until the travel distance reaches specification. Use pressure to cycle the aneroid a few times to be sure that the assembly is working smoothly without binding.

Servicing Diaphragm Transfer Pump Not only must the transfer pump be serviced to correct obvious mechanical defects, it must also be serviced to maintain the correct transfer-pump pressure.

DISASSEMBLY AND INSPECTION To disassemble a diaphragm transfer pump, first unscrew and remove the inlet and outlet unions. Withdraw the valve spring and the valve (see Fig. 29–45). After you have removed the screws, carefully separate the valve housing from the pump body. Remove the self-locking nut from the spindle and lift off the back plate, diaphragm, and front plate. Slide out the tappet spindle, body spring, and pressure spring. At this point, inspect the pump lever for any signs of wear on the shaft, the bearings, or the fork.

Check the valve seat in the valve housing for damage. Check the tappet body and the tappet pad for wear. Replace the valves and, when necessary, the diaphragm, pressure spring, and tappet spring. Check the joint face of the pump body and the tappet housing and, if necessary, lap the surfaces.

REASSEMBLY To reassemble the pump, slide the tappet body and tappet spring onto the lubricated spindle. Separately lubricate the tappet body and its bore. Then return the preassembled tapped body to the bore.

Compress the tappet spring and slide the pressure spring, spring collar, front plate, diaphragm, and back plate onto the spindle. Align the holes of the diaphragm with the pump body by inserting a few

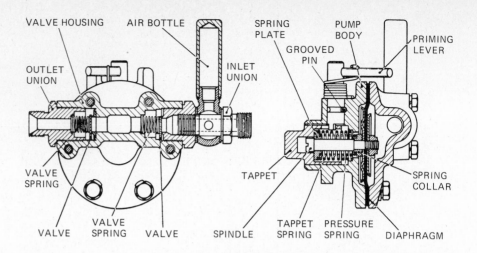

VALVE HOUSING · AIR BOTTLE · SPRING PLATE · PUMP BODY · PRIMING LEVER · OUTLET UNION · INLET UNION · GROOVED PIN · VALVE SPRING · TAPPET · SPRING COLLAR · VALVE · VALVE SPRING · VALVE · SPINDLE · TAPPET SPRING · PRESSURE SPRING · DIAPHRAGM

Fig. 29–45 Cross-sectional views of a fuel feed pump (type PL 16). *(CAV Ltd.)*

aligning screws. Then secure the assembly to the spindle with a new locknut.

Place the valve housing over the pump body. Tighten the screws evenly and alternately to assure equal screw force. Install the valve spring and the union.

Testing the transfer pump is outlined later in this unit, under Adjusting, Calibrating, and Testing Fuel-Injection Pump on Test Stand.

Injection-Pump Installation Before you install a new or reconditioned injection pump, make certain that the cause of the pump failure has already been corrected. Replace all the fuel filters and, when necessary, clean the fuel tank.

To position the No. 1 piston on the compression stroke when the firing order is 1-5-3-6-2-4, remove the rocker cover to visibly expose the valve action of the No. 6 cylinder. Turn the engine over in the direction of normal rotation until the intake valve of the No. 6 cylinder starts to open. This ensures that the No. 1 piston is on its compression stroke. Continue to rotate the camshaft until the engine timing marks line up.

Loosen the drive-coupling bolts. Align the timing marks on the injection pump and slide the injection pump into place. Align the pump to the coupling (use a feeler gauge). Then tighten the mounting and the drive-coupling bolts.

To recheck the injection pump to engine timing, crank the engine 720°. During the cranking, check the alignment of the drive couplings.

When you install and time an injection pump with a fixed drive gear or drive coupling, first time the engine and the injection pump, and then slide the pump into position (Fig. 29–46). Always install a new gasket or well-lubricated O ring to seal the injection pump. Support the pump to its mounting with the nuts or bolts, but do not tighten them because it may be necessary to rotate the pump toward or away from the engine in order to align the timing marks. Place the drive gear on the camshaft drive flange, and tighten the bolts to the recommended torque.

Check the engine timing marks and the timing marks of the injection pump. If the timing marks of the pump do not coincide, rotate the pump to bring them into alignment. Then tighten the mounting nuts or bolts to the recommended torque.

Some fuel-injection pumps are not timed by rotating the pump. Instead, they have an engine drive gear with slotted mounting holes, a drive gear which is held by friction onto the drive coupling, or an adjustable drive coupling.

When the drive gear has elongated mounting slots (Fig. 29–47), time the engine and the injection pump. Center the slots, and then install and tighten the injection pump to its mounting flange or base.

Take up the gear backlash before tightening the hex bolt of the drive gear. Rotate the crankshaft 720° to recheck the timing, and then secure the drive locks.

When you install an injection pump where the drive gear is held by friction to the drive coupling (Fig. 29–48), loosen the hex bolts which hold the drive gear to the tapered drive coupling. Make certain the gear can move freely. Time the engine and the injection pump. Mount the pump to the engine. Take out the gear backlash before tightening the hex bolts to the specified torque. Then rotate the crankshaft 720° and recheck the timing marks.

CAUTION When using a timing pin, do not forget to remove it before turning the engine over; otherwise the injection pump will be damaged.

When installing a fuel-injection pump having an adjustable drive coupling, loosen the two hex bolts that hold the two halves of the coupling together (Fig. 29–49). Time the injection pump and engine, and mount the pump to the engine. Then take out the gear lash, but make certain that you do not change the engine or injection-pump timing. Now tighten the two coupling bolts to specified torque and recheck the injection-pump timing.

Bleeding (Venting) Fuel-Injection System After the injection pump is timed to the engine and its mounting bolts are torqued to specification, connect the low-pressure fuel line, the injection lines, the throttle, and the shutoff control to the injection pump.

Bleed the air from the filters, fuel lines, and the injection pump to prevent lack of lubrication to the

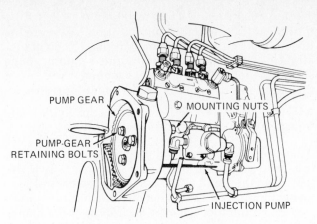

Fig. 29–46 Installing an injection pump having a fixed drive gear. (*J. I. Case Co., Components Div.*)

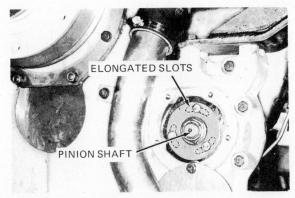

Fig. 29–47 Drive gear with slotted mounting holes. (*International Harvester Co.*)

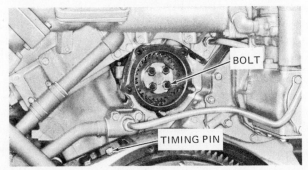

Fig. 29–48 Drive gear which is held by friction to the drive coupling. (*Caterpillar Tractor Co.*)

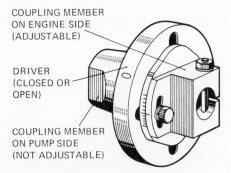

Fig. 29–49 Adjustable drive coupling.

pump elements and possible damage to the closely fitted parts.

To bleed the system, first check the level of the fuel tank. If it is below the injection pump, fill the tank.

Open the shutoff valve, the primary filter, the secondary filter, and the injection-pump vent screws. Allow fuel and air to flow from the vents until the fuel is free of air bubbles. Then tighten the vent screws in sequence, one after the other.

If you bleed the system by using the hand-priming pump, do not forget to lock it. If the pump is not locked, there will be no transfer-pump pressure because fuel will circulate within the transfer pump.

Bleeding Injection Lines   The high-pressure side of the injection lines is usually self-bleeding since the air is forced out through the action of the pumping element. However, when the cranking power is only moderate, when the engine fails to start, or when it continues to run rough, venting of the injection line is necessary.

To vent the injection line, set the throttle to the full-fuel position and the shutoff lever or the switch in the "run" position. Place the decompression lever in the "decompression" position. Loosen all the injection-line fittings at the injectors. Crank the engine until fuel squirts from the injection lines, and then tighten the connection .

ADJUSTING, CALIBRATING, AND TESTING FUEL–INJECTION PUMP ON TEST STAND

After a fuel-injection pump is serviced, it must be calibrated and the governor must be adjusted. Calibration is an adjustment of the pumping elements to ensure that each delivers the same specified quantity of fuel. The governor must be adjusted to ensure high-idle, low-idle, and maximum-fuel delivery.

Mounting Injection Pump to Test Stand   After the head clearance or plunger lift and the phase angle are adjusted, mount the injection pump to the test stand. Make certain that the drive couplings are aligned, and allow a small clearance between them.

Fill the injection-pump housing and the governor housing with lubricant to the specified level or connect an oil-supply pump to the recommended fittings. Install the recommended test injector and injection line.

With hoses, connect the transfer pump to the test-stand lift fitting. Connect the transfer-pump outlet to the fuel manifold of the injection pump.

Back out the high-idle and the low-idle adjusting screws. If equipped with a damper or buffer screw, back it out the specified distance.

If an aneroid is installed, carefully follow the manufacturer's instructions regarding removal or setting.

Open the bleed screw at the injection pump and injector holders. Select the correct test-stand rotation and, if an external oil supply is used, switch on the motor and adjust the oil pressure as specified.

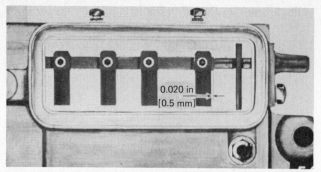

Fig. 29–50 Setting the first control fork to specification. *(CAV Ltd.)*

Fig. 29–51 Adjusting the No. 1 element to maximum-fuel output. *(CAV Ltd.)*

Fig. 29–52 Adjusting the remaining control forks.

Calibration First, adjust the fuel-control-rack travel according to specification. Using the Minimec fuel-injection pump as an example, set the first control-rod fork at 0.020 in [0.5 mm], from the square section on the control rod (Fig. 29–50). Next, start the test stand and operate it at the recommended speed to bleed air from the injection pump and from the injectors. When all air is removed from the pump and injectors, close the bleed screw and increase the operating speed to specified rpm to warm the fuel and injection pump to proper temperature.

Testing Operation of Transfer Pump First, set the throttle at the maximum-fuel position. Operate the injection pump at recommended rpm. Then gradually close the lift valve until the vacuum gauge shows the required minimum vacuum. Next, open the valve and check the transfer-pump pressure.

NOTE Do not proceed with any other tests if the transfer pump is not within specification.

Adjusting Maximum Fuel Delivery Back out the low- and high-idle adjusting screws. Move the throttle to the right. This will bring the control rack to maximum fuel. Now, operate the injection pump and adjust the fuel delivery according to the manufacturer's instructions. Again, using Minimec as an example, set the counter to the recommended stroke, and then check the delivery of the No. 1 pump element. If the delivery is lower than specified, adjust the maximum-fuel stop screw until the maximum fuel delivery from the No. 1 element meets specification (Fig. 29–51). Repeat the test several times to ensure proper adjustment. Then lock the fuel stop screw. Position the other control fork on the control rod so that the delivery from each element is the same as that from No. 1 (Fig. 29–52).

Electronic Phase-Angle Check When each element is adjusted, check the phase angle of the fuel-injection pump by switching on the electronic phase unit and the No. 1 selector switch. Rotate the degree wheel (Fig. 29–53) so that the flash accurately coincides with the zero on the wheel. Switch on the remaining selector switches. Note the flash and the angled position of all elements.

Governor Adjustment The next step is to adjust the high-idle stop screw so that the breakaway occurs at the specified rpm, say, 100 rpm above high-idle speed. Move the throttle to the full-fuel position. Then gradually increase the operating speed until the fuel rack starts to move to the no-fuel position. At this point, note the operating speed. When the breakaway exceeds the specified rpm, turn the high-idle adjusting screw clockwise to reduce the governor spring force. When the breakaway is less than specified, turn the high-idle stop screw counterclockwise, and then recheck the breakaway.

NOTE To prevent the engine from overspeeding, the speed (rpm) and the control-rack-travel position must be within specification.

Testing Idle-Speed and Cranking Fuel Delivery First, reduce the operating speed to the specified idle speed. Next, set the counter to the recommended strokes, and then check the fuel delivery. When the idle-speed fuel delivery is lower than specified, turn the idle-speed stop screw clockwise to increase governor spring force. Recheck the idle-speed fuel delivery.

To check cranking fuel delivery, operate the fuel pump at recommended speed and set the counter to specified stroke. Push the excessive fuel shaft inward. Move the throttle to the full-fuel position,

Fig. 29–53 Electronic phase-unit degree wheel.

and then check the fuel delivery. Remove the injection pump from the test stand. Plug all openings and lock-wire all adjusting screws.

Questions

1. Relate each mechanical and hydraulic action which takes place in the pumping element, beginning with the moment fuel is under transfer-pump pressure in the fuel manifold.

2. Why do some injection pumps use an upper-helix plunger?

3. Explain the mechanical action which takes place when the fuel rack moves to the full-fuel position. (Refer to Fig. 29–14.)

4. What is the purpose of a delivery valve?

5. List the steps you must take *before* disassembling an injection pump.

6. Identify the inspection or check points of the barrel and plunger, the delivery-valve assembly, the camshaft, and the control rack.

7. Briefly outline the procedure to measure the plunger travel on a PEP injection pump.

8. Explain in detail how to spill-time an injection pump having an upper helix.

9. When servicing a diaphragm transfer pump, which component parts require special attention?

10. List the essential steps or checks to be taken prior to installing an injection pump to an engine having an adjustable drive coupling.

11. When calibrating an injection pump, why is it necessary to use (a) only the specified test injectors, (b) only the specified high-pressure lines, and (c) only test oil which has been brought to specified temperature?

12. If, when checking fuel delivery on the test bench, you find that (a) at idle speed one pump element is lower than specified and (b) at maximum fuel the same pump element is correct (according to specification), what steps would you take to correct problem (a) above?

UNIT 30

Caterpillar Fuel-Injection System

As mentioned previously, although all port and helix-metering fuel-injection pumps operate on the same principle, they differ in construction, pump-element arrangement, method of phasing, governor design, and adjustment.

The components which make up one of Caterpillar Tractor Company's fuel-injection systems are shown in Fig. 30–1. It should be noted here, that the location of the bypass valve, the priming pump, or the pressure gauge may vary depending on the engine design or application. Furthermore, Caterpillar has designed not only two different port and helix-metering fuel-injection pumps but also a sleeve-metering fuel-injection pump. The latter pump is discussed separately in Unit 35.

Comparison Between Forged-Body and Compact-Housing Fuel-Injection Pumps The oldest Caterpillar fuel-injection pumps are the forged-body pumps. These injection pumps have individual pumping elements mounted on the top of the housing (Fig. 30–2). The pumping elements of the compact-housing injection pumps are located inside the housing, similar to the Robert Bosch or CAV injection pumps (Fig. 30–3). They are held in place by an external, threaded retaining bushing screwed into the housing.

The forged-body and the compact-housing pumping elements, though different in appearance have like components, and operate on the same principle (see Fig. 30–4).

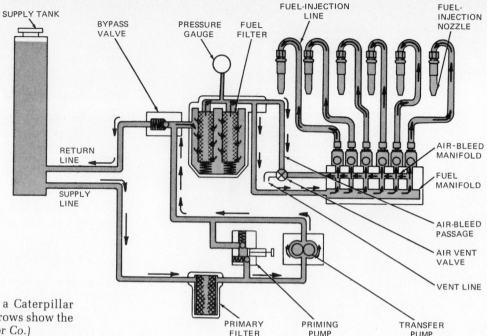

Fig. 30–1 Components of a Caterpillar fuel-injection system. (The arrows show the fuel flow.) (*Caterpillar Tractor Co.*)

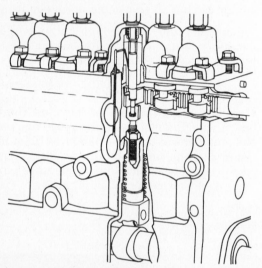

Fig. 30–2 Forged-body fuel-injection pump. (*Caterpillar Tractor Co.*)

Fig. 30–3 Compact-housing fuel-injection pump. (*Caterpillar Tractor Co.*)

The barrels of all Caterpillar pumping elements have only one fuel-inlet port; the delivery valve is a check valve with a displacement plunger. All Caterpillar plungers are of the constant-beginning and variable-ending design and have the gear segment bolted to the plunger. Their plungers have a second circumferential groove which is in line with the port in the barrel when the plunger is at TDC. This allows the fuel used for lubrication to return through a passage to the inlet manifold (Fig. 30–5).

The plungers of the forged-body pumping elements are hooked into a yoke which is screwed into the lifter tappet (Fig. 30–4a). The plungers of the compact-housing pumping element have a lower circumferential groove to lock the lower spring seat (washer) to the plunger (see Fig. 30–6.) The plunger-return spring is positioned between the barrel and the spring seat. This eliminates the necessity of the yoke arrangement since the plunger spring forces the plunger and the lifter assembly against the cam lobe. The lifter assembly is similar in design to the assembly of a Simms injection pump. The barrel and bonnet have a machined-out groove. Figure 30–7 shows the two locating dowels, which hold the barrel and bonnet in postion. When positioned, the barrel rests in the housing against a spacer which is used to change lifter height (port closure). The spacer rests against the shoulder in the pump housing and the assembly is held in the housing by the retaining bushing. The square O ring between the bonnet and bushing seals the assembly against fuel leakage. A felt seal is used to protect the assemblies against dust, dirt, etc.

NOTE The check valve design of the forged-body pump is the same as that of the compact-housing pump. The camshafts of all injection pumps are supported by bushings and are driven by an offset tang coupling, a split coupling, or a slip drive gear.

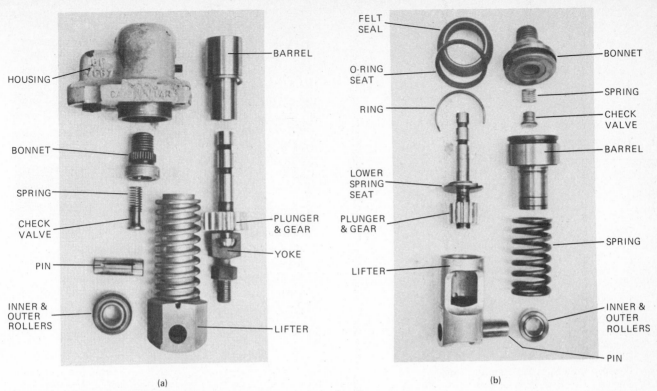

Fig. 30-4 *(a)* Disassembled forged-body injection pump; *(b)* disassembled compact-housing fuel-injection pump.

The fuel racks of the compact-housing injection pump have machined circular grooves rather than teeth like the forged-body injection pump.

FORGED-BODY FUEL-INJECTION-PUMP SERVICE

Removing Forged-Body Pump There is nothing unusual in the removal procedure of a forged-body or a compact-housing injection pump. However, the procedure does change according to the type of drive coupling used.

When the slip-type gear or the offset tang coupling are used, first remove the necessary fuel lines and accessories, and then remove the housing mounting bolts. Lift the pump housing from the engine with a suitable hoist. Either coupling will then slide out of mesh.

A brief insight into the servicing of a 769B truck engine (split coupling) follows, however, *you are cautioned to refer directly to the Caterpillar service manual when actually removing this pump for service.* This pump uses a split drive coupling. To remove the pump, loosen the bolts and slide the coupling toward the timing-gear housing. Attach a hoist to the pump housing. Slide the adapter housing toward the timing-gear cover to clear the ferrules

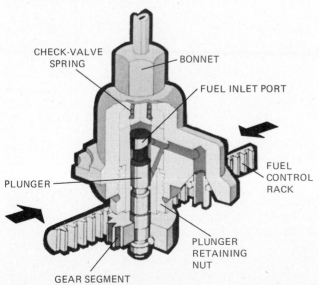

Fig. 30-5 Cutaway view of the pump, showing the fuel passages. *(Caterpillar Tractor Co.)*

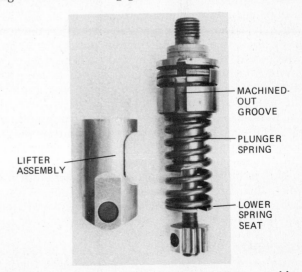

Fig. 30-6 Compact-housing pump and lifter assembly.

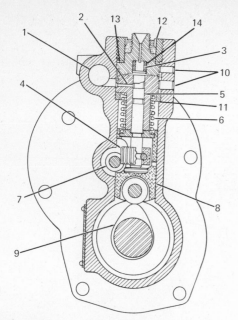

1. Fuel passage
2. Inlet port
3. Check valve
4. Gear segment
5. Pump plunger
6. Spring
7. Fuel rack
8. Lifter
9. Camshaft
10. Dowel pins
11. Spacer
12. Retainer bushing
13. Gasket
14. Check-valve spring

Fig. 30–7 Cross-sectional diagram of compact-housing fuel-injection pump. (*Caterpillar Tractor Co.*)

and the camshaft. Then remove the mounting bolts and lift the pump housing from the engine.

Disassembling Forged-Body Pump *Never* attempt to disassemble a Caterpillar forged-body fuel-injection pump without reference to the service manual. However, the usual precleaning of the work area, the unit and the tools, precedes disassembly of *any* fuel-injection pump.

Here is a review of most of the steps to disassemble this pump. Mount the clean pump housing in a swivel vise and remove the governor. Since the 769B engine has a servo-type governor, the rack limiter must first be removed (see Fig. 30–8). Remove the cover and the pin which connects the governor to the control link. Then remove the bolts which hold the governor housing to the injection-pump housing. Remove the governor housing by working it loose from the dowel pins. Next, remove the bolts and clamps that hold the injection pump to the housing. Lift the pump housing from its dowels. Slide the plunger out of its yoke, and remove it as a unit from the housing (Fig. 30–9). (Do not let the plunger fall from its barrel.)

If it is difficult to remove the plunger from the yoke, then lift the pump housing off and, with a clean magnet, lift the plunger out of the yoke and immediately insert it into the barrel.

CAUTION Do not manually lift the plunger because even the slightest dirt accumulation on the plunger will cause it and the barrel to wear.

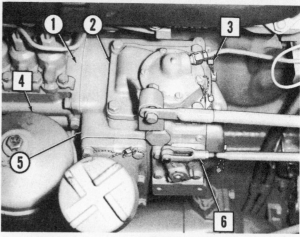

1. Fuel-injection-pump housing
2. Rack limiter
3. Seal
4. Front oil-filter assembly
5. Governor housing
6. Governor control rod

Fig. 30–8 Preparing to remove the governor housing. (*Caterpillar Tractor Co.*)

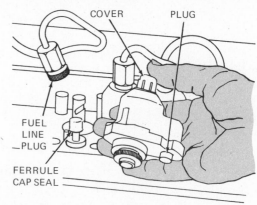

Fig. 30–9 Removing the forged-body fuel-injection pump. (*Caterpillar Tractor Co.*)

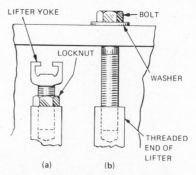

Fig. 30–10 (*a*) Removing lifter yokes and (*b*) raising lifters. (*Caterpillar Tractor Co.*)

To remove the camshaft, loosen the locknuts and screw out the yokes (see Fig. 30–10*a*). Use a 24NF ⅜-in bolt, 3 in long, to raise the lifters (see Fig. 30–10*b*). A large washer must be placed between the bolt head and the housing on which the bolt head can rest. Screw the bolt into each lifter to withdraw each one from the camshaft lobes.

To hold the camshaft when removing the drive-gear nut, place a piece of wood under a cam lobe near a bearing. When the nut is removed, lift off the plate, the spring, and the two pins, and slide off the drive gear (Fig. 30–11). Take note of the position of the spring since it is the connecting link from the camshaft to the governor drive gear. The drive gear can rotate about 33° without moving the camshaft. This assures a pulsation-free drive to the governor driven gear. Only two bolts need be taken out in order to lift the camshaft from the pump housing. Take care not to damage the camshaft bushings. If the lifters require servicing, remove the bolts, one by one, to release the lifter-spring tension and withdraw the lifters.

Inspecting the Forged-Body-Pump Components

Check all the mounting surfaces for nicks and burrs and all the threaded holes for damage. Measure the camshaft and fuel-rack bushings for wear with an inside micrometer and a hole gauge and then compare the measurement with the specification in the service manual. If the measurements indicate that the bushings should be replaced, push them out with a suitable adapter. When pressing the new bushings in place, make sure the oilholes are aligned and that the bushings are positioned in accordance with the service manual instructions.

Measure the camshaft-bearing journals, and then compare this measurement with the service manual specification. Check the cam lobes. If there is any sign of surface damage, the camshaft must be replaced.

The lifter body can be checked for wear by measuring its diameter. If the lifter roller shows any sign of rough spots or grooving, the assembly must be replaced. With a spring tester, measure the free length and test the energy of the lifter spring.

The Caterpillar service procedure for cleaning and checking the barrel and plunger or the delivery valve and spring is generally the same as that for similar components of the Bosch and CAV components. However, the Caterpillar Tractor Company recommends measuring the length of the plunger to determine possible wear (Fig. 30–12). If the measurement is not within specification or if the helix, plunger, or barrel is damaged, the barrel and plunger must be replaced.

Always check the yoke. It should never be allowed to deteriorate to the extent shown in Fig. 30–13.

Reassembling Forged-Body Pump

Lubricate the lifter and its bore. Slide the spring over the lifter body, and place it in the bore. Screw the retracting bolt far enough into the lifter to raise it sufficiently to allow installation of the camshaft. Repeat this procedure with the remaining lifters.

Lubricate the bearings and camshaft-bearing surfaces. Slide the camshaft into position, and bolt it to the housing. Make certain the camshaft rotates freely and the end play is within the specified limit before resting the lifters on the camshaft. Screw the locknut onto the yoke. Screw the yoke

Fig. 30–11 Governor-drive-gear removal. (*Caterpillar Tractor Co.*)

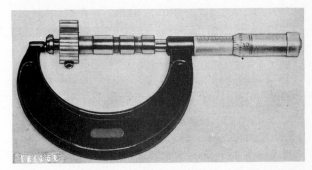

Fig. 30–12 Checking plunger length. (*Caterpillar Tractor Co.*)

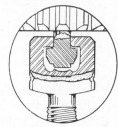

Fig. 30–13 Wear between yoke and plunger. (*Caterpillar Tractor Co.*)

into the lifter until the locknut bottoms against the lifter. Place the lubricated drive gear on the drive flange, and then install the dowel pins and spring. **NOTE** The drive spring must "wind up" when driving the governor gear. (See Fig. 30–11.)

Place the plate washer and the locking plate over the end of the camshaft, and tighten the nut to specification. Before installing the pump elements, measure and then, if necessary, adjust the lifter height.

Pump Lifter Setting (Off-Engine)

Setting the lifter height to the recommended specification can be compared with timing and phasing, except that the measurement is made with a depth micrometer rather than by using the spill-timing method. To check or to adjust the lifter height, install the degree wheel onto the drive end of the camshaft and the pointer onto the housing as shown in Fig. 30–14. Refer to your service manual for lifter-height dimension and to Table 30–1. In this example the lifter-height dimension is 1.88 in ±0.002 in [47.78 mm ± 0.05 mm].

Table 30–1 LIFTER SETTING IN DEGREES (OFF-ENGINE)

Lifter number (numbered consecutively front to rear)	Timing plate degrees
1	½
2	120½
3	240½
4	60½
5	300½
6	180½

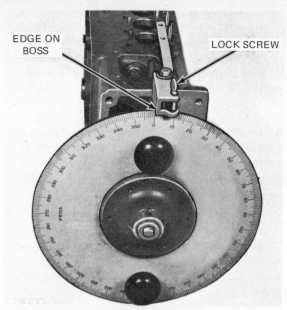

Fig. 30–14 Setting forged-body-pump lifter height (off-engine). *(Caterpillar Tractor Co.)*

With the degree wheel and the pointer in place, rotate the degree wheel so that the ½° mark is aligned with the left edge of the pointer. This brings No. 1 camshaft lobe and No. 1 lifter into position to be measured (and if necessary, adjusted). The No. 1 lifter is on the governor side of the pump housing, and the lifters continue (Nos. 2, 3, 4, 5, 6) to the left, toward the wheel.

Place the depth micrometer into the bore of the pump element. Make sure the surface of the housing is clean. Measure the distance from the surface of the housing to the yoke surface (A in Fig. 30–15). Using adjusting tools, raise or lower the yoke to bring the distance to specification. Then lock the yoke with the locknut, and remeasure the lifter height. If it is within specification, turn the degree wheel to 300½°. This will bring No. 5 camshaft lobe and No. 5 lifter into position to be adjusted.

Measure, adjust, and recheck the measurements of No. 5 lifter, then proceed to the next lifter (No. 3) and so on, until all lifter heights are adjusted.

Pump Lifter Setting (On-Engine) To set the lifters while the pump is installed to the engine it is necessary to position the injection-pump camshaft lobe to measure the lifter height. The camshaft-lobe position is found through the timing marks on the flywheel or by screwing a timing bolt into the flywheel (Fig. 30–16).

The flywheel has three threaded holes which align with the holes identified in Fig. 30–16 as No. 2, No. 3, and No. 4. When the timing bolt is in No. 2 hole and is threaded into the flywheel, two pistons (Nos. 1 and 6) are at precisely TDC (one piston is on compression and the other is starting its intake cycle). If the valves of No. 6 cylinder are open, No. 1 piston is on compression. The lifter of No. 1 injection pump is now ready to be measured, and if necessary, adjusted.

When No. 1 lifter is adjusted, remove the timing bolt. Turn the crankshaft 120°. Align the threaded bore in the flywheel with the No. 3 hole, and thread the timing bolt into the flywheel. The lifter of No. 5 injection pump is now in position to be measured (and if necessary, adjusted).

CAUTION If you overshoot the hole in the flywheel, turn the crankshaft at least 20° in the opposite direction, past the hole, then turn it back carefully to realign the hole. This will prevent gear lash.

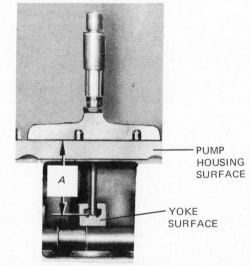

Fig. 30–15 Measuring forged-body lifter height (off-engine). *(Caterpillar Tractor Co.)*

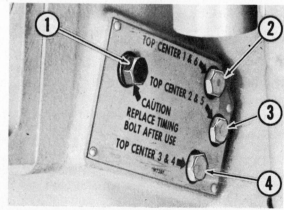

Fig. 30–16 Locating camshaft-lobe position by the timing-bolt method. *(Caterpillar Tractor Co.)*

Do not forget to replace the timing bolt in the No. 1 hole and to replace the three plugs.

NOTE Remember that the lifter-height specifications for setting the lifters "on" engines always differ from the setting of those "off" engines, and vary with engine application.

Installing Forged-Body Pump Before you install the fuel-injection pump, make certain that the housing surface is clean. Position the fuel rack so that the timing mark is in the center of the bore. Remove the plunger from the barrel and, with a magnet, lower it into position; that is, position it with the lower part of the plunger hooked into the yoke and the timing marks aligned (Fig. 30–17).

Next, install a new sealing ring. Use grease to hold it in the large recess of the injection-pump housing. Using extreme care, slide the barrel over the plunger.

To position the pump housing, align the dowels with the holes in the pump housing. Install the clamp and hex bolts, and torque the bolts to specification. Then install the remaining injection pump. When you use the rack-setting gauge to check the rack travel, leave out one pump element.

Servicing Rack Limiter First, remove the cover assembly, the locknut and nut, and then withdraw the adjusting screw, retainer plate, diaphragm, spring, and shaft (see Fig. 30–18). Check each component very carefully for damage and/or wear. Measure and test the springs. (The diaphragm usually is replaced.) Check the bell crank and shaft, particularly where the adjusting screw contacts the bell crank.

When you reassemble the rack limiter, make sure the diaphragm fits smoothly between the plates and lies flat between the cover surface and the limiter housing surface, then tighten the screws. The rack limiter on the 769B engine needs no spring preload adjustment. The spring is automatically correctly preloaded (adjusted) when the roller of the bell crank is positioned between the adjusting screw and the shaft, and the nut is torqued to specification.

Servicing Governor To service the governor, remove the cover, slide out the cylinder and its blocks, the piston assembly, and the springs (see Fig. 30–19). Slide the thrust bearing and sleeve from the spindle. Then pull out the spindle assembly, governor weights, drive gear, and bearing. If there is any doubt about the serviceability of the control-lever assembly, remove them also.

The speed limiter used with this governor is shown in Fig. 30–20. It is a device which limits the travel of the control lever and as a result the engine speed, until the oil pressure of the engine is about 10 psi [0.7 kg/cm²]. The oil pressure then forces the piston to retract against its spring force and the control rack is free to move to full-fuel position.

When you examine the governor components for wear, check particularly the governor weight assembly, the pins, rollers, spindle, drive gear, and

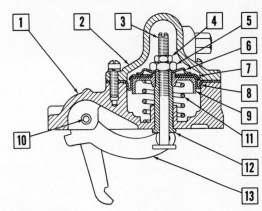

Fig. 30–17 Installing the forged-body fuel-injection pump. *(Caterpillar Tractor Co.)*

1. Limiter housing
2. Cover assembly
3. Adjusting screw
4. Locknut
5. Nut
6. Retainer
7. Retainer plate
8. Diaphragm
9. Plate
10. Shaft
11. Spring
12. Shaft
13. Bell crank

Fig. 30–18 Rack limiter. *(Caterpillar Tractor Co.)*

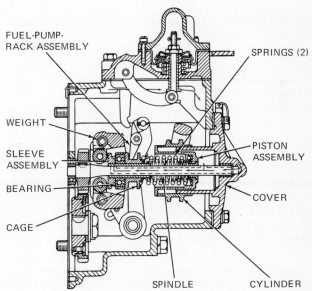

Fig. 30–19 Cross-sectional view of the governor. *(Caterpillar Tractor Co.)*

bearing. Along with the thrust bearing and sleeve, these are the most vital components of the governor (see Fig. 30–21). If any are worn they should be replaced.

Check the two speeder springs and measure their free length and energy. It should coincide with the

1. Stop   4. Drilled hole
2. Plunger 5. Oil-supply passage
3. Spring

Fig. 30–20 Speed limiter at operating pressure. (*Caterpillar Tractor Co.*)

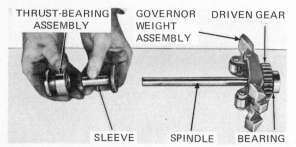

Fig. 30–21 Thrust-bearing assembly, sleeve, and spindle assembly. (*Caterpillar Tractor Co.*)

length and energy of the same color code shown in the service manual.

Check the torque spring and spacer. Compare their thicknesses with the service manual specifications (torque spring group).

Check to be sure that the pistons and the cylinder are free of any rough spots and grooves. (The piston must slide freely into the cylinders.) Check the grooves in which the blocks of the control lever are riding. The blocks must fit, without play, in the grooves and in the pins of the control lever.

Check the bushings, shaft, and pins of the rack-control lever for wear and make certain that the rack-setting adjusting screw is sufficiently loose so that it may be turned. Do not overlook the speed-limiter piston. It must move freely within its bore.

Reassembling Governor Wash all the component in clean fuel. Reassemble them in precisely the reverse order to that in which they were disassembled; however, leave off the rack-limiter housing, side cover, and speed limiter.

Place the governor on the dowel pins of the housing. Mesh the drive and driven gear, and tap the governor housing into place. Tighten and lock the governor mounting bolts, and then connect the control link to the control lever arm. Now position and tighten the rack-limiter housing to the governor housing.

Adjusting the Fuel Rack Using a Rack-Setting Gauge The fuel rack can be adjusted before or after the pump is installed to the engine. Two methods are used to measure the travel: the rack-setting-gauge method and the dial-indicator method.

The outer circumference of the rack-setting gauge is inscribed with marks, from 0 to 0 (Fig. 30–22). The longer lines represent 0.100 in and the shorter lines represent 0.050 in. The inner dial face (which is connected to the gear segment) has two scales, each line representing 0.005 in. One scale is used when the control rack moves to the right, and the other scale is used when the control rack moves to the left in its travel to full-fuel position.

When using the rack-setting gauge, one fuel-injection pump must be removed and the rack-setting gauge installed so that the timing marks on the gear align with the fuel-rack marks, as shown in Fig. 30–17. Check this alignment through the gauge hole. When the marks are aligned, the dial should read 0.

To check the rack adjustment, loosen the locknut and turn the rack-limiter adjusting screw clockwise until the bell crank can no longer restrict the fuel-rack movement.

Remove the speed limiter to allow unrestricted rack travel. Place (between the torque spring and the housing) a feeler gauge of the same thickness as that of the spacer, say, 0.100 in, to prevent the torque spring from bending during the adjustment. Now move the throttle to the full-fuel position. This movement moves the fuel rack and the dial of the gauge and bottoms the adjusting screw of the rack control lever against the torque spring. When this position is reached, read the dial on the rack-setting gauge. Say the dial reads 1.230 in, as shown in Fig. 30–22a, but according to the service manual specification, the rack setting should be 0.475 in (Fig. 30–22b). You must move the fuel rack from 1.230 in to 0.475 in. To do this, loosen the locknut and turn the adjusting screw clockwise (Fig. 30–23) until the

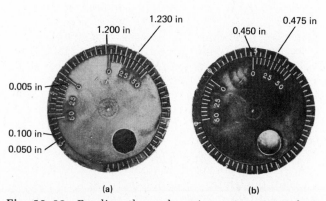

Fig. 30–22 Reading the rack-setting gauge. (*a*) Before fuel-rack adjustment. The 0 mark of the inner dial is just past the 1.200-in mark of the outer dial. The inner dial aligns with the outer dial at 0.030 in. The reading is 1.200 + 0.030 = 1.230 in. (*b*) After fuel-rack adjustment. The 0 mark of the inner dial is past the 0.450-in mark of the outer dial. The inner dial aligns with the outer dial at 0.025 in. The reading is 0.450 + 0.025 = 0.475 in.

dial coincides with the specified value. Then tighten the locknut.

Adjusting Rack Limiter   To adjust the rack limiter, remove the feeler gauge which was placed between the torque spring and the housing. During removal, the fuel rack will travel an additional 0.100 in [2.54 mm]. Recheck the dial reading because the torque spring will have become bent and will be resting against the housing. Loosen the locknut and turn the rack-limiter adjusting screws counterclockwise until the dial of the rack-setting gauge coincides with the setting specified in the service manual, and then lock the adjusting screw.

Recheck the adjustment by moving the throttle to the low-idle position and back again to the high-idle position. The dial reading should not have changed.

Adjusting Fuel Rack Using Dial Indicator   When you use the dial-indicator method to adjust the fuel rack and the rack limiter, remove the speed-limiter piston and turn the rack-limiter adjusting screw clockwise until the bell crank no longer interferes with the rack movement. Move the fuel rack forward. Slide the rack-centering gauge between the collar of the fuel rack and the housing. Position the throttle to the full-fuel position. This will hold the gauge in place.

Now, install the dial indicator as shown in Fig. 30–24, making certain that the adapter arm is in front of the fuel-rack collar. When in this position, zero the dial indicator and place a feeler gauge of 0.100 in [2.54 mm] between the torque spring and the housing. Remove the rack-centering gauge. When the gauge is removed, the fuel rack will move toward the full-fuel position until the adjusting screw of the rack control lever rests against the torque spring. The movement of the fuel rack

moves the adapter arm, and this motion is transmitted to the dial indicator. The travel of the fuel rack can now be read directly from the dial indicator.

Turn the rack-control-lever adjusting screw to obtain the specified rack travel of 0.475 in [12.06 mm]. To adjust the rack limiter, remove the feeler gauge, turn the rack-limiter adjusting screw counterclockwise until the dial indicator reaches the specified value, and then lock the adjusting screw.

NOTE   Always recheck adjustments.

Remove the rack-setting tools. Install the injection pump, cover, and speed limiter. Do *not* install the service meter at this time.

Installing Pump to Engine   Before you install the pump to the engine, turn the injection-pump camshaft until the timing dowel slides into the camshaft. This correctly positions the camshaft for timing the pump to the engine. Install the assembled pump to the engine in exactly the reverse order to which it was removed.

Position No. 1 piston on the compression stroke and install the timing bolt to the flywheel. (Make certain that No. 1 piston is on the compression stroke.) To take out the gear lash, slide the split coupling into position and tighten the bolts on the variable-timing shaft. Then turn the coupling toward you and tighten the bolts to the specified torque. The timing dowel should fit loosely in the injection-pump camshaft. Assuming that it does, remove the timing bolt from the flywheel, crank the engine 720°, and again install the timing bolt to the flywheel. In this position the timing dowel should slide into the camshaft without restriction. If it does, the injection pump is timed to 11° BTDC.

Spill-timing Caterpillar Pump   When you use the spill-timing method to check the pump timing, remove the fuel-injection line from the precombustion chamber of No. 1 cylinder. Remove the retainer nut, the injection-valve body, and the fuel nozzle.

Bring No. 1 piston to TDC on the compression stroke. Install a dial indicator with an extended

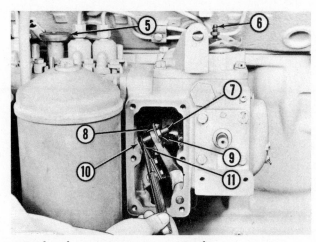

5. Fuel-rack-setting gauge
6. Rack-limiter-bell-crank adjusting screw
7. Rack-setting adjusting screw
8. Lock
9. Nut
10. Governor housing
11. Torque spring

Fig. 30–23   Adjusting the fuel rack. *(Caterpillar Tractor Co.)*

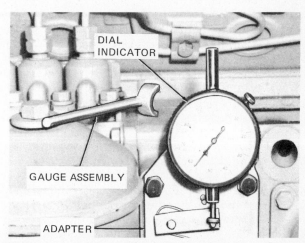

Fig. 30–24   Positioning the dial indicator before adjusting the fuel rack. *(Caterpillar Tractor Co.)*

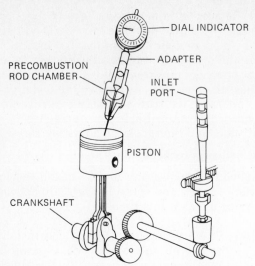

DIAL INDICATOR

ADAPTER

PRECOMBUSTION
ROD CHAMBER

INLET
PORT

PISTON

CRANKSHAFT

Fig. 30–25 Components and tools for spill-timing the engine to TDC. *(Caterpillar Tractor Co.)*

contact point to the precombustion chamber (Fig. 30–25). Rotate the crankshaft counterclockwise to determine whether the dial has a minimum of a 0.100-in [2.54-mm] travel distance. Turn the crankshaft slowly in a clockwise direction until the needle of the dial stops, that is, until the piston is in the TDC position. Zero the dial indicator.

To recheck the TDC position, turn the crankshaft in a clockwise direction until the dial indicator reads 0.020 in [0.050 mm]. Mark this point on the flywheel or on the vibration damper. Turn the crankshaft counterclockwise until the dial indicator again reads 0.020 in [0.050 mm]. Mark this point on the flywheel or on the vibration damper. True TDC position is midway between these two marks. Correct your dial-indicator setting accordingly.

When the TDC position is set on the dial indicator, connect a swan-neck pipe to No. 1 injection-pump housing. Bring the fuel rack to the full-fuel position. Turn the crankshaft counterclockwise until the dial indicator reaches the end of its travel. Pressurize the manifold 15 psi maximum [1.05 kg/cm²]. Use a hand primer or pressurize the fuel tank (or use a special tank), and spill-time the engine for port closure.

Port closure for Caterpillar injection pumps generally is established when 6 to 12 drops per minute leave the swan-neck pipe. An exception to this is the Caterpillar engine with a 5.4-in [137.16-mm] bore using a 3S1467 injection pump. Port closure for this engine generally is established when 12 to 30 drops per minute leave the pipe. However, there are exceptions, therefore always refer to your service manual.

When you have determined the port closure, read the dial indicator. It should read 0.072 in [1.83 mm], which is equal to 11° BTDC (Table 30–2). The piston travel toward TDC position measured in thousandths of an inch [hundredths of a millimeter] and the conversion to degrees are shown in Table 30–2. However, since the stroke on some engines differs, the piston travel varies in degree of crankshaft rotation. The travel is greater on an engine having a

Table 30–2a PISTON TRAVEL TO TOP DEAD CENTER (DIMENSION IN INCHES) (Degrees indicate crankshaft rotation from top dead center)

	D311	D311H D320A	D315 D318	D330A D330B D333A D333B 1673	D326 D337
3.0°					
3.5°					
4.0°					
5.0°	0.012	0.013	0.014	0.014	0.015
5.5°					
6.0°	0.018	0.018	0.019	0.020	0.021
7.0°	0.024	0.025	0.026	0.027	0.029
8.0°	0.031	0.032	0.035	0.036	0.038
8.5°					
9.0°	0.039	0.041	0.044	0.045	0.047
9.5°					
10.0°	0.048	0.051	0.053	0.056	0.059
11.0°	0.059	0.061	0.065	0.068	0.072
12.0°	0.070	0.073	0.078	0.080	0.085
13.0°	0.082	0.085	0.091	0.094	0.100
14.0°	0.095	0.099	0.106	0.109	0.115
15.0°	0.108	0.113	0.121	0.125	0.133

Table 30–2b PISTON TRAVEL TO TOP DEAD CENTER (DIMENSION IN MILLIMETERS) (Degrees indicate crankshaft rotation from top dead center)

	D311	D311H D320A	D315 D318	D330A D330B D333A D333B 1673	D326 D337
3.0°					
3.5°					
4.0°					
5.0°	0.30	0.33	0.36	0.36	0.38
5.5°					
6.0°	0.46	0.46	0.48	0.51	0.53
7.0°	0.61	0.64	0.66	0.69	0.74
8.0°	0.79	0.81	0.89	0.91	0.97
8.5°					
9.0°	0.99	1.04	1.12	1.14	1.19
9.5°					
10.0°	1.22	1.30	1.35	1.42	1.50
11.0°	1.50	1.55	1.65	1.73	1.83
12.0°	1.78	1.85	1.98	2.03	2.16
13.0°	2.08	2.16	2.31	2.39	2.54
14.0°	2.41	2.51	2.69	2.77	2.92
15.0°	2.74	2.87	3.07	3.18	3.38

larger stroke, and smaller on an engine with a shorter stroke.

Adjusting Idle and High-Idle Speeds Replace all covers, fuel lines and the necessary components which were removed to spill-time the injection

D326F D337F	D343 1693	641 650 651 657 660 666	D339 D342 D353 D8800 D13000	D17000	D364 D375 D379 D386 D397 D398 D399	D348 D346 D349	1673C 1674 D330C D333C D334	1676 D336 621 980	1140 1145 3145	1150 3150	1160 3160
								0.005			
								0.007			
							0.010	0.009			
0.016	0.016	0.016	0.019	0.020	0.019	0.016	0.015	0.013	0.010	0.011	0.013
								0.016			
0.024	0.023	0.022	0.029	0.028	0.028	0.022	0.022	0.019	0.014	0.016	0.018
0.032	0.031	0.030	0.039	0.038	0.038	0.030	0.029	0.026	0.019	0.022	0.025
0.042	0.041	0.040	0.051	0.050	0.049	0.040	0.038	0.034	0.025	0.028	0.032
								0.039			
0.053	0.052	0.050	0.064	0.063	0.062	0.050	0.048	0.043	0.032	0.036	0.041
								0.048			
0.066	0.064	0.062	0.079	0.078	0.076	0.062	0.060	0.053	0.039	0.044	0.050
0.079	0.077	0.075	0.096	0.094	0.092	0.075	0.072	0.064	0.047	0.053	0.060
0.094	0.092	0.089	0.113	0.112	0.110	0.089	0.086	0.077	0.056	0.063	0.072
0.111	0.107	0.104	0.133	0.132	0.129	0.105	0.101	0.090	0.066	0.074	0.084
0.128	0.124	0.121	0.155	0.153	0.149	0.121	0.117	0.104	0.077	0.086	0.097
0.147	0.143	0.139	0.177	0.175	0.171	0.139	0.134	0.119	0.088	0.098	0.112

D326F D337F	D343 1693	641 650 651 657 660 666	D339 D342 D353 D8800 D13000	D17000	D364- D375 D379 D386 D397 D398 D399	D348 D346 D349	1673C 1674 D330C D333C D334	1676 D336 621 980	1140 1145 3145	1150 3150	1160 3160
								0.13			
								0.18			
							0.25	0.23			
0.41	0.41	0.41	0.48	0.51	0.48	0.41	0.38	0.33	0.25	0.28	0.33
								0.41			
0.61	0.58	0.56	0.74	0.71	0.71	0.56	0.56	0.48	0.36	0.41	0.46
0.81	0.79	0.76	0.99	0.97	0.97	0.76	0.74	0.66	0.48	0.56	0.64
1.07	1.04	1.02	1.30	1.27	1.24	1.02	0.97	0.86	0.64	0.71	0.81
								0.99			
1.35	1.32	1.27	1.63	1.60	1.57	1.27	1.22	1.09	0.81	0.91	1.04
								1.22			
1.68	1.63	1.57	2.01	1.98	1.93	1.57	1.52	1.35	0.99	1.12	1.27
2.01	1.96	1.91	2.44	2.39	2.34	1.91	1.83	1.63	1.19	1.35	1.52
2.39	2.34	2.26	2.87	2.84	2.79	2.26	2.18	1.96	1.42	1.60	1.83
2.82	2.72	2.64	3.38	3.35	3.28	2.67	2.57	2.29	1.68	1.88	2.13
3.25	3.15	3.07	3.94	3.89	3.78	3.07	2.97	2.64	1.96	2.18	2.46
3.73	3.63	3.53	4.50	4.45	4.34	3.53	3.40	3.02	2.24	2.49	2.84

pump. Start the engine. Check the operation of the speed limiter. Bring the engine up to operating temperature. Remove the cover to gain access to the adjusting screws. Set the throttle at high idle and turn the adjusting screw (which rests against the control lever) to achieve the recommended high-idle speed. Then move the throttle to the low-idle position. Turn the low-idle adjusting screw until the idling speed meets the specification. Use a hand tachometer or a strobe light to measure the engine speed. Always cycle the engine to recheck adjustments.

COMPACT–HOUSING
FUEL–INJECTION–PUMP SERVICE

The highlighted procedures which follow refer to the compact-housing fuel-injection pump used on a 1674 truck engine.

Disassembling Compact-Housing Pump After you have cleaned the injection pump, mount it in a holding fixture. Disconnect the fuel-ratio control by removing its mounting bolts and sliding out the adjusting bolt from the fuel-rack stop collar. Remove the cover, the high- and low-idle adjusting screws, the cover plate, and the electric shutoff. Then slide out the cover assembly.

Next, remove the stop collar and the bolts which hold the governor housing to the injection-pump housing. Lift off the governor housing. The governor-activating mechanism must then be removed. To remove it, take out the three bolts which hold the cylinder, move the assembly to the left, and slide the valve from the fuel rack (Fig. 30–26).

Servicing Governor When checking the components of the hydraulic governor for serviceability, thoroughly check the valve, piston, sleeve, and cylinder. These components must be free of any surface damage. If it is necessary to clean their surfaces, use crocus cloth or tallow.

Check the end of the valve and the end of the piston which locks into the fuel rack. It should show no signs of wear or damage. It must not exceed the specified maximum clearance.

Check the thrust bearing and washer. They must be free of any grooving. The weight assembly must show no pin or bearing wear.

Check the fingers of the flyweights. They should be round, that is, without flat spots. Replace the coned washer and the spring if wear is visible. Check the spring seat and also the bolt which is the connecting link to the shutoff lever. It should show no wear.

Removing Compact-Housing Fuel-Injection Pump After you have removed the governor, remove the drive gear. The drive gear and the drive arrangement of this injection pump are similar to those of the forged-body injection pump.

With the special socket wrench, loosen the injection-pump retaining bushing about two full turns. Screw the extractor onto the bonnet. Press the extractor downward, but do *not* turn it while unscrewing the retaining bushing.

Lift the fuel-injection pump straight upward without rotating it so that the grooves on the bonnet and barrel will clear the two dowel pins on the housing (Fig. 30–27). Remove the remaining injection pumps. Next, remove the fuel rack, and lift out the spacers and lifters. Place all three components (as a unit and in the order in which they were removed from the housing) into clean fuel oil. **NOTE** Do not interchange the spacer or the lifter with the other pump elements because this may vary the lifter height.

Inspecting Compact-Housing Camshaft and Lifters The checks and measurements for the compact-housing injection pump (that is, the camshaft and its bushings, the drive flange, drive spring, and drive gear), do not differ from those of identical components in the forged-body injection pump. The lifter, however, should be checked for body, pin, and roller wear. Rough spots on the body surface or in the housing should be removed with 600-grit body paper. If the wear-washer is worn (Fig. 30–28), the lifter must be replaced.

Inspecting Injection Pump To remove the plunger, pull the plunger return spring from the barrel. Remove it and the spring seat from the plunger. Check the end of the plunger which contacts the wear washer. When worn, the assembly must be replaced.

Measure the length of the plunger. If the plunger is shorter than specified or if the gear shows wear,

1. Collar
2. Speed-limiter plunger
3. Lever
4. Seat
5. Governor spring
6. Thrust bearing
7. Oil passage
8. Drive gear (weight assembly)
9. Cylinder
10. Bolt
11. Spring seat
12. Weight
13. Valve
14. Piston
15. Sleeve
16. Oil passage
17. Fuel rack

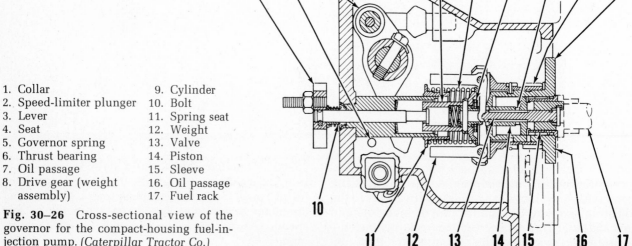

Fig. 30–26 Cross-sectional view of the governor for the compact-housing fuel-injection pump. (*Caterpillar Tractor Co.*)

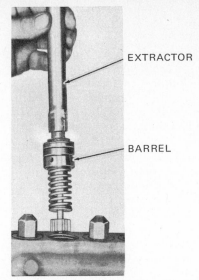

Fig. 30–27 Removing the compact-housing-pump element. *(Caterpillar Tractor Co.)*

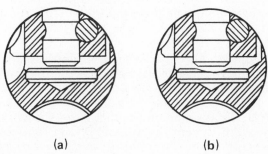

(a)   **(b)**

Fig. 30–28 Patterns of wear between washer and plunger. *(a)* Contact surfaces of new pump plunger and lifter washer; *(b)* flat end of new plunger making poor contact with worn lifter washer (resulting in rapid wear to both parts). *(Caterpillar Tractor Co.)*

the assembly must be replaced. Replacement is necessary because delivery (injection) begins later if the plunger is shorter. When the gear is worn, the plunger can rotate on its own, and this causes an alternation in fuel delivery.

If the assembly is reusable, clean the barrel and plunger assembly with tallow. Lap the bonnet, barrel, and check-valve surface to ensure effective sealing.

The fuel-rack grooves must show no wear and the rack must slide freely within its bushings but without excessive play. When in doubt, measure the bushings and the fuel rack and compare the measurements with those specified in the service manual.

Inspecting Pump Housing Check the threaded holes for damage. Do not forget to check the pump housing for nicks and burrs. Check especially the shoulder on which the spacer rests because its damage can distort the lifter-height measurement reading. Check for damaged locating dowel pins. Using an inside micrometer, measure the camshaft

bushing for wear, and compare with service manual specifications.

Installing Camshaft and Lifters Wash the pump housing in clean fuel and dry it with compressed air.

Lubricate the bushings and camshaft journals. Install the shaft to the housing and the drive gear to the drive flange.

CAUTION Do not damage the bushings during installation. Do not install the drive spring in the wrong direction.

With the shaft in place, lubricate the lifters and their bores. Return each lifter to its bore so that the roller of the lifter rests against the camshaft lobe. Then place each spacer on the shoulder of the bore. Make certain the lifters rest on the cam lobes before you install the fuel rack, or else the lifter may be damaged.

Measuring Fuel-Pump Timing Dimension Mount the degree wheel (timing plate) to the camshaft and the pointer to the housing, as shown in Fig. 30–29. Turn the degree wheel (in a counterclockwise rotation) so that the zero degree on the wheel aligns with the left edge of the pointer. This brings the No. 1 lifter into position to be measured. See Table 30–3 for the timing-plate setting in degrees.

Make certain that the lifter rests firmly on the cam lobe before placing the 2-in [50.8-mm] gauge block on top of the spacer. Use a depth micrometer with a 4- to 5-in [101.6- to 127.0-mm] rod, and measure the distance from the surface of the gauge block to the wear washer of the lifter. The micrometer should read, in this case, 4.390 to 4.394 in [111.50 to 111.607 mm] (see *A* in Fig. 30–30). If the microme-

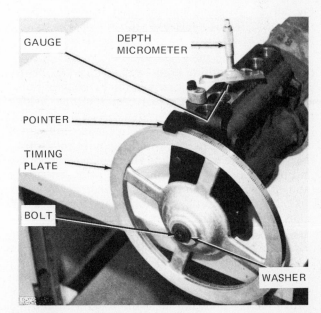

Fig. 30–29 Positioning components to measure compact-housing-pump timing dimension (off-engine). *(Caterpillar Tractor Co.)*

Table 30–3 LIFTER SETTING IN DEGREES (OFF-ENGINE)

Lifter number (numbered front to rear)	Timing plate degrees
1	0
2	120
3	240
4	60
5	300
6	180

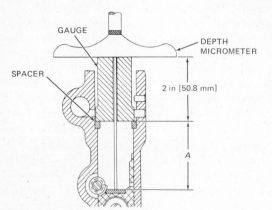

Fig. 30–30 Measuring the timing dimension. *(Caterpillar Tractor Co.)*

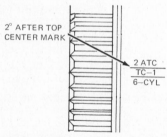

Fig. 30–31 Aligning timing marks in preparation to check pump timing (on-engine). *(Caterpillar Tractor Co.)*

ter reading is higher, say by 0.004 in [4.398 in], remove the spacer, measure its thickness, and select a spacer which is 0.004 in [0.101 mm] thinner to give the recommended lifter height.

When the No. 1 lifter is within specification, turn the degree wheel in a counterclockwise rotation so that 300° on the wheel aligns with the left edge of the pointer. This positions the No. 5 lifter. Measure (and, if necessary, change) the spacer to bring the No. 5 lifter within specification. Then measure and adjust the remaining lifters.

When checking the fuel-injection-pump timing while the pump is on the engine, bring the No. 1 piston on compression stroke and align the timing marks as shown in Fig. 30–31. Then measure the lifter height as outlined above. If the lifter height of a 1674 engine does not meet the specification, then the injection pump must be removed and an off-engine setting of all injection pumps must be made.

NOTE Follow, precisely, the instructions in your service manual.

Installing Compact-Housing Pumps Wash all the components of each injection pump in clean fuel. Place the spring seat in the groove of the plunger. Put the spring on the plunger. Carefully slide the plunger into the barrel and the spring over the barrel. **NOTE** The plunger must turn freely.

Next, place the check-valve spring and the valve into the bore of the bonnet. Position the barrel assembly on the bonnet. Clip the ring in the grooves of the barrel and the bonnet. Install a new O ring and then the retaining bushing on the bonnet. Screw the extracter onto the bonnet. Before you install the injection pumps, center the fuel rack by pushing the centering pin downward, holding it with the cover. Push the shoulder of the fuel rack against the centering pin. Turn the camshaft so that the lifter is on the base circle of the cam lobe. Align the notches of the bonnet and barrel with the slot of the gear segment (Fig. 30–32). Install the assembly into its bore. The barrel and the bonnet slot should engage smoothly with the dowel pins. The gear segment should engage smoothly with the fuel rack.

CAUTION Do not exert force. Unnecessary force will damage the gear segment and the fuel rack or will break pieces from the dowel pins.

With the assembly in position, press the extractor downward, locate the square O ring, and begin hand-tightening the retainer bushing until two threads protrude from the housing surface. Remove the extracter and tighten the bushing to 150 lb·ft [20.7 kg·m]. Install the remaining injection pumps.

Installing Governor Wash all the components of the governor in clean fuel and dry them with compressed air. Lubricate the components before reinstalling them. Carefully place the sleeve and the sealing ring into the cylinder and the valve into the piston. Hook the piston into the fuel rack. Make sure that the thrust bearing and spring are positioned properly, and that the coned side of the disk spring faces the rear, before installing the governor housing. Install the rack collar to the bolt, but make sure that the screw is in the groove of the bolt before tightening it. Install the torque-spring assembly in the order shown in Fig. 30–33. Do not install the

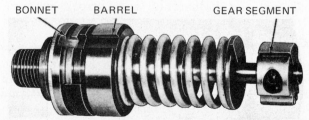

Fig. 30–32 Alignment of notches. *(Caterpillar Tractor Co.)*

spacer plate, speed limiter, or fuel-ratio control at this time.

Fuel-Rack Setting with Rack-Setting Gauge Install the rack-setting gauge over the front end of the fuel rack. Set the gauge to the specification recommended in the rack-setting information section of the service manual (Fig. 30–34). Now, move the throttle to the full-fuel position. The rack collar should barely contact the torque spring. If the torque spring becomes compressed or if the collar does not contact the torque spring, loosen the locknut and turn the adjusting screw until the collar just touches the torque spring. Then lock the screw. Now turn the rack-setting gauge slightly in a clockwise direction to reduce the rack travel. If the rack collar instantly moves away from the torque spring, the setting is correct.

Fuel-Rack Setting with Dial Indicator The dial-indicator method of measuring and adjusting fuel-rack travel is more precise than the rack-setting-gauge method. Another advantage is that the fuel-rack travel can be measured with the engine in either the running or the stopped position, or with the pump removed from the engine.

To use the dial-indicator method, install the gauge as shown in Fig. 30–35. This places the 1:1

Fig. 30–33 Compact-housing-pump torque-spring assembly. (*Caterpillar Tractor Co.*)

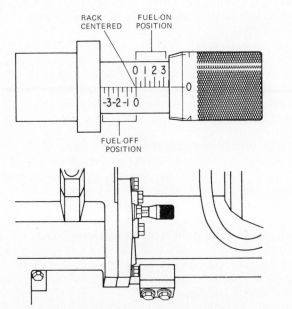

Fig. 30–34 Installed rack-setting gauge showing various rack-position readings. (*Caterpillar Tractor Co.*)

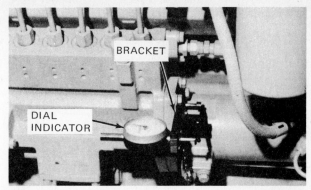

Fig. 30–35 Fuel-rack setting with a dial indicator. (*Caterpillar Tractor Co.*)

lever on the fuel rack and on the pointer of the dial indicator. Move the throttle lever to the fuel-off position. Depress the centering pin, move the cover over the pin, and then bolt the cover to the housing. Next, move the throttle lever to the full-fuel position to bring the fuel-rack stop against the centering pin. Zero the dial indicator. Recheck the setting by moving the throttle lever back and forth a few times. The gauge should remain at zero when the throttle is in the full-fuel position.

To determine the exact moment that the rack collar contacts the torque spring, connect one lead of the circuit tester to the insulated terminal; connect the second lead to an effective ground. After the test instruments are installed, remove the centering pin and move the throttle lever to the full-fuel position. The light should glow brightly. If it does not, loosen the locknut and turn the adjusting screw until the light becomes bright. Then move the throttle in the opposite direction until the light goes out. Again, (very slowly) move the throttle to the full-fuel position until the light barely glows. This indicates that the rack collar has just contacted the torque spring. At this point read the rack travel directly from the dial indicator. If necessary, continue to reset the adjusting screw until the dial indicator shows the correct setting while the light glows faintly.

Fuel-Ratio Control—Tests and Adjustments The fuel-ratio control coordinates the movement of the fuel rack with the inlet-manifold pressure to reduce exhaust smoke during acceleration. When the fuel-ratio control valve is serviced, it must be checked and tested before being reinstalled to the governor.

To test this valve, connect an air line with a pressure regulator to the inlet port of the chamber (see Fig. 30–36). Apply 5 psi [0.35 kg/cm²]. This should cause the diaphragm and bolt to move. Apply 16 to 20 psi [1.16 to 1.44 kg/cm²]. This should then cause the diaphragm to bottom onto the housing and the bolt to move fully outward.

To check the chamber for air leakage increase the pressure to 35 psi [2.46 kg/cm²], then close off the air supply and check the pressure drop. The pressure

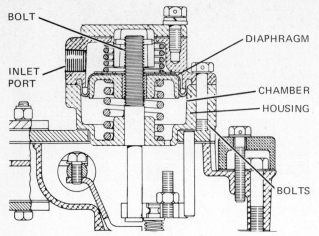

Fig. 30–36 Sectional view of the compact-housing-pump fuel-ratio control. *(Caterpillar Tractor Co.)*

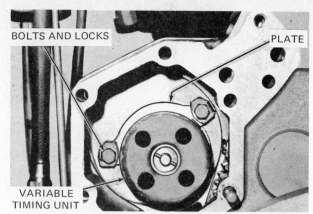

Fig. 30–38 Components visible when the accessory drive gear is removed. *(Caterpillar Tractor Co.)*

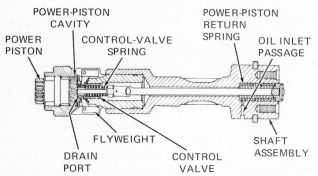

Fig. 30–37 Variable timing unit in low-rpm position. *(Caterpillar Tractor Co.)*

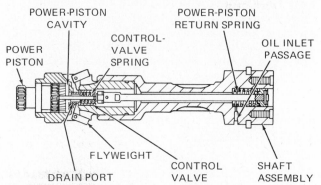

Fig. 30–39 Variable timing unit in high-rpm position. *(Caterpillar Tractor Co.)*

drop should not exceed 2 psi [0.14 kg/cm²] within 10 seconds. If the fuel-ratio control does not pass these two tests, it must be reserviced.

Fuel-Ratio Control—Installation and Adjustment Move the throttle lever to the fuel-off position. Slide the bolt into the notch of the rack collar. Install and tighten the hex bolts to specification. Connect the circuit tester as outlined earlier. Move the throttle to the full-fuel position; the test light should burn brightly. If it does not, turn the adjusting screw (with cover) clockwise until it does, then slowly turn it counterclockwise until the light barely glows. Assuming that the dial indicator is still connected, the dial-indicator reading should agree with the rack-setting specification. In this position, place an aligning mark on the cover and the housing. (See service manual for fuel-ratio control setting.)

Assuming that the fuel-ratio control setting is 0.020 in [0.508 mm], you must then retract the fuel rack by that amount. To make this reduction, turn the cover three holes from the alignment mark, counterclockwise. (The distance from hole to hole is equal to 0.008 in [0.20 mm] of rack travel.)

NOTE Always go to the next hole because this reduces the rack travel. In this case the reduction is 0.004 in [0.010 mm]. One complete turn of the

cover is equal to 0.050 in [1.27 mm]. Rack reduction can also be measured with a dial indicator.

When the fuel-ratio control is adjusted, install the bolts that hold the cover and the speed limiter, and then install the injection pump to the engine.

Variable Timing Drive When the fuel-injection pump is serviced, the variable timing drive must also be serviced since it is the drive link between the timing gear and the pump camshaft. Any malfunction in the variable timing drive will decrease the engine performance.

COMPONENTS OF VARIABLE TIMING DRIVE The components of the variable timing drive are shown in Fig. 30–37. As illustrated, the accessory drive gear is bolted to the right side of the shaft assembly. The shaft assembly is held in the housing by a plate (Fig. 30–38) and is supported by bushings. The power piston has straight splines on one end. These straight splines mesh with the straight splines of the camshaft. On the other end, the power piston has helical splines that mesh with the helical splines in the shaft assembly.

The control valve spring and the control valve are positioned over the rod of the power piston. The piston rod is connected to the followup rod by a pin.

The power piston return spring is situated between the bore in the shaft assembly and the spring seat. (The seat is pinned to the followup rod.) The two flyweights are fastened through pins to the drive-shaft assembly and the fingers of the weights rest in the groove of the control valve.

VARIABLE-TIMING-DRIVE OPERATION When the engine is running (below 1350 rpm), oil from the lubrication system is supplied to the front support bushing. From here it enters the inlet passage, flows to the control, the valve power-piston chamber, and then out through the drain port. This occurs because the flyweight force at this speed is lower than the spring force of the control valve (see Fig. 30–37).

As the engine speed increases, the flyweight force increases, moving the control valve to the right in Fig. 30–39, so that the drain annulus opening is reduced. This results in an increase in pressure in the power-piston chamber. The power piston is moved to the left, against its spring force. Because of the helical splines, the outward movement of the piston turns the camshaft slightly in the same direction as the variable timing drive.

The camshaft advances in relation to the drive-shaft assembly. The outward movement of the power piston has increased the control-valve spring force. This moves the control valve and increases the return annulus opening. As soon as the spring force and the weight force balance, the oil pressure in the power-piston chamber is constant. The greater the engine speed, the greater the advance until the power-piston flange butts against the camshaft. The spring and weight force and the helical spline govern the advance curve.

At any constant speed above 1350 rpm, the flyweight force and the spring force will balance each other.

Servicing Variable Timing Drive To remove the variable timing drive from the engine, remove the accessory drive gear, the plate, and then the mounting bolts which fasten the variable-timing-drive housing to the cylinder block. When these items are removed, slide the shaft assembly from its housing. Wash the components in clean fuel and dry them with compressed air. Check the housing bushings for damage. Measure them for wear. Check the bearing journals for scoring and grooving. Measure them for wear also.

To disassemble the unit, remove the pin of the power-piston-return-spring seat and then remove the spring seat and spring. Press out the flyweight pins and remove the flyweights. Slide out the power piston with the control valve, control-valve spring, and followup rod. Then remove the pin which locks the followup rod with the power-piston rod. Slide the control-valve spring and valve from the rod. When checking the components for serviceability, check the flyweight bores for wear and the weight fingers for flat spots. Check the control valve and its cylinder for scoring or excessive wear. Replace the control-valve spring and piston return spring. Check the helical splines of the power pis-

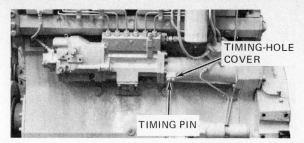

Fig. 30–40 Timing-pin location. *(Caterpillar Tractor Co.)*

ton, drive shaft, and camshaft. The splines must mesh smoothly within each other.

Reassembling Variable Timing Drive Install the drive assembly to the cylinder block, but do not tighten the accessory drive at this time. Use engine oil to lubricate the components during the assembly.

Installing Pump to Engine Position the No. 1 piston on compression stroke so that the timing bolt can be threaded into the flywheel, or, on recent models, set the position of the crankshaft according to service manual specifications. Install the timing pin to the variable-timing-drive-shaft assembly (Fig. 30–40).

Align the master spline of the injection-pump camshaft with the master spline of the power piston. Then slide the camshaft splines over the splines of the power piston until the mounting flange bottoms against the variable-timing-drive housing. Next, rotate the accessory gear to remove the slack from the drive gear, and tighten its bolts to the specified torque. Remove the timing bolt from the flywheel and the timing pin from the injection-pump drive shaft. Turn the engine 720°. Again install the timing bolt to the flywheel and check the timing. The timing pin must slide freely into the hole of the variable drive shaft.

High- and Low-Idle Adjustment To adjust the engine's high- and low-idle speed, bring the engine up to its operating temperature and remove the cover (Fig. 30–41). Position the throttle to the low-idle

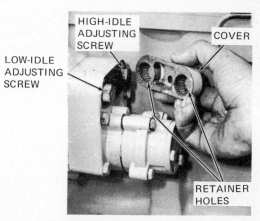

Fig. 30–41 Governor adjustments. *(Caterpillar Tractor Co.)*

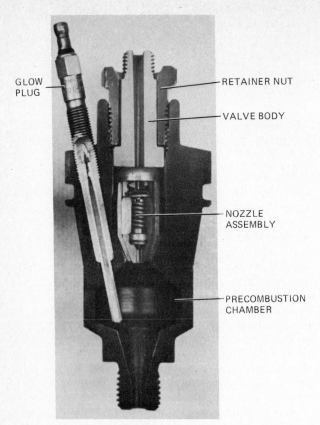

Fig. 30–42 Sectional view of a precombustion chamber with nozzle and glow plug. *(Caterpillar Tractor Co.)*

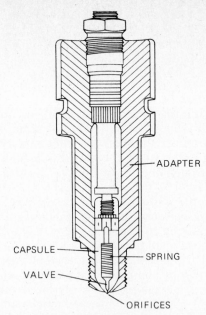

Fig. 30–43 Cutaway view of an adapter (direct-injection) with fuel-injection valve. *(Caterpillar Tractor Co.)*

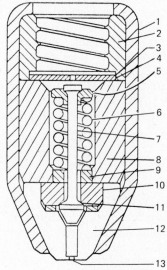

1. Shell
2. Collar
3. Screen
4. Screen support
5. Valve keeper
6. Spring
7. Valve
8. Spring guide
9. Lower spring seat
10. Valve-seat body
11. Valve seat
12. Nozzle tip
13. Orifice

Fig. 30–44 Schematic view of a capsule-type nozzle.

position. Turn the low-idling adjusting screw until the strobe light or hand tachometer indicates the recommended speed. Then move the throttle to the full-fuel position and turn the high-idle screw until the engine speed is up to specification.

If the exhaust smoke is excessive when accelerating the engine, check the operation of the variable timing, using a injection-timing tool (see Unit 51, "Troubleshooting and Tuneup") before you readjust the fuel-ratio control. When an adjustment is necessary, turn the cover counterclockwise (one hole at a time). Recheck the exhaust smoke.

CAUTION Make certain you have checked the air-intake system and the turbocharger boost pressure before adjusting the fuel-ratio control (see Unit 51, "Troubleshooting and Tuneup").

FUEL ATOMIZERS

Precombustion Chamber, Adapter, and Fuel-Injection Valves Most Caterpillar engines have a precombustion chamber with a capsule-type nozzle (Fig. 30–42). The 3200 Series engine (direct-injection) uses a Roosa Master injector. The 3400 Series truck engine is built as a precombustion chamber engine or as a direct-injection engine. When the engine is built as a direct-injection engine, the adapter replaces the precombustion chamber and the capsule-type injection valve replaces the capsule-type nozzle (Fig. 30–43). The precombustion chamber or adapter is screwed into the cylinder head and is sealed on the lower end with a copper-alloy gasket and on the upper end with an O ring.

NOTE Glow plugs are used only with precombustion chambers.

Precombustion Chamber or Adapter Removal and Installation If it is necessary to remove the precombustion chamber or the adapter, the cooling system must be drained and, on certain engines, the camshaft housing must be removed. *(Precise instructions are outlined in your service manual.)* After removing the valve cover, remove the injection line, the nut, and the body complete with nozzle assembly. Insert the special wrench adapter into the fine spline of the precombustion chamber or adapter. Loosen the precombustion chamber, or adapter, and unscrew it. If the interior or exterior of the chamber, or adapter, has not been damaged by corrosion or oxidation, glass-bead-peen the assembly to ensure adequate cooling. The chamber or adapter must be cleaned thoroughly with compressed air before it is reinstalled. Coat the O ring and groove and the chamfered portion of the cylinder head with the recommended grease (or soap). Apply an antiseize compound to the threads and, with grease, attach a new gasket to the bottom of the chamber or adapter.

Thread the precombustion chamber or adapter into the cylinder head and tighten it to 150 lb·ft [20.7 kg·m]. If a glow plug is used, make certain it is positioned in the "go" range. (When positioned in the "no-go" range, the rocker arm will damage the glow plug.) To change its position, change the gasket according to the service manual specifications.

Fuel-Injection Valve Function (Precombustion Chamber) The fuel-injection valve is a capsule-type nozzle. It consists of a threaded collar, a screen and screen support plate, a spring guide, a valve pressure spring, a valve with stem, a spacer, a valve seat, and the orifice body and shell.

When the fuel pressure in the area above the valve rises, pressure is exerted on the valve face and, when the force becomes higher than the spring force, the valve opens outward (toward the orifice) (see Fig. 30–44). Fuel is sprayed into the precombustion chamber until the plunger of the injection pump opens the inlet port. The pressure in the injection line quickly decreases and the spring force (of the valve) forces the valve onto its seat. As a result of the displacement plunger of the check valve, a predetermined pressure is maintained within the injection line and the area above the valve.

Servicing and Testing Nozzle (Precombustion Chamber) When the screen or the seating area of the nozzle is damaged, the nozzle must be replaced. If the orifice is blocked up with carbon, it can be cleaned with a drill having the same dimension as the orifice.

If the surface near or around the orifice is damaged, it can be lapped to restore its roundness and to make the corner sharp again. This is the only service possible on a capsule-type nozzle.

To test a capsule-type nozzle, mount it to an injector tester (Fig. 30–45). Check the opening pressure, spray pattern, and valve leakage (see Unit 27, "Fuel-Injection Nozzles and Holders").

The opening pressure for used nozzles should not be lower than 400 psi [28.1 kg/cm²] or higher than

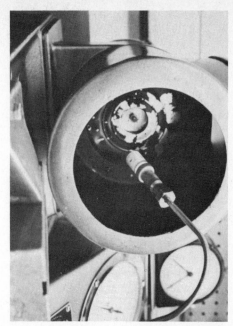

Fig. 30–45 Testing the capsule-type nozzle.

750 psi [52 kg/cm²]. (New nozzle opening pressure is 685 to 750 psi [48.2 to 52.7 kg/cm²]). The spray pattern must have a solid cone and be in line with the valve. A new nozzle should show no wetness at the orifice when pressurized to 450 psi [32.7 kg/cm²].

Installing Capsule-Type Nozzle Screw the nozzle (fingertight) to the body. Then check to make certain that the nozzle seat in the precombustion chamber is clean before you slide the assembly into the chamber. The fine body splines should engage smoothly with the splines of the precombustion chamber.

Tighten the nut to the specified torque but do not overtorque it. Overtorquing may cause nozzle leakage. The nut should not be undertorqued either because this can cause combustion leakage and insufficient cooling of the nozzle.

Questions

1. List the major differences between the Caterpillar fuel-injection system and the Detroit Diesel fuel-injection system.

2. List the differences in construction between a forged-body injection pump and a compact-housing injection pump.

3. What checks and measurements must be taken to properly check out the pump housing of a forged-body fuel pump?

4. What problems would arise if a new pump element were used with a worn yoke?

5. How could the engine timing and phase angle be affected if the micrometer accuracy were out 0.010 in [0.02 mm]?

6. Why is it important to zero the rack before installing the injection pump?

7. What is the purpose of a speed limiter?

8. Refer to Fig. 30–17. Draw a rack-setting gauge which indicates 0.115 in.

9. Outline the procedure to check the timing of a 1674 engine when using the spill method. (Port closing is at, say, 11° BTDC.)

10. What precautions are necessary when removing a compact-housing fuel-injection-pump element?

11. Why is it necessary to measure the plunger length of the fuel-injection pump?

12. What preliminary steps must be taken before measuring the lifter height of a compact-housing injection pump?

13. When setting or adjusting the fuel rack, why should you place (between the torque spring and stop plate) a feeler gauge of the same thickness as that of the spacer?

14. Why is it recommended to adjust the fuel-ratio control to the next hole reducing the rack travel rather than to the next hole increasing the rack travel?

15. What advantage does the variable timing drive afford an engine's performance?

16. Which parts of the precombustion chamber require detailed inspection?

17. Outline the mechanical and hydraulic action of the nozzle valve from the point fuel pressure increases to the end of injection.

18. Which checks and service work can be made on the fuel nozzle after it has been removed from the engine?

UNIT 31

Detroit Diesel Fuel-Injection System

COMPONENTS OF DETROIT DIESEL FUEL–INJECTION SYSTEM

The Detroit Diesel fuel-injection system is comparable to other fuel-injection systems (Fig. 31–1). Any differences which exist, however, occur in the unit injectors and their linkage to the governor.

The fuel-injection pump and the fuel-injection nozzle comprise one unit. Each cylinder is equipped with an individual unit injector. Each injector control rack (fuel rack) is snugly fitted to a rack-control lever which can be adjusted independently to the injector control tube to permit a uniform adjustment of all injector racks (Fig. 31–2). The injector control tube is connected to the governor by the fuel rod. The unit injector is held in its injector tube by the injector clamps. The follower (including the plunger), unlike the injection pumps previously outlined, is indirectly activated. Follower activation varies, however, on the 149 Detroit Diesel unit injector as opposed to the 53, 71, and 92 Series (Fig. 31–3).

The low-pressure side of the Detroit Diesel fuel-injection system (that is, the fuel tank, primary filter, transfer pump, and secondary filter) is comparatively similar to any other fuel-injection system.

Fuel Flow Refer back to Fig. 31–1. Fuel from the positive displacement gear-type pump is pumped through the secondary filter into the inlet manifold.

A manifold connector is screwed into the cylinder head to connect the inlet manifold with the inlet fuel pipe at one end, while the other end of the fuel pipe

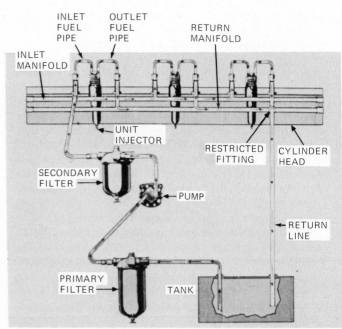

Fig. 31–1 Schematic view of a Detroit Diesel fuel-injection system. (The arrows indicate the fuel flow.) *(GMC Detroit Diesel Allison Div.)*

is connected with the filter cap of the injector. Fuel flows through the unit injector, cools, lubricates, and maintains a low pressure within the injector and also removes any air (when present). Fuel flows through the outlet filter cap, through the return fuel pipe and manifold connector and into the return manifold. The outlet manifold has a restricted orifice which raises the manifold pressure to the required specification. The pressure relief valve in the fuel pump prevents the pressure from exceeding specification.

Fuel Pump To check the efficiency of the fuel pump, you should connect a pressure gauge to the inlet manifold, disconnect the return line at a convenient place, and hold the open end over a measuring container. Start the engine and run it at 1800 rpm.

Assume your engine is a 71-T. The manifold pressure when using a 0.055-in [1.391-1mm] orifice should be 50 to 60 psi [3.52 to 4.22 kg/cm²] and the spillback should be 0.6 gal (U.S.)/min [2.27 l/min].

If the fuel pressure or the fuel spillback is lower than specified, check the fuel filter and the relief valve. If the filter and the relief valve are not defective, the trouble is in the fuel pump. Remove the fuel pump from the engine and cap all the fuel line openings.

Disassembling Fuel Pump Clean the pump externally and secure it in a soft-jaw vise before removing the cover. Take care not to damage the mating surfaces of the body or the cover.

Next, remove the drive shaft and gear and the driven shaft and gear (see Fig. 31–4). Remove the relief-valve assembly. Use a suitable puller to remove the seals.

Inspecting Fuel Pump Clean all components in fuel. Check the shafts for scoring and wear. Check the bearing bores in the body and cover. If the gears are worn, scored, or damaged, they must be replaced. The mating surfaces must be free of any nicks and burrs. Check the body flange and seal bores for damaged surfaces. When necessary, lap the surfaces to a smooth finish since no gasket is used to

seal against fuel leaks or pressure loss. If either the body or cover surface is damaged beyond refinishing, it must be replaced. If either the relief valve or relief-valve bore is pitted or corroded, a replacement is essential. Always replace the relief-valve spring and the seals.

Reassembling Fuel Pump Lubricate the seal lips and the seal case. Using the correct seal driver, install the seal. Always check your service manual for correct seal location and position.

If you are not using Detroit Diesel installation tools, take care that you do not insert the seals so deep that they cover the drain holes.

Press the gears onto their shafts until they are centered with the slot facing the pump cover.

NOTE Do not press the drive gear from the square end on the shaft. This will damage the gear and shaft seal surface.

Lubricate the gears, shafts, and seals. Carefully install the drive shaft, taking care that the seals are

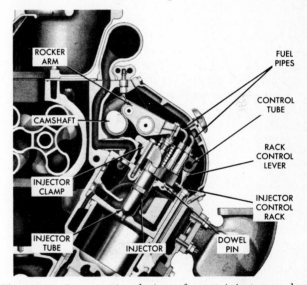

Fig. 31–3 Cross-sectional view of a 149 injector mechanism. *(GMC Detroit Diesel Allison Div.)*

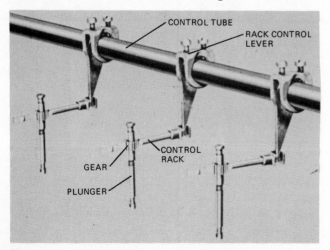

Fig. 31–2 Components that link the governor with the plungers. *(GMC Detroit Diesel Allison Div.)*

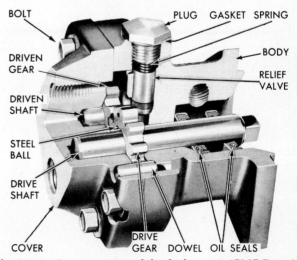

Fig. 31–4 Cutaway view of the fuel pump. *(GMC Detroit Diesel Allison Div.)*

not damaged. Install the driven gear and shaft. Place the pump cover on the dowel pins and install the cover bolts. Tighten the bolts alternately to the recommended torque. Check and make certain that the drive shaft rotates freely when turned by hand.

NOTE Detroit Diesel engines use left- or right-hand rotation fuel pumps depending on the engine rotation. Make certain when you install a replacement fuel pump that it has the correct rotation.

Locating Misfiring Unit Injector To locate a faulty or misfiring injector, remove the valve cover, start the engine, and force each follower down so that the injector cannot meter and inject fuel (Fig. 31–5). The sound of the engine should change as each follower is held down. When the sound does not change, you have located the faulty injector.

Removing Unit Injector Assume you have to remove a unit injector from a 149 Series engine. Start with the removal of the valve cover, fuel inlet, and outlet pipes. Instantly cover inlet and outlet filter caps and place the fuel pipes in a protected area. Then remove the injector-rack control-tube assembly (see Fig. 31–6). Remove the two bolts which hold the rocker-arm shaft to the camshaft-bearing caps. Lift the rocker-arm assembly from the cylinder head and lift the two exhaust-valve bridges from their bridge guides.

Remove the injector clamp. Lift the injector from its sleeve using an injector bar when necessary to break the injector loose.

Types of Unit Injectors Detroit Diesel 53, 71, 92, and 149 Series engines have the same unit-injector design but the components vary in size. The function of the unit injector is to time injection, meter fuel in relation to load and speed, and to pressurize and atomize the fuel.

The fuel-injection pumps mentioned earlier have nameplates which identify design, dimension, construction, etc. This is also true of the Detroit Diesel unit injector. Each engine (even within the same series) may use a different number of spray holes, different spray hole sizes, angles, plunger sizes, and

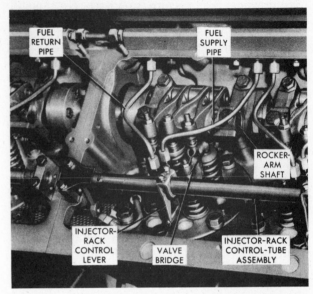

Fig. 31–6 Top view of the injector and valve mechanism. *(GMC Detroit Diesel Allison Div.)*

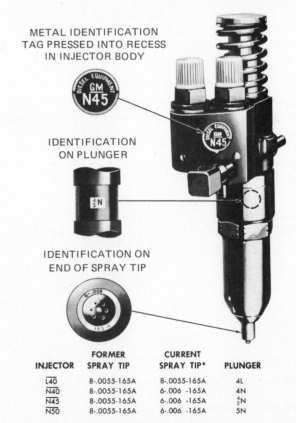

INJECTOR	FORMER SPRAY TIP	CURRENT SPRAY TIP*	PLUNGER
L40	8-.0055-165A	8-.0055-165A	4L
N40	8-.0055-165A	6-.006 -165A	4N
N45	8-.0055-165A	6-.006 -165A	$\frac{4}{5}$N
N50	8-.0055-165A	6-.006 -165A	5N

* First numeral indicates number of spray holes, followed by sizes of holes and angle formed by spray from holes.

Fig. 31–7 Injector identification. *(GMC Detroit Diesel Allison Div.)*

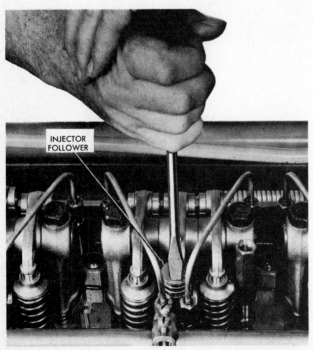

Fig. 31–5 Locating a misfiring injector. *(GMC Detroit Diesel Allison Div.)*

Table 31–1 CERTIFIED AUTOMOTIVE ENGINE MODEL INJECTOR

Engine family	Injector assembly		Engine timing		Tip assembly	Plunger-and-barrel assembly	
	Name	Calibration	Cam	Injector	Size*	Stamp	Tag color
53N and 53N	C40	42–46	Adv. †	1.460	6-0.006-165° A	4C	Gray
	C45	47–51	Adv.	1.460	6-0.006-165° A	45C	Red
	C50	50–54	Adv.	1.460	6-0.006-165° A	5C	White
L-71N(4V) and V-71N(4V)	71C5	50–54	Std. ‡	1.484	8-0.0055-165° A	5C	White
	C55	53–57	Std.	1.460	8-0.0055-165° A	55C	Orange
	C60	57–61	Std.	1.460	8-0.0055-165° A	6C	Blue
	C65	64–68	Adv.	1.484	7-0.006-165° A	65C	Brown
	C70	71–75	Adv.	1.484	7-0.006-165°	7C	Black
V-71(2V) and (4V) Coach	71C5	50–54	Std.	1.484	8-0.0055-165° A	5C	White
	C55	53–57	Std.	1.470	8-0.0055-165° A	55C	Orange
	C60	57–61	Std.	1.470	8-0.0055-165° A	6C	Blue
71T and 1T	C65	64–68	Std.	1.460	7-0.006-165° A	65C	Brown
	N70	71–75	Std.	1.484	7-0.006-165°	7N	Black
	N75	75–79	Std.	1.484	7-0.006-165°	75N	Gray

* First numeral indicates number of spray holes, followed by size of the holes (in thousandths of an inch) and angle formed by the spray from the holes.
† Advanced camshaft timing.
‡ Standard camshaft timing.

helices. Some information is engraved on the identification tag which is pressed into the recess of the injector body and some is written on the nut and plunger and on the spray tip. For example, referring to Fig. 31–7, the "N" on the tag identifies the valve design (needle valve); "45" is the identification for maximum fuel output in cubic millimeters per 1000 strokes; the number and/or letter on the injector nut and plunger indicates the plunger design and diameter; and the three numbers on the nozzle tip and the color tag identify the number of holes, the size of holes, and the spray angle (see Table 31–1).

The 1973 certified automotive injectors carry the letter "C" instead of "N" to identify the injector as a needle-valve injector with a variable-beginning and constant-ending plunger design. The letters "HV" shown on a Detroit Diesel unit injector identifies those units with a high valve injector nozzle.

Unit-Injector Construction Two unit injectors are shown in Fig. 31–8: (a) the older crown-valve unit injector and (b) the needle-valve (c-type) unit injector. They are similar in design. Each has a follower and a follower return spring held to the injector body by the stop pin. The stop pin, when installed, rests with its end in the groove of the follower to prevent it from rotating. The plunger hooks to the follower. The upper part of the plunger has a machined flat surface on which the gear is fitted. The control rack is supported in the injector body and meshes with the gear. The bushing (the barrel) has a lower and upper port. It also has another fuel passage, near the top, which allows the fuel that was used to lubricate the bushing and plunger to return into the area between the bushing and spill deflector. The bushing is positioned by one dowel pin to the injector body and, in conjunction with the gear retainer,

holds the gear in position. Inlet and outlet filter caps are screwed into the injector body to hold the wire-mesh filter in place.

Up to this point, the construction is similar to an APF fuel-injection pump. The features will now change because the fuel-injector nozzle is located directly below the bushing.

The crown-valve assembly is stacked onto the spray tip in the following order: check valve, valve cage, and valve seat. Inside the valve cage are the valve stop, spring, and injector valve.

The needle-valve assembly is stacked onto the bushing in the following order: check valve, check-valve cage, spring, spring seat, spring cage, and spray tip. Inside the spray tip is the needle valve and inside the spring cage are the spring seat and the valve spring.

In both cases the nut holds the assemblies and the bushings to the injector body. An O ring is used to seal the nut to the body. The locating dowel positions the injector to the cylinder head.

Cycle with Plunger in No-Fuel Position The pumping and metering principles of the Detroit Diesel unit injector are similar to any helix-metering fuel-injection pump except that the Detroit Diesel bushings have two offset ports (Fig. 31–9). The Detroit Diesel unit injector utilizes three types of plunger designs, each with a center hole and a cross-drilled hole (Fig. 31–10).

However, since 1973 most plungers used for 71, 92, and 149 Series engines are of the variable-beginning and constant-ending design. When the governor is moved to no-fuel position, it moves the control rack as well as the gear and the plunger, and thereby brings the plunger into a position where no fuel can be pressurized (Fig. 31–11). This occurs because the upper helix has a machined cutout recess which allows the upper port to stay open over

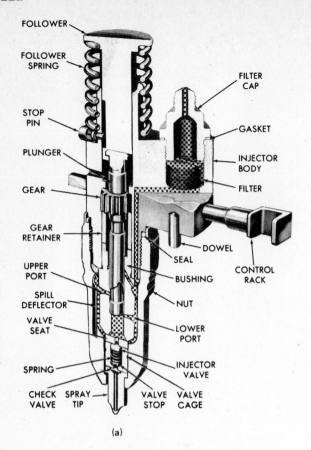

FOLLOWER

FOLLOWER SPRING

STOP PIN

PLUNGER

GEAR

GEAR RETAINER

UPPER PORT

SPILL DEFLECTOR

VALVE SEAT

SPRING

CHECK VALVE SPRAY TIP

FILTER CAP

GASKET

INJECTOR BODY

FILTER

DOWEL

SEAL

CONTROL RACK

BUSHING

NUT

LOWER PORT

INJECTOR VALVE

VALVE STOP VALVE CAGE

(a)

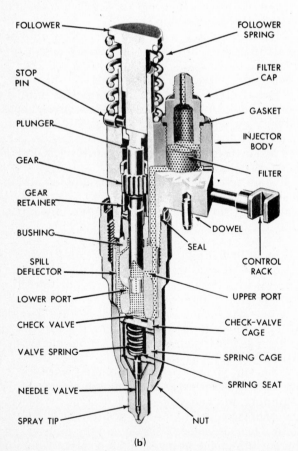

FOLLOWER

STOP PIN

PLUNGER

GEAR

GEAR RETAINER

BUSHING

SPILL DEFLECTOR

LOWER PORT

CHECK VALVE

VALVE SPRING

NEEDLE VALVE

SPRAY TIP

FOLLOWER SPRING

FILTER CAP

GASKET

INJECTOR BODY

FILTER

DOWEL

SEAL

CONTROL RACK

UPPER PORT

CHECK-VALVE CAGE

SPRING CAGE

SPRING SEAT

NUT

(b)

Fig. 31–8 (a) Sectional view of a crown-valve unit injector; (b) sectional view of a needle-valve unit injector (C60). (GMC Detroit Diesel Allison Div.)

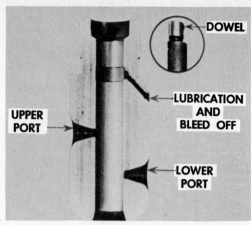

DOWEL

UPPER PORT

LUBRICATION AND BLEED OFF

LOWER PORT

Fig. 31–9 Cross section of a bushing. (GMC Detroit Diesel Allison Div.)

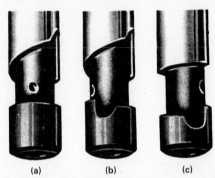

(a)   (b)   (c)

Fig. 31–10 Plunger designs. (a) Constant-ending type; (b) retarded type; (c) 50/50 type. (GMC Detroit Diesel Allison Div.)

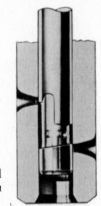

Fig. 31–11 Plunger in the no-fuel position. (GMC Detroit Diesel Allison Div.)

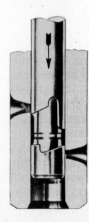

Fig. 31–12 Plunger showing the beginning of injection. (GMC Detroit Diesel Allison Div.)

a longer period of time and close just before the lower port is opened.

When the engine is cranked over, the fuel pump provides fuel, under pressure, to the cavities within the injector. When the camshaft is on its base circle, the follower spring has moved the plunger follower and the plunger upward, thereby opening the upper and lower ports. Fuel then pressurizes the helix area, below the plunger, the spring cage area, and the spray tip. The valve spring holds the needle valve on its seat, stopping fuel from entering the combustion chamber. As the camshaft rotates, the pivoting rocker arm forces the follower and the plunger downward, closing the lower port. The fuel displaced by the downward-moving plunger leaves via the center- and cross-drill holes out of the upper port and through passages within the injector to the outlet filter, filter cap, fuel pipe, outlet manifold, restricted orifice, and into the fuel tank. Just before the upper port closes, the lower port opens, allowing fuel to leave the lower port until the plunger movement stops, that is, as the camshaft moves from the lift flank to the cam nose. Further rotation of the camshaft allows the plunger to move upward. Fuel at fuel-pump pressure then fills the area voided by the retracting plunger, first through the lower port, them the upper port, and then both.

Metering and Injection The metering principle of the Detroit Diesel unit injector is the same as that of any other fuel-injection pump with a helix-designed plunger of variable-beginning and constant-ending. The governor spring and the weight force of the governor determine the plunger position and thereby change the effective stroke. When the control tube is moved by the governor, the rack-control lever moves the control rack, rotating the gear and the plunger. This rotation changes the upper-helix position in relation to the upper port. Assume for instance, that the plunger is at half fuel; when it moves downward, it closes the lower port. Fuel displaced by the plunger flows through the center- and cross-drilled holes and out of the upper port, until the upper port is closed by the plunger (Fig. 31–12).

From this point on pressure rises rapidly in the area below the plunger, helix, spring cage, and spray tip. The fuel pressure acting on the pressure taper of the needle valve increases and as the force of the fuel pressure becomes higher than spring force (3000 psi [210.93 kg/cm²]), the needle valve lifts off its seat and its upper shoulder comes to rest against the spring cage (needle lift). Fuel is sprayed in a predetermined pattern at a particular angle into the combustion chamber until the lower port opens and ends injection. The fuel at high pressure spills back out of the lower port, against the deflector sleeve. This reduces the pressure in the return passage within the injector. At the same time the lift force on the needle valve is reduced and the spring force quickly seats the needle valve.

On a C injector, the ending of injection is not related to helix position. However, the beginning of injection is advanced or retarded, depending on helix position (Fig. 31–13). The valve below the

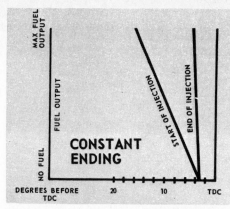

Fig. 31–13 Beginning and ending of injection as related to plunger position (C-type injector). *(GMC Detroit Diesel Allison Div.)*

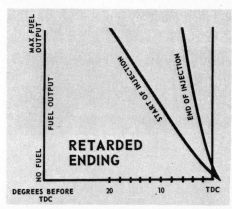

Fig. 31–14 Beginning and ending of injector as related to plunger position (retarded-type plunger). *(GMC Detroit Diesel Allison Div.)*

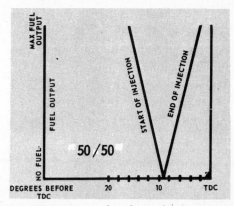

Fig. 31–15 Beginning and ending of injection as related to plunger position (50/50 -type plunger). *(GMC Detroit Diesel Allison Div.)*

bushing is a safety check valve to prevent combustion gases from entering the fuel passages in case the seat of the needle valve leaks.

Retarded and 50/50-Type Plungers Two other plunger designs are the retarded and the 50/50 type (Fig. 31–10b and c). Both are of the variable-beginning, variable-ending design. Both the upper and the lower helix of the retarded type have right-hand spirals, but the lower helix is more shallow than the upper.

NOTE The upper helix (beginning of injection) has a wide advance curve for greater fuel delivery (Fig. 31–14, p. 223). The lower helix also advances the ending of injection but to a lesser extent.

The lower helix of the 50/50-type plunger has a left-hand spiral running in a direction opposite to the upper helix. This also increases fuel delivery and advances injection (Fig. 31–15, p. 223).

By correctly selecting the plunger (diameter) size and helix design, you can increase fuel delivery and advance injection to accommodate different engine designs or their intended application.

Testing C60 Unit Injector Prior to Servicing Most fuel-injection service shops do not test the injectors prior to servicing unless it is to assist the mechanic in troubleshooting. Although servicing is not very difficult, it is extremely important that diesel mechanics pay particular attention to area and component cleanliness and follow service manual instructions exactly. Often, injector failure arises from improper tolerances, dirt and dust, faulty installation, or careless workmanship. To ensure good workmanship, the special servicing tools should be used.

SERVICING THE UNIT INJECTOR

Disassembling C60 Unit Injector Clean the exterior of the injector. Mount it and remove the filter caps. Inspect the threads and flared ends of the caps and discard the filters and gaskets. If the threads or the seats are damaged, also discard the filter cap. To withdraw the stop pin, push the follower slightly down with one hand and with pliers in the other hand, turn the stop pin 90°.

Tilt the vise about 145° to withdraw the follower, follower spring, and plunger. Place the plunger in clean fuel and the follower and spring in a separate container. Tilt the vise a full 180° and, using a deep socket to loosen the nut, unscrew and remove the nut. Take care not to displace the spray tip and valve components in the process! Separately remove the spray tip (taking care not to lose the needle valve), spring cage, spring seat, spring, check-valve cage and check valve (see Fig. 31–16). Place the components in individual containers. If the spray tip adheres to the nut so that it cannot be removed by finger force, place the nut on a wooden block and drive the tip through the nut using a tool which rests on the face of the tip. Remove the bushing, insert the plunger, and place the assembly in

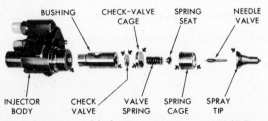

Fig. 31–16 Disassembled C60 unit injector. *(GMC Detroit Diesel Allison Div.)*

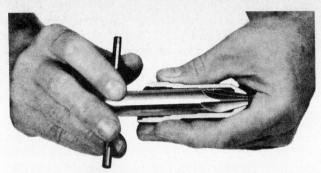

Fig. 31–17 Cleaning the injector-nut spray-tip seat. *(GMC Detroit Diesel Allison Div.)*

Fig. 31–18 Score marks on a plunger caused by one or more plugged spray holes. *(GMC Detroit Diesel Allison Div.)*

clean fuel. Tilt the vise to remove the gear container and the gear and place them in separate containers also. Finally, remove the control rack.

Cleaning and Inspecting Components Clean all individual parts separately using solvent, fuel, etc., as shop facilities dictate.

INJECTOR BODY Inspect the threads of the fuel-filter cap and the nut. Check the dowel pin. Resurface the ring face with a reamer. Then use a 0.375-in [9.37-mm] straight flute reamer to remove any nicks and burrs from the bore. **NOTE** Turn the reamer only in a clockwise direction to avoid damage to the flutes.

Check the contact surface of the bushing and if necessary, lap it to ensure a leak-free seal between the injector body and the bushing. Wash the injector body in clean fuel and place it in the assembly rack.

INJECTOR NUT Inspect the threads for damage and the O-ring groove and the sealing surface of the nut for corrosion. If any of these is defective, the nut must be replaced. Use one carbon removing tool to clean the spray-tip seat (Fig. 31–17). Then use another carbon-removing tool to remove the deposit from the tip bore. (Refer to the service manual for the correct tool type.) Failure to remove any carbon may cause the spray tip to misalign, may result in inadequate sealing (fuel leaks) between the tip and the spring cage, or may cause inadequate compression sealing between the nut and the injector tube.

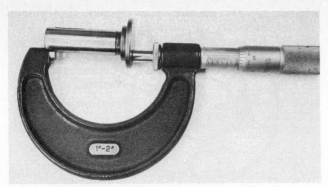

Fig. 31–19 Measuring follower length with a micrometer.

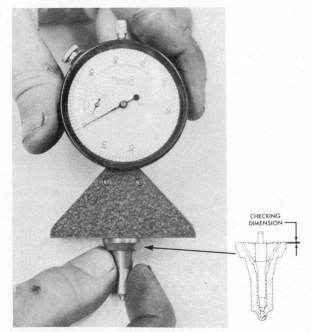

CHECKING DIMENSION

Fig. 31–20 Measuring needle-valve lift. *(GMC Detroit Diesel Allison Div.)*

BUSHING AND PLUNGER Insufficient fuel, dirty fuel, plugged or choked spray-tip holes often cause score marks on the injector bushing and plunger (Fig. 31–18). Inspect the plunger and bushing; use a magnifying glass to check for scoring or chipping. Closely inspect the helix. Check the flat surface on which the gear slides. When the surface cannot be made smooth with an oil stone, the assembly must be replaced. Check the ports in the bushing for carbon deposit, restriction, or damage. If the sealing surface of the bushing (which fits against the injector body) is damaged, the bushing assembly must be replaced. A faulty sealing surface allows fuel leakage and allows the fuel to leave the injector body at the control rack and contaminate the crankcase oil.

To clean the bushing and plunger, apply a moderate amount of mutton tallow to the plunger. Insert the plunger into the bushing. Gently move the plunger back and forth to remove wax or deposit from the bushing and plunger. Before cleaning the assembly, lap the lower seats, thoroughly clean the components, and then check for plunger freeness within the bushing. Until used, keep the assembly as a unit, in clean fuel.

DEFLECTORS When the deflector is out-of-round, pitted, or corroded, it must be replaced. If sharp edges or burrs are present, they should be removed with a 600-grit oil stone.

INJECTOR FOLLOWER SPRING Check the follower spring for corrosion and straightness. Measure the free length; it should be 1.504 in [38.00 mm], and the spring wire used should be 0.142 in [3.556 mm]. With a spring tester, test the spring compression. When compressed to 1.028 in [32.40 mm], it should read 70 lb [31.75 kg]. Do not use a spring which has less energy. It will affect metering because of the slower return action of the plunger.

FOLLOWER Check the contact surface of the rocker arm. If this area is slightly worn, lap it or use a valve grinder to restore the follower surface. **CAUTION** Do not remove more than 0.005 in [0.125 mm] of surface.

Measure the follower length using a micrometer or the special dial tester (Fig. 31–19). The follower length should be within 0.005 in [0.125 mm] of 1.650 in [41.65 mm]. If it is not, the follower must be replaced as it would otherwise adversely affect the beginning of injection.

SPRAY TIP The cleaning and service procedure for a unit-injector spray-tip assembly is the same as that for a multihole fuel-injection nozzle. However, before you clean and service any components, measure them with a micrometer. The spray-tip shoulder should not be less than 0.199 in [4.975 mm] thick. The gauge check valve must not be less than 0.163 in [4.075 mm]. The valve spring gauge must not be less than 0.602 in [15.05 mm]. The check valve not less than 0.022 in [0.049 mm].

If the spray-tip seat is damaged, the use of a polishing stick and polishing compound is recommended to restore the seat surface. **CAUTION** Do not smear the compound onto the lapped surface of the upper needle valve because this will increase the tolerance and may cause metered fuel to leak off, preventing it from being injected.

Check the freeness of the needle valve and valve lift in the same manner as that for a multihole or pintle-type nozzle valve (Fig. 31–20).

Check all other valve assembly components and lap the surface of those which are damaged (see Fig. 31–16). As a general rule, the valve spring should be replaced.

CONTROL RACK AND GEAR Check the control-rack-bearing surface and the teeth for damage, corrosion, and wear. Replace if necessary. Check the gear teeth for damage and the bore for roughness and wear.

Testing Needle-Valve Spray Pattern and Opening Pressure After the components are lapped and cleaned, assemble them (in sequence as shown from left to right in Fig. 31–16) onto a dummy injector body and bushing. Tighten the nut to 85 lb·ft [11.75 kg·m] torque, maximum. Place the protection shield over the assembly and then operate the hand pump of the tester until all air is removed from the assembly. Open the gauge valve slightly and operate the hand pump to raise the pressure until the needle valve lifts off its seat and fuel sprays from the nozzle. Observe and record the opening pressure on the gauge. The opening pressure should be within 2300 to 3300 psi [161.69 to 231.99 kg/cm²]; otherwise the valve spring must be replaced. However, an even opening pressure within 50 psi [3.52 kg/cm²] of 3000 psi [210.93 kg/cm²] is recommended for rebuilt unit injectors. To check the spray pattern, close the gauge valve and then operate the hand pump once every second with a sharp, even stroke. The spray pattern should be without distortion and without visible unatomized fuel. Each pattern should be evenly spaced from the other. Observe the chatter of the needle valve.

To check the needle valve seat for leakage, raise the pressure to 1500 psi [105.45 kg/cm²] and maintain this pressure for about 15 seconds. Although the spray tip should remain dry, a slight wetness is allowable. Next, check the spray tip for back leakage to ensure lubrication. It must be within the specified limit; otherwise too much metered fuel could be lost.

To make this test, bring the pressure to 1500 psi [105.45 kg/cm²] and time the pressure drop. The maximum allowable pressure drop within 5 seconds is 500 psi [35.1 kg/cm²]. If the pressure drop is higher, the spray tip must be replaced.

Reassembling C60 Unit Injector   Before you reassemble the unit injector, all components must be thoroughly cleaned, serviced, tested, and measured. Pay particular attention to the following points during reassembly:

1.   Install the filters with the slotted end upward.
2.   Use new gaskets and a torque wrench to tighten the filter cap.
3.   Use compressed air to remove fuel from the filter openings.
4.   Slide the control rack in place so that the drill marks are visible (A in Fig. 31–21). Place the gear in the injector body with its drill mark aligned between the two marks on the control rack (B in Fig. 31–21).
5.   Place the gear retainer on top of the gear.
6.   Positon the bushing dowel pins over the slot and position the bushing to the injector body.
7.   Assemble the valve and spray tip on the bushing in the following order: spill deflector, seal ring, check valve, check-valve cage, valve spring, spring seat, spring cage, and spray-tip assembly.
8.   Carefully guide the nut over the stacked assembly and hand-tighten the nut while turning the spray tip. When the spray tip cannot be hand rotated any further, use a torque wrench and deep socket to tighten the nut to 85 lb·ft [41.48 kg·m] maximum.
9.   Position the stop pin so that the tighter wound end of the follower spring, when placed over the injector body, rests on the flat side of the stop pin.
10.   Hook the plunger head into the follower. Align the follower slot with the stop pin and align the flat side of the plunger with the gear.

11.   Guide, but do not force, the plunger through the gear and into the bushing.
12.   With the plunger in position, push the follower downward and at the same time force the stop pin in place.
13.   Remove the injector from the vise. Bring the control rack to the full-fuel position to check the rack and plunger timing. The flat side of the plunger must be visible through the hole in the injector body. Place the unit injector in the concentricity gauge.
14.   Adjust the dial gauge to zero. Slowly turn the injector 360° and observe the dial movement. If the total runout of the spray tip is more than 0.008 in [0.2 mm], loosen the injector nut, reposition the spray tip, retorque the nut, and recheck the spray-tip concentricity (Fig. 31–22).

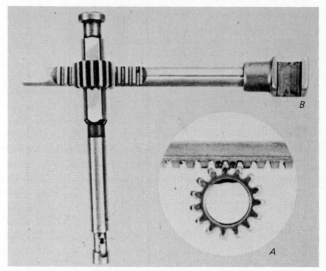

Fig. 31–21   Correct gear and control-rack position. (*GMC Detroit Diesel Allison Div.*)

Fig. 31–22   Checking spray-tip concentricity. (*GMC Detroit Diesel Allison Div.*)

Testing the Serviced Unit Injector

RACK FREENESS TEST To check the rack freeness to ensure smooth plunger movement, install the injector in the injector tester, and then force the follower downward. Move the control rack back and forth. A free rack indicates that the internal components of the injectors are free from damage and contamination.

HIGH-PRESSURE TEST With the injector installed, connect the test head connection to the filter cap by rotating the eccentric lobes. This forces the head connection against the filter cap. Operate the hand pump to pressurize the injector. Maintain a maximum pressure of about 2000 psi [140.60 kg/cm²]. Blow the injector dry, and then check for leaks at the filter cap, nut seal, and control rack. A slight fuel seepage at the control rack because of the higher pressure is permissible, but there must be no fuel leakage. If leakage occurs, it is an indication that the bushing-to-injector body is not sealing well.

VALVE-HOLDING PRESSURE TEST Pressurize the injector to 450 psi [31.63 kg/cm²] and time the pressure drop. A pressure of at least 250 psi [17.57 kg/cm²] should show on the gauge after a time lapse of 40 seconds. A higher pressure drop indicates the lapped surfaces are not seating properly, or the plunger to bushing has a high leak-by. It could also mean that the needle valve to spray tip has too much leak-by. This would have shown up had the needle valve leak-off previously been tested.

The spray pattern and valve-holding test should be repeated as a precautionary measure.

FUEL OUTPUT TEST When the unit injector has successfully passed all other checks and tests, install it to a fuel-flow calibrator or to a comparator.

When using the comparator to check the fuel output, make certain the injector is positioned properly and the handwheel has been tightly secured to the adapter. Always use test fuel. Make certain the filters are clean so that they do not restrict the fuel flow and that the fuel pressure is not too low.

Set the counter to 1000 strokes. Push the injector control rack to a no-fuel position. Switch on the motor which drives the camshaft and the fuel-transport pump.

When the fuel lines show no sign of air, push the control rack to the full-fuel position and hold it in this position. Press the fuel flow "start" button to direct fuel into the vial. At the same time the counter will start totaling the injection strokes and will automatically stop the flow into the vial after 1000 strokes. Take a reading and record it. Dump the fuel from the vial by turning the vial changer. Reset the counter and repeat fuel output tests. Take a second reading. It should be within specification (51 to 61 mm³).

NOTE Both fuel and water are attracted to the glass sides of the vial, thus forming a concave meniscus (crescent-shaped surface). Take your readings at the lower edge of the fluid surface (Fig. 31–23).

Installing Unit Injector To ensure good sealing, and cooling, and to prevent distorting the spray tip, the injector tube must be cleaned before the injector is placed in the tube. Use the injector bevel tube reamer and coat the flutes with grease. Turn the reamer in a right-hand direction to remove the carbon deposits from the tube. Do not use an injector nut reamer because it can remove metal and place the spray tip too far into the combustion chamber.

Clean the tube thoroughly. Remove the spray-tip protection cap, and place the injector into the injector tube. Align the locating pin with the hole in the cylinder head and push the injector into position. Position the rack control lever in the control rack. Install the injector clamp, washer, and bolt. Tighten the bolt to the recommended torque. After each injector is torqued into position, check the freeness of the control rack. Place the valve bridges over the bridge guides. Make certain they are positioned properly, then swing the rocker-arm assembly into position.

Install and torque the rocker-arm-shaft bolts to the recommended torque. Make certain the valve bridges are not dislodged during this procedure.

Remove the protection cap from the filter cap and fuel connector as you install each individual fuel pipe. Torque all fuel-pipe connections to the recommended torque to prevent leakage, twisting, or damage to the pipes.

When all fuel pipes are installed, attach a hand pump to the inlet manifold, and pump clean fuel through the injectors. When the manifolds and injectors are free of air, plug the return line, pressurize the system to 150 psi [10.55 kg/cm²], and then check for fuel leakage at the connection.

Engine Tuneup An engine tuneup is necessary on delivery when the engine is new or newly rebuilt, and again after about 100 to 150 hours of operation. It is also essential when the cylinder head, blower, governor, or injector have been serviced, or when the engine performance is not satisfactory even though the engine is in good mechanical condition.

An engine tuneup should be made after the coolant system, the lubrication system, the drive belts, and the fuel filters have been checked and the engine has been found mechanically sound. Tuneup involves a series of checks and adjustments to bring the components to the recommended specification.

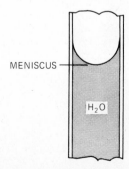

Fig. 31–23 Correct fuel-level reading. (*GMC Detroit Diesel Allison Div.*)

228

The effectiveness of an engine tuneup may be unsatisfactory unless the recommended procedure sequence is followed. The Detroit Diesel tuneup sequence is as follows:

1. Adjust valve bridge
2. Lash valves
3. Time injectors
4. Adjust governor gap
5. Adjust injector racks
6. Adjust no-load speed
7. Adjust idle speed
8. Adjust buffer screw
9. Adjust throttle delay

Valve Bridge Adjustment The 71, 92, and 149 Series engines utilize valve bridges. Valve bridge adjustment is required only after the cylinder head has been serviced, or when the valve bridge has been damaged or worn.

To make the adjustment, remove all the fuel pipes and cap the filter caps. Where the exhaust valves are closed, first remove the rocker-arm-shaft retainer bolts from the cylinder head, and then swing the assembly over center. Crank the engine over to position the other rocker arms, then remove these assemblies. Do not force the rocker arms over center.

Remove one valve bridge at a time. Place it in a vise to loosen the locknut and then place the bridge as shown in Fig. 31–24, turning the adjusting screw until it just touches the top of the valve stem. Add about one-eighth of a turn to compensate for thread looseness. Hold the bridge in a vise, and hold the adjusting screw with a screwdriver while torquing the locknut. Install the valve bridge on the same bridge guide and place a 0.0015-in [0.0375-mm] feeler gauge stock under each end of the valve bridge. When you press down on the valve bridge, the pull on each feeler gauge must be equal. Next, adjust the remaining bridges. Do not interchange valve bridges after they have been adjusted.

Reinstall the rocker-arm assembly and fuel pipe.

Valve Adjustment Crank the engine over. Make sure the governor is in the stop position when you bring No. 1 cylinder to TDC, that is, when the injec-

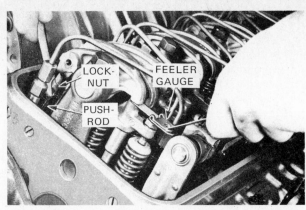

Fig. 31–25 Checking valve clearance (V71). *(GMC Detroit Diesel Allison Div.)*

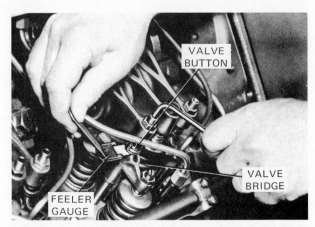

Fig. 31–26 Checking valve clearance (V149). *(GMC Detroit Diesel Allison Div.)*

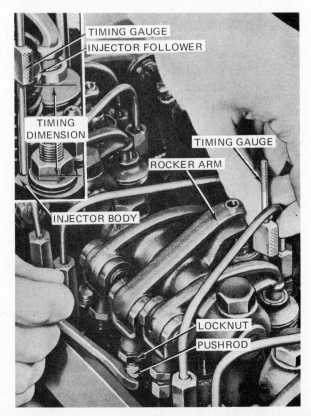

Fig. 31–27 Checking injector timing. *(GMC Detroit Diesel Allison Div.)*

Fig. 31–24 Valve bridge adjustment. *(GMC Detroit Diesel Allison Div.)*

tor pushrod has moved upward, depressing the injector follower spring.

Select the feeler gauge thickness designed for cold setting the exhaust valves. When possible, use a go–no-go gauge. For the 71 Series engine with two exhaust valves, slide the feeler gauge between the rocker arm and the valve stem. The adjustment is made by loosening the locknut and turning the pushrod until the feeler gauge has a slight drag when pulled and pushed between the surfaces. Then tighten the locknut and recheck the clearance.

For four-valve cylinder-head engines, the adjustment is made in a similar manner, except that the feeler gauge is placed between the rocker arm and the pallet surface of the valve bridge (Fig. 31–25).

The adjustment of the exhaust valves on the 149 Series engine is made by turning the adjusting screw to obtain proper clearance between the pallet of the valve bridge and the valve bottom of the adjusting screw (Fig. 31–26).

For the 53 Series engines, the adjustment is made in the same manner as for the 71 Series, except that the feeler gauge is placed between one valve stem and the surface of the pivot-mounted valve bridge.

After each adjustment, mark the cylinder so that you can keep track of the valves which have been adjusted. Assume you have a 6-71 engine, with a firing order of 1-5-3-6-2-4. You can adjust the exhaust valve of Nos. 1, 5, and 4 cylinders and the injector timing of Nos. 3, 6, and 2 cylinders when No. 1 cylinder is at TDC.

Injector Adjustment (Phasing) To ensure that each injector is timed correctly with regard to piston position, the height of the injector follower must be adjusted in relation to the injector body. Timing gauges of various dimensions are used. For example, when using a timing dimension of 1.484 in [37.69 mm], the beginning of injection is retarded, allowing the piston to reach a higher position within the cylinder before injection begins. This is essential because the plunger has to travel a further 0.024 in [0.6 mm] before the plunger closes the upper bushing port.

To check the injector timing or to make an adjustment, select the recommended timing tool and place its small end in the hole provided in the injector body (Fig. 31–27). You can start with either the No. 3, 6, or 2 injector. The timing is correct when the flat end of the gauge wipes off the oil on the top of the follower. If it does not, then adjust the pushrod in the same manner as you adjust the exhaust valve pushrod, until the proper follower height is obtained. Next, hold the pushrods while tightening the locknut to maintain the adjustment. Always recheck new adjustments and, when correct, mark the injector to identify that it has been adjusted.

After the exhaust valves and the injector adjustments have been made, turn the crankshaft 180° to bring the No. 6 cylinder to TDC. You can now adjust the exhaust valves of Nos. 6, 3, and 2 cylinders and the injector timing of Nos. 1, 5, and 4 cylinders.

The procedure to follow after the exhaust valves are adjusted and the injectors are timed depends on whether the engine has been running before, on whether it is a rebuilt engine, and on the type of governor used. Assuming the engine was rebuilt and a limited-speed governor (double-weight) is on the engine, positioning of the rack control lever would be the next step.

Positioning Individual Rack Control Levers To ensure that each injector control rack has maximum fuel output while the engine is running at full load, back out the idle-speed adjusting screw about ½ in [12.7 mm] beyond the locknut and back out the buffer screw about ⅝ in [15.87 mm]. Loosen all inner and outer injection-rack-lever adjusting screws. Hold the speed control lever with light finger pressure in the maximum-fuel position. Then turn the inner adjusting screw clockwise on the No. 1 injector rack control lever until you feel a stepup in effort and a light movement at the throttle shaft lever. Next, turn the outer adjusting screw clockwise until it bottoms lightly on the control tube, then alternately tighten both adjusting screws (Fig. 31–28). This should place the No. 1 injector in the full-fuel position. To determine if it is in this position, use a screwdriver or your fingertips and press down on the injector control rack. If the injector control rack returns with a springing action to its original position, but does not move the speed control lever, the control rack is probably correctly placed in relation to the full-fuel and governor positions.

The clevis pin which connects the fuel rod with the fuel-control-tube lever should also be checked for tightness. If the rack does not return to its original position, it is too loose. Loosen the outer adjusting screw and tighten the inner adjusting screw to get the proper control-rack action. When the lever has moved too much or the pin has become too tight, back off the inner adjusting screw and slightly tighten the outer adjusting screw.

If you are working on a V engine, remove the clevis pin from the fuel rod of the right cylinder bank before adjusting the No. 1 injector control rack on the left cylinder bank. When the No. 1 injector con-

Fig. 31–28 Adjusting the injector rack control lever on a V53 engine. (*GMC Detroit Diesel Allison Div.*)

trol rack is adjusted, remove the clevis pin from that side and place the pin in the fuel rod and control tube lever to connect the right cylinder bank with the governor. Then adjust the injector control rack of the No. 1 injector. When the fuel control racks of both injectors are evenly adjusted, the left cylinder bank's clevis pin should slide in place smoothly and both pins should drag equally when moving them within their bores.

This adjustment procedure must be followed precisely otherwise one bank receives less injector-control-rack travel than the other with the result that one cylinder bank receives less fuel.

Positioning Other Control Racks   After the No. 1 injector is adjusted, remove the clevis pin or pins (when working on a V engine) and manually hold the injector-control-tube lever so that the No. 1 injector control rack is in the full-fuel position. In this position, adjust the No. 2 control lever so that the No. 2 control rack is in the full-fuel position and is as equally responsive as that of the No. 1 control rack.

NOTE   Never adjust the No. 1 injector control rack.

Adjust the remaining control racks and make certain that all racks have equal resilience (bounce-back). When reconnecting the fuel rod to the control tube, the clevis pin must slide in without restriction.

Readjusting Exhaust Valve and Injector Timing   Position the valve cover(s) and make certain that all tools are removed from the engine before it is started. Warm the engine up to the operating temperature, 160° to 185°F [71° to 85°C]. At this temperature recheck and, if necessary, adjust the exhaust valves; recheck the injector timing. Now make certain that the valve lash and injector timing are precisely as specified. If necessary, again warm up the engine to maintain the operating temperature.

Governor Adjustment   Detroit Diesel engines have six different types of governors, each with its own adjustment peculiarity. Engines with mechanical governors require that the governor gap be adjusted prior to adjustment of the injector control rack. Also, the fuel rack must be adjusted before adjusting the hydraulic governor. Then the injector control rack is adjusted.

To adjust an engine with a limited-speed governor, (double-weight), start the engine and bring it to operating temperature. Next, stop the engine. Disconnect the throttle linkage. Remove the two attaching bolts. Remove the governor high-speed spring-retainer cover. Back out the buffer screw if you have not already done so. Start the engine and adjust the idling speed to that recommended on the unit option plate. (The option plate is attached to one valve cover. It lists precise information about engine design, additional equipment used, and recommended idle and high-idle speeds.) Then stop the engine. Remove the governor cover and lever assembly.

On in-line engines remove the fuel rod from the differential lever and the injector-control-tube lever.

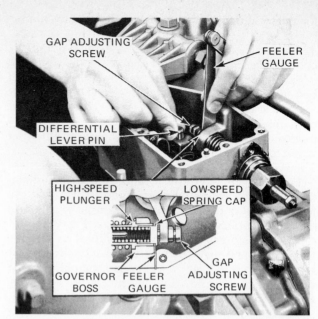

Fig. 31–29   Adjusting the governor gap. (GMC Detroit Diesel Allison Div.)

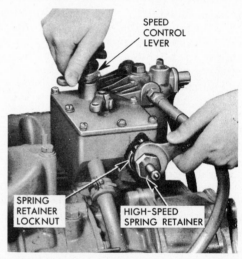

Fig. 31–30   Adjusting the high-idle speed. (GMC Detroit Diesel Allison Div.)

On V engines the fuel rod need not be removed. Hold the control-tube lever in one hand, start the engine and manually control the engine speed with the control-tube lever. Control the engine speed by controlling the differential lever.

With the engine at 800 to 1000 rpm, check the gap by placing a 0.0015-in [0.0381-mm] feeler gauge between the low-speed spring cap and the high-speed spring plunger. If the gap is correct, there will be a slight drag on the feeler gauge (Fig. 31–29). When adjustment is necessary, stop the engine and turn the gap adjusting screw to obtain the proper drag.

To recheck the gap setting, run the engine between 800 and 1000 rpm and force the gap closed with a screwdriver. When the setting is correct, the gap closing movement can be seen.

Another method of rechecking gap setting is to place a drop of oil on the gap area, and then slowly raise the rpm so that you can see the gap close.

However, there must be no mistake in the gap setting because this would affect the injector-control-rack adjustment. When gap setting is completed, stop the engine; replace the fuel rod and the governor cover. Do not reconnect the throttle linkage to the speed control lever at this time.

Readjusting Rack Control Levers When the governor gap has been adjusted, you must readjust the position of the rack control levers as outlined above.

Adjusting Maximum and Idle Speeds (Limiting-Speed Governor) With the engine at operating temperature, loosen the locknut and back off the high-speed spring retainer approximately five turns (Fig. 31–30). Start the engine and move the speed control lever to maximum-speed position, using a strobe light or hand tachometer to measure the rpm.

The maximum engine speed should not exceed the recommended no-load speed shown on the unit option plate. Turn the high-speed spring retainer until the engine runs at recommended no-load speed, and then tighten the locknut. Reduce the engine speed. Again bring the speed control lever to the full-fuel position, and recheck the maximum no-load speed.

To adjust the idling speed, bring the speed control lever to the idle position and turn the idling-speed adjusting screw until the engine runs about 15 rpm below the idling speed shown on the unit option plate. The engine should operate smoothly without hesitation and without a change in engine rpm. If it does not, all the adjustments must be rechecked.

Buffer Screw Adjustment When the idle speed is adjusted to that recommended on the option plate and the engine is at operating temperature, turn the buffer screw as shown in Fig. 31–31 until it just contacts the differential lever to smooth out engine idle speed. Do not increase the idle speed more than 15 rpm with the buffer screw. Always recheck the maximum no-load speed. If it has increased more than 25 rpm, readjust the buffer screw until the increase is less than 25 rpm.

Throttle-Delay Mechanism Some Detroit Diesel engines use a throttle-delay mechanism to reduce exhaust smoke during acceleration. This device—a piston which is part of a rocker-arm bracket—is connected over a link with the throttle-delay lever which is bolted to the injector control tube (Fig. 31–32).

Operation of Throttle-Delay Mechanism When the engine is running, oil is supplied from the rocker-arm bracket through a supply to the reservoir. However, the plug has a restriction orifice to reduce the oil flow into the reservoir. When the injector rack is in the full-fuel position, the injector tube has moved the throttle-delay lever, the link, and the piston to the left (in Fig. 31–32). With this action the piston blocks the oil opening in the reservoir. As the control rack is moved to the no-fuel position, the check valve opens and air can enter the

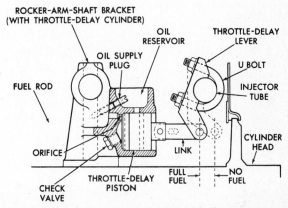

Fig. 31–31 Adjusting the buffer screw. *(GMC Detroit Diesel Allison Div.)*

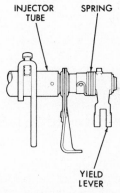

Fig. 31–32 Throttle-delay cylinder with the injector rack in the full-fuel position. *(GMC Detroit Diesel Allison Div.)*

Fig. 31–33 Yield lever and spring assembly. *(GMC Detroit Diesel Allison Div.)*

delay cylinder allowing the piston, and therefore the injector tube, to move freely to the right. As the piston uncovers the opening in the reservoir, oil enters the cylinder and displaces through the orifice above the check valve.

The governor and the fuel rod are permitted full accelerator travel because of a special yield lever and spring assembly instead of the standard injector-control-tube lever (Fig. 31–33). As the governor is moved to the full-fuel position, the spring assembly and yield lever transmit the motion from the governor to the injector tube onto the throttle-delay piston. The piston forces the oil through the orifice, delaying the rotation of the injector tube.

Throttle-Delay Adjustment When you have made all other adjustments, stop the engine, and place a (special) gauge between the injector body and the shoulder on the injector rack. Apply a slight pressure to the injector tube in the direction of the full-fuel position. Tighten the U-bolt nuts so that the delay piston is flush with the edge of the throttle-delay cylinder. Remove the tool. Check the freeness of the injector tube.

Start the engine and check the oil supply, the action of the yield lever and spring assembly, the check valve, and the combination air and oil bleed orifice [see section on engine run-in (Unit 51)].

Questions

1. Why is an outlet restriction fitting required in the return line of a Detroit Diesel fuel-injection system?

2. If the restriction fitting orifices were greater than specified, how would it affect engine performance?

3. What would occur if the fuel-pump seals were installed backward?

4. Outline two methods (other than those mentioned in this textbook) of locating a misfiring injector.

5. Explain how to remove a unit injector which is seized in the injector sleeve.

6. What does the engraved number on the nozzle tip signify?

7. Why aren't the spray angles of all nozzles equal?

8. List the differences in the metering and pumping principles between a unit injector pump and a Bosch injector pump.

9. Explain how each of the three plunger designs affects engine performance.

10. List the units of an N-type injector in their order of reassembly.

11. Why must the plunger-follower be within specified limits?

12. If the needle lift height of an N-type nozzle were greater than maximum specification, how would it affect the spray pattern and fuel flow?

13. What causes the rack on a unit injector to lose its freeness?

14. When working on a V engine, explain how to check and adjust the injector, adjust No. 1 injector control levers, and, check and adjust the governor gap.

15. Assume that an engine continues to run rough at idle even after you have increased (with a buffer screw) the idle speed by 60 rpm. What further steps (or repeated steps) must you take?

16. What would result if the orifice in the throttle-delay piston were clogged?

UNIT 32

Cummins Fuel-Injection System

The basic components of the Cummins PT fuel-injection system are shown schematically in Fig. 32–1. Superficially, this injection system appears to be the same as any other. In actual operating principles, however, the metering and injection operations differ considerably from others. To begin with, this system has a fuel pump instead of the regular injection pump. Injection is accomplished through a series of mechanical actions in the injector and the injector activating mechanism. The Cummins PT fuel-injection system, the metering of which is based on pressure and time, utilizes fixed openings (orifices) in the injectors and variable openings in the fuel pump.

The metering principle is based on the laws of hydraulics. It is commonly known that: (1) an increase in flow volume increases the pressure proportionately; (2) pressure is created by restricting the flow (adjustable restriction in the governor or the

regulator); (3) when a liquid flows through an orifice the volume and pressure decrease; (4) when the time is extended to allow the liquid to flow through the orifice, the volume increases.

The nameplate of the Cummins fuel pump and injectors, as with other fuel-injection systems, gives certain information required for servicing and adjusting the fuel pump and injector.

An example of a PTG fuel pump nameplate used with a KTA-2300 engine as shown below:

0179	994079
AR 40454	8230A

The current method of identifying pump application, design, and adjustment is as follows:

1. The first letter and number indicate the model, which determines the injectors, pistons, and camshaft to be used.

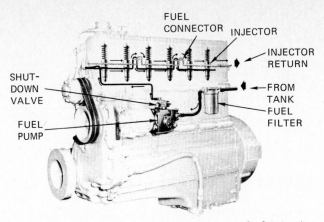

Fig. 32–1 Schematic view of a Cummins fuel-injection system (in-line engine, PTG pump). (Arrows show the fuel flow.) (*Cummins Engine Co., Inc.*)

2. The next set of numbers represents the pump series and the letter behind "L" or "R" indicates pump rotation.
3. The second line starts with "BM" or "AR," which is the assembly suffix letters.
4. The next four numbers give the calibration card number and suffix letter application.

PT FUEL PUMPS

PTR and PTG Fuel-Pump Components and Their Purpose Cummins Engine Company, Inc., builds three PT fuel pumps, (type R, type G, and the AFC type). Since type R is now used only for certain special engines, it will not be discussed in this textbook as fully as the PTG fuel pump. **NOTE** At this

1. Tachometer shaft
2. Filter screen
3. Fuel to injectors
4. Shutdown valve
5. Gear pump
6. Check-valve elbow
7. Fuel from tank
8. Pulsation damper
9. Throttle shaft
10. Idle adjusting screw
11. High-speed spring
12. Idle spring
13. Gear-pump pressure
14. Fuel-manifold pressure
15. Idle pressure
16. Governor plunger
17. Governor weights
18. Torque spring
19. Weight-assist plunger
20. Weight-assist spring
21. Main shaft
22. Bleed line

Fig. 32–2 PTG fuel pump. (The arrows show the fuel flow.) (*Cummins Engine Co., Inc.*)

writing, not enough adequate information is available to include the AFC fuel pump in this unit.

Externally, these three Cummins fuel pumps look alike, but the internal construction, the fuel flow, and the method of pressure control are quite different. The PTR pump has a return line and the current PTG fuel pump has a bleed line connected to the reservoir. The bleed line prevents excessive fuel temperature within the pump when the throttle is closed and the gear-pump speed (engine speed) is high due to downhill operation. The special check valve (in the return-line fittings) opens when the fuel pressure overcomes the spring force. This action permits excessive fuel to flow to the reservoir.

As previously mentioned, the PTR fuel pump is only used on special engines, therefore it will be compared only briefly with the PTG pump. Each of these fuel pumps uses a combined mechanical-electrical activation shutdown valve to stop the fuel flow to the injectors (see Fig. 32–2). The positive-displacement-type gear pump is driven directly by the main shaft to pump fuel through the fuel-injection system. The steel diaphragm-type pulsation damper smooths out pressure variations created by the gear pump. The opening in the throttle shaft controls the engine speed above idle by increasing or decreasing the throttle opening. This increases or decreases fuel flow to the injector to meet operating requirements under varying load and speed conditions. The throttle shafts of both fuel pumps vary with pump code.

The throttle shafts used with PTG and AFC fuel pumps have a restriction plug so that small pressure adjustments can be made (Fig. 32–3). The governor weight (flyweight) assembly is driven at increased speed from the gear on the main shaft. The gover-

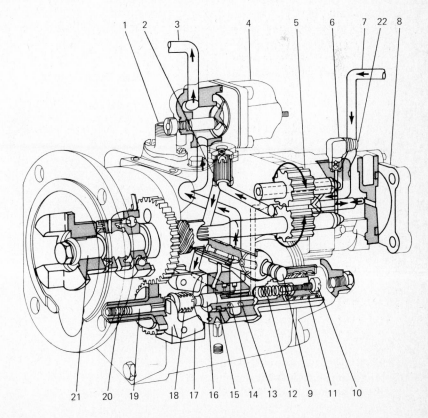

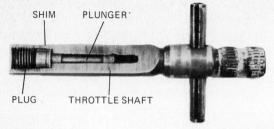

Fig. 32-3 Cutaway view of the throttle-shaft assembly.

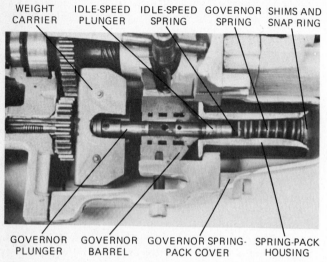

Fig. 32-4 Sectional view of a limiting-speed governor.

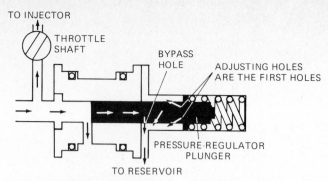

Fig. 32-6 PTR pressure-regulator action during downhill operation (throttle closed). (Arrows show the fuel flow.) (*Cummins Engine Co., Inc.*)

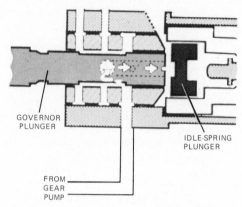

Fig. 32-7 PTG idle-spring-plunger operation. (Arrows show the fuel flow.) (*Cummins Engine Co., Inc.*)

nor plunger is held against the plunger guide on one end, and on the other end it is held by a snap ring in the spring pack housing.

The governor plunger, the weight carrier, the idle plunger, and the governor spring vary, depending on the pump type and fuel pump code (Fig. 32–4).

The governor plungers of the PTG and AFC pumps are center-drilled and have two cross-drilled holes of various sizes. The governor plungers of the PTR pump have only a single machined-out groove. PTG and AFC pumps use a weight-assist plunger, and some also use a torque spring (Fig. 32–3).

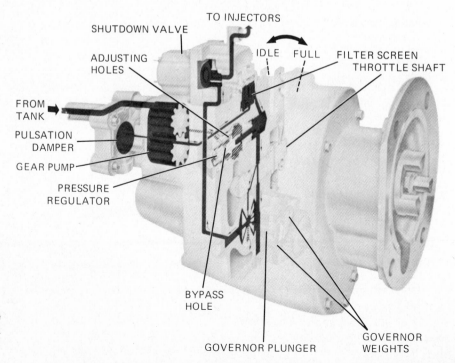

Fig. 32-5 Cutaway view of the PTR fuel pump showing the fuel flow. (*Cummins Engine Co., Inc.*)

PTR Fuel Flow and Pressure Regulation A schematic illustration of the PTR fuel pump is shown in Fig. 32–5. When the engine is running at idle speed, fuel under pressure is instantly available to the injector because the throttle is never fully closed. Also, a second passage is always open, thus keeping fuel lines and injectors filled with fuel. The fuel lubricates and cools the injectors, removes air, and assures quick acceleration. The governor weight force and spring force position the governor plunger to allow idle fuel to pass to the injectors. The pressure-regulator plunger, when forced to the right (in Fig. 32–5) by fuel pressure, opens the first bypass holes, thus reducing the fuel pressure. As the throttle is rotated, the opening in the throttle increases, allowing additional fuel to flow to the governor and injectors. As engine speed increases the gear pump and fuel pressure also increase. This increased fuel pressure moves the plunger of the pressure regulator further to the right. This uncovers the first adjusting holes allowing more fuel to bypass. With increased engine speed, the fuel pressure exerts an even greater force on the plunger. Then the torque holes are covered, and the pressure is again reduced.

The increased engine speed (from idle) increases the weight force causing the governor plunger to move against the high-idle spring in such a position that fuel is no longer restricted by the plunger. If the engine speed increases above high idle, the weight force moves the plunger even further to the right. This reduces the fuel flow to the injectors because another passage to the inlet side of the gear pump is opened. Should the engine speed increase even further, the fuel flow to the injectors shuts off (Fig. 32–6). At the same time, the fuel pressure increases, exerting a greater force on the pressure-regulator plunger. The plunger then opens the recirculating passage (dump holes) and returns fuel to the reservoir.

If it is necessary to increase or decrease the fuel pressure in order to come within specified pressure, shims must be added or removed from behind the pressure-regulator spring.

PTG Fuel Flow and Pressure Regulation When the engine is cranked over or is idling, fuel under pressure is readily available to the injector. This is because fuel from the gear pump has already traveled through the filter screen, the governor, the throttle shaft, and the shutdown valve into the fuel manifold. The idle-spring plunger rests against the governor plunger, preventing fuel from bypassing. The governor is positioned in its barrel by the idle spring and governor weight force (Fig. 32–2). This maintains the correct idle fuel flow through the idle fuel passage. As fuel flows through the center-drilled passage of the governor plunger to the idle-spring plunger, it forces the idle-spring plunger away from the governor plunger. Fuel is then allowed to escape into the pump housing to reduce the fuel pressure (Fig. 32–7).

Different size idle-spring plungers are used to control the amount of fuel that reaches the injectors. The larger the plunger number, the greater the area

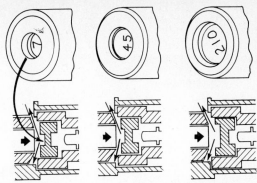

Fig. 32–8 PTG idle-speed plungers of different sizes for various fuel bypasses. *(Cummins Engine Co., Inc.)*

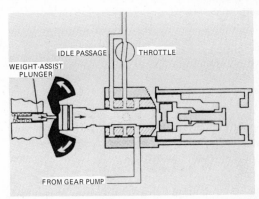

Fig. 32–9 Governor position at idle speed, showing the action of the weight-assist plunger. *(Cummins Engine Co., Inc.)*

on which the fuel pressure can act. Thus, a plunger with a large number allows more fuel to escape into the pump housing (Fig. 32–8). The throttle in the PTG fuel pump is never closed. Throttle leakage and fuel that passes through the idle fuel passage supplies fuel to the injectors (see Fig. 32–9).

To increase engine torque (fuel pressure) at low speed a weight-assist plunger and spring are used. The weight-assist plunger is forced, due to its spring, against the governor plunger, as indicated in Fig. 32–9. As a result, the governor plunger is forced against the idle-spring plunger. This increases the weight force and thereby reduces fuel bypass, but increases fuel pressure. Shims are used to extend or reduce the range of the weight-assist plunger on the governor plunger. The idle speed is adjusted by increasing or decreasing the idle-speed spring which, in turn, alters the force on the idle-speed plunger and the amount of fuel bypass.

The limited-speed governor controls idle speed, high-idle speed, and torque rise. The throttle position controls engine rpm between idle and high idle. If the engine speed increases, due to downhill operation, the weight force increases, causing the governor plunger to close the fuel passage to the throttle. At the same time, however, it opens another passage, allowing the fuel to pass into the pump housing and back to the inlet side of the gear pump. This results in a reduction of engine speed regardless of throttle position.

A standard pump can be adapted to suit many different engines by varying the governor weight force,

1. Air/fuel control valve
2. Tachometer drive assembly
3. Manifold air-line connection
4. Fuel out (to injector)
5. Shutdown valve
6. Filter
7. Fuel tank
8. Bleed off
9. Pulsation damper
10. Gear pump
11. "No air" adjusting screw
12. Throttle leakage adjustment
13. Idle spring
14. Governor spring
15. Adjusting screw
16. Fuel-pump housing
17. Restriction plunger
18. Weight carrier assembly
19. Weight-assist plunger
20. Throttle shaft
21. Drive coupling
22. Front cover

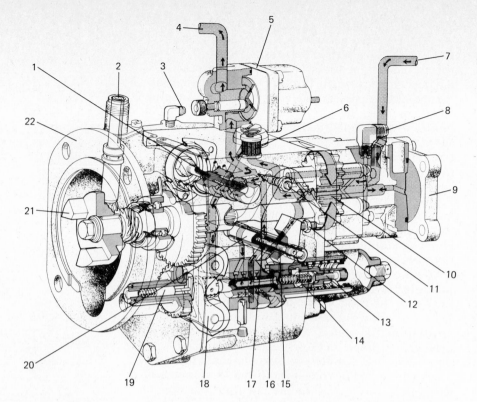

Fig. 32–10 APC fuel pump. (Arrows show the fuel flow.) (Cummins Engine Co., Inc.)

the governor spring force, or the idle-speed plunger to change the fuel pressure. The fuel pump code number indicates the specified combination for the correct fuel pressure and high-idle speed. However, if the high-idle speed is above or below the specified rpm, shims are used to increase or decrease the spring force. This, of course, increases or decreases fuel pressure because more or less fuel is bypassed at the idle-spring plunger. To increase the torque rise, a torque control spring is placed on the governor plunger. This modifies the weight force and the amount of fuel bypass. A specified combination may produce a 10 percent torque rise.

PTG fuel pumps using a variable-speed governor have only one fuel passage. This means that idle speed (idle fuel flow) and maximum speed (high-idle fuel flow) are governed through the same passage. To reset idle speed and high-idle speed, an idle-speed screw and a high-idle adjusting screw are used. They limit the travel of the adjustment lever, thereby limiting the spring force.

AFC Fuel Pump versus PTG Fuel Pump The AFC fuel pump is a modification of the PTG fuel pump. It is designed to control more adequately the exhaust emission and to make it tamperproof (Fig. 32–10).

To allow space for a new air/fuel control valve, the tachometer drive assembly has been relocated in the front cover. The air/fuel control valve is located in the area previously designed for the tachometer assembly.

The air/fuel control valve is similar in construction to the aneroid. It performs the same function except that although it restricts the fuel, it does not act as a bypass valve as does the aneroid. The throttle shaft has been redesigned and is a sealed unit. The adjustment of the restriction plunger is now made with an adjusting screw (instead of shims).

The adjusting screw is then permanently sealed. The throttle-leakage adjustment and throttle-shaft travel are still made with adjusting screws, but the adjusting screws are covered by a plate on which there is an inscription warning against tampering. The "no air" adjusting screw of the air/fuel control valve is located behind the throttle-shaft plate. Furthermore, the governor weight carrier is cast instead of being stamped from steel.

AFC Fuel Flow Refer again to Fig. 32–10. A comparison between the PTG and the AFC fuel-injection systems reveals an almost identical fuel flow. However, the fuel flow in the AFC is directed from the throttle shaft to the "no air" adjustment screw, to the air/fuel control valve and from there to the shutdown valve.

Removing Fuel Pump from Engine Never remove the fuel pump from the engine unless you are certain that it is the fuel pump that is causing the problem. The removal or installation of a PT fuel pump is very simple because the fuel pump is driven through a coupling from the compressor drive or fuel-pump drive. Timing or phasing is unnecessary because the fuel pump controls only the fuel pressure.

Before you remove the fuel lines and the throttle connection, clean the surrounding area thoroughly. Cap the hose and pump openings immediately to prevent contamination. Finally, loosen and remove the fuel-pump mounting bolts, and then lift the pump from the engine. Use a circular starter-type wrench to facilitate removal.

PTG Fuel-Pump Disassembly and Service Thoroughly clean the exterior of the fuel pump before mounting it to a swivel vice. You can then start the disassembly procedure from the drive shaft

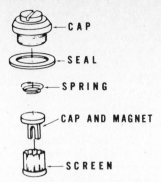

Fig. 32–11 Filter components. (Cummins Engine Co., Inc.)

Fig. 32–12 Removing the governor plunger. (Cummins Engine Co., Inc.)

side or from the gear pump side. You can also start by removing the shutdown valve from the pump housing, loosening and removing the filter screen cap, and lifting out the components in the order shown from top to bottom in Fig. 32–11.

NOTE Flat washers are used between the lock washers and aluminum components to prevent the lock washers from cutting into the soft metal.

Remove the bolts from the front cover and use a soft-faced hammer to tap the cover loose from its two dowels. Slide the assembly from the housing. Remove the governor plunger (Fig. 32–12). Handle the plunger with care because any small nicks will make it unusable. Remove the throttle-plate retainer and carefully slide the throttle out of its bore. When the tachometer drive requires servicing, remove its cover screws and carefully drive the assembly from the fuel-pump housing, using a brass punch. Next, remove the pulsation damper and the gear-pump mounting screws. Use a soft-faced hammer to lightly tap the gear pump loose from its dowels, then pull the pump from the housing.

To remove the governor-spring-pack assembly, remove the mounting screws of the spring-pack cover, and then lift it off. Remove the snap ring which holds the spring pack. Lift out the washer, shim, and high-speed spring from the spring-pack housing. Then remove the idle-speed plunger guide, washer, idle-speed spring, and idle-spring plunger. **NOTE** Some fuel pumps use an adapter. At this time do not remove the drive-shaft bush-

ings, the throttle sleeve, or the governor barrel because these may not require replacement. First, wash the housing using a suitable cleaning method. Then inspect the above-mentioned components for wear. If the governor weight shaft is loose in its bushings, or if the gear backlash between the drive gear and the drive-shaft gear is more than 0.009 in [0.23 mm], the front plate must be serviced.

Remove the weight-assist plunger, spring, and shims, if used. Heat the cover in an oven to about 300°F [149°C] and use a suitable puller to remove the weight carrier assembly. If the snap ring pulls away from the shaft, it will leave the bushing in the front cover. You must then remove the bushing with an internal engaging puller.

If the drive bearing is worn or damaged, remove the snap ring and the fuel-pump drive-coupling retainer screw and washer. Screw a ⅜-in 24-hex screw into the drive shaft and place the cover on the press. Press out the drive-gear assembly, and then press out the drive-shaft oil seal.

Fuel Pump Inspection and Service Inspect the pump housing for cracks, worn or damaged threads, studs, and threaded holes. When necessary, refinish the mounting surfaces of the front cover, shutdown valve, and governor-spring-pack cover.

Measure the wear of the drive-shaft bushing and shaft. If the bushing needs replacement, press the old one out. Apply a light coat of lubricant to the new bushing before pressing it into the housing, flush with the end of the bore. Remember, you must line-ream the bushing after installation to assure proper alignment and running clearance.

If the governor plunger is worn but the governor barrel is reusable, replace the plunger with one that is a class or two larger. Lap it to the old barrel with fine-grid lapping compound. Make sure that the replacement plunger has the same part number as the discarded plunger.

If the replacement plunger is the largest size (No. 5) and is still too loose in the barrel, or if the governor barrel is damaged, the barrel must be replaced. To remove it from the housing, first heat the housing in an oven to 300°F [149°C]. This prevents damage to the pump housing while being pressed out. Before you press the new barrel into the housing, scribe the centerline of the fuel passage on the governor barrel and the pump housing to ensure precise alignment and thereby an unrestricted fuel flow (Fig. 32–13). This is also a convenient time to check the spring-pack housing to determine if it is worn sufficiently to require replacement.

Place the lubricated barrel with the chamfered edge on the bore so that the pinhole is at the bottom and the scribe lines are aligned. Press the barrel into its bore until it rests totally against the spring-pack housing. The barrel-retaining pinhole must now align. Install the spring dowel with its slot facing the front of the housing.

If the throttle shaft or the throttle-shaft sleeve shows wear, a new shaft with a higher class number must be lapped to the throttle-shaft sleeve. However, when the sleeve is worn beyond a No. 5 shaft size, the housing assembly must be replaced.

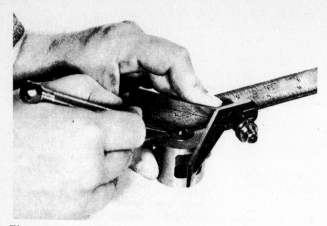

Fig. 32–13 Scribing the fuel-passage centerline on the governor barrel. (*Cummins Engine Co., Inc.*)

Be certain that the housing, the throttle sleeve, and the governor barrel are free of all lapping compound. If the governor plunger appears otherwise reusable, make sure that the thrust washer and the stop sleeve are not worn. If it is not reusable, remove the retainer pin and the governor plunger drive, and press the stop sleeve from the plunger.

NOTE If a torque spring is used, rotate it lightly when removing it from the governor plunger to prevent stretching.

When you do the reassembly, the two notches on the stop plate must face toward the barrel, and the chamfered side of the stop plate must be next to the drive.

There must be a clearance between the driver and thrust washer of at least 0.002 in (or not greater than 0.005 in) [0.05 to 0.127 mm]. Make sure that you install the specified torque spring. This specification information can be obtained from the pump calibration data.

Tachometer Drive If the tachometer drive requires servicing, remove the oil seal and press the shaft from the drive gear and bushing. Check the tachometer shaft for wear. If it is reusable, place the new bushing on the shaft so that the chamfered end faces the drive gear. Press the gear onto the shaft until the clearance is less than 0.005 in [0.127 mm] but the bushing still turns freely on its shaft (see Fig. 32–14).

To service the weight retainer, first remove the snap ring and slide the governor carrier bushing from the carrier shaft, then press the governor gear from the shaft. If there is any damage or wear on the weights or pins, the weight carrier must be replaced as a unit. The number on the replacement weight carrier must correspond with the fuel pump's code number.

When you reassemble the unit, press the new weight assembly onto the gear, making certain that you do not damage the weights or pins. Slide the new governor carrier bushing onto the carrier shaft and position the snap ring.

When you press the front and inner seals into the front cover, be sure that the lip of the front seal faces toward the drive coupling and the lip of the inner

seal faces toward the inside of the fuel pump. To prevent damaging the seals, use a seal-protecting sleeve, when pressing the drive-shaft assembly into the front cover. When it is in place, install the snap rings, and the drive-coupling key, and then press the drive coupling into position. During installation of the drive coupling, support the drive-shaft assembly with the governor drive gear. Heat the front cover in an oven. Coat the external area of the governor carrier bushing with a heat-resistant lubricant. Press the assembly (while meshing the gears) into its bore until the bushing fits firmly against the housing (Fig. 32–15).

If used, install the shims, spring, and weight assist plunger into the governor weight carrier shaft. Then measure the plunger protrusion, as shown in Fig. 32–16, or use a depth micrometer and subtract the weight-carrier-to-plunger dimension from the weight-carrier-to-housing-face dimension. Add or remove shims to achieve specified protrusion.

Gear-Pump Service Disassemble the gear pump by removing the six remaining Allen head screws. To remove the end cover from the pump body, screw two long screws into the body and drive it against the screws to separate the unit (the pump body and cover). Inspect the wear on the gears and shaft. Measure the gear width and compare it with the

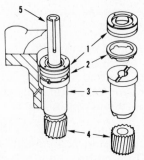

1. Drive seal
2. New drive-seal spacer
3. Drive bushing
4. Drive gear
5. Tachometer drive shaft

Fig. 32–14 Tachometer drive components. (*Cummins Engine Co., Inc.*)

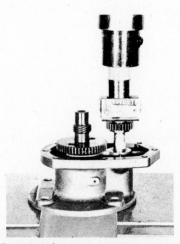

Fig. 32–15 Pressing the governor carrier bushing into the front cover. (*Cummins Engine Co., Inc.*)

specifications. When the gears or shafts are worn or scored, they must be replaced. If the gear body is damaged, or the wear marks pass 3 o'clock, or the cast-iron bearings are worn, the body must be replaced. If the end-cover surface is rough or damaged, it may have to be resurfaced by lapping it. The older gear pumps use needle bearings. When replacing these, press them out and press them in (do not "drive" them in). To remove the bearings from the body refer to the method outlined under Removing Antifriction Bearings (Unit 17).

NOTE When you press needle bearings into position, make sure that the end with the bearing identification is at the end on which the force must be placed, otherwise you could damage the needle bearings. If gears or body were lapped to restore proper running clearance, make sure that the gear is flush with the body surface. Measure the height of the gear pocket and the width of the gears. When you press new gears onto the shaft, heat the gear first and coat the shaft with a heat-resistant lubricant to prevent scoring the shaft. Make certain that the gears are positioned at the specified distance from the end of the shafts.

When you reassemble the gear pump, be sure to correctly position the notches that indicate pump rotation (see Fig. 32–17).

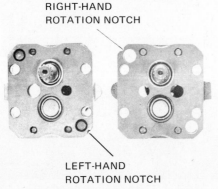

Fig. 32–16 Measuring weight-assist plunger protrusion. (*Cummins Engine Co., Inc.*)

RIGHT-HAND
ROTATION NOTCH

LEFT-HAND
ROTATION NOTCH

Fig. 32–17 Pump rotation notches. (*Cummins Engine Co., Inc.*)

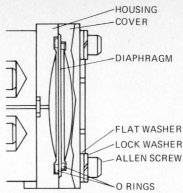

HOUSING
COVER
DIAPHRAGM
FLAT WASHER
LOCK WASHER
ALLEN SCREW
O RINGS

Fig. 32–18 Sectional view of the pulsation damper. (*Cummins Engine Co., Inc.*)

Lubricate the components and slide the shafts into the cover. Place a new 0.002-in [0.05-mm] gasket onto the cover and place the cover on the body. Align the holes and secure the pump assembly through its dowel before tightening the Allen head screws to the specified torque.

NOTE The pump must turn freely with finger force. If the fuel pump uses a cooling feature, install it at this point.

Pulsation Damper Remove the remaining Allen screws and separate the cover from the housing (Fig. 32–18). Clean the housing and the cover and, when necessary, lap their surfaces. Check the diaphragm. If it is bent, twisted or corroded, replace it. When reassembling the damper, coat the new O rings with grease and place them in the groove. Lay the diaphragm on the cover, then lay the housing on the cover. Do not forget to place flat washers between the housing and the lock washer before tightening the Allen screws. Then continue to tighten them until the lock washers are fully compressed.

Fuel-Pump Shutdown Valve Disassemble the valve by removing the four screws and the coil from the valve housing. Next remove the parts in the order shown from left to right in Fig. 32–19. Clean all parts except the coil with mineral spirit. Wipe the coil clean, and using an ohmmeter to measure the coil resistance, compare the measurement with the specifications. Inspect the valve and seat for wear, bonding failure, and corrosion. Replace all O rings. When you reassemble the valve, install a new O ring on the override shaft, lubricate it, and screw the shaft into the housing until it bottoms. Then set the depth micrometer to specification. Measure the distance from the face of the housing to the tip of the shaft. Adjust the shaft until its tip contacts the spindle of the micrometer. Press the knob onto the shaft until it is flush with the valve housing. Lubricate the housing O ring and place it in its groove. Place the spring washer on the valve with the concave side up, and with the washer piloted around the valve boss. Place the fuel shield on the valve housing. Tighten the screws to the specified torque.

Reassembling Fuel Pump Start your reassembly procedure by pressing the tachometer drive gear into

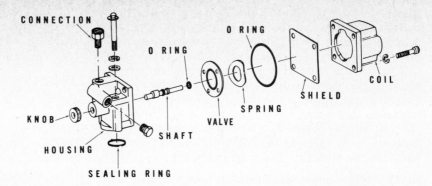

Fig. 32–19 Components of the fuel-pump shutdown valve. (*Cummins Engine Co., Inc.*)

its bore. Reassemble the other components in the order shown from bottom to top in Fig. 32–14. Install the gasket and tachometer drive cover to the fuel-pump housing. Next, install the filter screen. Position the seal and spring, and then tighten the cover to the specified torque. Select the specified idle spring plunger and high-idle-speed spring according to the fuel pump's code number. Assemble the parts as shown in Fig. 32–4. Install the spring-pack cover. Place a new gasket on the fuel-pump housing. Install the gear pump and pulsation damper. Tighten the Allen screws to the recommended torque. Install the fuel inlet and cooling fitting using a Teflon sealing tape. Be certain that you have the specified throttle shaft. When a restrictive plunger is used, install enough shims so that the plunger half-restricts the throttle opening. Next, tighten the plug to the recommended torque. Lubricate and install a new throttle shaft O ring using a protection sleeve. Slide the throttle plate over the shaft and insert the lubricated throttle shaft into the sleeve. Position the snap ring.

NOTE The counterbore of the throttle passage on PTG throttle shafts must face downward, whereas on PTR fuel throttle shafts they must face upward. If not done previously, install the throttle lever.

Lubricate the governor plunger, but make certain that it is numbered according to specification before sliding it into its barrel. Place a new gasket on the fuel-pump housing. Install the weight assist (shims, spring and plunger) into the carrier shaft. Lubricate the drive shaft and bushing. Rotate the weights and plunger drive tang to horizontal position. Now, carefully slide the front cover onto the fuel-pump housing while meshing the tachometer gears (Fig. 32–20). Engage the tang with the weight finger, then torque the mounting bolts to specification.

Calibrating PTG Fuel Pump Mount the fuel pump to the test stand according to the instructions of the test stand manufacturer. Make sure that you use the recommended test oil. Fill the gear pump and pump housing with test oil before connecting the fuel-pump inlet with hoses and before connecting the cooling bleed line to the test stand.

Remember to check that the nameplate properly describes the fuel pump to be calibrated. Select the calibration card. Adjust the throttle to the full-open position so that the port in the throttle shaft is indexed (in line) with the fuel passage in the throttle

Fig. 32–20 Installing the front-cover assembly. (*Cummins Engine Co., Inc.*)

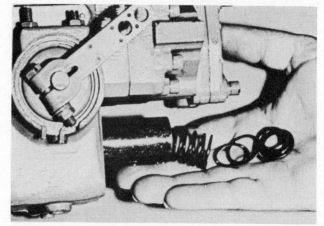

Fig. 32–21 Adding or removing shims to adjust cutout rpm. (*Cummins Engine Co., Inc.*)

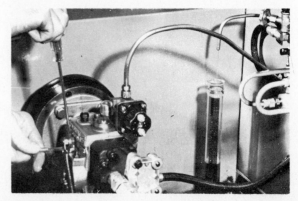

Fig. 32–22 Adjusting throttle leakage. (*Cummins Engine Co., Inc.*)

shaft sleeve. Manually turn the knob of the shut-down valve clockwise to fully open the valve.

Switch on the fuel heater. Open the main flow valve. Then close the idle and leakage valves. Select the correct pump rotation and operate the fuel pump between 800 and 1000 rpm. To prevent damage to the gear pump, make certain that it promptly pumps fuel. Operate the fuel pump at this speed until the test oil is at 90°F [32.2°C]. During this time check for external fuel leaks and for air bubbles in the flowmeter. Correct any leaks in the connections before the continuous warm-up period. Cummins Engine Company recommends adjusting the vacuum valves so that the gear pump draws 8 inHg [20.32 cmHg] during the warm-up period.

NOTE If you cannot obtain the specified vacuum, check that (1) the test stand filters are clean; (2) there are no restrictions in the inlet lines; (3) the main shaft seal, the throttle shaft seal, and the tachometer drive seal do not leak.

When the test fuel reaches the recommended temperature, operate the fuel pump at the rated rpm. Check rpm with a hand tachometer. Adjust the main flow valve until the correct (specified) flow for a pump under test registers on the flowmeter. (Refer to Cummins Engine calibration data sheet.) If the flow cannot be obtained, remove the shims from between the governor spring and retainer.

Setting Governor Cutout rpm To check governor cutout rpm, hold the throttle in the full-fuel position, open the fuel pressure valve, and then increase the pump rpm while observing the manifold pressure gauge. The rpm at which the pressure just begins to fall should coincide with the specified governor cutout rpm (see Cummins fuel-pump calibration values).

If the cutout rpm is lower than specified, add shims between the governor spring and retainer; if the cutout rpm is too high, remove shims (Fig. 32–21).

NOTE Before retesting the governor cutout rpm (after an adjustment), make certain that the flowmeter shows no air bubbles.

Adjusting Throttle Leakage (Limiting-Speed Governor) Operate the fuel pump at rated speed, move the throttle toward the gear pump and hold it firmly against its stop. Next, fully open the leakage valve and close the main flow control valve, then check the throttle leakage rate (in cubic centimeters per minute) indicated by the small flowmeter. **NOTE** Do not operate the pump under this condition any longer than is absolutely necessary.

The throttle-leakage flow rate must coincide with the Cummins fuel-pump calibration values. If the flow rate is not within specification, turn the front-throttle adjusting screw in clockwise or counterclockwise rotation to bring it within specification (Fig. 32–22).

NOTE This adjustment is very important because it affects the deceleration time of the engine. Do *not* forget to tighten the locknut.

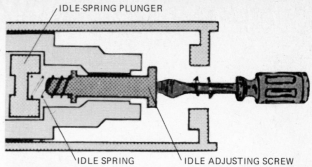

Fig. 32–23 Adjusting idle-speed pressure. *(Cummins Engine Co., Inc.)*

Idle-Speed Adjustment Stop the test stand motor. Close the main flow valve. Close the leakage valve, open the idle valve, and hold the throttle shaft firmly in idle position. Next, operate the fuel pump at the rpm recommended in the Cummins fuel-pump calibration values. The fuel-manifold pressure gauge should indicate the recommended pressure. If the pressure is too low, turn the idle adjusting screw in clockwise rotation to increase the fuel pressure. If the fuel pressure is too high, turn the idle adjusting screw in counterclockwise rotation to decrease the fuel pressure (Fig. 32-23).

Adjusting Fuel-Manifold Pressure with Internal Throttle-Shaft Plunger Place the throttle at the full-fuel position. Close the idling valve, open the main flow valve, and then operate the fuel pump at the rated speed indicated by the Cummins fuel-pump calibration values (under the heading "Manifold psi rpm"). Adjust the fuel flowmeter according to the fuel-flow calibration data and then check the pressure on the fuel-manifold pressure gauge.

NOTE Adjust the vacuum control valve so that the vacuum gauge reads 8 inHg [20.32 cmHg] during the test.

If the pressure is not within specification, stop the test-stand motor and remove the throttle shaft. Add shims to increase manifold fuel pressure or remove shims to decrease the pressure as required.

CAUTION After each adjustment, reset flow and vacuum control readings to the specified values before retesting.

If the throttle shaft has no restriction plunger, the manifold fuel pressure can be adjusted by turning the rear-throttle stop screw clockwise. This will reduce the opening of the throttle shaft and thereby reduce the fuel flow. The correct manifold psi and rpm is outlined in the Cummins fuel-pump calibration values.

Adjusting Final Fuel-Manifold Pressure with Rear-Throttle Stop Screw If the throttle shaft has a restriction plunger, operate the fuel pump at the speed indicated by the fuel-pump manifold-pressure data sheet. Then turn the rear-throttle stop screw clockwise until fuel pressure is reduced to the indicated specified value.

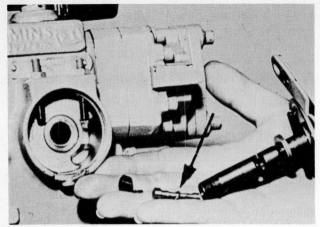

Fig. 32–24 Adding or removing shims to adjust manifold pressure. (*Cummins Engine Co., Inc.*)

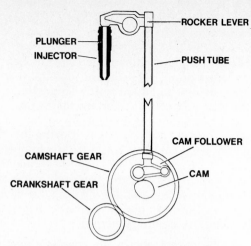

Fig. 32–25 Injector-activating mechanism. (*Cummins Engine Co., Inc.*)

Compare checkpoint pressure. Operate the pump at rated speed. Be sure that the fuel flow is set to specified value. Now, reduce speed to that specified under checkpoint 1 in the Cummins fuel-pump calibration values.

Check the manifold-pressure gauge reading. The check point pressure should be within the specified tolerance. If the readings are not within specification, check the torque spring to make sure it is properly seated. If a spring or shim is used, it must meet specification.

Weight-Assist Pressure Check If the fuel pump uses a weight assist, run the fuel pump at 800 rpm and check the fuel manifold pressure. It should be as specified in the calibration data. If the pressure is low, add shims below the governor weight-assist plunger. To decrease pressure remove shims (see Fig. 32–24).

NOTE If shims are added or removed behind the weight-assist plunger, the entire pump calibration must be rechecked.

Installing PTG Fuel-Pump to Engine Before you install the fuel pump, check the buffer or the spline coupling for wear or damage. Fill the gear pump and the pump housing with clean fuel. Then, using a new gasket, install the fuel pump to the mounting flange. Connect the inlet and return fuel lines to the fuel pump. Connect the fuel-supply line from the fuel-pump shutdown valve to the cylinder head (s). Connect the aneroid. Connect the electrical wire to the terminal of the fuel-pump shutdown valve.

Start the engine and bring it up to operating temperature. Then check the idle speed and, when necessary, adjust the idle-speed screw. Check the high-idle speed. You may have to add or remove shims to come within specification.

PT FUEL INJECTORS

Fuel injection on Cummins diesel engines is accomplished through a series of mechanical actions of the injector activating mechanism and the injec-

Fig. 32–26 Injector code number. (*Cummins Engine Co., Inc.*)

tor. The camshaft with its eccentric injector cam rotates at half the engine speed. The camshaft rotary motion is changed into reciprocating motion by the cam follower riding on the injector cam. This moves the push tube, causing the rocker lever to pivot and force the plunger down (see Fig. 32–25).

The PT injector is a simple but precision-built unit which meters, times, and injects fuel into the combustion chamber. The beginning of the injection is controlled by the camshaft, the quantity of metered fuel in the spray cup, and the injector adjustment.

The injector code number is located on the adapter (Fig. 32–26). The first three numbers in the code indicate the flow volume. The letter inscribed indicates the month and the number indicates the year in which the injector was manufactured. The next three digits identify the number of spray holes, the size of the holes (in thousandths of an inch), and the spray angle of the holes (in degrees). The inscription on the plunger coupling identifies its date of manufacture, the metering orifice size, and the plunger class size. For example, the inscription A 74 2 .027 would mean: January 1974, class size 2, and metering orifice 0.027-in [0.675 mm] (Fig. 32–27).

The spray cup also has on it the manufacturing date, the number of holes, the size and the degree of the holes.

Fig. 32–27 Identification on plunger coupling.

Fig. 32–29 Components of a PTD injector. (*Cummins Engine Co., Inc.*)

Fig. 32–28 Types of fuel injectors. (*Cummins Engine Co., Inc.*)

Basic Components and Operating Principles of PTB, PTC, and PTD Fuel Injectors These three types of injectors are similar in basic construction and operate according to the same metering injection principle (see Fig. 32–28). They are cylindrical and are positioned in the injector sleeve. The injector sleeve is located in the cylinder head and the injector is held by either a mounting plate or yoke. O rings separate and seal the inlet and drain (return) fuel manifolds: the lower manifold is the inlet, the upper manifold is the drain. Since the PTD injector has come into extensive use during the last few years, it will be examined in detail in this section.

Two types of PTD injectors are used: one with a 5/16-in [7.94-mm] plunger and one with a 3/8-in [9.53-mm] plunger. The PTD injectors are actually a modified type B and C injector which, through component design changes, provides more interchangeability among those components which are subject to wear.

The PTD injector consists of seven parts (Fig. 32–29); an adapter, a barrel, a plunger, a plunger link, a spring, a spray cup, and a spray-cup retainer. The type C injector, unlike the type D, is manufactured with the adapter or body and the barrel as one piece.

The new PTD Top Stop Injector, to the right in Fig. 32–30, provides a positive stop; that means that the plunger is limited in its upward travel resulting

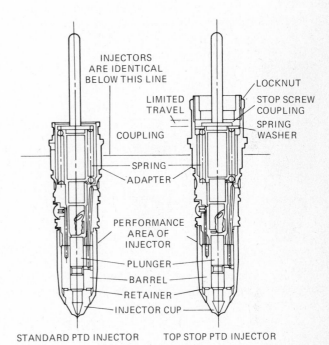

Fig. 32–30 Comparison of standard PTD injector and Top Stop PTD injector. (*Cummins Engine Co., Inc.*)

in better injector lubrication and less wear. In addition, the injector mechanism is not under strain (it is unloaded) and oil can flow into the balls, sockets, rocker arm, and follower areas.

PTD Adapter and Barrel The adapter has three external O-ring grooves by which the O rings seal and separate the inlet manifold from the drain manifold. It has an inlet opening (in which the balance orifice is threaded) and an outlet opening. Internal fuel passages connect the inlet and outlet ports with the fuel passages in the barrel and the manifolds. Two dowel pins (in the adapter) are used to position the barrel to the adapter. The barrel is short, allowing the plunger to be fitted more closely. It is made of hard-wearing resistant material, is lap-fitted to the plunger, and contains the ball check valve and the metering orifice.

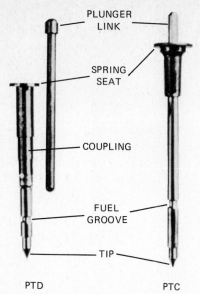

Fig. 32–31 Comparison of PTD and PTC plungers. (Cummins Engine Co., Inc.)

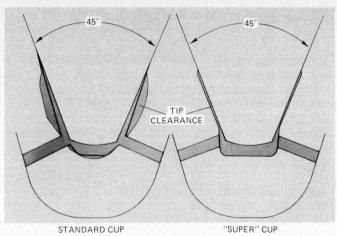

Fig. 32–33 Comparison of standard cup and "super" cup. (Cummins Engine Co., Inc.)

"drip" into the combustion area during metering is lessened, resulting in a cleaner burning engine.

Plunger Spring and Link The plunger spring which is positioned between the adapter and the coupling forces the plunger upward. The plunger link rests in the socket of the coupling and the rocker-arm adjusting screw. The long link lessens wear and reduces side thrust on the plunger and barrel because it can follow the arc of the rocker arm.

Comparison of PTC and PTD Injectors
1. The adapter and barrel of the PTC is one unit, whereas on the PTD they are separate units.
2. The PTC check valve is located in the top part of the injector body, but it is in the barrel of the PTD.
3. The PTC injector has four O rings and the PTD has three.
4. The metering orifice of the PTC is threaded into the lower end of the injector body and is interchangeable. The metering orifice of the PTD is fixed.
5. The PTC plunger is longer and is one piece whereas the PTD plunger is a two-piece unit.
6. The PTC retainer is shorter than the PTD and uses an O-ring seal.
7. The PTC plunger spring is longer but the link is shorter than the PTD.
8. The PTC differs from the PTD in that the diameter of its plunger is equal from the top to the tip, and the cup and the retainer are one piece.

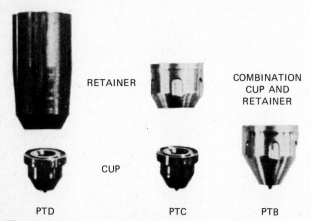

Fig. 32–32 Comparison of PTD, PTC, and PTB spray cups and retainers. (Cummins Engine Co., Inc.)

PTD Plunger The plunger is bonded to the coupling, and the tip diameter is narrowed and "fixed," regardless of plunger class (Fig. 32–31). The plunger diameter increases, but the seat area remains the same, so the amount of fuel metered does not change. The tapered plunger seat is lapped to the cup. To cool the injector during the combustion and exhaust stroke, a fuel groove is cut around the plunger which connects the inlet fuel to the drain passage when the plunger is seated against the cup.

Cup and Retainer The cup is lapped to the plunger seat. The spray holes assure good atomization and the spray-hole angle assures precise distribution of the droplets over the area required for combustion. The retainer nut holds the cup and the barrel to the adapter (Fig. 32–32). **NOTE** No seals or gaskets are used.

A newly designed "super" cup is shown in Fig. 32–33. You will notice from the illustration that the plunger tip clearance is reduced, leaving less fuel in the cup after injection. Therefore the chance of fuel

Fuel Flow and Metering through a PTD Injector A simplified schematic diagram of the PTD-injector fuel flow is shown in Fig. 32–34. The actual relationship between the camshaft actuating mechanism and the injector is shown in Fig. 32–35.

Late in the intake cycle of the compression stroke, the rotating injector cam lobe allows the follower roller to move from the outer base circle onto the retraction ramp and then onto the inner base circle. At the same time the injector spring lifts the plunger and pivots the rocker lever, forcing the pushrod and the follower roller down. The metering orifice is thereby uncovered. Fuel pressure controlled by the PTG fuel pump and fuel flow controlled by the throt-

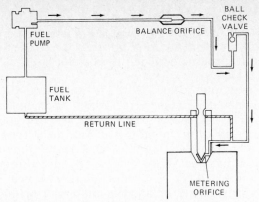

Fig. 32–34 Schematic diagram of PTD-injector fuel flow. (The shaded area shows the fuel flow when the plunger is seated.) (*Cummins Engine Co., Inc.*)

tle enter the injector adapter, flow through the balance orifice and around the ball check valve, through the metering orifice, into the cup. The follower roller remains on the inner base circle during the compression stroke to approximately 19° before TDC (Fig. 32–35*a*). At this point the injector ramp lifts the roller. This action lifts the follower and the push tube, pivots the rocker lever, and forces the plunger downward. The downward movement of the plunger closes the metering orifices and opens the drain outlet. Fuel under pressure which was metered by the throttle, balance orifice, and metering orifice is trapped in the cup (Fig. 32–35*b*). The amount of metered fuel which enters the cup depends on the time the metering orifice is open (which is proportional to engine speed).

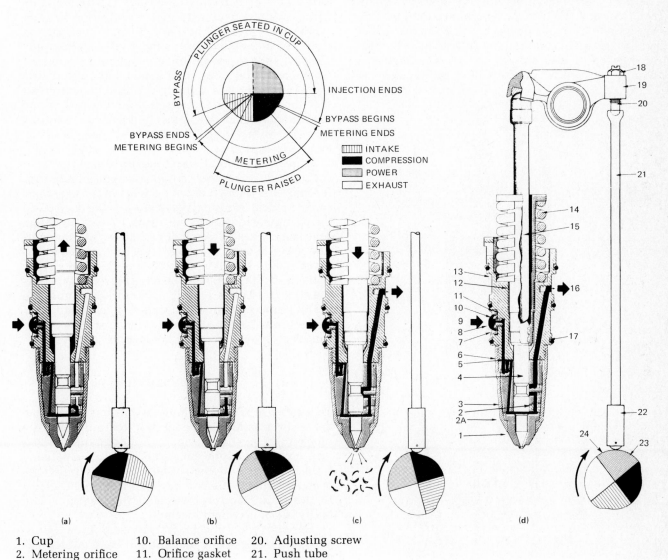

1. Cup	10. Balance orifice	20. Adjusting screw
2. Metering orifice	11. Orifice gasket	21. Push tube
2A. Cup retainer	12. Coupling	22. Cam follower
3. Barrel	13. Adapter	23. Cam lobe
4. Plunger	14. Spring	24. Injector ramp
5. Ball check valve	15. Link	
6. Gasket	16. Drain outlet	
7. Clip	17. O ring	
8. Screen	18. Nut	
9. Fuel in	19. Rocker lever	

Fig. 32–35 Relation of camshaft rotation during metering and injection. (*a*) Metering; (*b*) preinjection; (*c*) injection; (*d*) purging. (*Cummins Engine Co., Inc.*)

As the plunger moves downward the fuel pressure in the cup increases until fuel is forced through the spray holes into the combustion chamber (Fig. 32–35c). Injection ends when the plunger is seated against the cup, that is, when the cam follower roller reaches the nose of the outer base circle (Fig. 32–35d). The plunger remains on the cup seat during the power stroke, exhaust stroke, and the greater part of the intake stroke. The check ball is used to isolate hydraulic pulsation between the injectors and the inlet manifold.

Metering Function As illustrated in Fig. 32–35, the position of the cam determines when metering and injection should take place. The engine speed determines how long the metering orifice is open. The governor determines the fuel pressure in the manifold, and the throttle determines the fuel flow through the throttle's opening. The balance orifice size governs the amount of fuel that will enter the injector. It also provides a means for matching the flow for a set of injectors for a given engine so

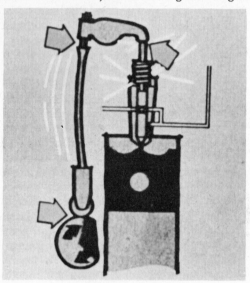

Fig. 32–36 Injector adjustment too tight. (*Cummins Engine Co., Inc.*)

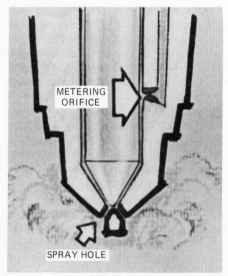

Fig. 32–37 Carbon and varnish buildup caused by loose adjustment. (*Cummins Engine Co., Inc.*)

that, regardless of any difference in the machining process or metering orifice, each injector will deliver a specified amount of fuel. The balance orifice also standardizes a single-model injector body for many different engines.

The metering orifice is the last point where the fuel is metered. The metering orifice, to change the fuel flow, is replaceable in the type C and the type B injectors, unlike the metering orifice in the type D barrel, which is fixed. In the type D barrel, the fuel flow for a given set of injectors is changed through the balance orifice. Different engines, however, may require larger or smaller metering orifices; therefore, the barrel and plunger must be changed to suit the engine's application.

Other Factors Affecting Metering There are other factors which affect metering and injection in addition to time, pressure, and orifice size. For instance, if the injector adjustment is incorrect, say too tight, the push tube exerts an abnormal amount of pressure on the entire injection mechanism and on the injector (Fig. 32–36). Of still more importance is the fact that the cam contacts the cam follower earlier and the plunger descends too low in the barrel, resulting in reduced metering time (reduced fuel per cycle) and early injection.

Loose adjustment has the opposite effect. It causes a loose plunger-to-cup seat, gives more time for metering (more fuel per cycle), and the cam must rotate several more degrees before it has any effect on plunger movement, resulting in late injection. The loosely seated plunger can allow hot combustion gases to enter the cup. This can cause a buildup of carbon which will reduce the size of the spray holes and the metering orifice (Fig. 32–37). Reduced spray-hole size will cause early ignition and will increase wear on the injection mechanism. Enlarged spray-hole size increases fuel droplet size and causes early injection and incomplete combustion.

INJECTOR SERVICE

To locate a misfiring (Cummins) injector, remove the valve cover and run the engine at the rpm where the misfiring is most noticeable. Hold one injector at a time on its seat to prevent it from metering and listen for a change of engine sound. If the cylinder has been misfiring, there will be *no* noticeable difference in the sound and operation of the engine. The faulty injector is then located.

To remove the injector for service, loosen the injector-adjusting-screw locknut and back out the adjusting screw. Push the rocker lever down (against the plunger spring) and swing the push tube from under the adjusting screw. Then tilt the rocker lever away from the injector. Remove the injector link, the injector hold-down bolts, and the hold-down clamp.

NOTE Different types of hold-down clamps are used for different engine models.

To remove an injector with the hold-down clamp shown in Fig. 32–38, thread a ⅜-in 17-hex screw into the threaded hole of the clamp and swing the clamp from its normal position. Turn the hex bolt down to lift the injector from its seat, and then lift the injector from its bore.

Disassembling PTD Injector Some service shops recommend that the injector be given a preliminary cleaning in a solvent and that then the injector be tested to determine the plunger-to-barrel leakage. Others suggest disassembling the injector without testing it.

When you disassemble the injector, remove the plunger and spring first, and then place them upright on a clean workbench. Next, remove the spring from the plunger. Remove the O rings from the adapter. Place the adapter in the special socket wrench and clamp the socket in a vise. Place an open-end wrench on the flat side of the adapter and loosen the retainer (Fig. 32–39). Hold the injector upside down, remove the retainer, and then remove the cup (Fig. 32–40). Remove the barrel from the adapter dowels, and then remove the check-valve ball from the barrel (Fig. 32–41).

NOTE It is usually necessary to lap both ends of the barrel, the spray-cup seat, and sometimes the adapter, before the components can be cleaned.

Clean all parts in a sonic cleaner (see Unit 47, "Shop Tools") or use a suitable cleaning tank. Some service shops additionally clean the spray cup by soaking it in Everite acid.

NOTE Do not interchange the plunger and barrel. Always slide the barrel onto the plunger immediately after they both are cleaned to minimize the possibility of a mixup of parts.

Inspecting PTD Injector

PLUNGER SPRING AND LINK Check the plunger spring for damage and test its energy with a spring tester. Examine both ends of the plunger link and the plunger coupling for wear or damage. Replace the pushrod if either end is worn.

PLUNGER Check the plunger-to-coupling bond for cracks and tightness of fit, as shown in Fig. 32–42. If the plunger cannot be twisted in the top by hand,

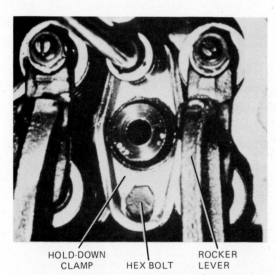

Fig. 32–38 Injector hold-down clamp. *(Cummins Engine Co., Inc.)*

Fig. 32–40 Removing the cup by hand. *(Cummins Engine Co., Inc.)*

Fig. 32–39 Loosening the cup retainer. *(Cummins Engine Co., Inc.)*

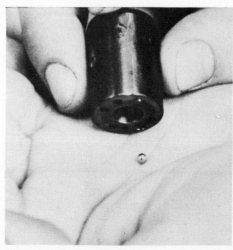

Fig. 32–41 Removing the check-valve ball from the barrel. *(Cummins Engine Co., Inc.)*

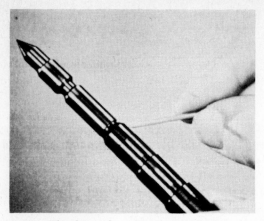

Fig. 32–42 Checking the plunger-to-coupling bond for cracks and tightness. *(Cummins Engine Co., Inc.)*

the unit is acceptable. When there is noticeable wear or fritting on the spring-contact area of the coupling flange, the plunger and the barrel must be replaced.

Check the plunger closely for scuffed or scored surfaces. (A shiny surface is acceptable at the top of the lapped plunger area and on the opposite side at the bottom of the plunger because this is a normal result of rocker-lever thrust.) A plunger can be reused if no surface damage is visible and if the plunger seat continuously covers 40 percent of the surrounding seating area. However, the injector must pass the injector leakage test.

BARREL Check the barrel surface at both ends for flatness. With a strong magnifying glass, check the fuel passages and orifices for damage and wear. If the metering orifice is damaged, the plunger and barrel must be replaced. Shine a light into the barrel and check for evidence of metal seizure or unusual wear. Do not neglect to check the check-valve ball seat and plugs for tightness. When necessary, lap the check-valve ball seat.

SPRAY CUP If the cup spray holes are damaged, elongated, or eroded or if there is evidence of excessive heat, the cup must be replaced. It must also be replaced if the cup-seat pattern is uneven, or if it is damaged due to carbon buildup (Fig. 32–43).

ADAPTER AND RETAINER Visually inspect the adapter at the O-ring area for nicks and burrs and for damaged retainer threads. Remove the balance orifice retainer clip and screen to check the balance orifice for damage or restriction (see Fig. 32–35). Make certain both fuel passages are clean.

Check the internal threads of the retainer and the cup seating flange. They must be free from nicks and burrs. Check the outside cone area. It must be damage-free. (Imperfect cone-seat area can cause improper injector-to-sleeve sealing.)

Reassembling PTD Injector After all parts have been inspected, make certain the new replacement parts (or the old ones) correspond with the injector code number and that all parts are perfectly clean.

Start your reassembly by placing the ball check valve in its barrel seat and then place a new gasket

on the adapter. Make certain that all holes match. **NOTE** Only the $^5/_{16}$-in [7.94-mm] injectors require a gasket.

Position the adapter on the barrel. Align the dowels with the holes in the barrel and guide them into place. Take care not to damage the gasket (Fig. 32–44). Place the cup on the barrel and screw the well-lubricated retainer finger-tight to the adapter, and then loosen the retainer about one-quarter turn. Immerse the injector plunger in test oil and slide it into the barrel (without the spring). Install the assembly into the holding fixture as shown in Fig. 32–45. Torque the adjusting stud to specification. This method of injector installation aligns the cup with the plunger. Use a crowfoot wrench, installed to the torque wrench, to tighten the retainer to the specified torque.

To check plunger-to-cup alignment, remove the plunger and dip it in clean fuel. Hold the injector vertically and allow the plunger to drip a few drops of fuel into the barrel. Insert the plunger about ½ in [12.7 mm] into the barrel, and then quickly force the plunger firmly against the cup seat (Fig. 32–46). Turn the injector upside down (cup up). The plunger should be free enough to drop out due to its own weight or when the injector is lifted quickly. If the plunger does not slide out, repeat the reassembly procedure.

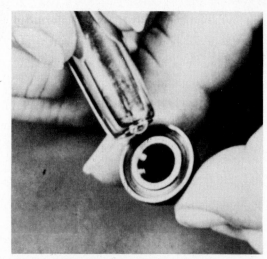

Fig. 32–43 Checking the spray-cup orifice and seating area using a flashlight. *(Cummins Engine Co., Inc.)*

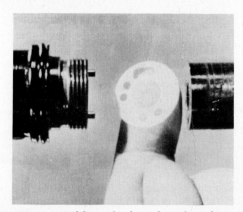

Fig. 32–44 Assembling the barrel to the adapter. *(Cummins Engine Co., Inc.)*

After sliding out the plunger, install the plunger spring and the new O rings into the proper grooves in the adapter. Install a new fuel-inlet screen and retainer clip.

Plunger-to-Barrel Leakage Test   Install the injector in the manner recommended by the Cummins shop manual to test plunger-to-barrel leakage. For example, install the injector into the adapter so that you can see the balance-orifice alignment through the burnishing tool hole. Then install and tighten the locating screw. Install and tighten the plug into the inlet of the adapter pot. Connect the pressure line to the drain passage of the adapter pot.

Select the specified spacer and place it on the adapter. Position the extension tool on the plunger coupling. Then place the assembly in the leakage tester according to the shop manual to check the barrel-to-plunger leakage.

Check-Valve Leakage Test   If the plunger-to-barrel leakage is within specification, the ball check valve must then be checked for leakage. To make this test, connect the pressure line to the drain passage of the adapter pot. Remove the plug from the inlet, and then hold the assembly vertically, and press the

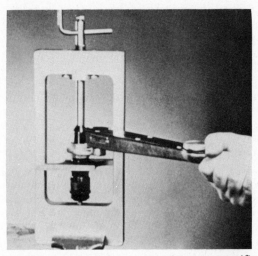

Fig. 32–45   Torquing the cup retainer to specification after the injector is installed to the holding fixture. *(Cummins Engine Co., Inc.)*

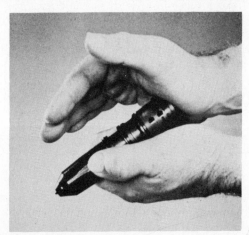

Fig. 32–46   Preparing and making a plunger drop test. *(Cummins Engine Co., Inc.)*

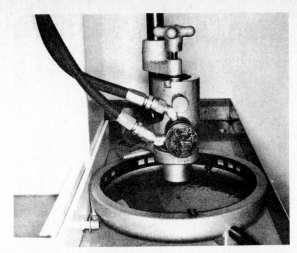

Fig. 32–47   Checking the spray pattern on an injector spray tester. *(Cummins Engine Co., Inc.)*

plunger firmly onto its seat. Apply about 150 psi [10.5 kg/cm²] pressure to the injector. At this pressure no fuel should leak from the balance orifice.

Spray-Pattern Test   When an injector has passed each of the above tests, remove the plunger and spring from the injector and plug the plunger bore with the correct size plug. (Make certain the plunger is protected from dirt or damage when removed.)

Select the correct seat spacer, according to the cup and specified seat, to obtain the correct spray angle. Then install the injector to the spray-pattern tester (Fig. 32–47). Next, select a target ring that corresponds with the spray holes of the cup. Pressurize the injector according to recommendations in the shop manual since this pressure varies with orifice size. Rotate the target ring so that one injection stream obviously hits the center of the No. 1 (enlarged) window. Closely check each of the other injection streams. Their spray must be confined to the inside of the window.

NOTE   The spray pattern (the cup) is acceptable if no more than one injection stream exceeds its target by the area of the No. 1 (enlarged) window.

After the spray-pattern test has been made, remove the injector from the tester. Make certain the plunger is clean before you slide the plunger (with its spring) into the injector body.

Calibrating and Testing PTD Injector   Before you install the injector to the calibrating test stand (Fig. 32–48), make certain that the test fuel meets the Cummins's specification, that the fuel level is as specified, and that the correct cam is installed.

Calibrating Test Stand   Index the timing marks to bring the cam in TDC position. Make sure that adequate air is supplied to the test stand. Place the load cell in the test stand (open the air valve to clamp it in place).

NOTE   Check the shop manual to determine if an orifice is required in the nose piece.

Fig. 32–48 Calibrating test stand.

Adjust the air pressure until the load-cell dial is within the marked black band, that is, 380 psi + 10 psi [26.6 kg/cm²], then lock the adjusting screw with the locknut.

Calibrating Test Stand with Meter Injector Select the recommended test-stand adapter and position the master injector. Lock it in place with the set screw. Fit the specified orifice into the nose piece. Select the specified stem extension and place it on the injector link. Open the air valve to clamp the injector in place. Make certain the injector link and the extension are positioned properly before you switch on the test-stand motor. Shortly after the motor is switched on, you should hear the injection and see fuel on the drain line. At this time, adjust the fuel pressure to 120 psi [8.44 kg/cm²] and set the counter to 1000 strokes. When the test fuel temperature is about 90°F [32.2°C], press the counter button to record the fuel delivery. Read the fuel delivery. Then dump the fuel from the vial and repeat the test several times to assure constant delivery. If, for example, the delivery of the master injector is 2 cm³ less than specified [132 cm³], do not change the fuel pressure. Instead, increase the counter setting to come within the fuel delivery. A change of seven to eight strokes on the counter setting varies the delivery by 1 cm³.

Repeat the test and adjust the counter as necessary to obtain 132-cm³ delivery at a pressure of 120 psi [8.44 kg/cm²]. Make three delivery tests. Reclamp the master injector, and then make three more delivery tests to assure maximum accuracy of test-stand adjustment.

Flow-testing PTD Injector Flow-testing an injector means to compare its performance and make any

necessary adjustments to reconcile the performance with that of the master injector. Master injectors are precisely calibrated on Hartridge equipment at the factory, by the Cummins Engine Company. Do not misuse your master injector.

When flow-testing a PTD injector, select the correct adapter and stem extension. Fit the specified orifice into the test-stand nose piece. Install and clamp the injector to the test stand. Mount the burnishing tool to the carrier block. Run through a test cycle and check the delivery against the test specification. If the delivery is too high, install a smaller balance orifice. If the delivery is too low, turn the burnishing tool knob until you can feel the needle enter the orifice. Lock the needle shaft to the large knob by turning the small knob counterclockwise, then turn the large knob clockwise until you feel positive contact. Continue turning the large knob clockwise according to the amount to be burnished (refer to your shop manual). (See Fig. 32–49.)

NOTE Take extreme care during the first burnishing operation because the initial reading may be false due to foreign matter or slight burrs in the balance orifice. Then remove the burnishing tool and run through a test cycle to check your fuel delivery.

Installing Injector and Making Injector Adjustment
Before you install the injector (in an order reverse to that in which it was removed), make certain that the injector sleeve is clean and that the O rings are well lubricated. When the injector hold-down bolts are torqued to specification, turn the crankshaft until the valve set mark correctly aligns with the pointer.

Tighten the injector adjusting screw until it bottoms in the push-tube socket, but make certain the adjusting screw turns freely. Tighten the injector adjusting screw about 10 psi [1.24 kg/m²] above the specified torque to force out any fuel which may be present in the cup. Loosen the adjusting screw and retorque it to specification. Then, while holding the adjusting screw, tighten the locknut (Fig. 32–50).

Injector Adjustment with Dial-Indicator Method
A far more efficient and more effective method of adjusting the injectors is with a dial indicator. This method prevents torque errors due to thread friction and thread looseness because the injector plunger travel is premeasured. It also provides uniform beginning of injection, improved fuel economy, better emission control and, due to improved plunger seating, longer injector life.

Checking Injector Plunger Free Travel Adjusting the injector by the dial-indicator method is only recommended when the engine has an instructional decal or when the free plunger travel is not greater than 0.205 in [5.20 mm]. Therefore, the plunger free travel of all injectors must be measured first. To do this, install the dial indicator with the dial-indicator extension resting on the top of the plunger coupling.

Position the dial indicator so that it is at the middle of its total travel, as shown in Fig. 32–51. Loosen the injector-adjusting-screw locknut. Turn the adjusting screw 1½ turns counterclockwise and re-

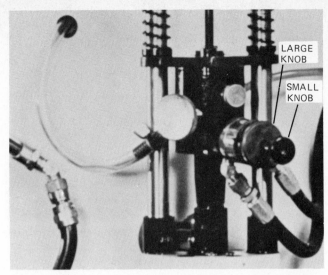

Fig. 32–49 Burnishing tool (installed). *(Cummins Engine Co., Inc.)*

Fig. 32–50 Tightening the locknut while holding the adjusting screw. *(Cummins Engine Co., Inc.)*

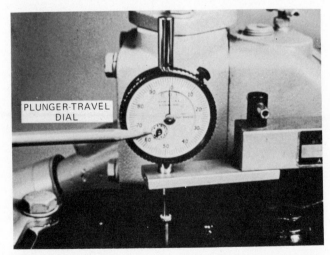

Fig. 32–51 Setting up the dial indicator to measure plunger free travel or to check injector adjustment. *(Cummins Engine Co., Inc.)*

tighten the locknut. Next, turn the crankshaft and note when the dial indicator reaches its lowest and highest readings. The difference between these readings equals the total plunger travel. Check all injector plunger free travel. When all are within

Table 32–1 INJECTOR AND VALVE SET POSITION

Bar direction	In pulley position	Set injector	Cylinder valve
Start	A or 1-6 CS	3	5
Adv. to	B or 2-5 VS	6	3
Adv. to	C or 3-4 VS	2	6
Adv. to	A or 1-6 VS	4	2
Adv. to	B or 2-5 VS	1	4
Adv. to	C or 3-4 VS	5	1

specification, use the dial-indicator method to adjust the injectors.

Injector Adjustment To check the injector adjustment, position the dial indicator as first explained. Then turn the crankshaft until one of the timing marks aligns with the pointer on the cylinder block. Table 32–1 shows the old and new pulley adjustment marks and identifies which injector and valve can be adjusted when in the "A", "B", or "C" position. But remember, when the engine is at the valve set mark "A" (or "1–6" on an old engine), then Nos. 1 and 6 cylinders are 90° ATDC; Nos. 5 and 2 cylinders are 30° BTDC; and Nos. 3 and 4 cylinders are 150° BTDC. However, only one cylinder (either No. 5 or 2) is on the compression stroke, and only one injector (either No. 3 or 4 is metering. When No. 5 piston is on the compression stroke, both valves (intake and exhaust) are closed and the injector of No. 3 cylinder is metering because No. 3 cylinder follows No. 5 cylinder in the firing order.

Position the dial indicator as explained above, under Checking Injector Plunger Free Travel. Next, place the rocker-lever actuating tool on the rocker lever and, with a box-end wrench, rotate the tool to force the plunger onto its cup. This action squeezes the fuel from the cup.

Repeat this action several times but be careful when you release the force from the rocker lever not to damage the dial indicator. Again force the plunger onto its seat and hold it there while zeroing the dial indicator. (Check the zero setting several times.)

Slowly allow the plunger to rise, and read the plunger travel on the dial. When the plunger travel is not within specification, loosen the locknut and turn the adjusting screw until the specified reading is obtained. Then tighten the locknut to the specified torque.

NOTE The locknut must be torqued to specification before you take the dial reading. The plunger travel must always be rechecked after an adjustment is made.

An important advantage of the dial-indicator method over the torque method is that it does not place a strain on the rocker-lever train and thus makes it possible for the valves to be adjusted more accurately.

Injector Timing It is well known that precise injection timing is essential to assure rated power and proper engine performance. Injector timing is necessary after any service work has been done on the timing gears, camshaft, or followers or when the push tubes have been replaced. This is necessary because the relation between the push-tube travel and the piston travel may have changed.

To obtain precise injector timing, you must measure both the push tube and the piston travel. The timing is changed by changing the camshaft key or by altering the thickness of the cam-follower housing which changes the relation between the cam lobe and followers.

To check injector timing, remove the rocker housing and the injector specified in the service manual. Install the piston-timing tool to the injector sleeve with the dial-indicator extension resting on the piston. Install the injector-timing tool so that the extension of the dial indicator rests in the injector push tube. Tighten the timing tool, using the injector mounting bolt hole and the special holding fixture (Fig. 32–52). Then turn the crankshaft in the direction of engine rotation. Bring the piston with the installed injector-timing tool to TDC on the compression stroke, if not previously done. At the point of maximum piston travel, zero the dial (above the piston). Continue to turn the crankshaft to the 90° ATDC position, and zero the dial indicator above the push tube (Fig. 32–53). Then turn the crankshaft 135° in the opposite direction, that is, to 45° BTDC. Again, turn the crankshaft slowly in the direction of engine rotation until the dial indicator above the piston reads specified piston travel (Fig. 32–54).

NOTE Both indicators should move in the same direction when the piston is on its proper stroke.

Assume you have to check the timing on a KT (A)1150 engine. The specification for this engine are as follows:

- Piston travel: 0.2032 in [5.156 mm]
- Push-tube travel: 0.108 in ± 0.002 in [2.743 mm ± 0.050 mm]

When the dial indicator above the piston reads 0.0032 in [0.0812 mm] before zero, this is actually 0.2032 in BTDC, since the indicator hand has gone around twice as the crankshaft was moved to 45° BTDC. In this position read the dial indicator above the push tube. It must be 0.108 in ± 0.002 in [2.743 mm ± 0.050 mm]. If it is not, the camshaft key must be changed in accordance with the shop manual directions.

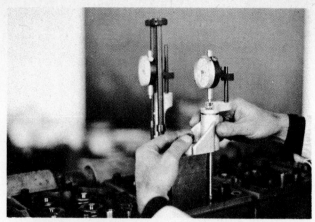

Fig. 32–52 Installing the injector timing tool. (*Cummins Engine Co., Inc.*)

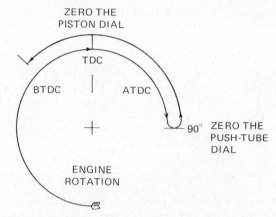

Fig. 32–53 Crankshaft rotation and position for setting the dial indicator to zero. (*Cummins Engine Co., Inc.*)

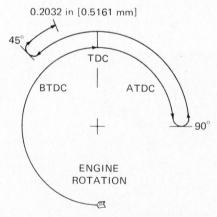

Fig. 32–54 Crankshaft rotation and position for checking injector timing. (*Cummins Engine Co., Inc.*)

Questions

1. List the major differences between the PT fuel-injection system and a helix-metering system in regard to their injection principles and their metering principles.

2. List the components which limit the fuel pressure in a PTG fuel pump.

3. Which component controls the engine speed between idle and high idle?

4. Explain how the governor components limit the high idle and the low idle.

5. How would engine performance be affected if the throttle shaft were worn or if the throttle shaft O ring allowed air into the fuel pump?

6. How would engine performance be affected if the gear pump were worn?

7. Which components must be adjusted to alter the cutout rpm? (Assume the fuel pump is on the test stand.)

8. If the throttle leakage were lower than specified, how would engine performance be affected?

9. Explain how to adjust throttle leakage.

10. When decreasing manifold pressure by less than 10 psi, why should you *not* use the rear throttle screw?

11. List the major differences between a PTD injector and an American Bosch injector.

12. List the differences between a PTD injector and a PTC injector. What advantages, if any, has the PTD injector over the PTC injector?

13. Outline the fuel flow through a PTD injector when metering and after injection has ended.

14. What four factors are involved in governing the quantity of fuel which enters the spray cup?

15. Which components in the PTD injector are used to change the fuel metering?

16. Outline in detail the procedure used to install an injector.

17. Outline how to adjust an injector using the torque method and using the dial-indicator method.

18. When using the dial-indicator method to adjust the injector (say of No. 1 cylinder), in what position is the No. 1 piston?

19. If an injector had been adjusted too loosely, how would engine performance be affected and how would the injector unit be affected?

UNIT 33

DPA and Roosa Master Distributor-Type Fuel-Injection Pumps

The Stanadyne Hartford Division (Roosa Master) and the CAV Limited, London (DPA) distributor-type injection pumps are described as "opposed plunger, inlet-metering distributor-type pumps." This type of injection pump uses only one metering valve to meter the fuel, either one or two opposed plungers to pump the fuel, and only one component, the distributor rotor, to distribute the metered fuel into the hydraulic head. The distributor pumps are of the variable beginning and constant-ending-type delivery. Although the distributor-type injection pumps of these two companies operate on the same principle, there are minor differences in the construction as well as in the governors and accessories.

The nameplates on the DPA and the Roosa Master distributor-type injection pumps, as with other pumps, contain abbreviated information required for servicing. For example, a Roosa Master DM nameplate may be designated DM 2 6 33 JN 2580. These letters and figures correlate with the pump series, plungers, cylinders, etc., as follows:

```
DM = D series pump
 2 = Number of plungers
 6 = Number of cylinders
33 = Abbreviation of plunger size (0.330 in
     [8.38 mm])
JN = Accessory code
2580 = Specification number
```

Components and Their Function Roosa Master and DPA distributor-type injection pumps are shown in Figs. 33–1 and 33–2. Although the injection pumps differ externally, a careful comparison of the pictures will reveal that the components responsible

for metering and distributing fuel, and for fuel and speed control, are similar in design (DPA).

Both injection pumps use an aluminium-alloy housing which contains the following major operating components: drive shaft, distributor rotor, transfer-pump blades, pumping plungers, internal cam ring, hydraulic head, pressure-regulator assembly, governor, automatic advance, and metering valve.

DRIVE SHAFT The drive shaft engages with the distributor rotor through a tang or through splines. It engages with the governor weight (flyweight) retainer and with the transfer-pump blade through the slots in the rear of the rotor. The shaft is supported by a bushing or a ball bearing.

The oil seals of the illustrated models vary in design as well as in their location. The lip of the forward seal faces forward to prevent engine oil from entering the pump. The rear-seal lip faces the pump and retains fuel.

GOVERNOR The inner flange of the weight retainer is splined through a tang slot, fastened to the rotor but the drive-to-governor weight retainer goes through a cushioning drive to reduce pulsation. (Cushioning drives vary from company to company and according to the models produced by the same company.)

The weights are slide-fitted in the weight-retainer pocket (Fig. 33–3). They are shaped to pivot on the outer edge. Each weight fits with its machined-out groove in the thrust sleeve.

One end of the governor control arm rests against the thrust sleeve and the other end is connected to

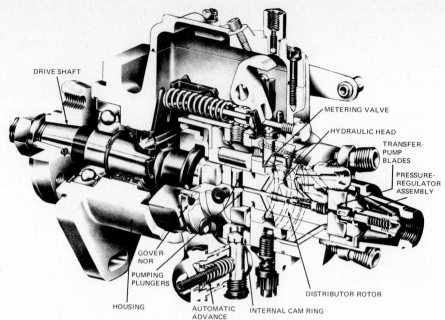

DRIVE SHAFT

METERING VALVE

HYDRAULIC HEAD

TRANSFER-PUMP BLADES

PRESSURE-REGULATOR ASSEMBLY

GOVERNOR

PUMPING PLUNGERS

DISTRIBUTOR ROTOR

HOUSING

AUTOMATIC ADVANCE

INTERNAL CAM RING

Fig. 33–1 Main components of a Roosa Master DM fuel-injection pump. (*Roosa Master Stanadyne/Hartford Div.*)

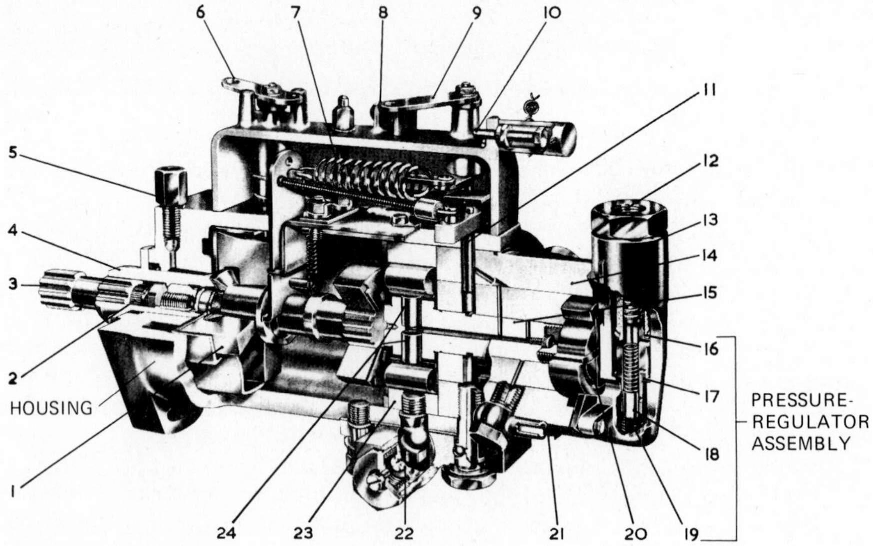

1. Governor weights
2. Drive-hub securing screw
3. Drive shaft
4. Drive hub
5. Back lead connection
6. Shutoff lever
7. Governor spring
8. Idle stop
9. Control lever
10. Maximum-speed stop
11. Metering valve
12. Fuel inlet
13. End-plate assembly
14. Hydraulic head
15. Distributor rotor
16. Nylon filter
17. Regulating valve sleeve
18. Regulating piston
19. Priming spring
20. Transfer pump
21. To injector
22. Automatic advance   23. Internal cam ring   24. Pumping plungers

HOUSING

PRESSURE-REGULATOR ASSEMBLY

Fig. 33–2 Main components of a DPA fuel-injection pump. (*CAV Ltd.*)

the governor spring and to the metering valve through a linkage hook. The control lever is connected to the shutoff lever and the fulcrum lever is connected to the governor spring.

DISTRIBUTOR ROTOR The distributor rotor is lap-fitted to the hydraulic head. On the drive end are two or four pumping plungers which move in opposite directions within a transverse bore. The plunger rests against the guide shoes in which the cam roller rotates (see Fig. 33–4). The cam ring fits over the cam rollers.

The outward movement of the roller shoes is limited by a leaf spring or by the adjusting plates. A radial passage (or passages) in the rotor connects the pumping chamber with the radial hole (or with a number of holes equal to the number of cylinders) to

the charging annulus and ports. A center-drilled passage from the pumping chamber leads to the discharge port. Some rotors have a delivery valve within this passage.

On the other end of the distributor rotor, rotor slots are machined for the transfer-pump blade of the positive displacement vane-type pump. On some DPA pumps the transfer-pump rotor is threaded into the distributor rotor.

HYDRAULIC HEAD Passages within the hydraulic head connect the transfer pump with the metering valve, the metering valve with the charging ports, the discharging ports with the discharge fittings, and the metering valve with the automatic advance (or the transfer pump with the advance) (see Figs. 33–1 and 33–2).

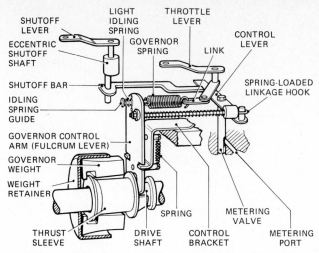

Fig. 33–3 Governor components of a DPA fuel-injection pump. *(CAV Ltd.)*

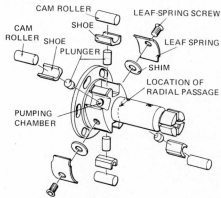

Fig. 33–4 Rotor assembly (Roosa Master DM4). *(Roosa Master Stanadyne/Hartford Div.)*

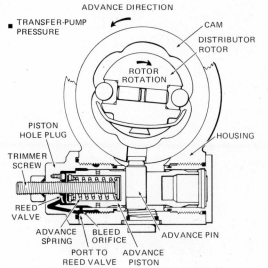

Fig. 33–5 Sectional view of the automatic advance. (Roosa Master). *(Roosa Master Stanadyne/Hartford Div.)*

END PLATE The end plate, which absorbs the end thrust and is the housing for the pressure regulator, covers the rear of the rotor and the hydraulic head. (Fig. 33–2). Its design and the design of the components of the pressure regulator vary according to the company and according to the model.

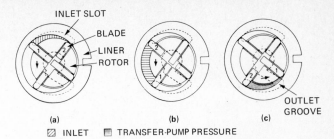

Fig. 33–6 Schematic illustration of transfer-pump operation (model DM). *(a)* Beginning of charging; *(b)* maximum charge; *(c)* discharging. *(Roosa Master Stanadyne/Hartford Div.)*

AUTOMATIC ADVANCE Although the components of the automatic advance vary in design, all models use fuel pressure to move the piston and the advance lever, and thereby the cam ring (Fig. 33–5). Fuel to the advance piston comes from the metering valve or directly from the transfer pump.

Fuel Flow (Tank and Transfer Pump) When the drive shaft rotates, the pair of rotor blades carried in the slot of the rotor rotate within the liner (Fig. 33–6a). The liner is eccentric from the axis of the rotor and the rotation of the blade increases the volume within the liner, creating a low pressure (Fig. 33–6b). Atmospheric pressure forces fuel from the tank through the filters, through the inlet filter screen, around the pressure-regulator assembly to the inlet side of the transfer pump. As the blade retracts, the volume between the blades and the liner is reduced (Fig. 33–6c). Fuel is then forced into the rotary retainer to the regulator and into the annulus in the hydraulic head.

The fuel flow of the Roosa Master (DM model) differs slightly. Fuel flows through the groove of the regulator assembly, past the rotor retainer and into an annulus on the rotor, leading to the annulus in the hydraulic head. This fuel-flow arrangement thoroughly lubricates the rotor and prevents it from seizing.

Pressure-Regulator Operation The pressure regulator controls the fuel pressure by maintaining a definite relationship between the fuel pressure and the engine speed. Some pressure regulators include compensation for viscosity changes.

The components of the pressure regulator vary between the different pump models, but their function is the same, that is, to control maximum pressure. As the pump speed increases, more fuel is pumped and the pressure increases progressively until the piston uncovers the regulator slot (Fig. 33–7). Fuel is directed through the slot to the inlet side of the transfer pump. Should the pressure rise beyond the capacity of the slot to reduce it, the increased pressure moves the piston still further, uncovering the high-pressure relief slot (Roosa Master DM).

Charging and Metering (Model DM) When the engine is being cranked over or when it is running, some of the fuel at transfer-pump pressure passes into an annulus behind the transfer-pump rotor and

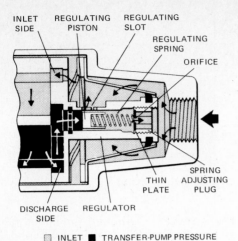

Fig. 33-7 DM pressure-regulator components and fuel flow. *(Roosa Master Stanadyne/Hartford Div.)*

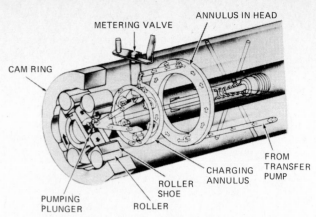

Fig. 33-9 DC metering (or charging) fuel flow. *(Roosa Master Stanadyne/Hartford Div.)*

through a connecting passage to the head locating screw and to the automatic advance (Fig. 33-8). The remainder of the fuel passes into the annulus in the hydraulic head.

The outlet side of the transfer pump is connected by a restricted air-vent passage to the pump housing. This air-vent passage serves two purposes: (1) it fills the injection-pump housing with fuel which is used to lubricate and to cool the injection-pump components (excessive fuel is returned to the fuel tank via the return line), and (2) when air has entered the transfer pump, it is carried off with the fuel and does not enter the annulus in the hydraulic head.

The fuel flow of a DC injection pump is slightly different. In that model fuel at transfer-pump pressure leaves the outlet port but leads through a passage in the bottom of the hydraulic head and proceeds forward to the annulus (Fig. 33-9).

NOTE The transfer-pump rotor of a DC pump is not as well lubricated as a transfer-pump rotor in a DM injection pump.

Fuel flow from the annulus is very similar in all distributor-type injection pumps because a drilled passage on the top of the annulus connects the fuel with the metering valve. Engine requirements and governor action then controls the amount of fuel that

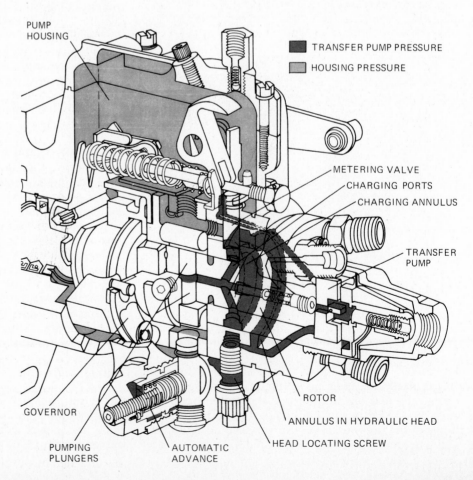

Fig. 33-8 DM metering (or charging) fuel flow. *(Roosa Master Stanadyne/Hartford Div.)*

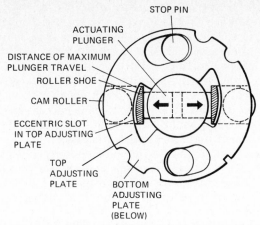

Fig. 33–10 Maximum-fuel adjustment components in the rotor assembly. *(CAV Ltd.)*

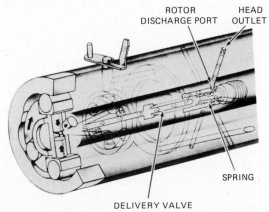

Fig. 33–11 Discharging (injection) fuel flow. *(International Harvester Co.)*

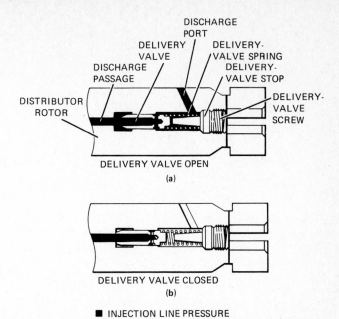

Fig. 33–12 Delivery-valve position *(a)* when open and *(b)* when closed. *(Roosa Master Stanadyne/Hartford Div.)*

flows past the metering valve and into the charging annulus. The number of charging ports in the annulus equals the number of cylinders. When the two charging ports in the rotor register with two of the charging ports in the charging annulus, metered fuel passes into the chamber of the pumping plunger. The outward movement of the plunger is governed by: (1) the opening of the metering valve, (2) the time available to charge the pumping chambers, and (3) the fuel pressure. However, the adjustable leaf spring(s) (Roosa Master) or the eccentric adjusting plates (DPA) limit the outward travel of the roller shoes and, therefore, the maximum separation of the plungers (see Fig. 33–10).

Fuel that passes into the slots or into the small grooves on the rotor (located on each side of one of the charging ports) lubricates the cam rollers and the governor components. It passes into the housing and, with fuel from the air vent, fills the entire pump lubricating the components within the pump and absorbing heat. It then returns through the return line to the fuel tank.

Discharging (Injection) Fuel Flow As the rotor continues to rotate, the charging ports of the rotor pass out of registry with the charging port in the hydraulic head. For a very short interval fuel is trapped in the chamber since the discharge port has not yet come into registry with the charging port in

the hydraulic head. Immediately before the cam rollers contact the cam lobes, the discharge ports come into registry with each other and, as the rollers force the plunger inward (toward each other), fuel is discharged from the pumping chamber (Fig. 33–11). Some injection pumps have a delivery valve in the discharge passage while others use a cam relief (cam retraction) to reduce the pressure in the discharge line for sharp cutoff of injection, to prevent secondary injection, and to prevent nozzle dribble. The cam retraction takes place just before the discharge port comes out of registry.

Delivery-Valve Action The delivery valve is a check valve with a displacement plunger. Its operating principle is the same as the check valve on the Caterpillar injection pump and it serves the same purpose. It pressurizes the delivery-valve spring chamber for a quicker beginning of injection (Fig. 33–12a) and quickly reduces the injection-line pressure when the roller passes the innermost point of the cam ring and begins to move outward, ending injection (Fig. 33–12b).

Residual injection-line pressure is achieved by the delivery valve or through the cam retraction, but it is maintained by the rotor as the discharge ports come out of registry.

Automatic-Advance Action Nearly all DPA and Roosa Master injection pumps use a simple automatic speed advance which provides control movement of the cam ring. In other words, it provides progressive or digressive retarding of injection as the engine speed increases or decreases the fuel pressure. (Some pumps use a combination load and speed advance.)

Fuel that passes by the second helix on the metering valve is directed to the head locating screw by the check valve and flows on to the pressure side of the advance piston (Fig. 33–13). The fuel pressure

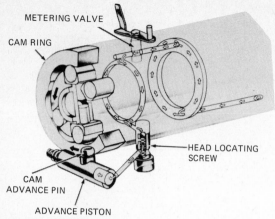

METERING VALVE

CAM RING

HEAD LOCATING SCREW

CAM ADVANCE PIN

ADVANCE PISTON

Fig. 33–13 Speed-advance operation. (The arrows show the fuel flow.) (International Harvester Co.)

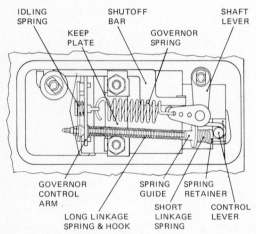

IDLING SPRING

SHUTOFF BAR

SHAFT LEVER

KEEP PLATE

GOVERNOR SPRING

GOVERNOR CONTROL ARM

SPRING GUIDE

SPRING RETAINER

CONTROL LEVER

LONG LINKAGE SPRING & HOOK

SHORT LINKAGE SPRING

Fig. 33–14 Top view of a DPA governor. (CAV Ltd.)

pushes the piston against the spring force and the dynamic injection loading on the cam ring in order to change the cam ring. The cam ring is moved in the opposite direction of the rotor rotation against the roller face. The check valve holds the fuel pressure in the chamber, preventing the cam ring from retarding. At low engine speed the fuel pressure is low, and the cam ring is held in the retard position by the spring and the roller force. As the metering valve allows increased fuel flow, the engine speed increases the fuel pressure, and the piston moves against the spring force. The maximum movement of the cam ring is limited by the length of the piston or the stop.

Roosa Master and DPA injection pumps use a trimmer screw or shims to change the preload on the advance spring and to change the beginning of the cam-ring movement.

DPA Mechanical Governor Action When the throttle lever moves to the full-fuel position, the governor spring fully compresses the idling spring, and the spring guide bottoms against the governor control arm (Fig. 33–14). The spring force pivots the control arm on the control bracket and forces the thrust sleeve to move, causing the governor weights to close. At the same time the governor control arm moves the spring-loaded linkage hook to the right

(in Fig. 33–14), rotating the control lever as well as the metering valve. Governor weight (engine speed) force and spring force determine the position of the governor control arm and thereby the metering-valve position. If the engine speed drops (due to added load), the flyweights move inward and the control arm moves the linkage hook, the control lever, and the metering valve. This increases fuel flow and maintains the selected engine speed. If the engine speed increases (due to decreased load), the governor weight moves outward and the governor control arm pivots in the opposite direction, moving the metering valve and reducing the fuel flow. The engine speed drops in response to the reduction in fuel flow.

When the throttle shaft is in the idle position, the tension from the governor spring is removed and the light idle spring assures sensitive idle-speed control because of the low weight force. The energy of the idling spring is set through the low-idle-speed adjusting screw. The energy of the governor spring is set by the high-idle adjusting screw.

To stop the engine, you must move the shutoff lever to the no-fuel position. The rotation of the shutoff lever moves the shutoff bar and the control lever. The control lever rotates the metering valve and closes off the fuel flow.

DPA Hydraulic Governor Action The components of a DPA hydraulic governor are shown in Fig. 33–15. The metering valve is lap-fitted to the governor housing and is center-and cross-drilled. The cross-drilled hole leads to an annulus which is in line with the metering port. The up-and-down movement of the valve varies the area of the metering port and, therefore, the fuel flow to the pumping chamber. The rack or the control sleeve fits loosely over the extended stem of the metering valve.

When the governor is assembled as shown in Fig. 33–16), the shutoff nut and washer pressurize the idle and governor springs, and thereby position the rack or the sleeve on the metering-valve stem. As the throttle moves to the full-fuel position, the throttle lever moves the rack or the sleeve downward (in Fig. 33–16) and the governor spring moves the metering valve, thus increasing the effective metering port area. As the engine speed increases, the fuel pressure increases, and the metering valve is forced upward until the forces on the metering valve and the governor spring are equal. Valve movement then stops, and the fuel flow through the metering port remains constant. If there is a reduction in fuel pressure (due to lower engine speed), the metering port opening increases because the spring force is stronger than the fuel pressure.

When the throttle lever is moved to the idle-speed position, it compresses the idle spring and moves the metering valve downward. A balance is reached when both forces are equal and the fuel flow to the metering port is constant.

A change in the idling speed accordingly increases or decreases the fuel pressure. The spring force increases or decreases the metering port opening until the fuel pressure and the spring force are again equal.

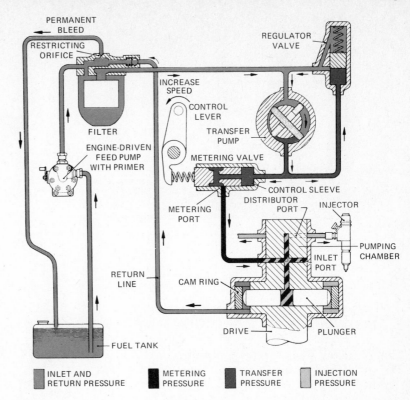

Fig. 33–15   Schematic illustration of DPA fuel-injection system having a hydraulic governor. *(CAV Ltd.)*

▮ INLET AND RETURN PRESSURE	▮ METERING PRESSURE	▮ TRANSFER PRESSURE	▯ INJECTION PRESSURE

Adjustment to the high or low-idle positions is made by turning the adjusting screws clockwise or counterclockwise, respectively. This increases or decreases the angle of the throttle lever and as a result, increases or decreases the spring force.

Injection pumps using a hydraulic governor may also use a combined load and speed-advance unit (Fig. 33–17). The outer piston is sensitive to engine-speed changes because it is moved by the action of the transfer-pump pressure against the spring force. The inner piston, not being subject to transfer-pump pressure, is sensitive to load; therefore, the spring force moves toward the advance.

Removing Distributor-Type Injection Pump from Engine   Thoroughly clean the injection pump, the injectors, the fittings, and the surrounding area to prevent dirt from entering the system when the fuel and injection lines are being removed.

CAUTION   Do not steam-clean the injection pump while the engine is running. (The temperature difference may cause serious damage to the pump.)

Crank the engine over to bring the No. 1 piston onto compression stroke and to allow the engine timing marks to coincide (Fig. 33–18). Turn off the fuel supply. Remove the timing window. If the pump is correctly timed, the timing line on the cam will be in line with the marks on the weight-retainer hub. Remove the injection lines. Take care not to bend or twist them. [See Removal of Fuel Injector (Unit 27).] Disconnect the throttle, the shutoff lever, the fuel-inlet line, and the fuel-return line. Cap all the openings promptly to keep out any dirt. Remove the mounting nuts and washers, and carefully slide the injection pump from the drive shaft.

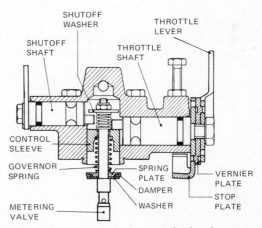

Fig. 33–16   Sectional view of a DPA hydraulic governor. *(CAV Ltd.)*

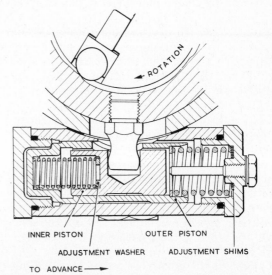

Fig. 33–17   Sectional view of a DPA combined load and speed-advance device. *(CAV Ltd.)*

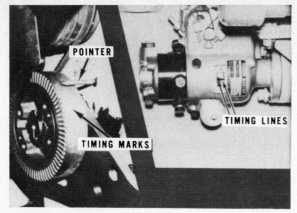

Fig. 33–18 Preparation for removing the pump. *(International Harvester Co.)*

Fig. 33–19 Removing a Roosa Master throttle shaft lever and throttle shaft assembly. *(International Harvester Co.)*

SERVICING DPA AND ROOSA MASTER DISTRIBUTOR–TYPE INJECTION PUMPS

The following basic service procedures apply to all DPA and Roosa Master distributor-type injection pumps. Because it is impossible to cover the precise service details for each model here, be sure to refer to the service manual of the specific pump you are working on.

Wash the injection pump with clean fuel, dry it with compressed air, and then mount it in a holding fixture. Remember that dirt is the fuel-injection pump's greatest enemy. Keep the disassembled components or parts in clean fuel until they are to be reassembled. Inspect each component or part carefully during disassembly; set apart those parts which are not reusable. (This step will speed up the reordering process.) Always discard used seals, gaskets, washers, and O rings. Discard bearings, if used.

Table 33–1 is a detailed inspection chart listing components and parts which require inspection for wear or deterioration.

Disassembling Distributor-Type Injection Pump

ROOSA MASTER To disassemble a Roosa Master injection pump, first remove the cover and gasket. Then, with a screwdriver, remove the horseshoe-like shutoff and throttle-shaft retainer. Slide out the throttle-shaft assembly and throttle-shaft lever from the shutoff lever (Fig. 33–19). Take note of the position of the throttle-shaft key in relation to the groove on the throttle-shaft assembly.

While removing the guide stud from the housing, remove the governor spring together with the idling spring, retainer, and guide (Fig. 33–20). Raise the governor linkage hook at the metering-valve end. Then disengage it from the governor arm and hang the linkage over the housing. Lift out the metering valve and spring. **CAUTION** Do not cock the valve when lifting it from its bore.

To remove the governor control arm, remove one cap nut from the pivot shaft, then withdraw the shaft, and lift out the control arm.

Fig. 33–20 Removing a Roosa Master governor spring. *(International Harvester Co.)*

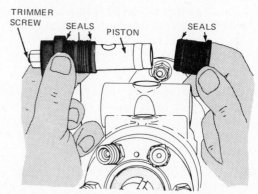

Fig. 33–21 Removing a Roosa Master DM advance mechanism. *(Roosa Master Stanadyne/Hartford Div.)*

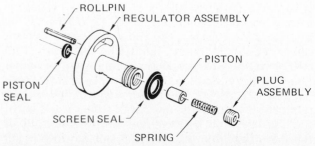

Fig. 33–22 Roosa Master transfer-pump regulator components. *(Roosa Master Stanadyne/Hartford Div.)*

Table 33-1 DETAILED INSPECTION CHART

Part group	Part	Excessive wear	Foreign material or rust	Nicks or chipping	Scratches or scores	Thread damage	Cracks	Distortion	Freedom of movement	Specifically inspect
Housing and drive	Housing	X	X	X	X	X	X	X		See Supplementary Inspection 1, page 4.2
	Drive shaft	X	X	X	X	X	X	X		
	Ball bearing	X	X	X	X				X	
Hydraulic head and rotor	Hydraulic head	X	X	X	X	X	X			
	Vent wire		X						X	See Supplementary Inspection 2, page 4.2
	Discharge fittings	X	X	X	X	X	X			Inside diameter of discharge fittings, sealing area for wear, nicks and scores
	Distributor rotor	X	X	X	X	X	X		X	See Supplementary Inspection 3, page 4.2
	Delivery valve	X	X	X	X		X		X	See Supplementary Inspection 4, page 4.2
	Plungers	X	X	X	X				X	See Supplementary Inspection 5, page 4.2
	Cam rollers and shoes	X	X	X	X		X		X	See Supplementary Inspection 6, page 4.3
	Leaf spring(s) and screw(s)	X	X	X		X	X			See Supplementary Inspection 7, page 4.3
	Cam	X	X	X	X		X	X		See Supplementary Inspection 8, page 4.3
	Governor weight retainer	X	X	X			X	X		Where weights pivot in retainer socket, also "E" ring area for wear, loose pins
	Governor weights	X	X	X			X		X	See Supplementary Inspection 9, page 4.3
	Governor thrust washer	X	X	X	X		X	X		Contact areas for excessive wear
	Governor thrust sleeve	X	X	X			X	X		Points of contact with governor arm for excessive wear
Transfer pump	End cap		X			X	X			
	Inlet screen		X				X	X		Screen and soldered area for breakage
	End-plate adjusting plug		X		X					Tightness in regulator, plugged orifice, loose plate
	Regulating piston	X	X	X	X				X	
	Regulator		X	X	X	X	X	X		Inside diameter of regulator
	Blades	X	X	X	X		X		X	See Supplementary Inspection 10, page 4.3
	Liner	X	X	X	X					Inside diameter in high-pressure area for wear
	Rotor retainers	X	X	X	X					
Governor	Pivot shaft	X	X	X	X		X			Chipped or worn knife edge
	Arm	X	X	X			X	X	X	Points of contact with thrust sleeve and pivot shaft for *excessive* wear
	Metering valve	X	X	X	X		X		X	See Supplementary Inspection 11, page 4.4
	Metering-valve arm	X	X	X	X		X	X		See Supplementary Inspection 11, page 4.4
Linkage	Linkage hook	X	X	X	X	X	X	X	X	Metering-valve pinhole
Advance	Piston	X	X	X	X				X	
	Pin	X	X	X	X				X	
	Reed valve			X	X		X	X		
	Plugs	X	X	X	X	X	X	X	X	Bore for excessive wear

SOURCE: Roosa Master Standyne/Hartford Division service manual—DM model.

To remove the advance mechanism and the hydraulic head from a Roosa Master DM pump, you first must remove the advance plug and end plugs. Next, lift out the advance pin. Slide out the advance components as shown in Fig. 33–21. Remove the head locating screw, the head locking screw, and the transfer-pump end cap.

To disassemble the transfer pump and regulator, remove the transfer-pump end cap, the filter screen, and the components in the order shown from right to left in Fig. 33–22. Next, lift the liner and the pump blades from the rotor and, with a small wire, lift off the transfer-pump cap seal.

When removing the hydraulic head from the housing, pull and swivel it with both hands.

CAUTION Do not drop the weights, thrust sleeve, or washer when removing the hydraulic head from the housing.

To remove the distributor rotor from the hydraulic head of a Roosa Master DC injection pump, remove the two retainers first, and then lift the hydraulic head from the rotor (Fig. 33–23a). Next, lift off the cam ring (Fig. 33–23b). Note the direction on the arrow located on the cam ring.

To remove the pumping plungers, remove the leaf-spring screws, the leaf springs, the cam rollers, and the shoes. Use a cork to hold the plungers in place until they are examined and cleaned. Before removing the delivery valve, you first must remove the plug with a hex wrench. Then tap out the stop, spring, and delivery valve (Fig. 33–24).

The weight retainer is removed from the rotor of a DM model *after* removing the retainer rings, retainer cushion, and cushion governor.

To remove the weight retainers from a DC model, remove the snap ring and press the rotor splines from the retainer splines while using two cam rings to support the weight retainer.

To remove the drive shaft from a model DM Roosa Master pump, use a puller arrangement and pull the gear hub from the drive shaft. Remove the retaining ring snap rings and the spring washer. Slide the drive shaft (with bearings) from the housing.

DPA To disassemble a DPA injection pump, first remove the throttle arm and shutoff lever nuts and lift the levers from the shafts. With light finger force press the throttle shaft downward while lifting the cover from the housing. With the cover removed, press out the shutoff shaft. Disconnect the governor spring from the link and idle-spring guide. Remove the governor spring, throttle shaft with link, idle spring and guide. Remove the cover studs. Lift out (as an assembly) the control bracket, governor arm, metering valve, and linkage hook assembly.

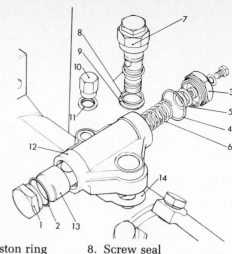

1. Piston ring	8. Screw seal
2. Seal	9. Washer
3. Piston spring cap	10. Cap nut
	11. Washer
4. Seal	12. Advance housing
5. Shim washers	13. Advance piston
6. Outer piston spring	14. Gasket
7. Head locating screw	

Fig. 33–25 Disassembling a DPA automatic advance. (*CAV Ltd.*)

Fig. 33–23 Removing a Roosa Master DC (a) hydraulic head and (b) cam ring. (*International Harvester Co.*)

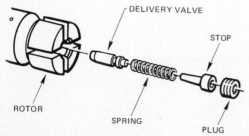

Fig. 33–24 Roosa Master DC delivery-valve components. (*Roosa Master Stanadyne/Hartford Div.*)

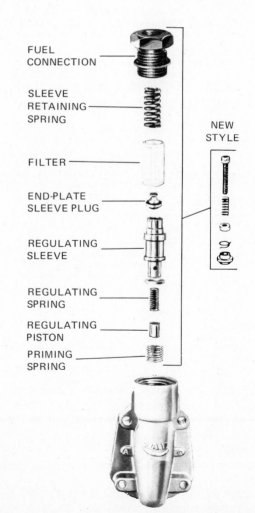

FUEL CONNECTION

SLEEVE RETAINING SPRING

FILTER

END-PLATE SLEEVE PLUG

REGULATING SLEEVE

REGULATING SPRING

REGULATING PISTON

PRIMING SPRING

NEW STYLE

Fig. 33–26 DPA regulator components. (*CAV Ltd.*)

Remove the piston plug, seal and piston, piston-spring cap, seal, shim washers, and spring. Then remove the head locating screw, seal, and washer (Fig. 33–25). Next, pry the advance housing (with piston) from the pump housing and with a special wrench, unscrew the advance screw. Remove the four end-plate screws. Lift the end plate and thrust plate from the hydraulic head. Remove the regulator components from the end plate in the order shown from top to bottom in Fig. 33–26. Before you remove the liner and pump blades, take note of the position of the slot on the liner.

The direction in which the transfer-pump rotor should be loosened is often indicated by an arrow. When the rotor is unmarked, loosen it in the direction of pump rotation.

CAUTION When loosening the transfer-pump rotor, use only the special wrench, otherwise the rotor may be broken.

After removing the transfer-pump rotor, remove the head locating screw and the vent-screw assembly, and then withdraw the hydraulic head.

To remove the rollers, cam shoes, and plungers, first lift the cam ring from the rotor. Remove the drive-plate bolts, drive plate, top adjusting plate, rollers, cam shoes, and the lower adjusting plate. Use corks to hold the plungers in their bores.

To withdraw the weight retainer from a DPA drive shaft, first remove the drive-shaft screw by holding the weight retainer in a special holding fixture. Then slide the drive shaft with its weight retainer from the pump housing. Lift the cushion-drive back plate, cushion-drive rubber insert, drive splined hub, and the weight retainer from the drive shaft (Fig. 33–27).

Supplementary Inspection

DRIVE SHAFT Measure the wear of the drive-shaft tang. The wear of the tang should not exceed 0.430 in [10.75 mm]. Check the drive-shaft seal area for nicks or scratches. Check the wear of the drive-shaft splines.

HYDRAULIC HEAD Make certain that the vent wire in the hydraulic head air-bleed passage is loose and that the passage is clean.

DISTRIBUTOR ROTOR Measure the tang slot or check the drive splines for wear. Check the contact area of the leaf spring(s) and all slots for wear. The charge and discharge ports must have no chipped edges or scratches. The rotor should be free of excessive scratches. Excessive rotor-to-hydraulic-head wear can only be determined when checking the cranking fuel delivery.

Check the delivery-valve retraction piston for corrosion and chipped edges. Always replace the stop and the delivery-valve spring.

Do not handle the rotor shank when the plungers are removed for inspection. Make certain that each plunger is correctly replaced in the end of the bore

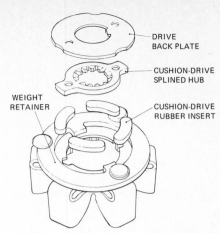

Fig. 33–27 DPA weight-retainer components. *(CAV Ltd.)*

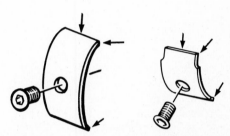

Fig. 33–28 Leaf-spring checkpoints (indicated by arrows). *(Roosa Master Stanadyne/Hartford Div.)*

from which it was removed, but do not use force when inserting the plungers into their bores. If the plungers are sticking in their bores but show no evidence of damage, clean them and their bores with acetone or a lacquer thinner. Hold the plungers (with cork) in their bores and slide the rotor into the hydraulic head. Immerse the unit in clean fuel until it is to be assembled.

Make certain that each cam roller is in its shoe and that each rotates smoothly. There should be no chipping or excessive wear on the top edge of the shoes where they are contacted by the leaf springs or adjusting plates.

Check the leaf-spring edges at the points shown in Fig. 33–28 or check the half-moon grooves of the adjusting plates. If they show wear, they must be replaced.

Carefully check each cam and retraction lobe and the flat surfaces of the cam. If there is evidence of wear, spalling, flat spots, or flaking out, a replacement is necessary.

Check the governor weight heels for flat spots and the toes for rough edges (Fig. 33–29). (The transfer-pump blades usually are replaced when the injection pump is serviced.)

If either the metering-valve arm or the pin is loose or if the valve is pitted or shows wear, the metering-valve assembly must be replaced. If the valve spring is distorted, replace it.

Reassembling Distributor Head Remove the hydraulic-head assembly from the fuel container, slide each roller into the end of its shoe and, install the shoes to the roller slot. Then position the leaf

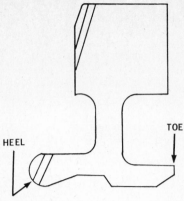

Fig. 33–29 Governor-weight checkpoints. *(Roosa Master Stanadyne/Hartford Div.)*

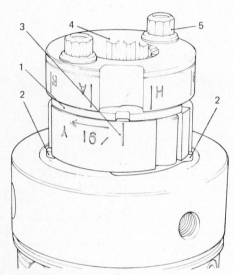

1. Top adjusting plate
2. Bottom adjusting plate
3. Alignment mark on rotor
4. Drive plate aligned between *A* and *H*
5. Plate screw

Fig. 33–30 Reassembled DPA drive hub. *(CAV Ltd.)*

Fig. 33–31 Checking roller centrality with a special dial indicator adapter. *(International Harvester Co.)*

Fig. 33–32 Transfer-pump liner position for clockwise rotor rotation. *(International Harvester Co.)*

springs or assemble the drive hub in the order outlined in the service manual (see Fig. 33–30).

Roller-to-Roller Measurement and Adjustment This adjustment must be made precisely to specification because the roller-to-roller dimension governs the maximum fuel injected per pumping cycle. To make this measurement, place the rotor in the special tool supplied by the Roosa Master manufacturer or place it in the hydraulic head. Connect an air-supply hose with a pressure regulator to one of the discharge fittings. Now, pressurize the head inlet 40 to 100 psi [2.81 to 7.03 kg/cm²] with clean filtered air. Rotate the rotor until the roller shoes bottom on the leaf spring(s) or adjusting plates. Measure the distance across the rollers with a micrometer and compare it with specification. Make certain that the micrometer does not force the roller inward.

To increase or decrease the distance across the rollers, you must adjust both leaf springs since only

one end of each leaf spring changes the shoe travel.

To ensure that each roller contacts the cam lobes simultaneously, the rollers must always be checked for centrality with a special dial indicator (Fig. 33–31). The total variation in roller extension must not exceed 0.003 in [0.07 mm]. You may have to change the leaf springs, rollers, or shoes to come within specification. However, if a change is made, you must remeasure the roller-to-roller dimension. Once the roller-to-roller dimension is adjusted, place the cam ring over the rollers so that the directional arrow points in the direction of pump rotation and the advance pin bore is located at the bottom.

Reassemble the transfer pump and pressure regulator in the reverse order of their removal.

NOTE Make certain the transfer-pump liner is installed according to rotor rotation (Fig. 33–32).

Turn the head assembly over and place the governor weights in the weight retainer. Insert the thrust washer (bevel edge up) and thrust sleeve (with its two grooves up) into the slots of the weights. Tilt

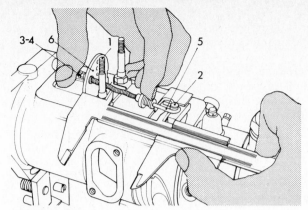

1. Cover stud
2. Metering valve lever pin
3, 4. Locknut and linkage nut
5. Hook linkage
6. Governor control arm

Fig. 33–33 Setting DPA governor linkage. *(CAV Ltd.)*

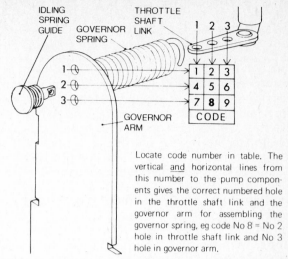

Locate code number in table. The vertical and horizontal lines from this number to the pump components gives the correct numbered hole in the throttle shaft link and the governor arm for assembling the governor spring, eg code No 8 = No 2 hole in throttle shaft link and No 3 hole in governor arm.

Fig. 33–34 Application of the setting code. *(CAV Ltd.)*

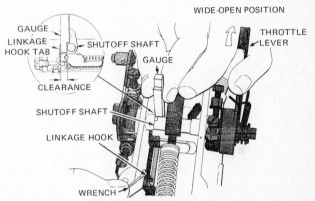

Fig. 33–35 Adjusting DM throttle linkage. *(Roosa Master Stanadyne/Hartford Div.)*

the weights and work both components into the correct position. Spin the weight retainer to ensure that there is no excessive restriction. Tilt the holding fixture, line up the holes for the head locating screws, and, with both hands, slide the assembly into the housing. (Use rotary motion but maintain alignment.) Realign the holes. Install the bottom head locating screw and the remaining head locating screws, then torque them evenly to specification.

Install the advance mechanism but make certain that the piston-hole plug is screwed into the correct side to ensure advance action (and not retarding action). (See Fig. 33–5 for correct position.)

Place the governor arm into position with the linkage-hook fork facing the rear. Insert the pivot shaft. Its sharp, milled edge must face toward the governor arm. Place the spring, the shim, and the metering valve on the arm assembly and insert the valve into the bore. (Make certain it rotates freely.)

Depress the metering-valve assembly. Install the guide stud and washer to the housing, and then tighten the guid stud. Pull back the governor linkage hook. Stretch the spring just enough to assemble the hook correctly to the fork on the governor arm. Position the opposite end of the hook over the pin on the metering-valve arm. Make sure the governor parts move freely. Assemble the governor spring as shown in Fig. 33–20.

Apply a light film of grease to the throttle, shutoff, and shaft seals. Partially assemble the throttle shaft assembly in its bore in the housing. Slide the throttle-shaft lever over the throttle shaft so that the projection in the throttle-shaft-lever bore engages with the rear keyway on the shaft. Position the fork end of the throttle lever so that it straddles the guide stud.

When you assemble the shutoff-lever assembly, use a slight rotary motion to prevent damage to the seals. If the throttle-shaft lever is correctly installed, rotate the throttle-shaft assembly toward the rear. It must compress the governor spring.

Throttle-Linkage Adjustment—DPA Injection Pump To check the governor linkage of a DPA injection pump, measure the distance between the large diameter of the governor-control-cover stud and the metering-valve-lever pin (Fig. 33–33). Make sure that you hold the vernier caliper parallel to the housing and that the metering valve is in the wide-open position. If an adjustment is necessary, loosen or tighten the linkage nut to increase or decrease the distance.

When you install the governor spring, first check the code number and then use the table in Fig. 33–34 to determine in which hole the idle-spring guide must be positioned and in which hole of the throttle-shaft linkage the governor spring must be hooked.

Throttle-Linkage Adjustment—DM Injection Pump To adjust the throttle linkage, back out the torque screw and hold the throttle lever in the wide-open position (Fig. 33–35). Place the linkage gauge between the rear of the shutoff shaft and the vertical tab on the linkage hook, and then check the clearance.

If an adjustment is necessary, loosen the adjusting screw and slide the linkage to its maximum opening. Insert the gauge, slide the linkage hooks together until the face of the tab is flush against the gauge, and then tighten the adjusting screw.

Install a new shutoff cam with the straight inner edge engaging the slot. Install a new throttle-shaft retainer with the straight inner edge engaging the slot. Using a new cover gasket, install the cover.

Pressure-testing Rebuilt Pump It is mandatory to pressure-test a rebuilt pump for external leakage before mounting the pump to the test stand. To make this test, connect an air hose with a pressure regulator to the return-line fitting. Immerse the pump in clean fuel, pressurize the pump to about 20 psi [1.41 kg/cm²], and check for air bubbles.

Mounting Pump to Test Stand First mount the mounting bracket to the test stand and align it within 0.005 in [0.127 mm] of the total indicator reading. With a dial indicator measure the face runout. It should not exceed 0.010 in [0.254 mm]. Mount the pump securely to the mounting bracket.

NOTE The test-stand drive shaft must not deflect the pump drive shaft. There must be 0.001 to 0.005 in [0.02 to 0.127 mm] clearance between the coupling surfaces.

With a hose, connect the test-stand feed pump to the transfer-pump inlet. With another hose, connect the return fitting to the test-stand-return fitting. Connect the test-stand pressure gauge (with a third hose) to the bottom of the transfer-pump endplate. On DM models, use the special pump connector and shutoff valve, and then install the advance test window. When necessary, check the opening pressure of the test injectors.

Select the specified injection lines and install them. Open the bleed-off valve on the injector holders. When all hoses and lines are connected, turn on the fuel supply and regulate the fuel pressure to 2 psi [0.14 kg/cm²]. **CAUTION** The inlet pressure must never exceed 10 psi [0.7 kg/cm²].

When an electric shutoff is used, energize the solenoid before starting the test-stand motor. Run the pump at its lowest test-stand speed. Back out the high-idle adjusting screw. (Also back out the torque screw, if used.) Move the throttle to the full-fuel position. Allow air to bleed from the pump and the injection line before you close the bleed-off valve at the injection holders. When all the air has been bled from the system, operate the pump at 1000 rpm to warm up the pump and the test fuel. During the warm-up period, check for fuel leakage and repair any leaks.

Transfer-Pump Vacuum Test The transfer-pump vacuum test is the first test you must make when checking for adequate fuel supply. Operate the injection pump at 2000 rpm, close the fuel-supply valve, and observe the combination pressure and vacuum gauge. When the gauge shows 18 inHg [45.72 cmHg], check the inlet connection for tightness before you examine the transfer pump. Insufficient vacuum can be caused by a damaged O ring under the inlet fitting, under the pressure-regulating-valve sleeve, or between the pump liner and the end plate. It may also be due to excess

clearance between the pump blades and end plate, to misalignment of the end plate with the pump liner, to damaged or worn blades, or to a worn pump liner.

Pressure Fuel-Flow Test To make the pressure fuel-flow test, change the inlet line fitting from "pressure fitting" to "lift fitting." Disconnect the return line from the test stand and direct it into a graduated calibrating vial. Operate the pump at its recommended speed and measure the return oil flow. It should be 100 to 140 cm³ in 1 minute.

NOTE Each of the following test specifications is merely for the purpose of illustration.

Transfer-Pump Pressure Test Operate the pump at 2400 rpm with the throttle wide open. Observe the transfer-pump pressure. It should be between 80 and 86 psi [5.63 to 6.05 kg/cm²]. If the pressure is not within specification, stop the test-stand motor and adjust the pressure-regulating-spring plug to increase or decrease the spring force. Then recheck the adjustment.

Cranking Speed Fuel-Delivery Test Operate the pump at 150 rpm. Set the counter to 1000 strokes, and then check delivery. It should be 43 cm³. If the fuel delivery is less than this (even though the transfer-pump pressure is as specified), replace the delivery-valve assembly and retest. If the delivery remains lower than specified, there is excessive leakage between the rotor and the hydraulic head. The head must be replaced.

Peak Torque Fuel-Delivery Test Operate the pump at 1600 rpm. Check the transfer-pump pressure. It should be between 64 and 67 psi [4.50 to 4.71 kg/cm²]. Check the pump delivery per 1000 strokes. It should be between 66 and 68 cm³. If the fuel delivery is lower or higher than this range, an adjustment of both leaf springs is necessary to either raise or lower the fuel delivery. Failure to equalize both springs can result in an uneven load on the pump roller.

Rated-Speed Fuel-Delivery Test (Droop Speed) Operate the pump at 2400 rpm. Check the transfer-pump pressure and measure the delivery per 1000 strokes. The transfer-pump pressure must be within specification. The average fuel delivery should be between 71 and 73 cm³. If the fuel delivery is not within specification, the opening of the metering valve must be increased or decreased. You can do this by turning the torque adjusting screw inward to decrease the opening and outward to increase the opening.

High-Idle-Speed Fuel-Delivery Test Operate the pump at (high idle) 2610 rpm. Measure the delivery per 1000 strokes. It should be between 24 and 26 cm³. If the delivery is higher or lower than this range, turn the high-idle adjusting screw outward or inward to change the governor-spring tension. After you have made the high-idle adjustment,

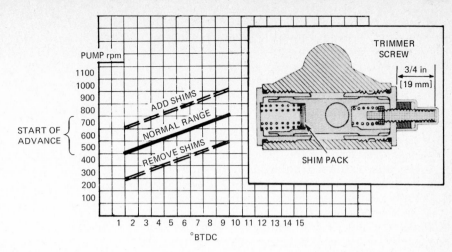

Fig. 33–36 Roosa Master speed-advance adjustment. *(International Harvester Co.)*

always recheck the rated-speed delivery. You may have to adjust the high-idle screw slightly to maintain the correct rated-speed fuel delivery.

Speed-Advance Test To check the speed advance of a Roosa Master injector pump, first move the throttle to the full-fuel position. Operate the pump at slow speed, then gradually increase the pump speed to high idle. During this time observe the transfer-pump pressure and the timing window. Note the pressure rise and the speed at which the line on the cam starts to move toward the advance position and also at what speed the cam movement is completed. In these circumstances the advance should start between 500 and 800 rpm and should be complete at 1600 to 1900 rpm. If the cam fails to advance at its specified low speed or fails to complete its full advance at the recommended speed, adjust the trimmer screw or remove a shim. If the cam advances at a lower engine speed, add shims, readjust the trimmer screw, or replace the springs (Fig. 33–36).

To check the speed advance of a DPA injection pump, install the advance checking tool as shown in Fig. 33–37. Zero the gauge by moving the scale relative to the pointer. When checking the advance mechanism, fuel pressure moves the advance piston and the cam. This piston movement will be measured by the pointer on the advance checking tool.

Low-Idle Fuel-Delivery Test Operate the pump at 600 rpm. Move the throttle to the shutoff position and then slowly (with light finger force) move it toward the fuel-on position until the throttle shaft fork contacts the idle-spring guide. When contact is made, the pump will start to deliver fuel to the injectors. Next, move the throttle about 10° toward the shutoff position. Use the idle adjusting screw to bring the throttle-shaft fork against the idle-spring guide. Check the fuel delivery.

To increase or decrease fuel delivery, increase or decrease the idle-spring tension by turning the adjusting screw clockwise or counterclockwise. When an electric shutoff is used, check it at the speeds indicated in the specification.

Air or Hydraulic Timing of Distributor-Type Pump After the injection pump has been flow-

tested and adjusted, it must be timed by using compressed air or hydraulic pressure. Air or hydraulic timing establishes the exact point at which the cam rollers contact the face of the cam lobes. This is the static timing point and is the start of injection (at peak torque delivery).

To air-time or hydraulic-time the injection pump, remove the timing window and replace it with a half window.

NOTE An old timing window, cut into two pieces (one long and one short) provides a good nonconductive straightedge. File the ends square, and cut the window to a length appropriate to the cam static position.

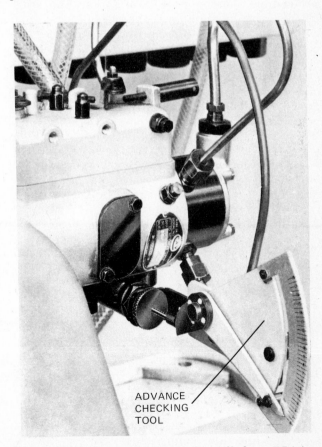

Fig. 33–37 Advance checking tool installed to a DPA injection pump. *(CAV Ltd.)*

Connect an air hose with a pressure regulator to the No. 1 discharge fitting, or connect the No. 1 discharge fitting to an injector tester. For a Roosa Master injection pump, pressurize the line to about 100 psi [7.03 kg/cm²]. For a DPA injection pump, pressurize it to about 426.0 psi [29.94 kg/cm²]. The applied pressure forces the piston outward. When the drive shaft is rotated (in the direction indicated on the nameplate), it positions the cam rollers so that they strike the faces of the cam lobes. As soon as you feel this resistance, hold the drive shaft firmly and, with an electric pencil, inscribe a line to coincide with the cam line on the weight retainer.

Upon completion of testing, adjusting, and timing the injection pump, use a sealing wire and lead seals to make the adjustment tamperproof. Then wire the throttle in the full-fuel position.

Installing Injection Pump to Engine   Before you install the injection pump to the engine, replace the drive shaft cup seals, if they are being used. Before you place the new seals on the installing tool, coat them with grease so that they will slide onto the shaft more easily.

Position the No. 1 piston on its compression stroke so that the timing pointer indicates the specified static time position. Remove the timing window and turn the drive shaft to align the timing lines, then slide the pump over the drive shaft onto its mountings. (You may have to rotate the pump to match up the spline or tang connection.) Use the seal compression tool to guide and protect the seals when sliding the pump over the seals.

Realign the pump timing lines by rotating the pump before you tighten the mounting stud nuts. Recheck the pump timing. Connect the fuel inlet and return lines. Install the injection lines but make certain you do not bend or twist the injection fittings. Always use new washers on each side of the banjo fitting.

Connect the throttle and shutoff lever. Bleed the injection system. Turn the ignition key to start the engine. If the engine does not start, loosen the injection line (at the injectors) and crank the engine until fuel leaves the loose connection. Retighten the connection and again turn the ignition key to start the engine.

When the engine is at its operating temperature, check the high- and low-idle rpm and make any corrections necessary.

CAUTION   Never change the timing while the engine is running because this could cause serious damage to the pump.

Questions

1. List the major differences between the three distributor-type injection pumps.

2. Trace the fuel (metering) of a DPA fuel pump from the inlet fitting to the point where fuel pressure forces the opposed plungers apart.

3. Describe the action of the delivery valve.

4. Describe the action of the Roosa Master governor when the throttle is in the full-fuel position and the engine speed is reduced due to increased load.

5. Describe the action within a DPA hydraulic governor when the engine speed is increased as the load is reduced.

6. Why should the fuel-pump timing be checked before removing the pump for service?

7. Describe how to measure and adjust the roller-to-roller dimension of a DPA pump.

8. What is the purpose of measuring and adjusting the throttle linkage?

9. What would result if either the cam ring or the transfer-pump liner were installed backwards?

10. Describe how to test the operation of the transfer pump while it is installed to the test stand.

11. Explain how you would increase the transfer-pump pressure of a pump having a regulator as shown in Fig. 33–22.

12. Outline the procedure to air-time a Roosa Master injection pump.

13. List the required steps to install a distributor-type injection pump to an engine having a static timing of 8°.

14. List the necessary checks or adjustments to be made after the pump is installed to the engine.

UNIT 34

American Bosch Distributor-Type Fuel-Injection Pumps

Although many companies manufacture distributor-type fuel-injection pumps, these pumps differ in construction as well as in metering principle. Each company's objective, however, is the same: to build a fuel-injection pump which will complement small-bore engines yet remain simple in construction and economical to service.

Some of the distributor-type injection pumps presently in use are no longer being produced; but new types are continually being introduced.

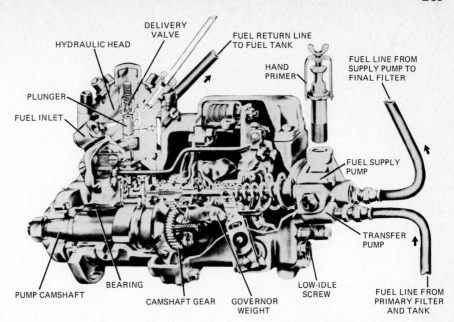

Fig. 34–1 Cutaway view of the model 100 injection pump. *(International Harvester Co.)*

American Bosch Company is one of the pioneers of the distributor-type injection pump. Their models have been modified over the years from the PSB, PSJ, PSM to the recent model 100 (Fig. 34-1). The construction of these four models is different but the pumping, metering, and distributing principles remain the same. All American Bosch models have these characteristics:

1. They have a constant-beginning and variable-ending delivery.
2. They are driven at engine speed.
3. They use a single plunger and barrel and only one delivery valve.
4. They use a mechanical governor and a gear-type transfer pump.
5. The plunger rotates through a gear train at half camshaft speed to distribute the metered fuel to the discharge outlet. There is a reciprocating motion for pumping action.
6. They use a sleeve-metering principle.

All PS series injection pumps (with the exception of the PSU) have a drive section, a pumping, metering, and distributing section, and a control section. (The location of the transfer pump varies between models.)

Drive Section The purpose of the drive section is to drive the governor, rotate the plunger, and move the plunger up and down. Its components consist of the drive coupling, camshaft, and gear (Fig. 34-2). When an automatic external mechanical timing device is used, the device is placed between the drive coupling and the camshaft. When a hydraulic automatic advance (intravance) is used, the camshaft and the intravance act as a single unit.

Regardless of the drive combination used, when the engine is cranked or is running, the timing gears transmit rotary motion to the drive coupling and onto the three- or four-lobe camshaft. When using an intravance unit, the rotation from the timing gear is transmitted to the drive coupling and to the three-

lobe camshaft (intravance). The gear on the camshaft drives the meshing gear on the governor drive shaft, and the end of the drive shaft drives the transfer-pump gear.

The governor weight assembly is driven over a friction clutch to reduce pulsation. The second gear on the governor drive gear drives the face gear and therefore the plunger.

Intravance Timing Device The intravance unit used with the model 100 is different to some extent from the unit used with the PSM injection pump. However, both advance units are located within the camshaft (see Fig. 34-3). Both units alter, hydraulically, the relation between the drive shaft and the camshaft. The degree to which the camshaft advances is governed by the sliding sleeve. The sliding sleeve has inner and outer longitudinal helical splines. The inner splines are engaged with the helical splines of the drive shaft and the outer helical splines with the camshaft. The governor fingers are linked with the servo valve and followup rod. The followup rod extends through the valve and weight assembly and holds the spring through a threaded

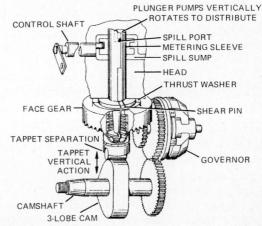

Fig. 34–2 Drive section. *(American Bosch AMBAC Industries, Inc.*

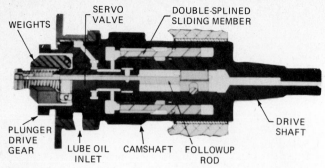

Fig. 34–3 Cross-sectional view of a model 100 internal timing device. (*International Harvester Co.*)

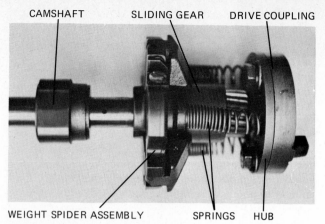

Fig. 34–4 TMF timing device.

spring seat against the servo valve. The other end of the followup rod is connected with the sleeve through a pin.

INTRAVANCE OPERATION (MODEL 100) When the engine is at idle speed or below 1000 rpm, the intravance spring force is greater than the weight force. Therefore, the oil-inlet port is blocked by the servo valve, and the sleeve is retracted to the right through the followup rod. As the engine speed increases, the weight force moves the servo valve to the right, opening the right oil annulus. Oil is supplied from the engine lubrication system to the rear bearing plate. The oil then travels by a passage to the servo valve, through the servo valve to the chamber between the drive shaft and the camshaft, and forces the sleeve to the left. The movement of the sleeve forces the followup rod to the left. This increases governor-spring force causing the servo valve to close the oil annulus to the chamber. Then the movement of the sleeve stops.

When the engine speed is increased, say to 2500 rpm, the weight force again moves the servo valves. The sleeve is moved even further to the left and against the drive shaft. The sleeve has advanced the cam to its maximum (14°) because of the helical angle of the sleeve, camshaft, and drive shaft. The maximum advance and the advance curve of the camshaft are governed by the helical angles, the spring rate, the preload, and the weight force.

When the engine speed decreases, the weight force decreases, and the spring force moves the servo valve to the left, opening the left annulus of the servo valve. This allows the oil from the chamber to drain through the shaft until the weight force and spring force balance, closing the drain annulus.

To maintain the camshaft in its advanced position during the pumping action, the oil-supply hole in the rear bearing plate comes out of alignment with the hole in the camshaft. The oil is trapped within the advance units, preventing the camshaft from moving to a retarded position.

TMF Timing Device The TMF is an external mechanical timing device which is bolted to the injection-pump housing. It uses a separate housing to locate and position the advance unit.

The components of the TMF are shown in Fig. 34-4. The drive coupling is keyed to the tapered shaft of the weight spider, and the hub is bolted to the injection pump camshaft. The spider and the hub have external helical splines, one curved left and the other curved right. The sliding gear has internal helix-angled splines which mesh with the weight spider and hub splines. Four springs are placed between the hub and the slding gear to force the sliding gear against the finger of the flyweights. A bushing and an adapter flange reduce the friction between the sliding gear and flyweight finger.

OPERATION OF TMF TIMING DEVICE When the engine is running, the drive coupling rotates the weight spider, sliding gear, hub, and camshaft. As the engine speed increases, the governor weights are forced outward against the force of the four coil springs. The weight force causes the sliding gear to move toward the hub. The movement of the sliding gear causes the camshaft to advance in relation to the weight spider because of the helical angle of the splines. The movement of the sliding gear stops as soon as the spring force equals the weight force. The spring rate and preload and the weight assembly and helix-angled splines are tailored specifically for this engine to compensate for its advance curve.

PSU Injection Pump The PSU injection pump is used on small high-speed engines. The PSU injection pump is actually only the hydraulic head of the model 100 or PSM pump placed in a special housing. The camshaft, the drive which rotates the plunger, and the governor are part of the engine rather than part of the pump.

Pumping, Metering, and Distributing Section The major components responsible for pumping, metering, and distributing the fuel are the camshaft and gear, the face gear and hub, the plunger, the hydraulic head, the sleeve, and the control unit (Fig. 34-5). Some internal passages in the hydraulic head lead from the reservoir which surrounds the bore to the fill ports and others lead from the delivery valve to the discharge fittings. One other passage from the lower end of the plunger leads back to the reservoir.

The plunger is center- and cross-drilled. Above the cross-drilled hole (fill port) is the distributor groove and slot. The plunger sleeve, connected to the control assembly, rides on the plunger. The plunger is connected to the face gear through the plunger drive guide and, as a result, when the face gear rotates, the plunger also rotates. The plunger spring rests at one end (against a washer) on the face gear. The other end is connected to the plunger through the spring seat and split retainer. The face gear is held to the hydraulic head by a gear retainer.

Control Section The components of a variable-speed governor are shown in Fig. 34-6. The basic governing principle is illustrated in Fig. 34-7. Here the fulcrum lever is connected to the throttle shaft (fulcrum-lever pivot point) at point A, to the control rod at point B, and to the control sleeve at point C.

When point A is moved to increase speed, points B and C move to the right (in Fig. 34-7). This moves the control rod, control assembly, and control sleeve. Since the control sleeve changes the effective stroke, more fuel is metered.

When point A is moved to the right while the engine runs at a constant speed, point B moves to the left. This increases fuel (since point C is "fixed") until the governor weight forces point C to the right, causing point B also to move to the right.

Maximum fuel (to develop maximum engine torque) is controlled by the position of the upper end of the fulcrum lever. When the throttle is in the full-fuel position and the engine speed is reduced because of the increased load, the cam nose comes to

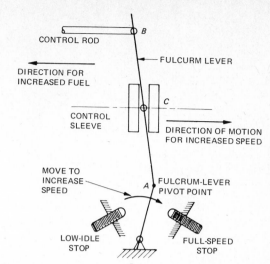

Fig. 34–7 Governing principle. (*American Bosch AMBAC Industries, Inc.*)

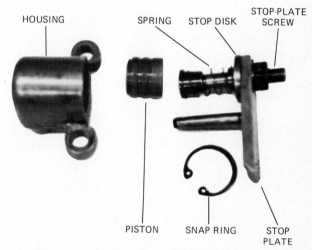

Fig. 34–8 Components of the excess-fuel device.

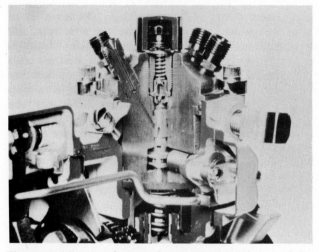

Fig. 34–5 Sectional view of pumping, metering, and distributing components.

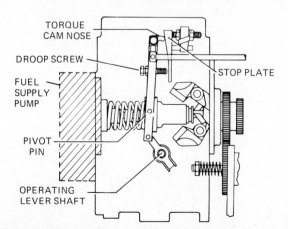

Fig. 34–6 Sectional view of the variable-speed governor components at maximum torque speed. (*American Bosch AMBAC Industries, Inc.*)

rest against the stop plate (see Fig. 34-6). This increases the effective stroke of the plunger.

The excess-fuel device gives the engine more fuel during starting because the spring force of the piston moves the stop plate to the left, allowing the control rod also to move to the left (see Fig. 34-8). As soon as the engine oil pressure builds up behind the piston, the pressure forces the piston to the right into the normal fuel-stop position.

Pumping, Metering, and Distributing Principle When the tappet roller rests on the base circle of the camshaft, the plunger is below the fill ports. At this time no discharge port is aligned with the distributor slot. Fuel from the reservoir area in the hydraulic head enters the chamber above the plunger through the fill ports (Fig. 34-9a). As the cam lobe lifts the tappet and the plunger button, it forces the plunger upward. Fuel displaced by the plunger leaves the fill ports until the top edge of the plunger closes the ports and traps fuel above the plunger (Fig. 34-9b). At the same time, the plunger rotates because the camshaft gear drives the governor gear as well as the face gear. When the fuel pressure overcomes the delivery-valve spring force,

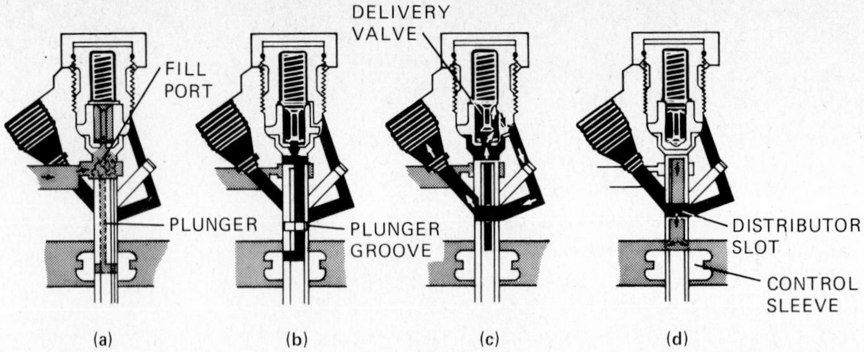

Fig. 34–9 Fuel pumping principle. (a) Intake; (b) beginning of delivery; (c) delivery; (d) end of delivery. (American Bosch AMBAC Industries, Inc.)

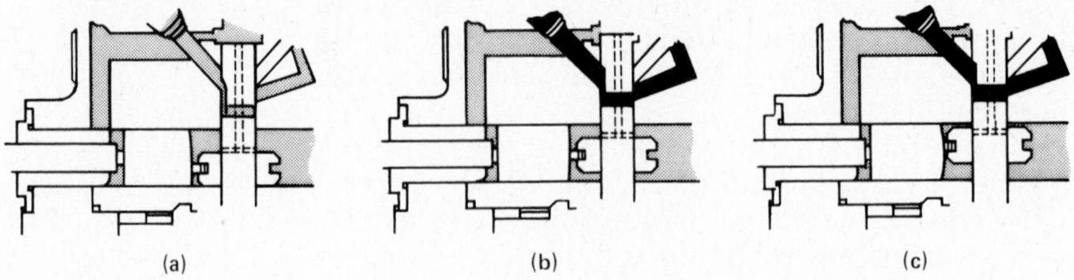

Fig. 34–10 Metering principle. (a) No delivery; (b) normal delivery; (c) maximum delivery. (American Bosch AMBAC Industries, Inc.)

the delivery valve is forced upward. As the retraction piston leaves its bore, fuel can pass around the valve, downward through the discharge passage, around the plunger groove, and up to the distributor slot (Fig. 34-9c). The distributor slot is then aligned with one discharge port. Fuel from the distributor slot discharges through the discharge port, into the discharge passage, to the discharge fitting, and into the injection line.

ENDING OF DELIVERY When the upper edge of the cross-drilled holes (spill ports) passes the upper edge of the control sleeve, fuel at high pressure spills back through the center- and cross-drilled holes into the reservoir area of the hydraulic head. When the spill ports open, the delivery valve snaps onto its seat, reducing the injection-line pressure. A fraction of a second later the distributor slot comes out of registration with the discharge port. The plunger closes this port and holds the pressure in the injection line (Fig. 34-9d).

As the plunger moves downward, fuel enters the spill ports and fills the area vacated by the plunger until the spill ports are covered by the sleeve. For a short interval no fuel can enter the area above the plunger until the fill ports are again opened by the plunger.

The plunger is pressure-balanced to minimize side force on it. The Oldham-type drive coupling reduces any tendency to side-load the plunger.

METERING When the control rod rotates the shaft of the control assembly to the no-fuel position, it moves the control sleeve downward uncovering the spill ports (see Fig. 34-10a). At this position no fuel can be trapped above the plunger. When the control sleeve moves upward to the fuel-on position, it covers the spill ports (Fig. 34-10b). The position of the sleeve on the plunger in relation to the spill ports determines the effective stroke (fuel delivery) (Fig. 34-10c).

Servicing Model 100 Injecton Pump After you have removed all excess grease and dirt from the outside of the pump, wash it with clean fuel or with a noncaustic solvent, and dry it with compressed air. Drain the fuel and lubrication oil from the pump and mount the pump in a holding fixture. Remove the drive-hub nut and the drive hub, using the appropriate holding tool and puller. Do not use a jaw puller because it will damage the mounting flange.

Next, remove the fuel-return valve, the control-unit cover, the control-unit plate, and the spacer. Carefully engage the control rod while sliding out the control unit. Loosen, but do not remove, the delivery-valve-holder nut.

To remove the hydraulic head, first loosen the four Allen-head screws. If the head does not rise, turn the camshaft to force it up. The head can then be turned to allow removal of the screws without damaging the discharge fittings. Lift out the hydraulic head, tappet guide, and roller using a pair of needle-nose pliers. Before you remove the transfer pump, loosen the regulator valve plug, then remove the mounting bolts. Using two bars, pry the pump housing from the governor housing. Lift from the governor drive shaft, the governor spring, spring seat, and spacer. Remove the excess-fuel device *very carefully* because it is connected to the oil-supply tube.

After the governor mounting screws are removed, tap the governor with a soft-faced hammer to break the gasket loose from the injection-pump housing and dowels. Tip the housing slightly to prevent the control rod and oil-supply tube from bending while lifting the governor from the pump housing.

If the camshaft requires servicing, remove the front-thrust-washer retaining plate and the rear bearing plate, and push the camshaft to the rear to remove it from the housing.

Servicing Drive Section Wash the components of the drive section in clean diesel fuel and dry them

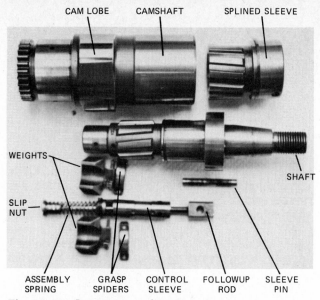

Fig. 34–11 Intravance unit components.

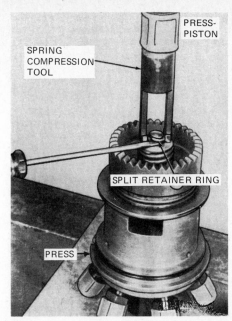

Fig. 34–12 Removing the split retainer ring. *(American Bosch AMBAC Industries, Inc.)*

with compressed air. Check the injection-pump housing for cracks and damaged surfaces. Check the threaded holes for damaged threads. Use compressed air to clean the lubrication oil passage.

Check and measure the front bushing in the pump housing for wear. If the front bushing is worn beyond specification, the pump housing must be replaced. If the rear bearing plate is damaged or the bushing is loose within the plate, or if the bushing is worn beyond specification, the plate must be replaced.

Check the tappet sleeve for wear. If the clearance between it and the tappet is more than 0.001 in [0.025 mm], the pump housing must be replaced.

Check the drive shaft for damage and/or scored taper. Check the keyway for size. If the camshaft lobes are worn, pitted, or scored, or if the bearing journals are damaged, grooved, or worn, the intravance must be replaced. (Disassembling and servicing the unit is not recommended.)

To clean the intravance, immerse the assembly in clean diesel fuel. Holding the drive shaft, move the followup rod in and out while rotating the camshaft. This action will clean the servo valve and also the splines of the drive shaft, sleeve, and camshaft (see Fig. 34–11).

Disassembling Hydraulic Head To disassemble the hydraulic head, first remove the retainer ring and the plunger button. Then place the hydraulic head in a press as shown in Fig. 34–12. Compress the plunger spring and remove the split retainer ring. (You may have to lift the plunger with a screwdriver in order to remove the retainer ring.) Slowly release the force and lift the lower spring seat, spring, and washer from the plunger.

To remove the plunger, remove the face-gear retainer, face gears, and plunger drive guide. Lift the thrust washer and the plunger from the head (Fig. 34–13). Slide out the control sleeve and immediately place it on the plunger. Unscrew the delivery-

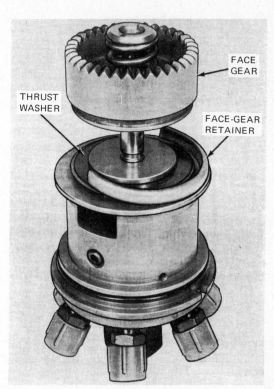

Fig. 34–13 Removing the face gear. *(American Bosch AMBAC Industries, Inc.)*

valve holder. Lift out the components in the order shown from right to left in Fig. 34–14. (You may need to use a wire to lift out the spacer.)

Servicing Hydraulic Head Wash the components of the hydraulic head separately in clean diesel fuel to prevent foreign particles from contaminating the head passages and grooves.

Inspect the head for damage and cracks and for damaged or leaking discharge fittings. Check for scored or damaged fill ports and outlet holes. If the

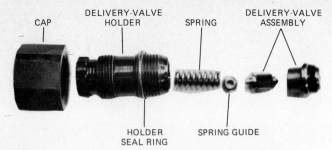

CAP • DELIVERY-VALVE HOLDER • SPRING • DELIVERY-VALVE ASSEMBLY

HOLDER SEAL RING • SPRING GUIDE

Fig. 34–14 Disassembled delivery-valve assembly.

thrust-washer contact surface is badly scored, the head assembly must be replaced.

Check the set screw and the ball check for damage and leakage. If the plunger or the bore of the sleeve show scratches or scuff marks, or if the spill port and/or the fill ports and distributor slot show wear, the assembly must be replaced.

Check the control unit for damaged, loose, or bent levers. Make certain the shaft rotates freely, that the plunger sleeve pin is not worn, and that the clearances are within specification.

Reassembling Injection Pump After the components are washed and dried with compressed air, lubricate the bearings, journals, and bushing. Next, slide the camshaft into the housing.

Install the O ring on the oil tube and position it in the rear bearing plate. Place the thrust washer on the bearing plate so that the bronze side faces toward the camshaft and the tang is locked into the bearing plate.

Install the rear bearing plate with the large cutout facing up (12 o'clock). Make certain that the oil tube and its O ring are correctly placed in the oil duct and that the rear-bearing thrust plate is correctly seated in the plate, as shown in Fig. 34–15. Then install and tighten the screws to the recommended torque.

Place the shims on the front housing and the thrust washer on the retainer plate (Fig. 34–16). (The bronze side must face toward the camshaft and the tang must be in the plate slot.) With the retainer plate in the correct position, gradually tighten the screws (alternately) to the specified torque.

NOTE Turn the camshaft while tightening the screws. This will prevent retainer flange distortion caused by misaligned tangs and slots or by insufficient shims.

After the screws are torqued to specification, measure the end play. If the end play is not within specification, add or remove shims of suitable thickness until you obtain the correct end play.

To check the clearance between the tappet guide and the roller, turn the camshaft so that the keyway faces up (12 o'clock). Lubricate the roller and place it onto the cam parallel with the camshaft. Place a strip of Plastigage on the bronze tappet-guide bearing and insert the guide (into its bore) on the tappet roller. With a soft punch, lightly tap the guide to flatten the Plastigage. Remove the tappet guide and measure the clearance. If the clearance is larger

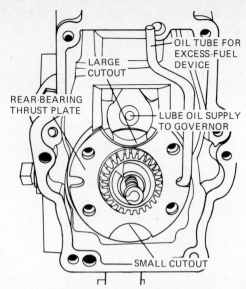

Fig. 34–15 Rear bearing plate position for installing the camshaft. (*American Bosch AMBAC Industries, Inc.*)

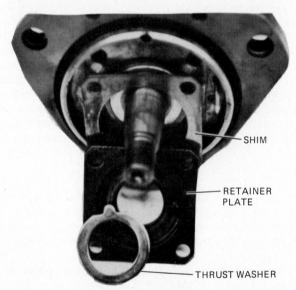

Fig. 34–16 Installing the front thrust washer and retainer plate.

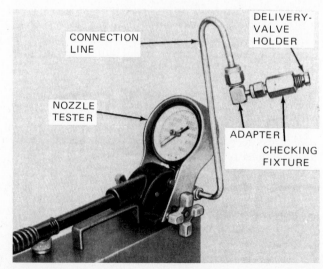

Fig. 34–17 Checking delivery-valve spring pressure. (*American Bosch AMBAC Industries, Inc.*)

than specified, replace the tappet guide and/or the roller.

If the delivery valve or seats are not in perfect condition, the complete assembly should be replaced. However, it is sometimes permissible to lap the valve seat to its body seat. If the delivery-valve seat in the head is damaged, replace the assembly.

CAUTION Never put lapping compound onto the relief piston or onto the fluted stem.

Carefully check the plunger drive guides and the button for cracks, grooving, or uneven wear.

Check the face gear for worn or damaged teeth, the thrust surface for wear, and the plunger guide for damage. If the plunger spring shows wear or has flat spots, or the spring ring is bent, or the spring seat is damaged, replace the damaged component. Always replace the thrust washer.

Testing Delivery Valve The delivery valve of the American Bosch distributor-type injection pump must be tested for leaks as well as for its opening pressure. To make this test, install the valve assembly (without the plastic spacer) in the test fixture. Connect the assembly to an injector tester (Fig. 34–17). Slowly increase the pressure until the valve opens and then record the opening pressure. The opening pressure should be 1150 to 1450 psi [80.85 to 101.94 kg/cm²]. If the opening pressure is not within this specified range even though a new delivery valve spring is used, change the spring guide to raise or lower the opening pressure. (Spring guides are available in various thicknesses.)

Reassembling Hydraulic Head First install the delivery-valve spacer, then install the remaining components in the order shown from right to left in Fig. 34–14. Make sure before installing the valve-holder

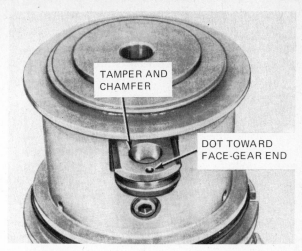

Fig. 34–18 Positioning the metering sleeve for reassembly. *(American Bosch AMBAC Industries, Inc.)*

seat that it has no nicks and burrs. Tighten it finger-tight, then place the head into the holding fixture. Place the lubricated thrust washer on the bottom of the head. Slide the piston sleeve (the drill marks should face upward) into position, as shown in Fig. 34–18, and carefully, by rotating the plunger, slide it into the sleeve. Next, assemble the plunger drive guide, face-gear retainer, and face gear.

Check to make sure that the sleeve and plunger do not stick. If they stick, totally immerse the assembly in clean fuel. Rotate and push the plunger and move the sleeve up and down until the plunger and sleeve move freely. Leave the assembly in clean fuel until it is to be installed.

Servicing Transfer Pump The transfer pump must be disassembled and checked to ensure that the pump can supply the proper quantity of fuel and can pump against a pressure of 40 psi [2.81 kg/cm²]. When the pump is disassembled, Fig. 34–19, check

1. Relief-valve screw
2. Relief-valve spring
3. Relief valve
4. Pump body
5. Pumping gears
6. Cover
7. Plugs
8. O ring
9. Hand primer

Fig. 34–19 Exploded view of the transfer pump. (The numbers indicate the order of component installation.) *(American Bosch AMBAC Industries, Inc.)*

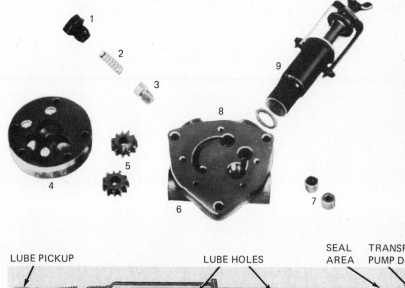

Fig. 34–20 Governor drive shaft. *(American Bosch AMBAC Industries, Inc.)*

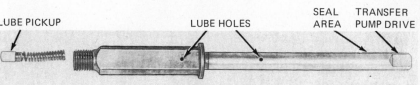

the internal bores for wear and grooving. Check the idler shaft and the governor drive shaft for wear, paying particular attention to the contact surface of the transfer-pump seal area (Fig. 34–20). If the pump cover is damaged or grooved or the primary check valves show wear, the cover must be replaced. Small nicks and burrs should be removed by lapping the cover and body surface. If the gear teeth indicate wear, the gears should be replaced. Always replace the governor-shaft seal and the relief-valve spring. When pressing the new double-lip seal into place, use a sealing compound on the outer surface and push it flush with or slightly below the edge of the seal bore.

If the pump body and the gears are reusable, measure the gear clearance; it should not exceed the specification. Don't forget to service the hand primer.

Check the barrel for out-of-roundness, damage, and corrosion. The plunger must be free of score marks and must slide freely (without the O ring) in the barrel. Check the threads of the barrel and the plunger for damage.

Reassembling Transfer Pump Dip all components in clean diesel fuel. Place the idler gear and the drive gear so that the chamfer faces toward counterbore of the body. Apply a thin coat of sealant to the sealing surface of the cover. Place the cover on the body. Misalignment is impossible as the bolt holes in the cover and body will line up in one position only. Do not forget to torque the screws to the specified torque.

Lubricate the relief valve and install the components in the order shown in Fig. 34–19. When the transfer pump is assembled and the hand primer installed, the pump should be tested for leakage as shown in Fig. 34–21.

Servicing Governor The governor components must be thoroughly checked for wear. If the weight

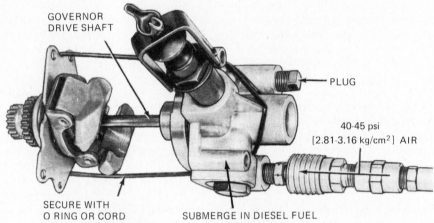

GOVERNOR DRIVE SHAFT

PLUG

40-45 psi [2.81-3.16 kg/cm²] AIR

Fig. 34–21 Checking the transfer pump for leakage. (*American Bosch AMBAC Industries, Inc.*)

SECURE WITH O RING OR CORD

SUBMERGE IN DIESEL FUEL

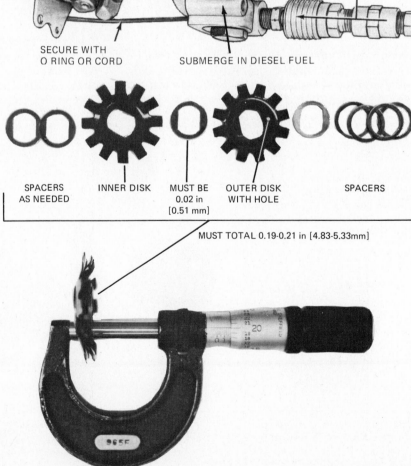

SPACERS AS NEEDED   INNER DISK   MUST BE 0.02 in [0.51 mm]   OUTER DISK WITH HOLE   SPACERS

MUST TOTAL 0.19-0.21 in [4.83-5.33mm]

Fig. 34–22 Measuring the thickness of the friction drive.

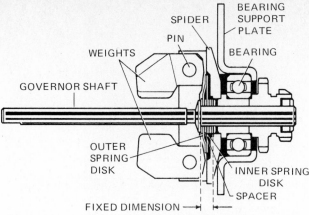

Fig. 34–23 Friction-clutch weight assembly.

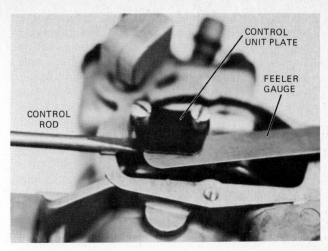

Fig. 34–24 Checking the clearance between the control rod and the control unit plate. (*American Bosch AMBAC Industries, Inc.*)

assembly (weight spider), the levers, or the pins have a loose fit, or if the weight fingers have flat spots, they should be replaced.

Replace the thrust bearing and thrust washer if they show signs of wear. If the sleeve is scored, the bushing worn, or the guide slot rough, replace the whole sleeve. It is essential that the sleeve slide freely on the weight shaft and on the fulcrum pivot pins.

If the throttle-linkage holes are out-of-round or too loose, replace the operating lever. The recommended procedure is to replace the low- and high-idle spring and the spring seat or spacers if they are worn or damaged. If the spring disk or the friction drive spiders are bent, broken, or cracked, they should also be replaced. Always replace the ball bearings and bearing bushing. Check the governor drive gear for damage and chipped teeth. If the governor drive shaft is bent, scored, or worn (especially at the transfer-pump seal area), it should be replaced. Replace the lube pickup when the tip surface is worn (Fig. 34–20).

Reassembling Governor To assure a smooth pulsation-free governor drive, the assembled friction drive must be measured and the pack must be within specification (see Fig. 34–22). When you have completed this, install the components to the drive shaft as shown in Fig. 34–23.

NOTE When pressing the bearing into position, press on the outer race only, and be careful not to distort the plate.

When the unit is assembled, measure the slippage torque. To do this clamp the weight assembly (spider) in a soft-jawed vise. Lubricate the contact area of the friction drive spider with engine oil. Place a 15/16-in [8.00-mm] socket with an inch-pound torque wrench over the governor drive gear. Rotate the drive gear at 30 to 50 rpm through several complete revolutions using the torque wrench, and continuously observe the torque scale. It should be between 30 to 40 lb·in [0.345 to 0.455 kg·m].

Reassemble the remaining governor components with the exception of the transfer pump. When you install the governor to the injection pump, apply the recommended gasket sealer to both sides of the new governor gasket. Place the gasket on the governor housing, then install the governor to the injection-pump housing. Be careful not to bend or damage the control rod or the lubrication-oil-supply tube. Torque the new long lock screws to specification. Install the excess fuel device.

Installing Hydraulic Head Rotate the camshaft so that the keyway faces up (12 o'clock). Place a well-lubricated square O ring into the bore of the hydraulic head. Align the location slot in the flange of the hydraulic head with the tongue of the locating plate. Then rotate the face gear until the timing line on the gear is aligned with the raised timing mark in the timing window. Now, carefully slide the head into the pump housing. Be especially careful not to damage the O ring. Rotate the head slightly out of position to install the fourhead retainer screws. Tighten these screws finger-tight.

Realign the slot with the locating plate. Realign the line mark on the face gear with the raised mark in the timing window. While tightening the head-retaining screw, rock the camshaft back and forth. This assures a smooth meshing of the face-gear teeth with those of the governor gear. Remember that the keyway must be at the 12 o'clock position and the line mark must align with the mark in the timing window when the head screws are torqued to specification.

When installing the control unit make sure that the O ring is well lubricated. The plunger sleeve and the plunger-sleeve pin must be in the lowest position but the pin dot must face up.

To check that the plunger sleeve pin is positioned properly, rotate the arm of the control unit. When it rotates 360°, the pin is in place.

Slide the control-rod pin into the bore of the control arm and install the control-unit plate. Now, measure the clearance between the control rod and plate (Fig. 34–24). If the clearance is larger than specified, use a larger offset plate to bring it to specification. Do not forget to lock-wire the two control-unit-plate screws.

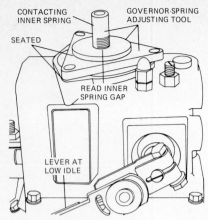

Fig. 34—25 Checking the high-speed (inner-spring) gap. *(American Bosch AMBAC Industries, Inc.)*

Installing and Adjusting Governor Springs Turn the injection pump so that the transfer-pump end faces you. Move the control lever to the low-idle fuel position. Next, check that the governor weights are closed and that the thrust washer and bearings are seated flat on the governor sleeve. When this is done, place the inner spring (high speed) on the sleeve, the adjusting spacer and spring seat on the spring, and position the governor-spring adjusting tool (Fig. 34—25).

When the spring gauge is positioned properly and the inner spring gauge rests lightly against the spring guide, read the distance from zero to the last visible quarter millimeter gradation. This is the measurement of the inner spring gap. If necessary, remove or add spacers to meet the gap specification. Check with your service manual or injection-pump test specification for the required measurements.

When you have adjusted the inner spring gap, place the outer spring spacer and the outer spring (low speed) in the sleeve bore, and then place the spring seat on the spring.

Reinstall the governor-spring adjusting tool. Be sure there is no space between the inner face of the tool and the spring seat. Record the millimeter reading on the inner spring gauge. Press the outer spring gauge against the governor housing and reread the gauge. The compression of the outer spring is the difference between these two readings. If adjustment is necessary, remove or add spacers in order to increase or decrease the outer compression rate of the spring.

Installing Transfer Pump Place the outer spring seat in the bore of the transfer pump and a new O ring in the groove of the pump body. Lubricate the O ring, the end of the drive shaft, and the double-lip seal before you place the pump on the governor housing. Turn the governor drive gear slowly to engage the flat side of the governor drive shaft with the flat side of the drive gear. Do not use force or tilt the pump housing during the alignment because this will damage the seal. The screws must always be torqued to specification. When this has been done, turn the governor drive gear to check that the transfer-pump gears turn freely.

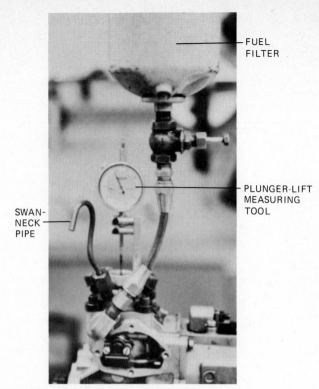

Fig. 34—26 Checking plunger lift to port closure.

Spill-timing for Port Closure and Measurement of Plunger Lift to Port Closure When the hydraulic head is installed, the port closure and the plunger lift to port closing must be checked and the timing remarked on the drive hub. To check the timing (port closure), remove the delivery-valve assembly and install the plunger-lift measuring tool in place of the delivery valve (see Fig. 34—26). Install a swan-neck pipe to No. 1 discharge fitting and a fuel filter to the inlet fitting. Also, install the port-closure locking plug if used.

Align the timing mark of the face gear with the raised mark in the timing window. This should bring the keyway on the camshaft to the 12 o'clock position. Next, place the throttle in the idle position. Turn on the fuel to remove air from the reservoir and passage within the hydraulic head. When only fuel leaves the swan-neck pipe, move the throttle to the full-fuel position. Rotate the camshaft slowly in the direction of normal rotation until you have found the point of port closure, that is, when one to five drops in 10 seconds drip from the swan-neck pipe. In this position, zero the dial indicator. Turn the camshaft in the opposite direction until the tappet roller is on the base circle of the camshaft and read the dial indicator. The dial-indicator reading is the plunger lift to port closure. It should be 0.60 to 0.75 mm (0.123 to 0.209 in).

NOTE When you check the port closure, make certain that the intravance is fully retarded, that is, that followup rod is all the way out.

If the plunger lift to port closure is not within specification, you must remove the hydraulic head and change the plunger button to increase or de-

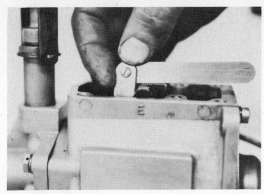

Fig. 34–27 Measuring cam nose angle.

crease the plunger lift. Repeat the spill-timing, and then remeasure the plunger lift.

When plunger lift to port closure is within specification, spill-time the injection pump again and inscribe a line on the drive hub to coincide with the pointer. Make sure that the old mark is removed.

Pressure-testing Injection Pump   The injection pump must always be tested for fuel leaks. Leaks can occur at the top of the housing and head flange, at the control unit, or at the screw of the index plate. To make the test, connect an injector tester to the inlet fitting and plug the return fitting with a 1/4-in NPT plug. Pressurize the hydraulic head to 400 psi [28.12 kg/cm²] and check for any external leakage. If the pressure drop is rapid but no external leaks are noticeable, the cause of the quick drop may be leakage by the lower O ring or excessive wear of the plunger and barrel. It may also be due to a cracked housing.

Mounting Pump to Test Stand   Secure the injection pump to the test stand, making certain that the pump is properly aligned and that there is a slight clearance between the drive couplings of the test stand and of the injection pump. Connect the lubrication-oil-supply pump to the injection pump. Switch on the lubrication-oil-supply pump and adjust the pressure to 40 psi [2.81 kg/cm²]. There must be no oil leakage. (Lubrication oil when mixed with the test fuel will change the fuel viscosity.)

Connect a hose from the lift fitting on the test stand to the inlet side of the transfer pump. With another hose, connect the outlet side of the transfer pump to the 2.5-µm (micron) final-stage filter. Two more hoses will be required in order to connect the filter to the hydraulic-head inlet fitting and the overflow valve to the return fitting on the test stand.

Be certain that you have the correct test injector and that the opening pressure is adjusted properly. Connect the injection lines to the injector holder and to the discharge fitting on the hydraulic head. The first injector holder corresponds to the first cylinder in the firing order and so on, respectively. Make sure that you use the specified length and the correct inside diameter of the injection lines.

Checking and Adjusting Injection Pump   Assuming the test stand is in good working order and the

test fuel meets specification, measure and adjust the torque cam nose. To do this, remove the governor cover and turn out the droop screw and the high-idle adjusting screw. Rest two feeler-gauge blades against the cam nose, as shown in Fig. 34–27. Then withdraw the gauge and measure its angle with a protractor. All test specifications are given as an example only.

If an adjustment is required, use the support block to support the fulcrum while loosening the cam-nose retaining nut. After making the necessary adjustments, install the special governor cover so that you can observe the governor and the action of the excess-fuel device.

Start the lubrication-oil-supply pump. Readjust the pressure, if necessary. At the same time, check the action of the excess-fuel device. Within 25 seconds the stop plate must move to its full outward position. Next, bleed the air from the injection pump by using the hand primer to open the bleed screw at the injector holders. Switch on the drive motor, and set the throttle to the low-idle position. With the injection pump camshaft turning at low speed (500 rpm), set the throttle to the full-fuel position. When all the air has been removed from the injection lines, close the bleed screw on the injector holder and warm up the injection pump.

Flow-testing Torque Check Speed   When the test fuel temperature is about 110°F [43.3°C] set the counter to 1000 strokes, set the throttle to the full-fuel position, and increase the test-stand rpm to torque check speed (1800 rpm). At this pump speed, the cam nose should be in contact with the stop plate. Now, trip the counter and observe the fuel rate. The average fuel delivery of all six injectors should be within specification (86 cm³). If the fuel delivery is lower or higher than this, you must adjust the stop-plate screw inward or outward to reduce or increase the travel of the cam nose. When making this adjustment, be careful not to bend or twist the stop-plate pin or the guide. Recheck the fuel delivery at the torque check speed.

Flow-testing Rated Speed   To check the fuel delivery at rated speed, increase the test-stand speed to the rated speed (2400 rpm). Trip the counter and measure the fuel delivery. The average fuel delivery should be 81 cm³. If the average delivery is higher than specified, the governor spring gap must be rechecked.

To adjust the high-idle adjusting screw, increase the pump speed by 80 rpm, and then turn the high-idle adjusting screw until the cam nose just moves off the stop plate.

To check the governor cutoff speed, increase the test stand to 2800 rpm and trip the counter. The average fuel delivery must not exceed 4 cm³.

Droop-Speed Adjustment   To measure fuel delivery at droop speed, reduce the test-stand speed to droop check speed (2200 rpm), then trip the counter. The average fuel delivery should not exceed 82 cm³. If the fuel delivery is incorrect, loosen the locknut of the droop adjusting screw and turn the screw in or out to reduce or increase droop-speed fuel delivery (Fig. 34–28). Recheck fuel delivery at all speeds before checking the intravance.

Checking Intravance To check the operation of the intravance, run the test stand at idle speed, switch on the pressure phase switch, and set the degree wheel of No. 1 injector to zero. Gradually increase pump speed to high idle. Observe when the advance starts and at what speed the camshaft has fully advanced (14°). If the advance curve does not coincide with the specification, remove the intravance for servicing.

Installing Injection Pump to Engine Position the engine timing mark (with the No. 1 piston on the compression stroke) and align the pump hub line with the pointer. Then mount the injection pump to the engine. Place the drive gear on the drive hub while meshing the teeth of the gear with the teeth of the timing gears. **NOTE** No timing of the gears is required.

Holding the camshaft, install and tighten the drive-gear bolts. To recheck the timing, rotate the engine crankshaft about 45° counterclockwise, and then slowly rotate it in a clockwise direction until the engine timing marks are aligned. Next, check the drive-hub timing.

Install the fuel-supply and fuel-return lines, lubrication lines, and injection lines. Also install the throttle and the components, covers, etc., which were previously removed.

Bleed the system. Leave the injection lines at the injector holder loose. Crank the engine until fuel leaves the injection line, and then tighten the connection. Start the engine and bring it to its operating temperature. Check the high- and low-idle speed, making any necessary adjustments.

Questions

1. List the major differences between a Roosa Master injection pump and an American-Bosch distributor-type injection pump.

2. Explain how the intravance changes the beginning of injection.

3. What is the major difference (in regard to the operation) between the intravance and the TMF timing device?

4. Describe the mechanical action within the injection pump during camshaft rotation.

DROOP ADJUSTING SCREW

Fig. 34–28 Adjusting droop-speed fuel delivery.

5. Explain the pumping, metering, and distributing action from the point where the camshaft forces the plunger upward to the point where the tappet roller again rests on the base circle of the camshaft.

6. How would you increase the end play of the camshaft?

7. Why is there a specified clearance between the tappet guide and roller?

8. Why is it so essential for the opening pressure of the delivery valve to be within specification?

9. What checks must be made and what precautions must be taken when you reassemble the hydraulic head?

10. How do you check for leaks in the check valves of the transfer pump?

11. What would occur if the slippage torque of the friction drive was less than specified torque? What would happen if it was more than specified torque?

12. Outline the procedure to install the hydraulic head.

13. Describe how to spill-time and measure plunger lift to port closure.

14. If the average delivery tested out lower than specification during a flow test for rated speed, what checks and adjustments would then be required?

15. How is droop speed adjusted to increase fuel delivery?

Caterpillar Sleeve-metering Fuel-Injection System

A. Fuel passage
B. Fuel passage
1. Speed limiter
2. Inlet for engine oil pressure to speed limiter
3. Plunger of speed limiter
4. Injection pump
5. Sleeve control shaft
6. Housing for fuel-injection pumps
7. Transfer pump
8. Lever on governor shaft
9. Drive gear of transfer pump
10. Governor housing
11. Governor spring
12. Thrust collar
13. Carrier and governor weights
14. Sleeve lever
15. Camshaft
16. Bypass valve and spring

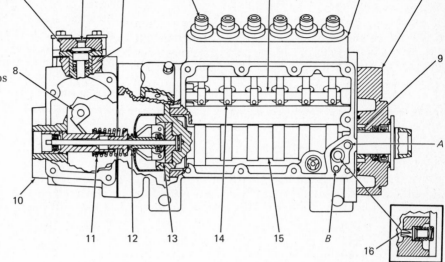

Fig. 35–1 Sectional view of the sleeve-metering injection pump. *(Caterpillar Tractor Co.)*

The sleeve-metering fuel-injection pump made by the Caterpillar Tractor Company is very similar to the American Bosch distributor-type injection pump discussed in Unit 34. It is also a constant-beginning and variable-ending delivery design. The main differences between these pumps are in the mechanical construction of the governor and the connecting links to the control sleeves. Since the Caterpillar pump has a sleeve-metering pumping element for each cylinder, it uses a standard camshaft. This injection pump is lubricated and cooled by diesel fuel.

Fuel Flow to (and within) Injection Pump   Filtered fuel from the fuel tank enters at the rear of the governor housing (see Fig. 35–1). Passages within the governor housing and injection-pump housing direct the fuel to port B, through the side-cover passages into port A, and on to the gear-type transfer pump. A fuel filter can be mounted on the side cover so that the fuel is filtered on leaving port B. When the engine is being cranked over or is running, the transfer pump pressurizes the fuel contained in the injection pump and governor housing. The maximum pressure at full load is 2.1 kg/cm^2 ± 0.4 kg/cm^2 (30 psi + 5 psi) and is limited through a bypass valve. When the bypass valve opens, it directs fuel through a passage to the inlet side of the pump.

Pumping Element Design   Each pumping element consists of a retainer, bonnet, check valve, spring, barrel, plunger and sleeve, and plunger-return spring with washer (lower spring seat) (Fig. 35–2). These components, too, are similar in design to the American Bosch distributor-type pump. The

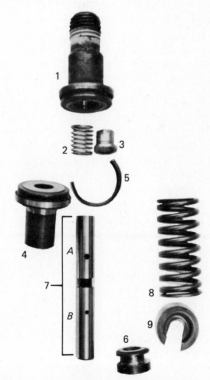

1. Bonnet	6. Plunger sleeve
2. Spring	7. Plunger (A, fill port;
3. Check valve	B, spill port)
4. Barrel	8. Plunger return spring
5. Retainer	9. Washer

Fig. 35–2 Exploded view of one pumping element. (The numbers indicate the component assembly order.)

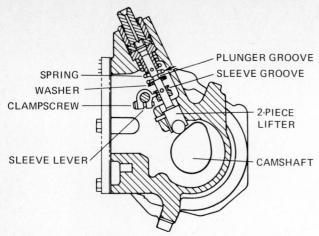

Fig. 35–3 Sectional view of a pumping element (installed).

plunger is also center-drilled with two cross-drilled holes (the fill and spill ports). The precision-lapped sleeve is again similar to the American Bosch type, except that the sleeve is fitted on the lower end of the plunger. However, the barrel has no passages or ports.

When the pump is positioned in the housing, the washer (lower spring seat) hooks into the plunger groove and the spring fits over the barrel (Fig. 35–3). The groove on the sleeve rides in the lever which is connected to the sleeve control shaft by a clamp screw. The two-piece lifter rests against the camshaft lobe because the plunger spring forces the plunger downward.

Pumping Principle When the lifter is on the base circle of the cam lobe, the fill port of the plunger is below the barrel, and the sleeve covers the spill port (Fig. 35–4a). The internal fuel pressure pressurizes the passage within the plunger and the area above the plunger. When the throttle is in the fuel-off position, the sleeve is moved by the sleeve control shaft and lever below the spill port (Fig. 35–4b). The area above the plunger remains pressurized by the pressure within the pump. As the plunger is forced upward through the action of the cam lobe and lifter, the barrel closes the fill port. [**NOTE** The spill port remains uncovered (Fig. 35–4b).] Fuel in the barrel is displaced by the plunger and released out of the spill port until the

plunger is at TDC. As the plunger is forced down by the plunger spring, fuel flows through the spill port and fills the area vacated by the plunger.

When the throttle is being moved to the idle position, the half-fuel, or the full-fuel positions, the sleeve control shaft and level move the sleeve to cover the spill port (Fig. 35–4c). At idle position, the sleeve's upper edge barely covers the upper edge of the spill port (Fig. 35–4d). With an increase in throttle setting, the sleeve is moved further upward, thereby increasing the distance from the upper edge of the sleeve to the upper edge of the spill port. This causes the effective stroke to increase.

In the fuel-on position, regardless of the position of the sleeve, only the fill port is open (Fig. 35–4a). As the plunger moves upward and the fill port closes, the passage within the plunger and the area above it are pressurized (Fig. 35–4c and d). This increase in fuel pressure lifts the check valve off its seat. This, in turn, increases the pressure in the injection line until the force on the nozzle valve overcomes the spring force. As soon as the upper edge of the spill port passes the upper edge of the sleeve, the fuel spills back through the spill port (Fig. 35–4e). This reduces the pressure, causes the check valve to seat, and ends injection. The plunger continues ascending until it reaches TDC during which time the displaced fuel leaves through the spill port (Fig. 35–4e). As the plunger moves downward, fuel can enter through the spill port until it is closed by the sleeve. There is a short distance of plunger travel when no fuel can enter the plunger area. This is the period between the closing of the spill port and the opening of the fill port.

Governor Operating Principle The mechanical governor used with the sleeve-metering injection pump differs somewhat from other mechanical governors (See Fig. 35–5). But the governor spring and weight assembly, measure the engine speed and change the position of the various levers in order to increase or decrease the fuel quantity injected per cycle.

As the throttle is moved to the full-fuel position, the force on the governor spring increases. Lever 10 (in Fig. 35–5), however, resists the movement of the thrust collar, and only a part rotation is made to the load stop lever (6). This load stop lever moves the

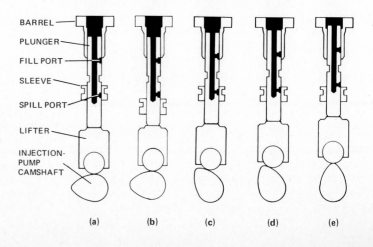

Fig. 35–4 Pumping element when the throttle is in various positions. (a) Plunger in metering position, sleeve in fuel-on position; (b) plunger in metering position, sleeve in no-fuel position; (c) plunger at beginning of injection, sleeve in fuel-on position; (d) plunger near end of injection, sleeve in fuel-on position; (e) plunger at end of injection, sleeve in fuel-on position.

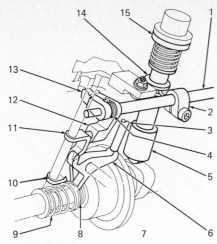

1. Sleeve control shaft
2. Sleeve lever
3. Load stop pin
4. Torsion spring
5. Lifter
6. Load stop lever
7. Weight assembly
8. Thrust collar
9. Governor spring
10. Lever
11. Lever (connects to thrust collar and lever 12)
12. Lever (slips on shaft)
13. Lever (pressed on shaft)
14. Sleeves
15. Injection pumps

Fig. 35–5 Fuel-injection components showing the mechanical governor. *(Caterpillar Tractor Co.)*

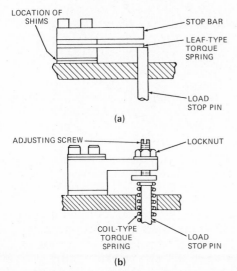

Fig. 35–6 Fuel-system settings (adjustments) with (a) leaf-type torque spring and (b) coil-type torque spring. *(Caterpillar Tractor Co.)*

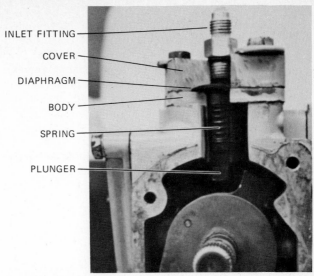

Fig. 35–7 Sectional view of the diaphragm-type speed limiter.

Fig. 35–8 View of the check valve and bypass valve.

load stop pin and lever 12. This movement causes the sleeve control shaft to rotate, moving with the sleeve lever (2) and the sleeves. Any changes in the spring force or weight force are directly transmitted to the load stop lever (6). Its maximum allowable travel is reached when the load stop pin compresses the torque spring (Fig. 35–6a) or butts against the adjusting screw (Fig. 35–6b).

The speed limiter used with this governor is a diaphragm type (Fig. 35–7). It limits the movement of the governor lever through the action of the speed-limiter plunger and lever (10 in Fig. 35–5) until the oil pressure is about 0.4 kg/cm² (6 psi).

One major difference in this (sleeve-metering) mechanical governor is that its weight assembly is bolted to the camshaft. Therefore, it is driven at camshaft speed.

Servicing Sleeve-metering Fuel-Injection Pump The tests and checks outlined under Fuel-Injection Pump Service (Unit 29) apply to the sleeve-metering pump. The pump should never be removed for service unless the failure is specifically diagnosed as being in the injection pump. After you conclude that the problem is in the pump, thoroughly clean the pump and mount it in a holding fixture. Remove the side cover, the check valve, and the bypass valve (Fig. 35–8).

NOTE If not thoroughly experienced with the Caterpillar pumps, refer directly to your service manual before proceeding further. Highlights of how you will proceed will now be given.

Remove the pumping elements by first loosening the bushing. The lever will slide out of the sleeve as you lift out the elements. At this time, do not remove the sleeve from the barrel and do not disassemble the unit. Place the assembly in clean fuel.

When you disassemble the element, remove the sleeve by carefully pulling on the spring to remove it and the plunger from the barrel. Take care not to nick the plunger during removal.

Remove the retainer to separate the barrel and the bonnet. Lift out the check valve and spring. If the plunger shows any sign of surface damage on the areas where it contacts either the sleeve or the barrel, or on the edges of the fill or spill port holes, the assembly must be replaced.

If the components appear to be reusable, measure and test the plunger spring, and lap the mating surfaces of the bonnet, barrel, and check-valve seat. Clean the plunger and the barrel following the procedures outlined for in-line injection pumps.

Wash all components in clean fuel. Assemble them in the order indicated in Fig. 35–2.

NOTE The groove on the sleeve must face toward the fill port; no other alignment or timing is necessary.

Keep the elements in clean fuel oil until you are ready to install them.

Servicing Transfer Pump To remove the transfer pump, use the timing pin to hold the camshaft, and then thread the bolt into the drive sleeve. Tighten the bolt and pull the sleeve off the shaft (Fig. 35–9).

Remove the mounting bolts and lift the transfer-pump assembly from its housing. If the wear marks from the gears (on the inlet side of the pump) are past 2 o'clock, the body must be replaced because this indicates the shaft or the bores are worn beyond their limit. If the injection-pump-housing surface is only worn slightly, refinish the surface with an oil stone. However, if the damage is significant, the pump housing must be replaced.

NOTE Always replace the seals. The lip of the inner-lip seal should face toward the gear. The lip of the outer-lip seal should face outward.

Servicing Governor When the governor requires service, remove the governor bolts which attach it to the pump housing, and then lift off the governor housing. Remove the cover, torque spring assembly, seat, and spring from the shaft (to the left in Fig. 35–10). Pull the shaft with the lever upward to remove it from the housing (to the right in Fig. 35–11). Remove the dowel from the housing and slide the load stop lever from the dowel. Remove

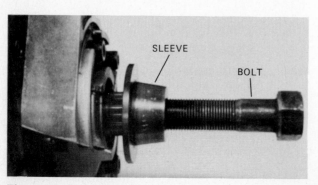

Fig. 35–9 Removing the drive sleeve.

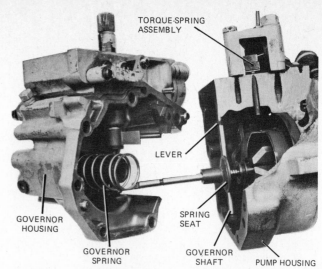

Fig. 35–10 Removing the governor housing.

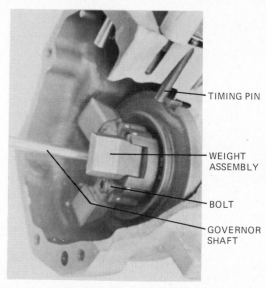

Fig. 35–11 Removing the bolts that hold the weight assembly to the camshaft. (Partial cutaway view.)

the governor-weight cover. Place the timing pin in the camshaft to hold it firm while removing the bolts which fasten the governor-weight assembly to the camshaft (Fig. 35–11). If the throttle shaft requires service, remove the lock, spring, spring guides, plates, pin and ball from the shaft. Then lift out the shaft, hub, and lever (Fig. 35–12). Always replace the seals in the housing end cover.

When inspecting the governor components for serviceability, check the speed-sensing components, that is, the thrust collar, bearing, washers, and weight assembly (Fig. 35–13). Do not overlook the lever which transmits the motion from the thrust collar to the control sleeve.

Sleeve-metering Injection-Pump Housing and Camshaft The pump housing and the camshaft must be checked and measured on a sleeve-metering injection pump, just as it is on any other injection pump.

Reassembling Sleeve-metering Injection Pump Wash the injection-pump housing, camshaft, and

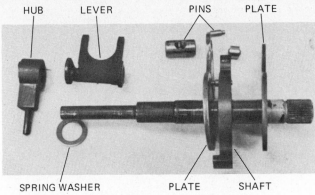

Fig. 35–12 Throttle shaft components.

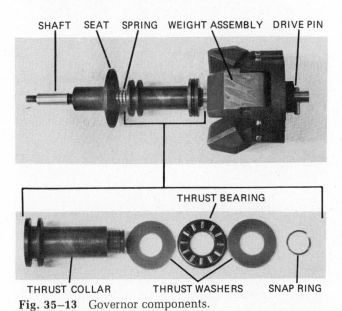

Fig. 35–13 Governor components.

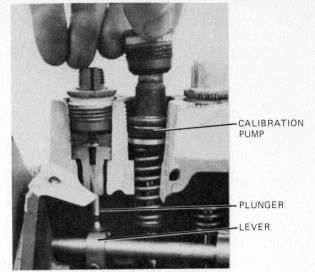

Fig. 35–14 Installing the calibration pump. (Partial cutaway view.)

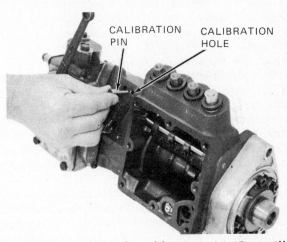

Fig. 35–15 Installing the calibration pin. (*Caterpillar Tractor Co.*)

components of the governor and transfer pump, and dry them with compressed air. Generously lubricate the bushings and the camshaft journals. Place the camshaft in the housing, turning it to ensure that it rotates freely. Install the key and the drive gear on the shaft. Place a new O ring in the groove of the transfer-pump housing. Install the idler gear. Lubricate the interior of the transfer-pump housing and the seals before you place the transfer-pump housing on the pump housing.

Place the timing pin in the camshaft. Install and tighten the sleeve to the specified torque. Recheck to make sure that the camshaft rotates freely. If it does not, recheck the transfer pump. Measure the end play. It must be within specification.

Make certain that you lubricate the bushings, shafts, seals, pins, and thrust bearing before you reassemble the governor. Do not install the side cover on the injection pump or the cover of the governor.

Fuel-Pump Calibration Before you install new or serviced pumping elements, the individual levers on the sleeve control shaft must be adjusted to the height specified in the service manual. This ensures correct positioning of the sleeve in order to inject the correct amount of fuel.

To measure the height of the lever, install the calibration pump into the bore with the plunger's flat side toward the lever (Fig. 35–14). Turn the calibration pump to engage the plunger tang with the lever. When it is engaged, tighten the bushing to the recommended torque. Install the calibration pin in the calibration hole (Fig. 35–15), and hold it with the cover in place. Insert the timing pin into the oil-supply inlet of the speed limiter. This will move the lever out of the way and allow the throttle to move to the full-fuel position.

NOTE It is not necessary to position or time the camshaft, but it is essential to measure the lever height. The beginning of delivery is timed by the precision-machined ports on the plunger and the height of the barrel, lifters, and cam lobes.

Set the dial indicator on the microgauge, bring it to zero, and lock it in position. Now place and hold the dial indicator firmly on the top of the calibration pump.

Suppose, for example, the specification states that the dial indicator should read "zero" and an adjustment of the lever is necessary. Loosen the lever screw and swivel the lever on the shaft (up or down) to attain a zero dial-indicator reading (Fig. 35–16).

Fig. 35–16 Adjusting (calibration) of an injection pump. *(Caterpillar Tractor Co.)*

Installing Pumping Elements Turn the camshaft until the lifter roller is on the camshaft-lobe base circle. Place the throttle in the full-fuel position. This brings the sleeve lever into a higher position to aid installation. Remove the pump elements from the fuel container. Slide the element into its bore. Engage the sleeve with the lever.

Place a new sealing ring on the bonnet. Tighten the bushing (using the special wrench) to the recommended torque. Cap the bonnet opening.

Install the remaining elements and then place the bypass valve, spring, and cover in position. When you install the cover on the pump housing, make certain that the bypass-valve spring is in its groove.

Fuel-System Setting (Rack Setting) This involves checking or adjusting the sleeves for the maximum-fuel position, that is, when the load stop pin contacts the torque spring or the load stop pin contacts the adjusting spring. (Refer to Fig. 35–19 for locations of the torque spring and load stop pin.)

First, remove the cover. Install the zero pin, and hold it in place with the spring and bolts. Next, insert the timing pin into the speed limiter oil-supply hole. Push the pin down to relieve the speed limiter (Fig. 35–17). This puts the sleeves in the full-fuel position. Install the dial indicator as shown in Fig. 35–18 and zero both dials. Connect one lead of the circuit tester to a good ground and the other to the load stop contact as shown in Fig. 35–19.

Slowly loosen the bolt (to allow the governor spring to move the stop pin) until the light just glows. Read the dial indicator. When both the dial indicator and the rack-setting information are the same, the fuel system is adjusted correctly. When the fuel-injection system is not adjusted properly, the dial-indicator reading will differ from the rack-setting specification.

To bring the dial-indicator reading in line with the rack-setting specification, you must add or subtract shims to lower or raise the torque spring (Fig. 35–6a). The number of shims to be added or removed will be determined by the difference between the dial-indicator reading and the rack-setting information.

Fig. 35–17 Installing the bolt to relieve the speed limiter. (Partial cutaway view.)

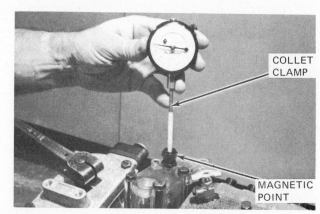

Fig. 35–18 Installing the dial indicator. *(Caterpillar Tractor Co.)*

Fig. 35–19 Circuit tester connections.

When a coil-type torque spring is used (Fig. 35–6b), loosen the locknut and turn the adjusting screw until the dial-indicator reading corresponds with the dimension given in the rack-setting specification.

Remove the fuel-system setting tools. Replace the cover. Check all the bolts and nuts. Turn the injection-pump camshaft to allow the timing pin to be installed.

Idle- and Low-Idle-Speed Adjustment After you have installed the injection pump, bleed the low-pressure system. Start the engine and bring it up to its operating temperature.

To check the high idle speed, move the throttle to the full-fuel position. Measure the engine rpm with a strobe light or a tachometer. When the engine speed is above or below specification, make the necessary adjustment by turning the high-idle adjusting screw (Fig. 35–20). When checking the low-idle speed, move the throttle to the low-idle position and then, if necessary, change the engine rpm by turning the low-idle adjusting screw.

NOTE To assure proper engine speed, change the engine speed after each adjustment by moving the throttle.

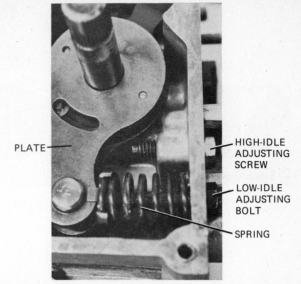

Fig. 35–20 View of the high- and low-idle adjusting screw and bolt.

PLATE — HIGH-IDLE ADJUSTING SCREW — LOW-IDLE ADJUSTING BOLT — SPRING

Questions

1. List the major differences between the American Bosch distributor pump and the Caterpillar sleeve-metering injection pump.

2. Describe the mechanical action which takes place when the throttle is moved from the no-fuel position to the full-fuel position.

3. Which component changes the effective stroke?

4. If the sleeve were placed on the plunger with the narrow flange downward, what effect would it have on metering and on the beginning of injection?

5. What precaution must be taken when you remove the pumping element?

6. Which part of the governor requires particularly close scrutiny when checking it for serviceability?

7. Why is it that the Caterpillar sleeve-metering pump does not require a phase-angle adjustment?

8. Explain how to calibrate a Caterpillar sleeve-metering injection pump. List the components which you actually adjusted or measured to accomplish calibration.

9. If you inadvertently omitted to screw the timing pin into the speed limiter and then attempted to adjust the fuel-rack setting to specification, would you be setting the fuel rack to more fuel or less fuel?

10. Explain how you would adjust the fuel-rack setting on a fuel pump with a coil-type torque spring, when the dial indicator reads 0.127 mm less than specification.

SECTION 5

Electrical System

UNIT 36

Electricity and Magnetism

ELECTRICITY

Electron Theory   An understanding of the electron theory and of magnetism is mandatory in order for you to service electrical components efficiently. As a brief review of these subjects you should look at the following facts:

All matter is composed of small parts called *molecules*. These molecules are made up of two or more atoms. Atoms are made up of even smaller particles called *electrons*, *neutrons*, and *protons*.

All matter has an atomic weight. It is expressed by an atomic number which also represents a total number of electrons (or protons). For example, copper has an atomic number of 29. It has 29 electrons that orbit in four rings around a nucleus of 29 protons and 29 neutrons.

Electrons revolve only in a certain orbit or energy level around the nucleus. The nucleus has the same number of protons and neutrons as it has electrons (Fig. 36–1).

The greater the atomic number, the more numerous are the electrons circulating around the nucleus. The nucleus, which has the greatest mass and weight, keeps the electrons in orbit around itself by gravitational pull. The electrons which are closest to the nucleus are more strongly attracted (pulled) to it than the electrons which are farthest away. The electrons in the outer orbit are partially

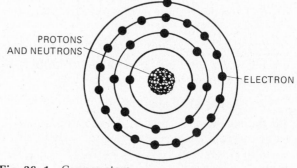

Fig. 36–1   Copper atom.

free from the pull, and therefore can sometimes be forced off their orbit. They are called *free electrons*. Those in the inner orbit are called *bonded electrons*.

Normally the atom is electrically balanced. If a state of imbalance should develop through the loss of electrons, the atom becomes positive in charge. It then tends to attract other electrons in order to regain its balance. This electrical imbalance (potential difference) is the basic principle underlying electric current because imbalance causes electrons to flow from atom to atom.

Conductors   A conductor may be defined as matter made up of atoms whose free electrons are dislodged easily and whose orbits overlap, allowing easy electron flow from atom to atom.

The atomic structure governs the conductivity of an element, or the extent to which it resists the movement of electrons. Elements such as silver, copper, or aluminium are good electron carriers because their free electrons are not very closely tied to the atom. Copper is presently the most common carrier of current (electrons) because it is commercially economical, bends easily, and offers little change in resistance with changes in temperature.

There are two other factors in addition to temperature which can change the resistance of a conductor. They are the cross-sectional area and the length. As the cross-sectional area is increased, the electron-carrying capacity increases because there are more atoms available to carry the current and so the resistance is reduced. When a conductor is lengthened without increasing its square area, more electrons have to be dislodged. This causes heat and increases resistance.

Insulators (Nonconductors)   An insulator is composed of one or more atoms with an atomic structure of more than four electrons bonded in a compact outer orbit. The structure will not allow the movement of electrons because there are no free electrons present. The most common insulators used for electrical wiring are natural or synthetic rubber, varnish, mica, Bakelite, and various fibers.

Insulators are essential to the conductors of any electrical device because they prevent electron losses, avoid a short or grounded circuit, and can separate two conductors. They also protect the conductor from external damage, oil, dirt, water, and heat.

Semiconductors   Semiconductor materials (elements) are neither good conductors nor good insulators because they are materials which have an atomic structure which has three or five electrons in their outer orbit. However, these materials can be altered to fulfill a useful purpose, that is, to act as an electrical device which can act as a conductor under certain conditions and as a nonconductor under other conditions. (See the section on semiconductors in Unit 44 for more details.)

Resistance   Resistance is any force tending to hinder motion. Each atom resists the removal of its free electrons due to gravitational pull of the nucleus. Resistance in a conductor, semiconductor, or insulator is the opposing force which the atoms give to the movement of electrons.

NOTE   The resistance of insulators or special resistors usually decreases with increase in temperature.

Electron Flow   Electrons will flow from atom to atom when there is a difference in magnitude and when there is a path which connects the concentration of electrons on the negative terminal with positive terminal, which lacks electrons. When the terminals are connected (by a conductor), an electric circuit is made and the free electrons flow from atom to atom (from terminal to terminal) at a speed of 186,000 nmi/s (nautical miles per second) [344,472

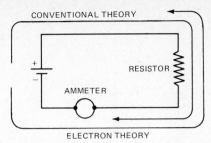

Fig. 36–2   Demonstrating the electron theory and the conventional theory.

km/s (kilometers per second)]. However, it has been proven scientifically that they travel only a few inches per minute in the conductor.

Current Flow   There are two theories, the conventional theory and the electron theory. The conventional theory states that current flows from the positive terminal of the power source to the negative terminal of the source. The electron theory states that free electrons flow from negative to positive (see Fig. 36–2). This electron flow, or *current*, is the rate of electron flow per second past a given point. The way to measure this flow is by using an ammeter. Regardless of whether the electrons move from positive to negative or from negative to positive, they have to pass the ammeter which measures the flow in amperes.

NOTE   Whenever there is a current flow, all electrons that leave the negative terminal will reenter the positive terminal. So, whether you measure the current flow at the negative or at the positive side of a circuit, the ampere reading always will be the same.

DIRECT CURRENT AND ALTERNATING CURRENT   *Direct current* refers to the flow of electrons in one direction only. This means the voltage source does not change its polarity. When the flow of electrons changes direction (cycle), it is called *alternating current*.

Voltage or Electromotive Force   Voltage is the difference in the concentration of electrons between the negative and the positive terminals. When the two points of different potential are connected, the potential difference at these two points causes the electrons to move because of their forces of attraction and repulsion. The potential difference is called *voltage* or *electromotive force* (emf), and the unit of measurement is the *volt* (V). The greater the concentration at one terminal, the greater will be the voltage or potential energy to do work. Work is the combined effort of voltage and current expressed in watts (W).

Ohm's Law   Ohm's law states that the intensity of the electric current in a circuit is proportional to the voltage (emf) and inversely proportional to the resistance. In other words, the current flow can be changed by changing the voltage or by changing the resistance in the circuit. Note, however, that current has no effect on resistance or voltage, but each has an effect on current.

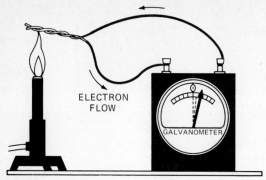

Fig. 36–3 Creating voltage through heat by means of a thermocouple (*United Delco-AC Products, General Motors of Canada, Ltd.*)

If you know two of the three factors [*I* (amperes), *V* (volts), or *R* (resistance)], then you can apply Ohm's law to find the third factor.

Voltage Source   There are many ways to produce voltage including friction, heat, light, pressure, chemical energy, and magnetism.   This textbook will be concerned only with voltage produced by heat, chemical energy, and magnetism.

Voltage through Heat   You may associate the principle which creates voltage through heat with such devices as thermocouples, heat indicators, or temperature gauges.   In each instance heat is used to produce voltage.   When two wires or plates made of different metals twisted or welded together are heated, the junction transmits the heat to their opposite ends, causing a potential difference (see Fig. 36–3).   The greater the difference in temperature between the opposite ends of the wires or plates, the greater the voltage, and therefore the greater the current flow.   The voltage produced through heat is very small, but the current flow is sufficient to energize the coil of a small gauge.

MAGNETISM

Theory of Magnetism   Each molecule in iron or in iron alloy has a north and south pole.   When not magnetized, the molecules are oriented randomly within the material.   An iron bar can be magnetized through magnetic induction by striking a magnet over the bar to magnetize it, by placing the iron bar in a strong magnetic field, or by electromagnetic induction.   These actions cause the molecules within the bar to align themselves with all north poles pointing in the same direction (Fig. 36–4).   It is now known that these magnetic lines of force are essentially electrical in nature.   They arise from the interaction between the spinning and orbiting electrons.   When these electrons align so that their forces are combined, the iron bar is magnetized.

Basic Facts about Magnetism
1.   A magnet has a north and south pole and the magnetic lines of force leave the north pole and enter the south pole.   This can be proven by placing a

SOUTH POLE ■□ NORTH POLE
IRON MOLECULE

UNMAGNETIZED IRON

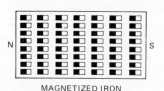

MAGNETIZED IRON

Fig. 36–4 Unmagnetized iron and magnetized iron. (*United Delco-AC Products, General Motors of Canada, Ltd.*)

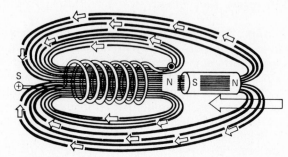

Fig. 36–5 Reducing magnetic lines of force. (*United Delco-AC Products, General Motors of Canada, Ltd.*)

compass near a magnetic bar.   The compass needle will align itself so that the south pole of the compass needle points to the north pole of the bar.
2.   The concentration of magnetic lines of force is heaviest at the pole ends and decreases in density midway between the poles since magnetic lines have "like" polarity and therefore repel each other.
3.   Like poles (charges) repel each other, and unlike poles (charges) attract each other.
4.   If you cut a north pole from one end of a magnet, the result is a second short magnet with a north and south pole.
5.   To date there is no known insulation against magnetic lines of force   However, magnetic lines can be reduced by touching an iron core to one pole piece.   This results in magnetic lines of force splitting in two separate parts, reducing the lines of force of the left magnet (Fig. 36–5).
6.   Magnetic lines of force can penetrate certain materials more easily than others.
7.   Permanent magnets can be either natural or an alloy such as nickel-iron, or aluminium-nickel-cobalt (magnetized) because these alloys have an atomic structure similar to iron.   An alloy with a loose molecular structure allows magnetic lines of force to pass through itself more easily and is used as a temporary magnet.

Basic Facts about Electromagnetism
1.   A straight wire which carries current creates a magnetic field around itself.   The lines of force form concentric circles around the wire.   The right-hand

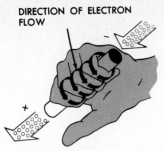

DIRECTION OF ELECTRON FLOW

Fig. 36–6 Left-hand rule (electron theory) for determining the direction of electron flow. (*United Delco-AC Products, General Motors of Canada, Ltd.*)

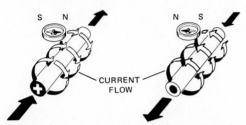

CURRENT FLOW

Fig. 36–7 Symbol indicating current flow and direction of magnetic lines of force. (*United Delco-AC Products, General Motors of Canada, Ltd.*)

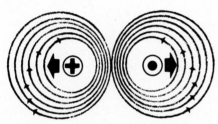

Fig. 36–8 Strong magnetic field between conductors. (*United Delco-AC Products, General Motors of Canada, Ltd.*)

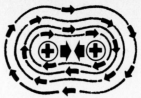

Fig. 36–9 Magnetic field between conductors canceled. (*United Delco-AC Products, General Motors of Canada, Ltd.*)

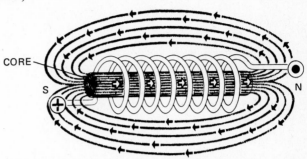

CORE

Fig. 36–10 An iron core placed inside the coil increases the number of lines of force (field strength). (*United Delco-AC Products, General Motors of Canada, Ltd.*)

rule to determine the direction of electron flow as well as the north and south poles is used when the conventional current theory is applied. The left-hand rule is used when the electron theory is applied (Fig. 36–6). The magnetic field of a current-carrying conductor behaves in the same manner as the magnetic field of a permanent magnet, as long as current flows.

2. When the direction of current flow is changed, the direction of the lines of force also changes. The symbol used to indicate current flow is shown in Fig. 36–7.

3. When two conductors carrying an equal current flowing in opposite directions are placed side by side, the lines of force are also in an opposite direction and are more concentrated between the conductors than on the outside (see Fig. 36–8). The current-carrying conductors move apart to relieve the imbalance in the magnetic field.

4. If there is an equal current flow in the same direction through two or more parallel conductors, each conductor alone creates a circular field of force and the fields of force move in the same direction (Fig. 36–9). The lines of force between the conductors move in the opposite direction. Because the strength of each field is the same, the total magnetic effect between the conductors is canceled. The con-

ductors will tend to move toward each other causing the magnetic effect to increase as the lines of force from each conductor join and surround the conductor.

5. An electromagnet is produced when a straight current-carrying conductor is formed into a single loop. All lines of force enter the inside of the loop on one side and leave on the other side.

6. When the number of loops (coil) is increased, the magnetic field of force increases. When using the right-hand rule to determine the polarity of the coil for the conventional current theory, the lines of force travel from the south pole to the north pole.

7. Air is a very poor conductor of magnetic lines of force. The smaller the air gap, the greater the number of lines of force.

8. When a soft iron core is inserted into the coil, it forms a true electromagnet (Fig. 36–10). The magnetic flux (lines of force) will increase as much as several hundred times, because the permeability of rough iron may be 2500 times that of air.

9. When the current flow is increased, say from 10 to 50 amperes (A), the magnetic lines of force also increase in the ratio of 1:5.

10. The same magnetic force will be produced by an electromagnet having 1000 turns of wire carrying 10 A as that of an electromagnetic force having 200 turns of wire carrying 50 A. The number of turns times the amperes equals 10,000 ampere-turns in either case.

Questions

1. Define matter, conductor, and insulator.

2. List four reasons why it is necessary to insulate conductors.

3. List the four most common causes of insulation failure.

4. Give the term used for one-way flow of electrons.

5. If you break a neutral magnet in half, would you have two neutral magnets? Why or why not?

6. Refer to Fig. 36–5. Explain why the magnetic lines of force have changed their path.

7. How can you transform unmagnetized material into magnetized material?

8. Explain why a magnetic field has a north and south pole after conductor has been formed into a coil.

9. Give three methods by which the magnetic strength of a coil conductor is increased.

UNIT 37

Electric Circuits and Test Instruments

Electric Circuits An electric circuit is basically a path for current flow. It includes a voltage source, a resistance unit (field coil, light bulb, etc.), and conductors (wires), which form the path for the current.

Electrical Failures No matter how well an electrical system is built or how well an electrical component (resistor unit) is designed, it cannot last or operate indefinitely. There are four types of electrical component or conductor failures: a short circuit, an open circuit, a grounded circuit, and a circuit having high resistance.

The term *short circuit* indicates that part of the circuit resistance is removed (Fig. 37–1a). A short circuit can be caused by a coil winding touching an adjacent winding.

The term *open circuit* indicates that the circuit is broken either by the conductor or by the resistor unit (Fig. 37–1b). This term also is used when the conductor is removed from the terminal or is loose and without conductivity.

The term *grounded circuit* indicates that the circuit is partly isolated because the conductor has found another ground (whether accidentally or intentionally) (Fig. 37–1c).

High resistance refers to the resultant added resistance caused by a conductor with a poor, loose, or corroded connection.

Test Instruments In order to make an electrical diagnosis quickly and precisely, accurate electrical measuring instruments such as an ammeter, voltmeter, or ohmmeter are required. Equally important, is the mechanic's ability to connect and read the instruments correctly (see later sections in this unit on the various types of circuits).

The three test instruments mentioned above have common basic working parts which include a permanent horseshoe magnet and a movable coil. Current flowing through the movable coil reacts with the permanent magnetic field causing the coil to rotate. The relative movement of the coil is in proportion to the amount of current flow through the coil

windings. A pointer attached to the coil moves across a calibrated scale indicating the number of volts, amperes, or ohms. The ohmmeter, additionally, has its own voltage source and one or more resistors which are directly related to the scales.

VOLTMETER A voltmeter is used to measure voltage and/or voltage differential (voltage drop), that is, the voltage between two points of a circuit. It must always be connected to parallel (across the circuit) because of the high meter resistance (Fig. 37–2). A voltmeter for use with diesel engines should have more than one scale. The lowest scale should be calibrated in tenths of a volt.

AMMETER As shown in Fig. 37–3, a shunt of low resistance is connected parallel with the measuring coil winding. This shunt carries most of the cur-

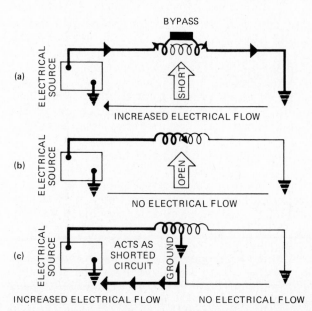

Fig. 37–1 Circuit failure. (*a*) Short circuit; (*b*) open circuit; (*c*) grounded circuit. (*United Delco-AC Products, General Motors of Canada, Ltd.*)

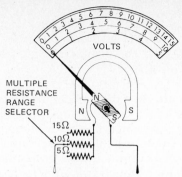

Fig. 37–2 Basic construction of a voltmeter. *(United Delco-AC Products, General Motors of Canada, Ltd.)*

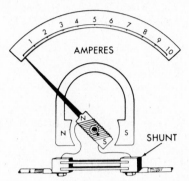

Fig. 37–3 Basic construction of an ammeter. *(United Delco-AC Products, General Motors of Canada, Ltd.)*

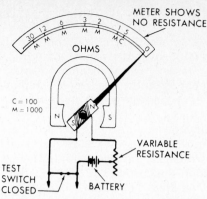

Fig. 37–4 Basic construction of an ohmmeter. *(United Delco-AC Products, General Motors of Canada, Ltd.)*

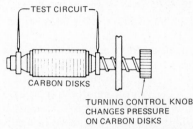

Fig. 37–5 Basic construction of a carbon pile. *(United Delco-AC Products, General Motors of Canada, Ltd.)*

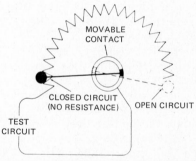

Fig. 37–6 Basic construction of a rheostat. *(United Delco-AC Products, General Motors of Canada, Ltd.)*

rent. Only a small current passes through the coil winding. An ammeter measures electron flow from the voltage source to the resistance and back to the voltage source. Therefore, it must be connected in series to the circuit. However, it may be placed on the positive side or the negative side of the circuit.

For a generator-regulator tester, an ammeter should have scales calibrated from 1 A to no less than to 100 A and as high as 1000 A for a battery-starter tester.

OHMMETER An ohmmeter, when connected, supplies its own test voltage by forcing current through the coil winding (tester) to the electrical components or to the circuit and back to its own voltage source (Fig. 37–4). The ohmmeter scale is directly related to and calibrated to the resistor within the ohmmeter to indicate the circuit or component resistance value in ohms.

An ohmmeter should have a selection of scales with the lowest scale calibrated to 10 (ohms) and the highest not less than 1000.

CAUTION Do not connect a voltmeter in series, an ammeter in parallel, or an ohmmeter to a live (hot) circuit. Any one of these procedures could damage the test instruments.

CARBON PILE A carbon pile is used to vary the resistance of a circuit. This type of resistor can absorb and dissipate reasonably high current flow without damage. A carbon pile consists of a stack or pile of carbon disks which can be forced together or sepa-

rated by a screw-type thread, thereby increasing or decreasing circuit resistance (Fig. 37–5). When the screw separates the disks, no current can flow through the carbon pile and while connected to a circuit, the circuit remains open. A control knob turns the screw and forces the disks together, thereby gradually reducing the resistance. When the disks are forced tightly together, all resistance is removed and a conductorlike effect is produced.

RHEOSTAT A rheostat is used to vary the resistance of a circuit though limited to current absorption and heat dissipation. It consists of a resistance wire wound around an insulator, over which a movable contact can slide (Fig. 37–6). The right side of the resistor wire is connected to the movable contact and the contact slip ring is connected to one circuit lead. The left side of the resistor is connected to the other circuit lead.

When the rheostat is connected in series to a circuit and the movable contact is in the position shown in Fig. 37–6, no resistance is inserted into the circuit.

However, when the contact is moved in a clockwise direction over the resistance wire, the resistance increases. The resistance inserted into the circuit depends on the position of the movable contact and the resistance of the coil wire.

Series Circuit Electric circuits are classified according to the manner in which the electrical components (resistors) are inserted into the circuit. In a series circuit the current has only one path through which it can flow. Therefore the total circuit resistance is equal to the sum all individual resistors.

Figure 37–7 illustrates a 12-V series circuit with four resistors. The total resistance for this circuit is $2\ \Omega + 5\ \Omega + 4\ \Omega + 1\ \Omega$, or $12\ \Omega$. The current, calculated in accordance with Ohm's law, is $I = V/R = 12/12 = 1$ A.

When an ammeter is connected to this circuit, it should read 1 A. However, when the current flow is higher than 1 A, it is apparent that the circuit has less resistance. An ohmmeter can be used to detect the components with the lower resistance. However, in order to do this, you need to know the value of the resistor and you must remove the battery connection.

A more expedient way of identifying a component with a lower or higher resistance is to use a voltmeter because the voltage drops when the current passes through the resistor.

When using the voltmeter method, you must know the voltage, the value of each resistor, and the current flow. As noted above, in Fig. 37–7 the voltage is 12 V, the values of the resistors are 2, 5, 4, and 1 Ω, and the current flow is 1 A.

To calculate the voltage drop across the resistor $R1$, multiply the amperes by the resistor value: $I \times R = 1 \times 2 = 2$-V drop. Subtracting 2 V from 12 V equals 10 V. Therefore, when the voltmeter is connected at point B and to ground, it should read 10 V. When it is connected at points A and B, the voltmeter should read 2 V.

When the voltmeter is connected at point C and to ground it should read 5 V because the resistor $R2$ has a 5-Ω resistance. ($I \times R = 1 \times 5 = 5$ V.) When there are 10 V at point B and the voltage drop is equal to 5 V, to arrive at the voltage at point C, you must subtract: $10\ V - 5\ V = 5$ V. When the voltmeter is connected at points C and D, it should read 4 V because the resistor $R3$ has a 4-Ω resistance. ($I \times R = 1 \times 4 = 4$ V.) To arrive at the voltage at point D, you subtract:

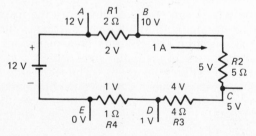

Fig. 37–7 Series circuit. (*United Delco-AC Products, General Motors of Canada, Ltd.*)

$5\ V - 4\ V = 1$ V. When the voltmeter is connected at point E and to ground, the voltmeter should read 0 V. When connected at points D and E, it should read 1 V; but after connecting the voltmeter from A to E, it should read 12 V.

Parallel Circuit In a parallel circuit the current has as many paths as there are resistors in parallel. These paths are sometimes called *branch circuits* (Fig. 37-8). The sum of the branch circuits' current is equal to the total current flow in the circuit. Current flow through each branch will differ as the resistance differs. The total current flow should be 12 A.

$$I = \frac{V}{R} = \frac{12}{6} = 2\text{ A} \qquad \text{in branch one}$$

$$I = \frac{V}{R} = \frac{12}{3} = 4\text{ A} \qquad \text{in branch two}$$

$$I = \frac{V}{R} = \frac{12}{4} = 3\text{ A} \qquad \text{in branch three}$$

$$I = \frac{V}{R} = \frac{12}{4} = 3\text{ A} \qquad \text{in branch four}$$

Therefore, the total current flow is

$$2 + 4 + 3 + 3 + 3 = 12\text{ A}$$

Another formula to calculate the total resistance R_T of a parallel circuit is

$$R_T = \frac{1}{R1} + \frac{1}{R2} + \frac{1}{R3} + \frac{1}{R4} = \frac{1}{6} + \frac{1}{3} + \frac{1}{4} + \frac{1}{4}$$
$$= 0.17 + 0.33 + 0.25 + 0.25 = 1\ \Omega$$

When an ammeter is connected to the circuit and the current flow shows a reading lower than 12 A, it may be due to additional resistance in the circuit (in the form of a loose or poor connection) either on the live or ground side of the circuit, or it may be the result of one or more resistors yielding higher resistance due to a damaged coil. When the current is higher than 12 A, one or more resistors may have short circuited or grounded.

To locate the cause of high or low current flow in the parallel circuit shown in Fig. 37–8, you would first connect a voltmeter at points A-B, A-C, A-D, and A-E. A battery voltage of 12 V would occur each time because the voltage drop across each resistor would be the same. However, if the voltage at any of these points is lower, it would be due to additional resistance after points B, D, and E (including battery connection).

To locate the resistance, connect the voltmeter at B to battery ground terminal, and then connect it at C, D, and E. If voltage readings occur at any of these points, there are probably loose or poor connections or partly damaged conductors.

Series-Parallel Circuit A series-parallel circuit is one major path and two or more branches through which current flows (see Fig. 37–9). The total cur-

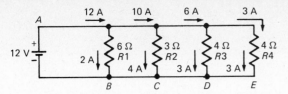

Fig. 37-8 Parallel circuit. (*United Delco-AC Products, General Motors of Canada, Ltd.*)

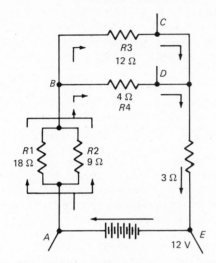

Fig. 37-9 Series-parallel circuit. (*United Delco-AC Products, General Motors of Canada, Ltd.*)

rent flow in these circuits is also based on Ohm's law: $I = V/R = 12/12 = 1$ A.

You should notice that the 18- and 9-Ω resistors and the 12- and 4-Ω resistors are in parallel and the 3-Ω resistor is in series with them. To arrive at the total circuit resistance, the two parallel resistors must be reduced to series resistance, using the formula

$$R_{eq1} = \frac{R1 \times R2}{R1 + R2} = \frac{18 \times 9}{18 + 9} = \frac{162}{27} = 6 \ \Omega$$

$$R_{eq2} = \frac{R3 \times R4}{R3 + R4} = \frac{12 \times 4}{12 + 4} = \frac{48}{16} = 3 \ \Omega$$

$$R_T = R_{eq1} + R_{eq2} + R5 = 6 + 3 + 3 = 12 \ \Omega$$

$$I_T = \frac{V}{R_T} = \frac{12}{12} = 1 \ \text{A}$$

where R_{eq1} = equivalent resistance of R1 and R2, Ω
R_{eq2} = equivalent resistance of R3 and R4, Ω
R_T = total resistance, Ω
I_T = total current, A
V = voltage, V

When the ammeter indicates that the current flow in this circuit is more or less than 1 A, use a voltmeter to measure the voltage drop and thereby locate the point of higher or lower resistance.

Questions

1. Briefly explain Ohm's law.

2. Why must a voltmeter be connected in parallel?

3. Why must an ammeter be connected in series?

4. What is the most significant difference between a carbon pile and a rheostat?

5. Basic meters use which of the following principles (a) revolving coil and magnet, (b) revolving coil and rotor, (c) permanent-magnet moving-coil meter movement, or (d) none of these.

6. A voltmeter is used to measure which of the following (a) watts, (b) intensity of current, (c) lack of pressure, or (d) electromotive force.

7. In which way is a voltmeter connected (a) in parallel to the load, (b) in series with the load, (c) not used in connection with a load, or (d) none of these.

8. In which way is an ammeter connected (a) in parallel to the load, (b) in series with the load, (c) across the load, or (d) none of these.

9. An ohmmeter is used to measure which of the following (a) current, (b) voltage, (c) resistance, or (d) none of these.

10. Give the abbreviations for the following: pressure (volts), volume (amperes), and resistance (ohms).

11. When electrical devices are connected so that the same current flows through all of them, then we have a (a) parallel circuit, (b) series circuit, (c) parallel-series circuit, or (d) series-parallel circuit.

12. When electrical devices are connected so that the same voltage is applied to each, we have a (a) parallel circuit, (b) series circuit, (c) parallel-series circuit, or (d) series-parallel circuit.

13. Three resistors of 2, 2, and 4 Ω are connected in parallel. The resistance of the combination as connected is (a) 0.014 Ω, (b) 0.8 Ω, (c) 8 Ω, (d) 16 Ω, or (e) 0.008 Ω.

14. Voltage drop is also called (a) VR/1 drop, (b) VR drop, (c) VI drop, (d) IR drop, (e) V $\times$ IR drop, or (f) none of these.

15. The resistance in a circuit is 6 Ω if the voltage is 12 V. What is the amperage? (a) 2 A, (b) 4 A, (c) 1.75 A, (d) none of these.

16. Current flow in a circuit depends on (a) voltage only, (b) resistance only, (c) a combination of voltage and resistance, or (d) none of these.

Wires and Terminals

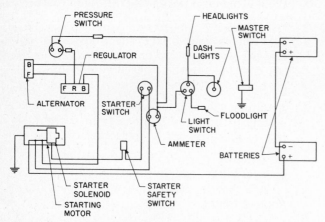

Fig. 38–1 Typical wiring diagram. *(Allis-Chalmers Engine Div.)*

Wiring Diagrams The schematic wiring diagram illustrated in Fig. 38–1 is similar to those illustrated in service manuals. The purpose of wiring diagrams is to help the mechanic trace and service the wiring system. The manuals also include separate wiring diagrams for each individual circuit to allow a more detailed layout of the conductor connections. The conductors are color coded, and the electrical components are indicated by a symbol or are illustrated schematically.

Some manufacturers prefer a two-wire system; others use the frame as the second wire (the ground wire). SAE recommends the following color code for a tractor-trailer combination: black for a hot wire (live wire), white for ground, brown for license and taillight circuit, yellow for stop and left-turn light circuit, green for stop and right-turn light circuit, and blue for the auxiliary circuit.

Conductor Size Selection The conductor size (wire number) used to connect the various electrical components is governed by the maximum allowable circuit resistance (voltage drop).

The maximum allowable circuit resistance (including connections and switches) must not exceed the values in Table 38–1.

Table 38–1 MAXIMUM ALLOWABLE VOLTAGE DROP AND CIRCUIT RESISTANCE FOR A 12-, 24-, AND 32-V CRANKING-MOTOR CIRCUIT

Circuit voltage, V	Maximum voltage drop, V/100 A	Maximum resistance, Ω
12	0.12	0.0012
12 (high-output starting motor)	0.075	0.00075
24–32	0.2	0.002

Deduct the following from the total circuit resistance recommended before determining wire sizes for a given length:

- Each connection = 0.00001 Ω
- Each contactor = 0.00020 Ω

NOTE Make certain when you make the cables up that the resistance of each soldering connection is not higher than 0.00001 Ω and that each contact resistance in the circuit is not higher than 0.002 Ω.

Cranking-Motor Circuit The cranking-motor circuit supplies the high current to the motor. The selection chart shown in Fig. 38–2 gives the resistance in ohms of various size battery cables, according to their length.

Cranking-Motor Control Circuit The cranking-motor control circuit is the one that permits switching. To stay within the maximum allowable voltage drop, the resistance of the cranking-motor control circuit, that is, the resistance between the battery and the solenoid switch, should not exceed 0.007 Ω for a 12-V starting circuit or 0.030 Ω in a 24-V starting circuit. The selection chart shown in Fig. 38–3 is indicative of the wire size required to maintain resistance within these values.

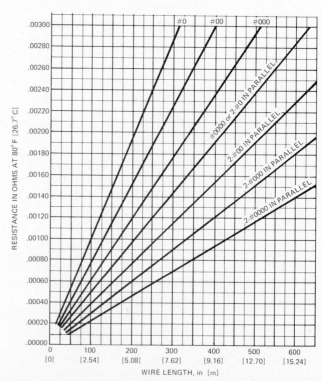

Fig. 38–2 Battery-cable selection chart. *(Delco-Remy Div. of General Motors.)*

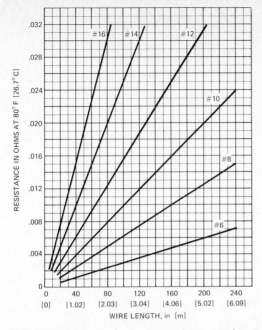

Fig. 38-3 Standard-wire selection chart. (Delco-Remy Div. of General Motors.)

Table 38-2 BATTERY-TO-ALTERNATOR WIRE SIZES

12-V system		24 to 32-V system	
Amperes (max.)	AWG* size	Amperes (max.)	AWG size
53	8	25	14
85	6	40	12
125	4	60	10
205	2	100	8

* American Wire Gauge.

Battery to Alternator When the total wire length from the battery to the alternator does not exceed 16 ft [4.88 m] the recommended wire sizes are as shown by Table 38-2.

Accessory Circuit Resistance The wire size for all accessory feed circuits should be No. 12 or No. 14 if the total length does not exceed 15 ft [4.62 m]. All other control circuits should use No. 14 wire (except the glow-plug circuit). When an insulated ground circuit is used, No. 10 wire is recommended.

TERMINALS

The electrical system is probably the most common problem area of all automotive or diesel equipment. Often the problem results from loose or corroded terminals or incorrectly selected terminals. Sometimes it is simply that the terminals are not properly fastened to the wire (conductor).

Terminal Ends Some terminal ends (wire ends) are shown in Fig. 38-4. Before selecting a terminal end, you must decide where crimping or soldering is to be used. Crimped terminal fastenings are acceptable for some applications while others require the security of solder.

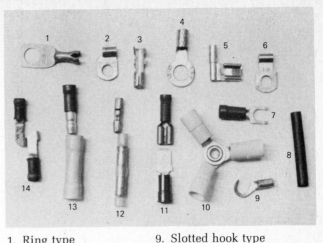

1. Ring type
2. Roll type
3. Female snap-on
4. Lug type
5. Female slide connector
6. Roll type
7. Slotted-flange bay type
8. Insulator
9. Slotted hook type
10. Three-way connector
11. Male and female slide connector
12. Male and female plug connector (bullet connector)
13. Male and female plug connector (bullet connector)
14. Knife disconnector

Fig. 38-4 Terminal ends.

NOTE A soldered terminal is more effective than a crimped fastening because the crimped fastening may connect only three-quarters of the strands with the terminal. This will reduce the current flow and may cause overheating.

Fastening Terminal Ends Before you crimp or solder the terminal end to the wire, you must remove part of the insulation. Use a wire stripper and cut and remove insulation equal to the length of the terminal barrel.

CAUTION To avoid cutting conductor strands, use the correct cutter opening.

FASTENING THROUGH CRIMPING To fasten a terminal end through crimping, first insert the wire into the terminal barrel until it bottoms or is flush with the barrel end. In this position, secure the wire by placing the crimping tool over the barrel (make certain to use the correct crimping opening), squeeze the barrel by forcing the handles together.

NOTE Do not use a pair of pliers, a side cutter, or a vise to crimp the terminal barrel because they will not produce an adequate connection.

FASTENING THROUGH SOLDERING When you use solder, insert the stripped wire into the terminal barrel. Apply a small amount of soldering paste to the open end of the barrel and hold the soldering iron in contact with the barrel and on the edge of the wire rosin solder. Allow sufficient melted solder to flow into the barrel, then remove the soldering iron from the terminal to allow the solder to bond with the barrel and wire. Clean the soldering connection and push the insulation onto the barrel.

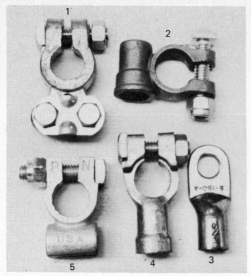

1. Clamp type
2. Close-barrel type (90°)
3. Lug-type close barrel

4. Straight close barrel
5. Flat type

Fig. 38–5 Battery terminals.

When you solder a lip-type terminal to a wire, position the wire so that the insulation is flush with the holding tang, then roll the tang around the insulation. Use a crimping tool to fasten the wire to the terminal. Bend the lips over onto the wire. Hold the soldering iron on the bent lips and hold the wire rosin solder in contact with the wire between the bent lips and tang. When the soldering iron reaches a high enough temperature, the solder will melt and draw toward the lips, bonding the wire and terminal.

Replacing Battery Terminals Figure 38–5 illustrates five popular battery terminals. Regardless of which terminal is selected as the replacement for the worn one, it must be soldered to the battery cable.

When you replace a battery terminal, cut the cable as close as possible to the worn terminal. Use a 32-tooth hacksaw blade to ensure that the cable maintains its roundness. Strip away the amount of insulation that will expose the cable by the exact length of the terminal barrel.

Hold the terminal (with the barrel up) in a vise. With an acetylene or propane torch, apply low heat to the cable end and terminal. Tin the cable end by holding the rosin solder against the copper wire until all strands are saturated with the rosin.

Apply soldering paste to the inside of the barrel and let the solder flow into the barrel.

Place the cable end over the terminal opening. Maintain low heat to the terminal and the cable end until the cable slides in the opening. Solder may run out of the opening at this point, however, continue to apply low heat to ensure a good bond.

When the solder on the edge of the opening is drawn into the barrel, the joint between the barrel and cable end will be well soldered. When there is no solder on the edges of the barrel, more should be applied to the terminals to ensure joint effectiveness.

After removing the torch, hold the cable steady until the solder is set.

Clean the terminal and tape part of the barrel and insulation to prevent corrosion.

Pointers for Connecting Wires
1. When you connect a terminal to a component, make certain that the contact surfaces are clean and not pitted.
2. If possible, use an internal-external lock washer. If this is not available, place a plain washer below the plain lock washer.
3. Do not allow the terminal to touch the component.
4. Do not allow the terminal stud bolt (post) to turn as you tighten the nut.
5. When you have a slide or bullet connector, tape the unit to prevent its separation.
6. Be certain the terminal screw is not too long because this may ground out the terminal.
7. Return the wire as it was originally routed and clamped.
8. When you install an additional circuit, use the existing wire or cable route, clamps, and grommet.
9. If you are forced to reroute the cable in a new direction, use rubber grommets or short pieces of rubber hose to prevent the wire from being cut by the sheet metal or by the sharp corners of the frame.
10. Secure the wire in enough places so that it does not hang loose.

Questions

1. What factors govern the selection of battery cables?

2. If you had a starting circuit of 24 V and the wire length from key switch to the solenoid was 24 ft, what should the wire size be?

3. Under the circumstances outlined above, what would be the result if you had selected a conductor smaller than that stated in your answer to the above question?

4. List three factors that govern the selection of wire ends.

5. How is a wire end fastened to a No. 10 wire?

6. Why should battery terminals be restricted to the soldering type?

7. Why should the existing cable routes be followed and the existing clamps be used when repairing or installing an additional circuit?

8. Why should a ring (closed) type wire end be used on diesel engines?

9. When should a flag ring-type wire end be used?

10. Explain how to join two wires by soldering.

11. After replacing a battery terminal, what steps or precautions should be taken before connecting the cable to the battery?

Relays, Switches, and Solenoids

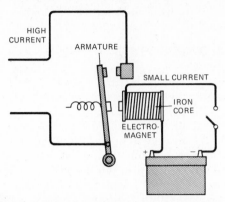

Fig. 39–1 Relay and circuit connections. (*United Delco-AC Products, General Motors of Canada, Ltd.*)

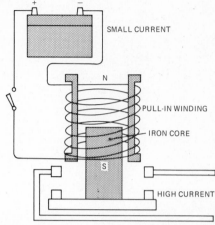

Fig. 39–2 Magnetic switch and circuit connections. (*United Delco-AC Products, General Motors of Canada, Ltd.*)

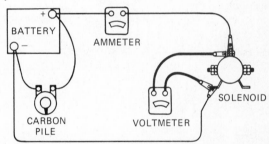

Fig. 39–3 Testing a magnetic switch or solenoid. (*United Delco-AC Products, General Motors of Canada, Ltd.*)

Applications of the various relays, switches, and solenoids are almost limitless. Some are designed to lift or push hundreds of pounds [kilograms], others are refined to push or lift only a fraction of an ounce [gram]. Some control a high current flow, while others are used to control low current.

Relays Relays, generally speaking, have many different applications. However, on diesel engines they are only used to close the field circuit to the generator when the engine is operating and/or to prevent the cranking-motor circuit from being energized when the engine is running.

All relays operate on the principle that a small current controls a large current flow (see Fig. 39–1). The small current is controlled by a toggle, push-button, or key switch which closes the electric circuit of the coil winding. The electromagnet pulls the armature toward its core, closes a set of contact points, and thereby closes the electric circuits with the higher current flow.

Magnetic Switches A magnetic switch is a mechanical switch that is operated electromagnetically. It is used on a starting circuit having a cranking motor with an inertia drive.

The main difference between a relay and a magnetic switch is that the iron core of a relay is stationary whereas the core of a magnetic switch moves within a tube over which the pull-in coil winding is wound (Fig. 39–2). The contacts of a relay close indirectly; those of a magnetic switch close directly.

Testing Magnetic Switch When a magnetic switch fails to pull in or fails to close the electric circuit, the actuating circuit may be defective. Other causes of failure may be an open, short, or grounded pull-in winding, or the contacts may have a high resistance. Several checks should made to locate the problem area: Check the activating circuit with a voltmeter.

Check the resistance of the pull-in winding or make the same connections of the test instruments and test as outlined under Testing Solenoid (Fig. 39–3).

To check the contact and disk for high resistance, close the circuit of the pull-in winding and connect an ohmmeter to the two large terminals. There should be no resistance.

Solenoids A solenoid is an electromagnet with one or two coil windings wound around an iron tube which also serves as the bushing of the movable iron core (Fig. 39–4). When current is passed through the coil windings, the iron core is pulled in or pushed out. This movement shifts a transmission or a cranking-motor drive-in mesh, or shuts off the fuel supply, or moves the fuel rack to the shutoff position.

Testing Solenoid When the action of a solenoid is too slow or fails to operate, the problem may be an open, short, or grounded pull-in winding, a poor connection, or high resistance of the movable iron core. To check the solenoid electrically, use an ohmmeter and connect the test probe to the two terminals. The resistance (in ohms) of the coil wind-

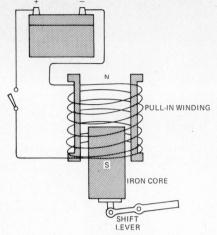

Fig. 39–4 Solenoid and circuit connections. *(United Delco-AC Products, General Motors of Canada, Ltd.)*

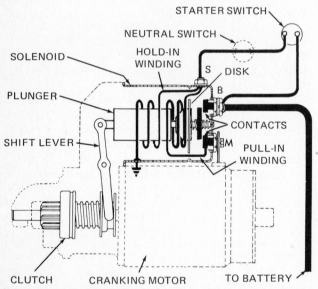

Fig. 39–5 Solenoid switch and circuit connections. *(United Delco-AC Products, General Motors of Canada, Ltd.)*

ing must coincide with the manufacturer's specification. A check can also be made by connecting a voltmeter, ammeter, and carbon pile as shown in Fig. 39–3. Close the coil circuit and adjust the carbon pile to the specified voltage. Read the amperage and compare the reading with the specification. A high ampere reading indicates a short-circuit or grounded winding and a low reading indicates excessive resistance.

Solenoid Switches A solenoid switch combines features of the solenoid and the magnetic switch. It does "work" and it also closes the electric circuit.

A solenoid switch is used on cranking motors which employ an overrunning clutch-type cranking-motor drive. It has two coil windings wound over a hollow cylinder and paired off at the S (starter) terminal (Fig. 39–5). The coil winding with the smaller gauge wire has its own ground and is called the *hold-in winding*. The *pull-in winding* (of heavier gauge copper wire) is connected to the cranking-

motor terminal and is in series with the field coil, armature, and the ground brushes. The iron core plunger) is connected at one end to the shift lever and at the other end to a spring-loaded copper disk.

Two insulated heavy-contact terminals are fastened in the end of the solenoid-switch housing. The terminal marked "battery" is connected to the battery, and the terminal marked "cranking motor" is connected to the cranking motor.

Operation of Solenoid Switch The starting circuit, which energizes the solenoid switch, consists of a starter or key switch, a neutral switch, and the necessary conductor wires (Figs. 38–1 and 39–5). When starter switch is closed, current flows from the battery terminal of the solenoid switch, through the starter and neutral switches to the S terminal of the solenoid switch. From this point on the current flows through the hold-in winding to ground and to the pull-in winding to the cranking-motor ground (ground brush). Both windings create a south pole at the right side, pulling the plunger to the right (in Fig. 39–5). The plunger pivots the shift lever and causes the cranking-motor drive pinion to mesh with the ring gear. Just before the meshing is completed, the spring-loaded disks contact both B and M terminals, thereby making a direct connection from the battery to the cranking motor, and cranking takes place. At the same time the current flows to the cranking motor, the pull-in winding is short-circuited because the contact disk has connected both ends of the coil winding to the battery. Only the magnetism of the hold-in winding holds the plunger, which in turn, holds the pinion in mesh and the disk in contact.

When the starting circuit is open, the hold-in winding is deenergized. However, for a split second, current from the battery flows from terminal B, through the disk, to the M terminal, then in a reverse direction through the pull-in winding to the S terminal, and finally in normal direction through the hold-in winding. This causes the pole polarity to change in the pull-in winding and to oppose the pull-in coil polarity. The magnetism is canceled out. The plunger-return spring forces the plunger to the left, opens the cranking-motor circuit, and moves the cranking-motor drive pinion out of mesh.

Solenoid-Switch Failures Under normal circumstances a solenoid switch seldom causes cranking trouble. However, when the solenoid switch will not pull in or will not stay in mesh, the trouble could be due to a faulty or discharged battery or to a poor battery connection. The problem may also be a defective starter or neutral switch, or the cranking-motor drive may be damaged. Complications can also arise from an open, short, or grounded hold-in or pull-in winding, or from contacts and disks which are burned, pitted, or misaligned.

Testing Starter Switch When tracing the cause of cranking trouble, start with inspection and testing of the power source, that is, the battery and its connections. Next, check and test the starter switch to determine if there is a poor or loose connection or a

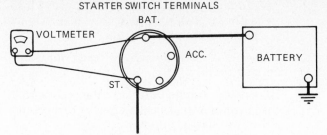

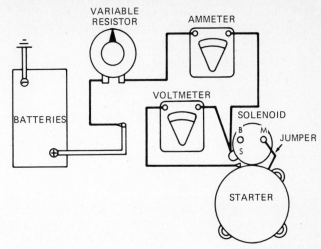

Fig. 39–6 Testing the starter switch. (*United Delco-AC Products, General Motors of Canada, Ltd.*)

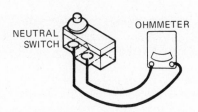

Fig. 39–8 Instrument connections for testing the solenoid switch. (*J. I. Case Co., Components Div.*)

Fig. 39–7 Testing the neutral switch. (*J. I. Case Co., Components Div.*)

loose terminal. Use an ohmmeter to measure resistance when checking for damaged or burned contact points. Alternatively, connect a voltmeter as shown in Fig. 39–6 and measure the voltage drop.

Testing Neutral Switch A neutral switch that is ineffective may be misadjusted or the electrical wiring may be damaged. To check misadjustment and/or the electrical condition of the switch, connect an ohmmeter as shown in Fig. 39–7. Then move the shift lever slowly into the neutral position. When the shift lever is in the neutral position, the ohmmeter should register "no resistance."

Testing Solenoid Switch A solenoid switch can be tested while the cranking motor is installed or after the motor has been removed. In either case, however, the test specification for the solenoid switch must be known before this test can be made:

TEST SPECIFICATION

Current draw at 10 V, 80°F [26.7°C]
Both windings... 70.5 –77.8 A
Hold-in winding... 18 –20 A

To test the solenoid switch while it is on the cranking motor, first remove the connector between the switch motor terminal and the motor terminal. Next, remove the battery cable and starter-switch wire. Connect the test instruments as shown in Fig. 39–8 and, with a variable resistor, slowly close the circuit until the voltmeter reads 10 V. The ammeter should read 70.5 to 77.8 A.

To check the hold-in winding, disconnect the jumper lead which grounds the pull-in winding. Adjust the variable resistor to obtain the specified 10 V. The current draw for the hold-in winding should be 18 to 20 A. If the solenoid switch does not meet specification, the switch should be repaired or replaced.

To check the resistance between the contact disk and the contact terminals, connect an ohmmeter to terminals B and M. Push the plunger so that the disk closes the circuit. The ohmmeter should read "no resistance."

Servicing Solenoid Switch Once the coil windings of a solenoid switch are short-circuited, open, or grounded, a replacement of the solenoid coil is necessary. More common problems, however, are broken connections, a burned disk, burned terminal contacts or burned insulation. Therefore, when you disassemble the switch, be careful not to break the connections or damage the insulation. Use a smooth file to clean up the disks and contacts or replace them if required. Never use damaged insulators, washers, or grommets. When you reassemble the switch, use the service manual as a guide to position the terminals and insulators in their correct places. Make sure the switch assembly is completely sealed to prevent dirt or water from entering. Always repeat the testing procedures previously outlined to verify the operation of the solenoid switch.

Questions

1. Explain how a simple relay works.

2. What is the main difference between a relay and a magnetic switch?

3. Trace the flow of current through the solenoid switch shown in Fig. 39–5 after the key has been turned to the start position.

4. Refer to Fig. 39–5. Why is there no current flow in the pull-in winding, when the disk bridges the B and M contacts?

5. Why is the contact disk spring-loaded and flexible?

6. What is the purpose of the hold-in winding?

7. If the pull-in winding is short-circuited, what result could it have on the operation of the solenoid switch?

8. If the contacts of the starter or neutral switch are pitted, what effect would there be on the operation of the solenoid switch?

9. Briefly outline how to test the operation of a relay.

10. If the air gap between the armature and the iron core is greater than specified, how does it affect the operation of the relay?

11. List several reasons why a relay may not operate properly.

12. List several causes why a solenoid switch could chatter if the key switch were turned and held in the start position.

UNIT 40

Batteries

BATTERY CONSTRUCTION

Function of Battery A battery is an electro-chemical device that stores electrical energy in chemical form which can be released as electrical energy. The battery, regardless of its size, capacity, construction, or type, is the heart of any electrical system. Just as the whole body suffers if the heart flutters in any way, so the electrical system will not function properly if the battery is faulty.

Battery Types There are two types of batteries: primary batteries, which are used in test instruments, and secondary batteries, which are used to (a) store electrical energy in chemical form, (b) stabilize voltage, and (c) supply electrical energy to the system.

Primary batteries (sometimes called *dry batteries*) irreversibly convert chemical energy into electrical energy. In other words, they cannot be recharged. The nickel-iron and nickel-cadmium batteries, however, can be recharged because they are primary *and* secondary batteries. In most primary batteries the electrolyte used is a pastelike substance that affects certain materials.

The most common primary batteries are the carbon-zinc, the mercury, and the silver-oxidized. The secondary batteries with which you are most concerned are those that convert chemical energy into electrical energy by a process that can be reversed. There are three different types of secondary batteries: lead-acid, nickel-iron alkaline, and nickel-cadmium. The lead-acid battery, developed by Gastone Planté in 1859, is the most economical. It produces the highest voltage per cell (2.2 V) and the lowest cost per watt-hour for its capacity. The cells withstand fairly high charge and discharge ranges.

Lead-Acid Battery Generally speaking, the internal and external construction of lead-acid batteries are similar regardless of trade name. Apart from very minor changes, only the number and the area of the plates differ. Two types of plates (negative and positive) form the basic unit. The Planté-type bat-

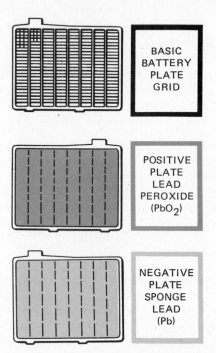

Fig. 40–1 Battery plate construction (lead-acid battery). *(United Delco-AC Products, General Motors of Canada, Ltd.)*

tery, the active material is electrically formed of pure lead by repeated reversal of the charging current. In the Faure type, the active material of the positive and negative plates is formed by applying a paste (largely of lead oxide) to the individual plate grids. The plate grids are made from lead-antimony to give them adequate strength and stiffness. Then, during the manufacturing process (precharge), the active material of the negative plates is converted to a grey sponge lead (Pb) and the positive plates are converted to lead peroxide (PbO_2), which is chocolate brown (see Fig. 40–1). The positive and negative plates are welded (lead burned) to the terminal post or plate strap casting to form a plate group (Fig. 40–2). As you can see from the illustration, the negative plate group has one more plate to cover the positive plate. This improves the chemical action during charging and discharging.

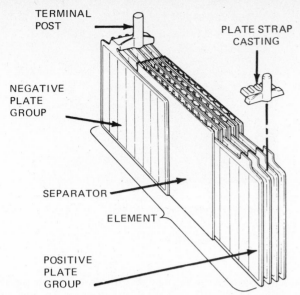

Fig. 40–2 One battery element (lead-acid battery). *(United Delco-AC Products, General Motors of Canada, Ltd.)*

Separators are placed between the negative and positive plates to prevent contact between them. The material used may be microporous plastic, rubber with or without plastic ribs, fiber-glass mats, resin-impregnated fiber with or without fiber-glass mats, and cellulose fiber impregnated with resin.

The two groups of plates shown in Fig. 40–2 are called an *element*. One element makes up a battery cell. The number and the area of the plates governs the amount of energy [ampere-hours (A•h)] that can be stored.

The voltage (emf) of a fully charged cell is about

2.2 V with specific gravity (electrolyte) of 1.275. With a higher specific gravity, say, 1.285, the voltage rises to about 2.3 V. With a specific gravity of 1.265, the voltage is 2.15 V.

The one-piece battery case is made of a hard rubber or plastic compound to withstand the temperature changes, stress, shock loads, the sulfuric acid. It is partitioned to separate the cells. Sediment chambers on the bottom allow the active material which wears off during charging and discharging, to settle without interfering with the plates. Such interference would cause a short circuit.

The cells are placed in the battery case and secured to supports. When a one-piece battery cover is used, cell connectors are welded to the intermediate terminal posts to connect the cells in series (Fig. 40–3a).

When individual cell covers are used, one cover is placed over each of the six cells (Fig. 40–3b). A positive intermediate terminal and a negative battery post, supported by lead bushings, protrude through one end cover. A negative intermediate terminal and a positive battery post protrude through the other end cover. An intermediate terminal connected to the positive plates and an intermediate terminal connected to the negative plates protrude through the center cover.

The cell connectors are welded externally so that the negative internal post of one cell is joined with the positive post of the adjacent cell.

Two advantages to a one-piece battery cover are that it provides an acid-tight seal and that it has shorter connections between the cells which reduce voltage loss. To prevent battery cables from being installed in inverse polarity, the positive and the negative battery terminals are standard. They vary

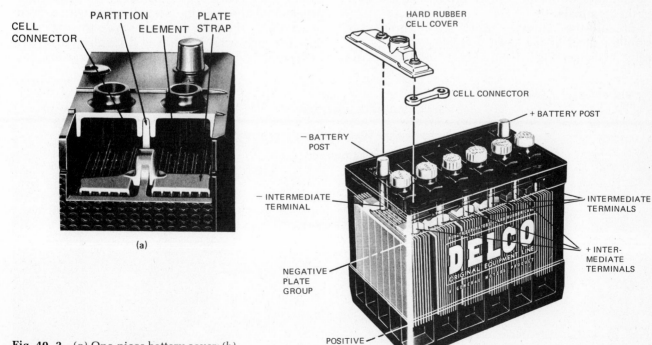

Fig. 40–3 *(a)* One-piece battery cover; *(b)* battery using individual cell covers (lead-acid battery). *(United Delco-AC Products, General Motors of Canada, Ltd.)*

in diameter but not in taper and length. The positive post has a top diameter of 0.6875 in [17.46 mm] and the negative post has a top diameter of 0.625 in [15.87 mm].

When the individual covers are installed and the cells are connected, the covers are sealed with a special bituminous substance that withstands temperature changes. This substance can be removed. The immovable one-piece cover is sealed with a permanent resin sealer.

The cell covers have vents of different construction. Their cover openings are closed off by the vent cap. The vent cap serves a dual purpose: (1) it allows the gases which are formed during charging and discharging to escape and (2) it closes the openings through which electrolyte is checked and water is added.

Sealed-Type Battery (Delco 1200)

Sealed-Type Battery (Delco 1200) In an effort to extend the battery life, to reduce the maintenance requirements and to increase its dependability, a new type of battery has been developed (see Fig. 40–4). The external appearance of this new Delco 1200 battery is somewhat similar to that of the conventional battery except that it has no filler caps nor battery posts and it includes a state-of-charge indicator. (The state-of-charge indicator gives you instant read-out on the charge condition of the battery.)

Internally, the sealed-type battery is much different from the conventional battery. It is constructed as a completely sealed unit and has the inherent advantage of a lifetime supply of electrolyte (1.265). There is, therefore, no need to check the electrolyte level. So the cells cannot be under- or overfilled and there is no danger of using unsuitable water. The battery terminals are made of stainless steel. The battery-cable connection is not made through the terminal bolts but through the lead base of the terminals.

Because the battery terminals are constructed of stainless steel, it has been possible to modify the cable ends (see Fig. 40–5). Short No. 2 cables are

Fig. 40–5 Battery post and battery-cable terminal connector (sealed-type battery). *(United Delco-AC Products, General Motors of Canada, Ltd.)*

used to connect the battery terminals with the main battery cable. A larger gauge cable No. 3-000 or 4-000 is spliced to the No. 2 gauge cable which connects the battery with the cranking motor. The terminal ends are covered with plastic and the protruded surfaces of the connectors ensure a good electrical contact. Additionally, stainless steel crown nuts are used which provide further protection against a poor terminal connection.

INTERNAL CONSTRUCTION The sealed battery was made possible by the removal of antimony from the plates. As a result this battery has greater counterelectromotive force (cemf) than the conventional battery, thus preventing overcharging which greatly reduces gassing. The gases which are formed during the charging and discharging cycle rise upward and are trapped in the liquid/gas separators. As the gases cool, they transform to liquid and reenter the electrolyte through special passages. The cooling is accelerated through the sealed cover which is located above the gas separator. Any internal pressure that may arise is released through a small vent hole in the side cover (called a *flame arrestor*).

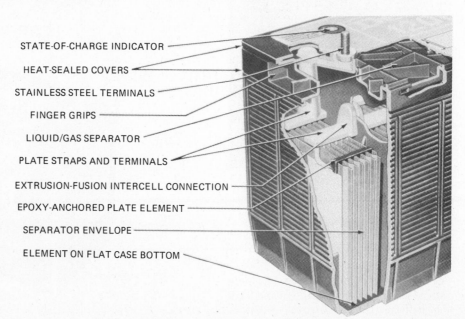

STATE-OF-CHARGE INDICATOR
HEAT-SEALED COVERS
STAINLESS STEEL TERMINALS
FINGER GRIPS
LIQUID/GAS SEPARATOR
PLATE STRAPS AND TERMINALS
EXTRUSION-FUSION INTERCELL CONNECTION
EPOXY-ANCHORED PLATE ELEMENT
SEPARATOR ENVELOPE
ELEMENT ON FLAT CASE BOTTOM

Fig. 40–4 Cutaway view of a Delco 1200 battery. *(United Delco-AC Products, General Motors of Canada, Ltd.)*

Table 40–1 CCA RECOMMENDATIONS

Manufacturer	Model	Displacement	CCA	Watts
Caterpillar	1674	638	1500	18,000
Mack Maxidyne	675	672	1460	17,520

The plates which are enclosed in envelopes, act as separators and sediment collectors and are bonded together with a special epoxy. As a single unit (called *an element*), it can rest directly on the case bottom, giving the plates added resistance to vibration (see Fig. 40–4).

The plates are extrusion-fusion-welded to the plate straps. The intercell connection is made through extrusion-fusion-welding of the plate straps. Thus the newer battery has increased terminal voltage because of this shorter electrical path.

STATE-OF-CHARGE INDICATOR When the state-of-charge indicator becomes dark with a green dot in the center, the battery is charged sufficiently and may be tested. When the green dot is not visible, the battery is undercharged and must be charged until the green dot appears.

CAUTION The battery must not be charged more than 60 A·h (ampere-hours). If the state-of-charge indicator is light, do not attempt to charge or test the battery. The battery must be replaced.

Battery Classification and Ratings The amount of current (capacity) that a battery can deliver depends on the number, size, and weight of its plates, and the volume of acid present. The emf, that is, the open-circuit voltage of a fully charged cell with a specific gravity of 1.265 is 2.1 to 2.15 V regardless of the number, size, or weight of the plates. But the voltage on discharge is influenced by the size of the cells—the larger the plate area, the longer the cranking voltage remains the same (under any given condition) when compared with a smaller battery plate area.

To crank an engine, you need electrical power. The power demanded from a battery varies with the engine displacement and compression ratio. Engine manufacturers test their engines to establish starting power, that is, the energy (in watts) required to crank an engine at 0°F (−17.8°C). From this test factor the recommended minimum cold-cranking amperes (CCA) for a given engine is established. For example, suppose that 18,000 W are required to start the engine, and that the cranking-motor voltage is 12 V. To arrive at the cold-cranking amperes, divide 18,000 by 12. The answer is 1500 CCA. If the cranking-motor voltage were 24 V, only 750 CCA would be required. Two engine manufacturers' recommendations are listed in Table 40–1.

To standardize the CCA requirements for the great number of diesel and other internal-combustion engines, battery manufacturers in cooperation with SAE have classified batteries with a number and letter system. This system distinguishes battery voltage, battery dimension, reverse capacity, cold-cranking amperes at 0 and −20°F [−17.8 and −28.5°C] overcharge, and charge acceptance in amperes.

The classification and ratings for five battery groups recommended for use on diesel engines are listed in Table 40–2.

BATTERY SERVICE

Electrolyte To utilize a battery of the conventional design, you must first bring it to life by filling it with electrolyte. The electrolyte supplies the sulfation and acts as a current (electron) carrier between the positive and the negative plates. The plate grids and the cell connectors carry the current running to and from the active material to the negative and positive battery posts.

The electrolyte is a solution of 65 percent water and 35 percent sulfuric acid by weight or about 76 percent water and 24 percent sulfuric acid by volume. This solution has a specific gravity of 1.265 at 80°F [26.7°C] since water has a specific gravity of 1.000 and sulfuric acid has a specific gravity of 1.835.

To increase the life of the battery, most manufacturers produce electrolyte with a specific gravity of 1.265 when it is to be used in a moderate climate. However, if the battery is to be used in a tropical climate, the specific gravity may be reduced to 1.250. If the battery is to be used in an arctic climate the specific gravity may be increased to 1.285 (see Table 40–3). As indicated in the table, when the strength of the electrolyte changes, the open-circuit voltage changes also.

Filling Cells with Electrolyte Electrolyte usually is shipped in a plastic container. Use a plastic funnel when filling the individual cells to the recommended level. Make sure that the battery and electrolyte are both at room temperature.

Table 40–2 BATTERIES RECOMMENDED FOR DIESEL ENGINE SERVICE

SAE	Group	Voltage, V	Reserve capacity, min	Cold-cranking amperes At 0°F [−17.8°C]	At −20°F [−28.5°C]	High discharge, 15 seconds, A
6T3A	7D	6	430	900	650	450
20T4A	4D	12	285	640	450	450
20T6A	6D	12	350	800	555	450
20T8A	8D	12	430	900	650	450
Delco 1200	4D	12	130	475	375	230

Table 40-3 ELECTROLYTE-STRENGTH VARIATIONS

Specific gravity	Opening voltage, V	Freezing temperature, °F [°C]
1.285	2.30–2.40	−100 [−73.3]
1.275	2.20–2.30	− 88 [−66.0]
1.265	2.10–2.20	− 79 [−61.5]
1.250	1.90–2.00	− 62 [−52.2]

CAUTION Handle the container with care. Do not spill the electrolyte during filling operations. It is an acid and will cause blisters if it contacts the skin directly. It will also damage clothes and deteriorate iron and paint, etc.

If electrolyte is accidentally spilled, immediately wash away the acid with water, sprinkle baking soda on the contacted area, or wash the area with a solution of water and baking soda.

After filling, recheck the electrolyte level, since the plates react like sponges and quickly reduce the level of the first fill. Always clean away any spillage before you charge the battery.

If through necessity you are forced to mix your own electrolyte from a concentrated sulfuric acid, do so with extreme care; wear rubber gloves, goggles, and an apron. Mix the solution in a plastic or rubber container.

CAUTION Mix by pouring pure acid very slowly *into the water.* (Pouring water into the concentrated acid will cause an explosion.) Use a glass, plastic, or wooden stick for mixing. Do not let the temperature exceed 150°F [65°C] while mixing.

To arrive at a specific gravity of 1.265, start with 1 qt [2.8308 l] water and a little less than $\frac{1}{4}$ qt [0.9436 l] acid. After adding the last sulfuric acid, allow the temperature of the solution to cool to about 90°F [32.2°C] before you take a hydrometer reading.

Take care when adding the final acid not to exceed the desired 1.265 specific gravity.

Battery Inspection The batteries of a diesel engine are sometimes so neglected that they will no longer start the engine. The proper procedure is to inspect and service the battery regularly to maintain battery life and to prevent battery failure. When you inspect the batteries, look for cracks caused by overly loose or overly tight hold-down clamps. Inspect the battery compartments for foreign matter.

Check for signs of wetness or leakage on the top of the batteries which might be the result of a missing filler cap, overfilling, or overcharging.

Check the battery terminals for corrosion. Excessive corrosion is an indication of high resistance (poor connections) at the terminals or within the battery itself.

To service the batteries efficiently, remove them to the service area. This will minimize battery damage during servicing and permit unobstructed access to the battery compartment.

Battery Maintenance To simplify reinstallation, make a mental note of the position of the terminal posts before you remove the batteries. Disconnect the ground cable first. This will prevent a short circuit which could damage the batteries or other components. It also may prevent your hands from being burnt, especially if you are wearing a ring.

Remove battery hold-down clamps (covers). When they are damaged or corroded, replace them. If they are in good condition, clean them thoroughly and paint them with acid-resistant paint. Use two wrenches when loosening the terminal bolt nut to prevent loosening or breaking the terminal post from the cell connector. Pull the terminals from the battery post with a battery puller. Do not use a pair of pliers or a screwdriver because either of these may damage the post connection or cell cover. Use a carrying strap to lift and carry the battery to the service area. Do not use a pair of pliers or a vise grip to lift the batteries from the compartment. Do not carry the batteries in your hands.

Before you clean the battery, tape or block the vent holes. Use a wire brush to remove excessive dirt or oil from the battery. Wash away corrosion and electrolyte from the battery and terminals with a baking soda solution, then rinse with clean water. If necessary, use steel wool or a terminal-cleaning brush to clean the battery posts. Use compressed air to dry the batteries. Make sure the sealer has not broken away from the case because this could allow acid to be drawn from the battery.

Battery Leakage Test The first test you should make is the electrical leakage test, since electrical leakage will drain the battery continuously. Use a low-reading voltmeter or adjust the voltmeter to its lowest scale. Rest one prong against the positive post. With the other prong resting lightly over the cell covers (or battery cover), slide it from the positive to the negative side of the battery and watch the voltmeter's reaction. When the prong rests on the last cell cover near the negative post, the voltmeter should not read more than 0.1 V. If the reading is higher, check for cracks or a loose intermediate post or a loose battery post.

Hydrometer Design Hydrometers are devices to measure specific gravity of a liquid. They operate on Archimedes' principle that buoyancy or lift (buoyant force) equals weight of liquid displaced.

A typical hydrometer consists of a sealed weighted glass vial, a glass barrel, and a bulb syringe to draw up a sample of the electrolyte (Fig. 40–6). The vial stem may have a scale calibrated in terms of specific gravity etched on it, or it may have a paper scale inserted in the stem.

A small thermometer and a correction scale are built into most hydrometers to allow for temperature effects when determining the correct specific gravity reading (Fig. 40–7). When the temperature of a liquid rises, the liquid density decreases; conversely, when the temperature decreases, the liquid shrinks and its density increases. Therefore, temperature changes cause the buoyancy to increase or de-

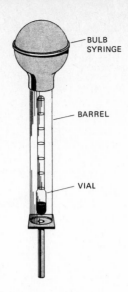

Fig. 40–6 Hydrometer. *(J. I. Case Co., Components Div.)*

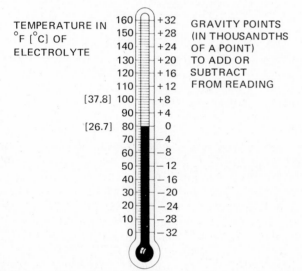

TEMPERATURE IN °F [°C] OF ELECTROLYTE

GRAVITY POINTS (IN THOUSANDTHS OF A POINT) TO ADD OR SUBTRACT FROM READING

160	+32
150	+28
140	+24
130	+20
120	+16
110	+12
[37.8] 100	+8
90	+4
[26.7] 80	0
70	−4
60	−8
50	−12
40	−16
30	−20
20	−24
10	−28
0	−32

Fig. 40–7 Hydrometer temperature correction scale. *(J. I. Case Co., Components Div.)*

once or twice or turn the headlights on for a few minutes.

NOTE Do not take a hydrometer test if water has just been added or if the engine has been cranked too long.

To ensure an accurate reading, warm up the hydrometer float to that of the electrolyte temperature by drawing electrolyte in and out of the barrel. Draw up enough electrolyte so that the float moves to the middle of the barrel. Hold the hydrometer at eye level without tilting it and allow the float to float. Do not take the fluid level at the float stem or at the barrel since the fluid level is high at both of these points due to surface tension of the fluid. At the same time, take the temperature reading.

Assume the hydrometer reading is 1.240 and the electrolyte temperature is 100°F [37.8°C] (see Fig. 40–7). Add 0.008 points specific gravity to account for the temperature rise above 80°F [26.7°C]. So, 1.240 + 0.008 = 1.248, the correct specific gravity. The cell is then about 90 percent charged and should have a voltage of 2.05 V.

Check all the cells of one battery and compare their specific gravity. A difference of more than 50 points (0.050 specific gravity) between the cells is generally an indication that the plates are deteriorating. However, the difference may be caused by an internal short circuit, a loss of electrolyte due to leakage, or an excessive charging rate. A battery with a high variation should be recharged and retested. If the variation persists, the battery should be replaced.

When battery cells have a specific gravity of less than 1.215, the battery's service life is questionable. It should be recharged and retested, or replaced.

Open-Circuit Voltage (emf Voltage) Another method of determining if a battery is in a fully charged condition, if it requires recharging, or if it is defective is by the use of a low-reading voltmeter with a 2.3-V scale in a 0.10-V division.

Place the test prong of the voltmeter across each battery cell as shown in Fig. 40–8. The individual cell voltages should be within 0.05 V of each other. When there is a greater difference (or when any cell voltage is 1.95 V or less), the battery should be recharged and retested. A high-discharge test (capacity test) should then be made.

NOTE Batteries with a one-piece cover cannot be tested with a voltmeter. A cadmium probe tester must be used to test their voltage (see Fig. 40–9).

Light-Load Test A light-load test is more reliable than an open-circuit voltage or a hydrometer test because the light-load test requires the battery to convert chemical energy into current flow. For this test the low-reading voltmeter, or the cadmium probe tester may be used.

When you use a cadmium probe tester to perform the light-load test, place a carbon pile across the battery and draw about 15 A from the battery. Alternatively, when the batteries are installed, turn on the

crease. Figure 40–7 shows that the "zero" point of the correction scale is at 80°F [26.7°C]. To arrive at the correct specific gravity when electrolyte temperatures exceed 80°F [26.7°C], you must add 0.004 for every 10°F [5.6°C] over 80°F [26.7°C]. When the electrolyte temperatures drop below 80°F, you must subtract 0.004 for every 10°F [5.6°C] below 80°F [26.7°C].

Hydrometer Test The state of charge of a battery may be measured with a hydrometer by testing the specific gravity of the battery fluid (electrolyte). As a cell discharges, sulfuric acid from the electrolyte enters the battery plates and the active material gradually changes to lead sulfate, thereby lowering the strength of the electrolyte. The specific gravity of the electrolyte varies directly in proportion to the strength of charge of each individual battery cell.

Before you check the specific gravity, make sure that the gases have escaped, that the sediments have settled, and that the surface charge is removed. To remove the surface charge, crank the engine over

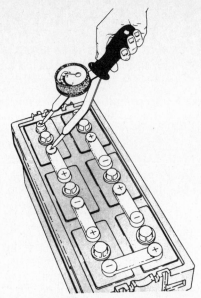

Fig. 40–8 Checking the battery with a voltmeter. *(Allis-Chalmers Engine Div.)*

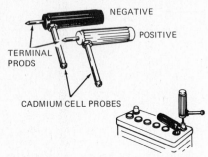

Fig. 40–9 Checking the battery using cadmium cell probes. *(Allis-Chalmers Engine Div.)*

headlights or other electrical components. This will equal a 15-A draw. With the current flowing from the battery, insert one cadmium probe in one cell and the other probe in the adjacent cell. Observe the scale reading.

To check the next cell, remove the outer probe from the cell, insert it into the third cell, and observe the reading. Continue until all cells are tested. If the cell voltage is 1.95 or higher with no voltage variation, the battery is in good condition. When the voltage is lower, but without variation, the battery requires recharging and retesting. When the voltage variation is more than 0.5 V, recharge and retest the battery. Also perform a high-discharge test to be certain the battery is reusable.

High-Discharge Test and Cold-Cranking Test The battery tests with which you will be most concerned are the high-discharge test, the cold-cranking test, and the reserve-capacity test. These tests are the most reliable because, under maximum and minimum strain, they measure the battery capacity to change chemical energy into current flow.

To perform the high-discharge or cold-cranking test, an ammeter, voltmeter, and carbon pile (variable resistor) or battery-starter tester are required. Connect the voltmeter and ammeter (variable resistor) as shown in Fig. 40–10.

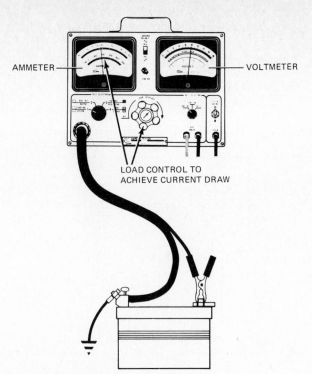

Fig. 40–10 Connections for a battery-capacity test (high-discharge test). *(Allis-Chalmers Engine Div.)*

NOTE To perform a cold-cranking test, the temperature of the battery should be brought to 0°F [−17.8°C], and that is not always possible. A high-discharge test is used to simulate the cold-cranking test. When the instruments are connected, measure the open-circuit voltage. When the open-circuit voltage is 6 V or lower for a 6-V battery (12 V or lower for a 12-V battery), or if the specific gravity is lower than 1.220, recharge the battery before testing; otherwise the battery plates could be damaged during testing.

One more requirement before making the actual high-discharge test is that you remove the surface charge in order to obtain a true battery-capacity reading. This is done with the help of a carbon pile. The amount of current draw from the battery needed to remove the surface charge is fixed for each battery group. For instance, batteries recommended for diesel engines, such as groups 7D, 4D, 6D, 8D, and Delco 1200, would require a current drain of 300 A for 15 seconds.

To perform a high-discharge test on groups 7D, 4D, 6D, or 8D batteries, draw 450 A from the battery for 15 seconds. When testing a Delco 1200 battery, draw 230 A for 15 seconds. At the conclusion of the test the voltage in all cases should register not less than 9.6 V for a 12-V battery and not less than 4.8 V for a 6-V battery, assuming that the ambient temperature is 70°F [21.1°C] or more. If the ambient temperature is less than 70°F, the minimum acceptable voltage is less, as indicated in Table 40–4.

When the load from the battery is removed (the variable resistor is open), the voltmeter reading should be within 0.2 V of the open-circuit voltage reading before the test. If the voltage is lower than

Table 40–4 AMBIENT TEMPERATURE AND MINIMUM VOLTAGE

Temperature, °F	70 and above	60	50	40	30	20	10	0
Minimum voltage, V	9.6	9.5	9.4	9.3	9.1	8.9	8.7	8.5

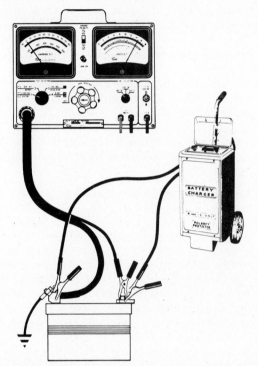

Fig. 40–11 Connections for a 3-minute charging test. *(Allis-Chalmers Engine Div.)*

the 9.6 or 4.8 V referred to above, or if it did not recover to within 0.2 V of the previous reading, the battery is in poor condition and should be replaced.

To perform a true cold-cranking test, place the fully charged battery in a cold room to reduce the temperature to 0°F [−17.8°C]. When the battery reaches the recommended temperature, connect the same test instruments you used to perform the high-discharge test to the battery to be tested. Now use the carbon pile and discharge the battery for 30 seconds at the same rate as its cold rating. At the end of 30 seconds, the individual cell voltage must be 1.2 V or greater, that is, the terminal voltage must be 7.2 V for a 12-V battery and 3.6 V for a 6-V battery.

Reserve-Capacity Test The reserve-capacity test procedure is similar to that of the old 20-A·h test. That is, a 100 A·h battery fully charged must deliver 5 A for 20 hours at a temperature of 80°F [26.7°C], and the cell voltage at the end of the test must be 1.7 V or higher.

For the reserve-capacity test you also test a fully charged battery at 80°F [26.7°C], but you draw 25 A from the battery. When the individual cell voltage reaches 1.75 V, or the terminal voltage for a 12-V battery reaches 10.5 V (5.25 V for a 6-V battery), you begin withdrawing amperes until the battery reaches the voltage specified above. The reserve-capacity test measures the length of time (in minutes) from the moment of ampere withdrawal to the moment cell voltage reaches 1.75 V.

Three-Minute Charging Test The result of a 3-minute charging test indicates the condition of the plates as well as the condition of the connections within the battery.

To make this test, connect a fast charger and a voltmeter to the battery as shown in Fig. 40–11. Set the voltage selector switch to 6 or 12 V (depending on battery voltage). Set the timer for 3 minutes. Start the charger and adjust the charging rate to 10 percent of the CCA of the battery. Observe the voltmeter while charging. The charging voltage should not exceed 15.5 V for a 12-V battery, or 7.75 V for a 6-V battery throughout or at the conclusion of the 3-minute charging time.

When the voltage is higher than specified, the battery is sulfated, damaged, or has high resistance in its connections. When the voltage is lower than specified, check the individual cell voltage. If the variation is not greater than 0.1 V, the battery should be recharged and the high-discharge test repeated.

Chemical Action in Battery when Discharging While a battery is discharging, the sulfuric acid within the electrolyte is absorbed by the sponge-lead negative plates and by the positive lead peroxide plates, and thereby both plates gradually are changed into lead sulfate. The greater the similarity between the positive and the negative plates, the lower the voltage within the cell, as voltage depends on the difference between the two materials.

Chemical Action in Battery during Charging While the battery is being charged by a generator or charger, current passes through the battery in a direction opposite to that of discharge. As the sulfuric acid leaves the plates and returns to the electrolyte, the negative plates change gradually to lead and the positive plates change to lead peroxide. During the charging action, the negative and positive plates give off a hydrogen and oxygen gas. The nearer the battery comes to the fully charged state, the greater the gassing action becomes.

CAUTION Do not approach the battery area with an open flame. Do not smoke when checking or testing batteries. The gases are highly explosive; therefore, the possibility of an explosion is always present.

Battery Charging Four major battery chargers are used to recharge batteries: (1) a dc or ac generator, (2) a constant-potential (voltage) charger, (3) a constant-current charger, and (4) a trickle charger (used to maintain test stand and storage batteries in a charging state).

Before you charge any battery, you should clean it thoroughly, test it, and adjust the liquid to the correct level. When you adjust the electrolyte levels, add only water that is as pure as drinking water. Do not use electrolyte.

When possible, use the liquid level indicator to ensure proper liquid height. If this indicator is not available, fill the cells to the lower edge of the cell opening.

NOTE Add electrolyte only when you accidentally have spilled some of this solution.

Battery Charging with Fast Charger Fast chargers are commonly of the constant-potential (voltage) type. The batteries to be charged must be of the same voltage or connected to the same voltage and then connected to the charger in parallel. The number of batteries that can be connected depends on the capacity of the charger. In all cases, follow the manufacturer's instructions.

Batteries which are in good condition will not be damaged (by a fast charger) when the charging temperature does not exceed 125°F [51.6°C] and the voltage selector switch is set to the correct battery voltage. Adjust the charging current to 10 percent of the CCA of the smallest battery.

NOTE A battery with badly sulfated plates cannot be fully charged with a fast charger.

Connect the batteries first, before switching on the charger. When the charger is switched on, you will notice that you have to adjust the current selector switch very high to achieve the desired charging current. Since electrolyte expands due to heat and gassing bubbles, as the batteries are warming up, reduce the current selector setting and, if necessary, remove some electrolyte to prevent it from spilling over. (Store the removed electrolyte for later replacement when the battery is charged.) You will notice that as the battery voltage (emf) increases, the charging current gradually decreases.

NOTE When the temperature is close to 125°F [51.6°C], reduce the charging rate.

When all cells are gassing freely, the battery is nearing its fully charged state. After this point take the hydrometer reading frequently. When the specific gravity ceases to increase, stop charging. When the battery has cooled off, check the electrolyte level.

CAUTION Switch the charger off before you disconnect the batteries to prevent an explosion. Avoid overcharging the batteries because this may corrode and bend the positive plates. The excessive gassing bubbles will remove active material from the plates and may increase the acid level which is harmful to the plates and separator.

Battery Charging with Slow Charger If time is available or if the condition of the battery plates is unknown, a constant-current charger (slow charger) should be used to prevent damage to battery plates. When being charged with a constant-current charger, the batteries must be connected in series.

Note that a 6 and a 12-V battery may be placed in the same line.

The most common charging rate is 1 A per positive plate of the smallest cell connected to the charger. For example, if the smallest cell has 11 plates, then it has 5 positive plates. Therefore, the selector switch of the charger should be set to allow a current flow of 5 A into the batteries.

When using a constant-current charger at normal rate, you will notice that the terminal voltage gradually increases from 2.14 to 2.3 V and then increases rapidly to 2.5 to 2.6 V. This latter interval is known as the *gassing period*. When this period is reached, you should reduce the charging rate in order to avoid unnecessary erosion of the plates.

Battery Installation Before you install a battery, make certain the battery compartment is clean and preferably repainted. Make sure all battery cables and terminals (including the battery-terminal bolts and nuts) are in good condition and that the battery hold-down clamps or covers have been repaired or replaced.

NOTE To prevent future electrical problems, install only fully charged batteries. When your equipment requires more than one battery: do not install one new battery alongside older batteries; do not install batteries of different capacity; do not install batteries from different battery manufacturers because the counterelectromotive force of the batteries differs and the capacity or plate area also varies. This last point is especially important when a series-parallel switch is used. [See cranking-motor circuits (Unit 41).]

Before you install the batteries in the battery compartments, arrange them on the shop floor in a sequence corresponding to the location and direction they will take in the compartments. This step not only speeds up installation, but also prevents the necessity of reinstalling because of insufficient cable length, wrong polarity, or insufficient voltage. Two common battery connections are shown in Fig. 40–12.

After you lift the batteries into place with a lift strap, make certain they are level and flat in their compartments. Tighten the battery hold-down bolts uniformly so that the cover rests evenly without cracking or distorting the battery case. Apply a light coat of grease on all battery posts and terminals

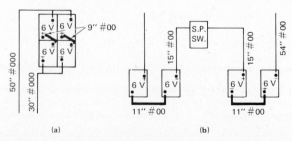

Fig. 40–12 Battery connection for a 12-V starting system; (b) battery connection for a 24-V starting system using an S. P. S4. (*Cummins Engine Co., Inc.*)

to reduce corrosion. Push the terminal ends onto the battery posts. Do not drive them on with a hammer because this will damage the battery. When necessary, use a terminal spreader to increase the diameter. Use two wrenches to tighten the battery bolts to prevent damaging the cell cover and connections.

Before you connect the ground terminal to the battery post, check the cranking-motor connection, the ground connections from the engine to the frame and from the frame to the battery. Use a voltmeter to determine if your equipment has an unwarranted ground (electric current drain).

Make certain all accessories are shut off. If there is no leakage to ground, the voltmeter will read zero. You can then connect the ground cable. Start the engine and observe the charging system. Check if the charging voltage is too high or too low.

Battery Capacity versus Temperature Do not expect 100 percent effectiveness from a fully charged battery when the temperature is near the freezing level because at low temperatures the chemical action of the battery is reduced and more power is required to turn over the cold engine. To overcome starting difficulties at extremely cold temperatures, remove the batteries, drain the lubricating oil and coolant, and store them at room temperature. When you are ready to start the engine, fill the cooling system with coolant and the crankcase with oil. Install the batteries. This will warm up the engine, give the batteries more cranking power and, at the same time, will reduce component wear.

Booster batteries are another way to aid starting when the temperature is very low or when batteries are partly discharged.

Booster Batteries to Aid Starting When you use booster batteries to assist starting, connect only boosters with the same voltage as the batteries of the truck, tractor, etc., which is supplying the voltage. Follow the steps outlined below to prevent battery explosion.

Connect one booster cable end securely to the tractor battery live post, and the other end of the booster cable to the booster battery post having the same polarity. **CAUTION** Reverse polarity could damage the regulator, generator, radio, and other accessories.

Connect one end of the second booster cable to the booster battery ground post and the other end of the booster cable to a good ground. Do not connect it to the truck battery because this could cause battery explosion.

After the engine has started, remove the booster cable from the ground connection. Then remove the cable from the live posts of the booster battery.

Questions

1. State the basic purpose of the lead-acid battery.
2. Name the main components of the lead-acid battery.
3. Briefly outline how to fill a dry-charge battery.
4. Why is a hydrometer correction scale needed?
5. What factors influence battery capacity?
6. What is meant by a 20-hour rating at 80°F?
7. List obvious defects which would indicate that a battery needs to be repaired.
8. List four battery conditions which a light-load test might determine.
9. Explain how to make a battery-capacity test using a variable resistor (carbon pile), a voltmeter, and an ammeter.
10. What information will a 3-minute test reveal with respect to battery condition?
11. Write a short paragraph on the differences between a fast charger and a slow charger.
12. What precautions must be taken when charging batteries, removing batteries, and installing batteries?
13. List three advantages of a sealed-type battery over a conventional battery.
14. List the major differences between an element of a sealed-typed battery and an element of a conventional battery.
15. What would you do if the state-of-charge indicator was dark but the green dot was not visible?

UNIT 41

Electric Starting (Cranking) System

The three most common diesel engine starting systems are the electric starting system, the air starting system, and the hydraulic starting system. Though each system has individual advantages or disadvantages, their common purpose is to rotate (crank) the engine so that it will start. This unit discusses electric starting systems; Unit 42 discusses air and hydraulic starting systems.

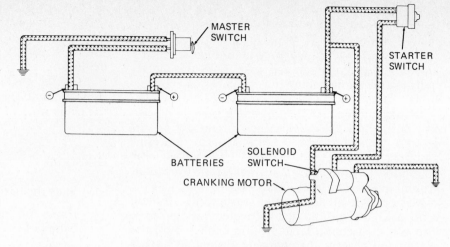

Fig. 41–1 Schematic drawing of a typical 12-V starting circuit. *(Allis-Chalmers Engine Div.)*

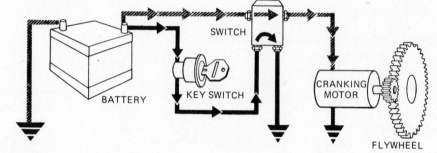

Fig. 41–2 Simplified starting circuit. *(United Delco-AC Products, General Motors of Canada, Ltd.)*

ELECTRIC–STARTING–SYSTEM COMPONENTS

The electric starting system consists of a battery (or batteries), the cranking motor with a solenoid switch, a starter switch, and the complementary wires and cables (Fig. 41–1).

Some electric starting systems have additional components. These can include the following:

- A *master switch*, as shown in Fig. 41–1, which is a safety switch to disconnect the batteries
- A *neutral switch*, which prevents the starting circuits from closing when the transmission is in gear
- A *series-parallel switch*, which connects the battery in series for starting and connects them in parallel when the engine is running
- A *sensing relay*, which automatically opens the starting circuit when the generator is charging

If any of these components are in poor condition electrically or mechanically, the cranking motor will not crank the engine effectively.

Cranking-Motor Principle The first electric self-cranking motor was invented by Charles F. Kettering. Its operating principle involves the effect of a current-carrying conductor in a magnetic field.

A cranking motor receives direct current from the battery and converts it to mechanical energy (rotating motion). The mechanical energy is transmitted from the armature in the cranking motor to the pinion and onto the flywheel ring gear, causing the crankshaft to rotate (Fig. 41–2).

The conductor is wound over an iron core and placed in a magnetic field. When current flows in the conductor, the lines of force are distorted. A strong field is produced on one side of the conductor and a weaker field on the other. The conductor moves to the weaker side, causing the loop (conductor) to rotate in a clockwise direction (Fig. 41–3). However, the rotation will stop when the loop reaches the center position of the north and south pole. To extend and equalize the force on the loop (conductor), the pole shoes are curved. The clearance between the iron core and the pole shoes is kept small to increase the magnetic field.

To make continuous rotation possible, the loop ends are connected to a pair of sliding contacts (the commutator segments). A set of brushes is placed in a fixed position on the commutator to connect the battery with the armature loop (see Fig. 41–3). When the circuit is closed, the armature turns in a clockwise rotation and coasts through the static neutral point (commutation). At the same time, the

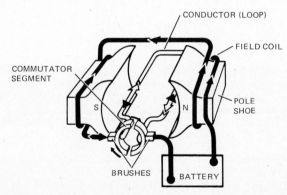

Fig. 41–3 One-loop cranking motor. *(United Delco-AC Products, General Motors of Canada, Ltd.)*

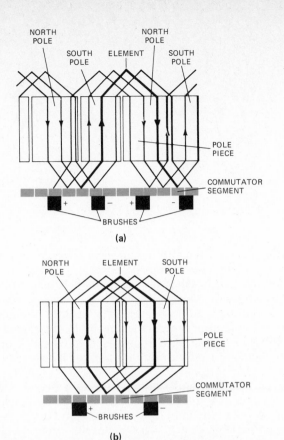

Fig. 41-4 (a) Wave-winding connection for a four-pole four-brush cranking motor; (b) lap-winding connection for a four-pole two-brush cranking motor. (*United Delco-AC Products, General Motors of Canada, Ltd.*)

brushes slide onto the other commutator segments thereby changing the direction of current flow in the armature loop resulting in continuous rotation. In order to make the motor operate smoothly and more powerfully, more loops are placed into the cylindrical iron core and their ends are connected to the segmented commutator.

Armature Assembly   An armature assembly consists of the armature shaft, a laminated iron core, the commutator, and the armature loops.   The armature shaft has machined bearing surfaces, drive splines, and devices to hold the laminated sections and the commutator to the armature shaft.   Thin steel sections, insulated from each other, reduce eddy current and keep the armature temperature low.   Eddy current develops as the induced voltage (emf) causes the free electrons in the laminated sections to swirl about.   The insulated armature loops (conductors) are placed in the laminated-iron-core grooves to transmit the force from the conductor to the iron core and onto the armature shaft.   The armature loops are made from heavy gauge wire or heavy copper ribbon and are soldered either in a wave-winding or lap-winding pattern to the individual commutator segments (see Fig. 41-4).

The ends of the wave-wound armature coil for a four-pole motor are connected to the commutator at an angle of approximately 180° (Fig. 41-4a), while

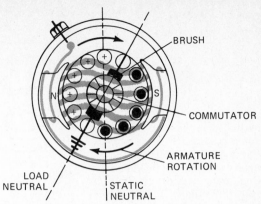

Fig. 41-5 Brush position. (*United Delco-AC Products, General Motors of Canada, Ltd.*)

on the armature of the six-pole motor the angle is approximately 120°.   The brushes contacting the ends of an armature loop are of the same polarity and the coil is short-circuited in the commutation position.

The ends of the lap-wound armature coils are connected to the adjacent commutator segment.   Since both ends of a coil contact a brush at the same time, the coil is short-circuited in the commutation position.   The wave-wound armatures have only two paths through which the current can flow, whereas the lap-wound armatures have as many paths as there are poles (Fig. 41-4b).   The commutator segments are fastened to the armature shaft and are insulated from the shaft and from each other.

The static neutral point of a cranking motor is always in the center between the north and south poles (see Fig. 41-5).   When the armature loops carry current, they create a magnetic field which distorts the lines of force between the poles.   As a result, the static neutral point moves against the armature rotation.   The brushes of the cranking motor, therefore, are positioned about ½ to 1½ commutator segments before the static neutral point, as shown in Fig. 41-5, to ensure an arcless (brush) commutation.   The greater the current flow, and the greater the armature speed, the greater is the distortion.

Field-Frame Assembly   The field-frame assembly consists of two or more field-coil windings, with the same number of pole shoes, one or more terminal connections, and the iron frame.   The field-coil windings are made of heavy copper ribbon to reduce resistance to the high current flow.   A four-pole, four-coil cranking-motor circuit is made by winding the field coils around the cast-iron pole shoes.   Two field coils are wound in a right rotation to make one pair of the pole shoes north in polarity, the other two are wound in a left rotation to make the other pole shoes south in polarity.   The pole shoes (surrounded by their field coils) are mounted alternately within the field frame.   The field frame is the mounting surface for the commutator end frame and the drive housing.   The pole shoes and the field frame concentrate the magnetic flux and its path.

Cranking-Motor Circuits   To increase torque or the speed of a cranking motor, the field-coil windings are connected in combinations of series, series-

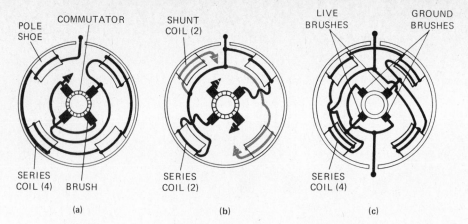

Fig. 41–6 *(a)* Series-coil cranking-motor circuit; *(b)* series-coil cranking-motor circuit using two shunt coils; *(c)* four-coil insulated cranking-motor circuit. *(United Delco-AC Products, General Motors of Canada, Ltd.)*

parallel, or parallel. Torque can also be increased by adding more field coils or by increasing the ampere-turns of the field coil.

A straight series circuit is used when high top speed is required (Fig. 41–6*a*). One or two shunt coils limit the top speed to prevent the armature loops from flying out of their grooves (Fig. 41–6*b*).

When a cranking motor runs, it generates voltage (emf) that cannot be shut off and, as the armature speed increases, the generated voltage increases. The induced voltage, which opposes the battery voltage, is called *counterelectromotive force (cemf)*. It is because of the cemf that a cranking motor has a larger current flow when starting than when running at full speed.

When a shunt coil is used to limit high armature speed, it creates the same magnetic strength as the other field coils. It maintains its strength because it has its own ground. Therefore, at high armature speed it revolves through the strong magnetic field and creates a greater countervoltage, reduces the current flow and, subsequently, the armature speed.

To increase the current flow (cranking power), and create a stronger magnetic field, the field-coil current is separated into two paths or split into three separate paths. Each separate field-coil circuit has its own live and ground brush (Fig. 41–6*c*). However, some cranking-motor circuits have the current flow split into four separate paths. Two of the field

coils use one live and one ground brush. You may notice when examining the cranking-motor circuits that all live brushes are connected to each other. The jumper leads equalize the voltage at the brushes to prevent arcing (burning). The ground brushes of cranking motors used in marine applications or used with series-parallel switches also have the ground brushes connected through a jumper lead.

Drive Housing and Commutator End Frame The drive housing and the commutator end frame, which are bolted to the field frame, enclose and support the armature (Fig. 41–7). To prevent the armature shaft from bending, most large cranking motors have an additional bearing support between the field frame and the drive housing. The drive housing encloses and supports the armature, shields the cranking-motor drive, and is the means by which the motor is bolted to the engine. In most cases, the commutator end frame also is the mounting surface for the brush holders and the terminal bolt.

CRANKING-MOTOR DRIVES

Although there are many different drives used to engage or disengage the cranking motor from the engine, to reduce motor speed, and to increase cranking torque, there are two major types. They

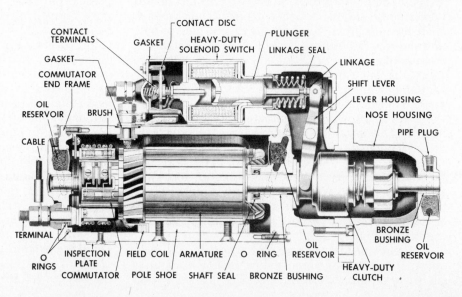

Fig. 41–7 Sectional view of a cranking motor with a heavy-duty sprag clutch. *(GMC Detroit Diesel Allison Div.)*

are the inertia drives and the drives which require a solenoid switch and a shift mechanism. The inertia drives, which rely on inertia for their pinion engagement, are used mostly on small engines. They employ a magnetic switch to close the cranking-motor circuit. The friction-type clutch (inertia drive) is used exclusively on larger engines.

An advantage which the cranking-motor drives requiring a shift mechanism and a solenoid switch have over the inertia drives is that their pinion engagement is made prior to the closure of the cranking-motor circuit. This ensures a smoother pinion engagement than that of the inertia-drive pinion which is spun into mesh.

Inertia Cranking-Motor Drive Though all inertia drives operate on the same basic principle, there are differences among them. Some designs engage and disengage more smoothly than others and have

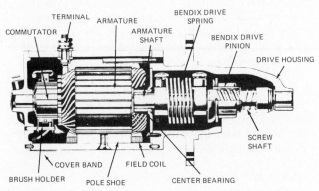

Fig. 41-8 Sectional view of a standard Bendix drive cranking motor. (*United Delco-AC Products, General Motors of Canada, Ltd.*)

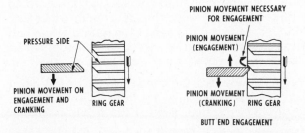

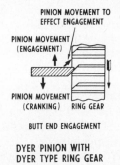

Fig. 41-9 Action of Dyer pinion in engaging with flywheel teeth of different chamfers. (*United Delco-AC Products, General Motors of Canada, Ltd.*)

better cranking-motor protection. However, all automatically engage when the cranking circuit is closed, and all automatically disengage when the pinion is driven faster than the armature.

Inertia-Drive Operation When the starting circuit is closed, the armature starts to rotate. This rotation carries through to the drive spring, or to the rubber cushion, or to the multiclutches, depending on which inertia drive is being used (see Fig. 41-8). It continues to the screw shaft or to the shaft assembly, which fits loosely over the armature shaft. The loosely fitted pinion, either because it is unbalanced or because of its inertia, does not revolve with the screw shaft or the shaft assembly. Therefore, the screw shaft or shaft assembly rotates within the pinion, causing the pinion to move longitudinally to engage with the ring gear. The engagement is made easier due to design of the pinion and ring-gear chamfer (Fig. 41-9). When the pinion is in mesh, the shock load, as well as the torque, is transferred first by the drive spring or the rubber cushion, then by the drive lugs. When the engine is running and when the ring gear drives the pinion faster than the speed of the armature, the pinion moves longitudinally on the screw shaft out of mesh, since its speed is higher than that of the armature.

Folo-Thru Drive The Folo-Thru drive shown in Fig. 41-10 or a Folo-Thru using a rubber cushion in place of a drive spring have three additional features over the standard or the pinion-and-barrel-type drives (Fig. 41-8). It has a spring-loaded detent pin (antidrift pin) that locks the pinion with the screw sleeve when disengaged to prevent the pinion from moving into mesh during engine operation. A second spring-loaded detent pin moves into a notch cut in the screw sleeve, and holds the pinion engaged until the pinion speed creates enough centrifugal force on the detent pin to overcome its spring. The pinion is then free to move longitudinally out of mesh. To prevent the armature from being driven by the pinion, a dentil clutch of a ratchet-tooth design is used.

As shown in Fig. 41-10, the drive-spring anchor plate locks over the dentil clutch and the teeth of the clutch transmit the torque to the teeth of the screw sleeve. When the ring gear drives the pinion faster

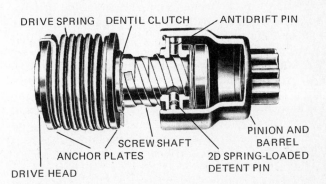

Fig. 41-10 Folo-Thru Bendix drive. (*United Delco-AC Products, General Motors of Canada, Ltd.*)

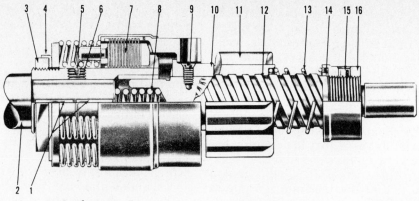

1. Woodruf keys
2. Armature shaft
3. Adjusting nut
4. Lock washer
5. Lock ring
6. Head screw
7. Clutch assembly
8. Meshing spring
9. Back stop screw
10. Back stop
11. Pinion
12. Screw shaft
13. Antidrift spring
14. Stop-nut ring
15. Stop-nut pin
16. Stop nut

Fig. 41–11 Sectional view of a friction clutch drive. *(United Delco-AC Products, General Motors of Canada, Ltd.)*

than the speed of the armature shaft, the teeth over-run until the detent pin is engaged.

Servicing and Troubleshooting Folo-Thru Drive
When a Folo-Thru drive will not stay in the engaged position, one of the following defects may be the problem: a worn armature shaft or bent shaft, a worn commutator end frame or drive-housing bushing, a damaged or worn pinion (ring gear), or a broken drive spring.

A Folo-Thru drive may be prevented from meshing or may engage unsatisfactorily because of a damaged or worn ring gear and pinion, a bent armature shaft, a dirty or worn screw sleeve, a worn or seized antidrift pin, a damaged drive spring, a worn or damaged dentil clutch, or a worn or damaged pinion guide.

When servicing the drive, give special attention to the pinion guide, detent pins and springs, dentil clutch teeth, bushings, and pinion teeth. Make certain when replacing the drive spring that it is replaced with a drive spring wound in the same direction.

When you reassemble the drive, make certain that the components are clean and free from oil, and that the snap rings are properly seated. Use graphite powder to lubricate the parts which slide or rotate. Do not use oil; it will reduce engagement efficiency.

Friction Clutch Drive
The friction clutch drive is used exclusively with large cranking motors. The major differences between this type and the other inertia drives is that the shock load and torque are transmitted through multiclutches (7 in Fig. 41–11). To prevent pinion engagement during engine operation, an antidrift spring is positioned between the pinion and the stop-nut ring. To reduce pinion and ring gear tooth wear and to assist in engagement, the pinion is chamfered and a meshing spring is placed between the pinion and the clutch housing.

Servicing Friction Clutch Drive
All inertia drives have the common problems of the pinion not staying in the engaged position, not engaging properly, or disengaging improperly. However, the friction clutch drives are often subject to the additional problems of ineffectively transmitting torque transfer to the armature or not transmitting it at all. In either case, a worn or improperly adjusted clutch pack is usually the cause.

In order to adjust the drive clutch so that its drive is positive but can nevertheless slip sufficiently to absorb the shock load, the cranking motor must be removed from the engine. The drive housing must also be removed in order to bend the locking plate lip back to turn the adjusting nut.

PROCEDURE Turn the adjusting nut clockwise either one-half turn or one turn, depending on the amount of slippage. The increased spring tension increases the force on the clutches. Lock the adjusting nut. Reassemble the cranking motor. Install the motor and recheck cranking-motor operation. If the clutch continues to slip, repeat the adjustment procedure previously outlined but adjust the nut only one-half turn at a time. This trial-and-error method must be followed until cranking-motor efficiency is attained.

NOTE Do not turn the adjusting nut to the point where the clutch springs are fully compressed. If this point is reached, service the drive assembly.

When the drive assembly is rebuilt, adjust the friction clutch, in the following manner. Install a dummy shaft with key in the lock ring, and hold the shaft in a vise. Install a deep socket on the pinion, and use a torque wrench to check the breakaway of the friction clutch. Adjust the nut until the breakaway of the clutch corresponds with the stalled torque test specification.

Overrunning Clutch
The ball or roller overrunning clutches are used with smaller cranking motors (see Fig. 41–12). They are operated with a solenoid switch and a shift mechanism.

Inside the shell is the outer part of a wedge arrangement which wedges the rollers or balls onto the external pinion surface. The sleeve has splines which fit loosely on the mating surface of the

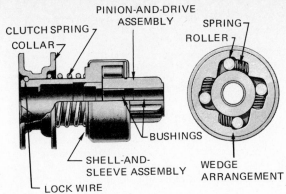

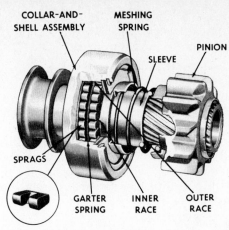

Fig. 41–12 Overrunning clutch. *(United Delco-AC Products, General Motors of Canada, Ltd.)*

Fig. 41–13 Sprag clutch. *(United Delco-AC Products, General Motors of Canada, Ltd.)*

straight or spiral splines of the armature shaft. The collar fits snugly over the shaft but can move on the sleeve. The pinion is permanently secured within the shell. An internal bushing supports the pinion-and-drive assembly at the front on the armature shaft. Coil or accordian springs hold the rollers or balls in their tapered cut recessed area. The balls or rollers lightly contact the pinion and shell surfaces.

Overrunning Clutch Operation When the solenoid switch is engaged, the shift lever moves the collar over the sleeve and compresses the clutch spring. The clutch spring then moves the shell, sleeve, and pinion into mesh with the ring gear. Should the teeth not mesh at this time, the collar further compresses the clutch spring, and as soon as the armature turns, the pinion engages. As the armature revolves the shell revolves also, transmitting torque because friction forces the balls or rollers to roll out of their grooves and wedge the pinion to the shell. When the engine is started and the pinion runs faster than the armature, the friction of the pinion surface forces the balls or rollers back into their grooves. The connection between the pinion and the armature is broken and the pinion can overrun the armature.

Overrunning Clutch Failure Overrunning clutch failure is often brought about by worn bushings which cause misalignment between the pinion and the shell and, therefore, prevent the balls or rollers from locking the pinion.

The overrunning clutch cannot be serviced. When it fails to operate properly, it must be replaced.

Sprag Clutch Cranking-Motor Drive In order to have an overrunning clutch that would accommodate larger cranking motors, the ball-and-roller overrunning clutch had to be redesigned. This meant a change from balls and rollers to sprags. The sprag-clutch assembly is beefed up to carry a greater torque to the ring gear and to ensure reliable engagement and disengagement. Furthermore, the assembly is now serviceable.

The collar-and-shell assembly is a one-piece unit (with internal splines) splined to the armature shaft

(see Fig. 41–13). The bushing, pressed in the sleeve, supports the armature shaft at the front and at the rear, the shaft is supported in the shell. The sprags are located between the shell surface and the sleeve surface and are held against these surfaces by a garter spring. The sleeve and the sprags are held in the shell by a snap ring. Bronze washers on both sides of the sprags are used to reduce friction. A seal protects the sprag assembly, preventing the entry of water or dirt. The pinion is spirally splined to the sleeve and is held by the meshing spring against the pinion stop.

Sprag Clutch Operation When the solenoid switch activates the shift lever, it moves the entire sprag clutch on the spiral armature shaft, causing the assembly to rotate about 10° and ensuring a smooth pinion engagement. Should the pinion butt against the teeth of the ring gear, the sleeve movement will stop. This also causes the shift lever to stop in its travel because the two retainer cups butt against each other. Since the cranking-motor circuit cannot be closed by the solenoid switch, the pinion and ring gear are protected from damage and the armature shaft cannot bend or twist.

The starter circuit must be opened to accomplish pinion engagement. When the solenoid switch closes the cranking-motor circuit, the armature rotates, and with it, the collar and the shell. The friction from the shell surface causes the sprags to tilt, locking the sleeve to the shell. When the ring gear drives the pinion faster than the armature, the pinion friction tilts the sprags in the opposite direction, allowing the pinion to overrun.

Servicing Sprag Clutch The sprag clutch should last the normal life of the cranking motor without requiring servicing. However, when the cranking motor is serviced, the drive assembly should also be cleaned and checked. When you disassemble the sprag clutch, take note of the position of the sprags before removing them. Check the collar for grooves which may have been caused by the shift lever. Check the drive splines of the shell and sleeve for damage and wear.

NOTE When the bushing in the sleeve is worn, you must replace the sleeve, otherwise the sprag may not lock or may not disengage properly.

Make certain the bronze thrust washers are smooth (without grooves). Check the pinion for worn spiral splines. Check the teeth and the chamfer for wear or damage. Replace the retainer cups if excessive wear is noticeable. Check the sprags and the lock surfaces of the pinion and shell for brinelling or rough surfaces. Lubricate the sprag-clutch assembly with light engine oil and the splines with graphite powder to promote longer, troublefree service life.

CAUTION Do not forget to lock the pinion stop, otherwise the whole assembly could come off the armature shaft.

Dyer Drive Since Dyer drives have now been replaced by the Positork drive, a discussion of them in this text will be brief. Their advantage over the sprag clutch or the inertia drive is that once the pinion is disengaged, it locks and cannot be reengaged until the solenoid switch is reenergized.

Dyer Drive Operation When at rest, the pinion guide is locked into the notch of the armature spline and the pinion spring prevents it from releasing (Fig. 41–14). The shift lever bottom is held in the cutout area of the shift sleeve. When the solenoid switch is energized, the shift lever moves the shift sleeve to the right, causing the pinion guide and the pinion to move longitudinally. This causes clockwise rotation and engagement of the pinion with the ring gear.

Should the pinion butt against the ring-gear teeth, the pinion guide rotates the pinion until the teeth align with the ring gear. The pinion spring forces the pinion into mesh. At this point the shift lever has traveled its full stroke and the solenoid switch has closed the cranking circuit. As the armature starts to rotate, the friction between the shift sleeve and the pinion guide causes the shift sleeve to rotate. The shift lever bottom then comes out of its notch, and the spring forces it to the left. When the pinion is driven faster than the armature, the pinion and pinion guide rotate on the armature shaft, disengage the pinion, and lock the pinion guide into the milled-out notch of the armature spline. The shift lever, however, remains in the same position. It will only return to its rest position when the solenoid switch (starting circuit) is opened. The shift lever then retracts through the coil spring to its rest position, and its bottom resets in the shift sleeve.

Positork Drive The Positork drive, shown in Fig. 41–15, is used with larger cranking motors and has replaced the Dyer drive. The Positork drive provides a positive engagement with the pinion remaining in engagement until the activating circuit is deenergized. If the pinion does not mesh with the ring-gear teeth, the cranking motor cannot be energized and the teeth of the pinion and of the ring gear, therefore, cannot be damaged.

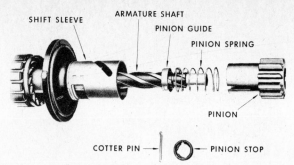

Fig. 41–14 Dyer-drive components. (*United Delco-AC Products, General Motors of Canada, Ltd.*)

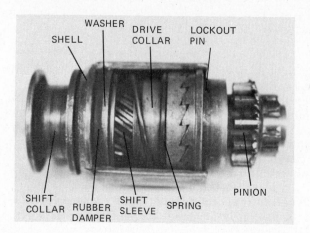

Fig. 41–15 Positork drive.

The components of a Positork drive are shown in Fig. 41–16. The shift lever is supported at the front by its bore on the armature shaft and at the rear by short helical splines that mesh with the helical splines of the armature shaft. The shift collar is fastened to the left side of the shift sleeve with a snap ring. The shell, rubber damper, and washer are placed over the shift sleeve. The drive collar has internal helical splines which mesh with the external splines of the shift sleeve. A flat washer and a cone washer are placed against the inner shoulder of the drive collar and a three-piece lockout device is loosely pinned to the interior of the drive collar. A snap ring holds the assembly to the shell.

Positork Drive Operation When the solenoid switch is deenergized, the spring of the solenoid disengages the Positork drive from the flywheel ring gear. At the same moment the spring in the Positork drive has forced the drive collar in mesh with the pinion, causing the coned washer to force the three-piece lockout device against the shift sleeve. The teeth of the pinion can then mesh with the teeth of the drive collar. When the solenoid-switch circuit is energized, it causes the shift lever to move, which in turn moves the pinion teeth in mesh with the teeth of the ring gear. A smooth engagement is assured because the shift lever is forced to turn on the armature shaft as the shift lever moves the shift sleeve. Should the pinion not mesh with the ring gear, the pinion rebounds and is cushioned on the rubber damper. However, the drive collar is forced to turn as it and the pinion move against the damper.

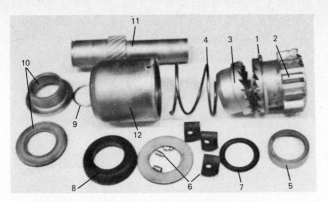

1. Snap ring
2. Drive pinion
3. Drive collar
4. Spring
5. Coned washer
6. Lock pin and lockout
7. O ring
8. Rubber damper
9. Snap ring
10. Shift collar
11. Shift sleeve
12. Shell

Fig. 41–16 Disassembled components of a Positork drive. (The numbers indicate component disassembly order.)

Effective engagement results as the spring forces the drive collar and pinion to the right. When the pinion is fully engaged, the cranking-motor circuit is closed (by the solenoid switch) causing the armature and the Positork to rotate as a unit.

When the engine starts, the pinion is rotated faster by the ring gear than the remaining parts are driven by the armature and a demeshing of the pinion with the drive collar occurs. The three-piece lockout device now holds the pinion in a demeshed position from the drive collar until the pinion reaches the speed of the armature or until the cranking motor is deenergized. The lockout prevents a reengagement of the pinion with the drive collar whenever the pinion is engaged with the ring gear and the engine is running.

Servicing Positork Drive The Positork drive can be disassembled and serviced by drilling a $^3/_{16}$ in [4.76 mm] hole in the shell to remove the snap ring. When the snap ring is removed, the components can be lifted from the shift sleeve in the order shown in Fig. 41–16.

When servicing the components, you will sometimes find that the cone washer has machine-grooved the three lockout pieces at their cone ends and/or that the pinion teeth are damaged. Other than this you should find no damage to the other components. The reason for the grooves in the lockouts is that the cranking motor was energized too long (or for short periods, too often) while the engine was running. Grooving can prevent proper lockout, causing pinion and ring-gear wear, as well as wear of the drive collar and pinion teeth.

CRANKING-MOTOR SERVICE

Let us assume that you are faced with a cranking motor that does not turn the engine or one that turns it too slowly. Before removing the cranking motor, you should test and check the batteries, the motor connection, and the starter, neutral, and solenoid switches. Check that all components are mechanically and electrically in good working order, that the engine is not seized, the cranking-motor drive is not locked with the ring gear, and that the ring gear is not loose and has no broken teeth. If, through this process of elimination, it is obvious that the trouble lies in the cranking motor, it will then have to be removed.

Removing Cranking Motor When you remove a cranking motor, always remove the ground battery cable first. If it is necessary to raise the vehicle to gain access to the cranking motor, make certain the vehicle is safely supported.

Remove the battery cable, starter, and ammeter wires from the cranking motor. When an insulated cranking motor is used, also remove the ground cable. Remove the mounting bolts, and then carefully remove the motor from its mounting.

CAUTION Do not overestimate your strength when lifting a large cranking motor from an engine. Always follow the prescribed safety rules.

After the motor has been removed from its mounting, check the condition of the ring-gear teeth. With a prybar, rotate the flywheel 360° so that all ring-gear teeth can be examined.

Disassembling Cranking Motor Before you make any attempt to disassemble the cranking motor clean it with a wire brush and solvent.

CAUTION Always wear safety glasses when disassembling and assembling a cranking motor.

With a center punch, mark the commutator end-frame, drive housing, and the field-frame assembly (Fig. 41–17). These alignment marks will ensure proper brush and drive-housing position when reassembling.

Remove the solenoid switch from the field-frame assembly. Remove the hex bolts which hold the drive housing and commutator end frame. Remove the dust cover and screws which connect the field coils to the brush holders. You can now remove the commutator end frame and the drive housing.

To remove the armature, remove the cotter pin and slide the pinion stop, pinion, pinion spring, pinion guide, shift sleeve, cup washer, and spacer washer from the armature shaft. When the intermediate bearing (bushing) is worn or damaged, remove it from the field frame.

Inspecting Armature Visual inspection, prior to cleaning and testing the armature, can often disclose the cause of trouble. If the armature is greasy, it should be cleaned with wood alcohol, otherwise compressed air cleaning is sufficient.

If the armature is polished on one end, the shaft may be bent, the bushings worn, or the pole shoes

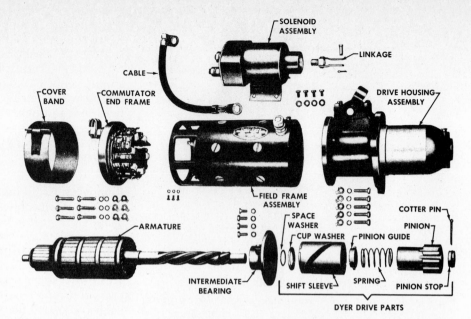

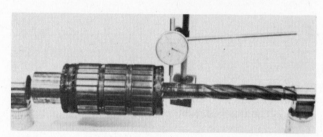

Fig. 41–17 Disassembled components of a heavy-duty Dyer drive cranking motor. (GMC Detroit Diesel Allison Div.)

loose. If there are signs of wear on the armature-coil insulation, the coils are probably loose in the grooves of the armature core due to overspeeding of the armature. The cranking-motor drive and ring should then be checked.

If one or more commutator bars are burned, one or more armature coils have an open circuit. Check the coil-to-commutator connection. If the commutator is excessively worn or out-of-round, it is possible that the armature shaft is bent, the bushings are worn, the brush-spring tension is too high, or the wrong brushes are being used.

If the mica (insulation between the commutator segments) is higher than the bronze commutator bars, the brushes are too soft or the mica should have been undercut.

If you find thrown solder, it is an indication that the cranking motor was overheated. Advise the operator to use the cranking motor for 30 seconds only, then allow it to cool before reusing it again.

Testing Armature Shaft and Commutator for Runout Large armatures should be checked for commutator and armature-shaft runout. When the commutator has to be serviced, test the armature-shaft runout only. Later when the commutator is resurfaced, test the commutator runout.

To test the armature-shaft runout, place the armature between two V blocks and position a dial indicator on the armature-shaft center-bearing surface (Fig. 41–18). Rotate the armature 360° and note the runout. If the runout is higher than 0.003 in [0.076 mm], the shaft must be straightened. If the bearing surface is damaged or worn, the armature must be replaced.

Testing Armature for Grounded Circuit An armature growler is used to test the armature loops for grounded, open, or short circuits. To test the armature for ground, place the armature growler as shown in Fig. 41–19, and switch on the test-probe circuit. Hold one test probe on the armature shaft or

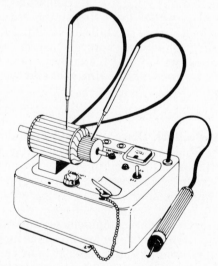

Fig. 41–18 Testing armature-shaft runout.

Fig. 41–19 Armature ground test. (J. I. Case Co., Components Div.)

iron core and slide the other probe over and around the commutator. The test light will go on when one or more loops or the commutator are grounded to the armature shaft. A replacement of the armature is usually necessary.

Testing Armature for Short Circuit Have the armature in the same position as in the previous test, but turn on the switch to energize the growler coil winding. The current flow through the coil windings creates a magnetic field which induces voltage into the armature loops. Rest the steel blade, supplied with a tester, parallel with the armature loops, and slowly rotate the armature at least 360°. The steel blade will vibrate at 60 cycles per second when one or more armature loops touch and short-circuit each other. The armature must be replaced if you cannot locate the short-circuited loops.

Testing Armature for Open-Circuit Loops (Armature-Loop Balance Test) The purpose of an armature loop balance test is obvious from its name. Position the armature as in the previous test. Adjust the test probes so that they are the center of the commutator with one probe on one commutator segment and the other probe on the adjacent commutator segment. When the ac power is turned on, the current flowing through the growler coil creates a magnetic field and induces ac voltage in the armature loops. This induced voltage is picked up by the test probe resting on the commutator segment. It is directed to the gauge and back to the other test probe.

Adjust the gauge to a full number, say 1, by adjusting the variable resistance, then turn the armature one commutator segment, and again hold both probes against a commutator segment. The gauge should once again read "1." Repeat the procedure until all loops are tested. Minute variations may occur due to the probe position; however, abnormal readings during testing indicate that there is a short circuit or a poor connection. When there is no reading at all, one or more armature loops have an open circuit. When the readings are uniform the armature is electrically in good condition.

Resurfacing Commutator When the commutator is rough, grooved, tapered, or out-of-round, it must be resurfaced. This can be done by machining and then polishing the commutator in an armature lathe or a small machinist's lathe.

To machine the commutator, first support the commutator end in a live center or bushing. Grind the cutting tool for bronze cutting (not undercutting). Then adjust it a fraction below the center of the commutator. Remove only enough material to obtain a smooth surface.

UNDERCUTTING MICA The mica of some commutators does not need to be undercut. Let your service manual guide you in this regard.

Mica can be undercut with a special cutting tool supplied with the armature lathe, with a fine cutting saw fastened to a hand drill, or with a hacksaw blade. When using a hacksaw blade, grind the blade parallel (not tapered) to the width of the mica. To achieve a good cut, pull the hacksaw blade toward you to remove the mica. Do not push it. Undercut the mica to the depth of the distance between the two commutator segments. When the mica is sawn, the cut should resemble D in Fig. 41–20.

After the commutator is undercut, reinstall the armature in the lathe, and polish the commutator with a 600-grit body paper until the surface is mirror-finished. Blow the cutting and grinding material

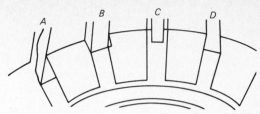

Fig. 41–20 Undercutting the mica. Points A, B, and C show how an undercut mica should not look; point D shows how an undercut mica should look. (*GMC Truck and Coach Div.*)

Fig. 41–21 Checking the field coils for open circuit. (*GMC Truck and Coach Div.*)

from the armature with compressed air. Retest the armature to confirm that you have not damaged it during servicing.

NOTE Do not forget to polish the bearing surface and to clean the straight or spiral splines. Do not forget to check the keyways and snap-ring grooves (if used).

Inspecting Field-Frame Assembly Very seldom will the field-frame assembly prevent the cranking motor from operating. However, if this assembly is the cause of cranking-motor trouble, the reason may be a cracked housing, the field-coil insulation burned by a short-circuited or grounded field-coil circuit, or the insulation damaged during assembly or disassembly.

Shiny pole shoes indicate that the armature shaft is bent, the bushings are worn, or the pole shoes are loose. The motor terminal should be partly removed so that the insulation, washers, and grommet can be examined.

After you make this examination, blow the assembly clean with compressed air and, if necessary, wash the field coils with alcohol before testing them.

Testing Field Coils A test light or an ohmmeter can be used to test the field coils for continuity and ground. Field coils cannot be tested for a short circuit because of their low resistance.

When testing for continuity, hold the test probes to each end of the field-coil windings, one set at a time (Fig. 41–21). When the test light fails to glow, or the ohmmeter does not move to zero, the field-coil circuit is open.

When testing for a grounded coil circuit, hold one probe to ground and the other probe to the field-coil lead. The test light should not glow.

When a motor terminal has been replaced, recheck it for continuity and ground. You may have created a ground when soldering or soldered the joint poorly.

When you replace field coils, use the correct tools, as shown in Fig. 41–22, to ensure proper installation. Make certain that the field coils fit properly on the pole shoes and that one coil is wound left and the adjacent coil is wound right, then alternately left and right, so that adjacent poles are in opposite polarity.

When you solder the coil ends together, take care not to burn the insulation. Also make sure you have soldered the joint well.

Testing Commutator End Frame Visually inspect the ground terminal, insulation, and terminal bolt threads. Check the brush springs, and replace any which are corroded or damaged. Gently tug each brush holder to determine if any are loose.

The commutator end frame can be checked electrically with a test light or an ohmmeter. Let us assume that the live and ground brush holders of the cranking motor are insulated. In this case when testing for ground, hold one probe to ground and the other probe to a brush holder. Proceed systematically to each holder until all six brush holders have been tested. Do not forget to test the insulated ground terminal.

Servicing Brushes When the cranking-motor brushes are worn to half of their original length, they must be replaced since the spring tension is then greatly reduced. If the brushes are reusable, remove

POLE-SHOE SCREWDRIVER

POLE-SHOE SPREADER

Fig. 41–22 Removing and installing field coils using a pole-shoe spreader and a screwdriver. (GMC Truck and Coach Div.)

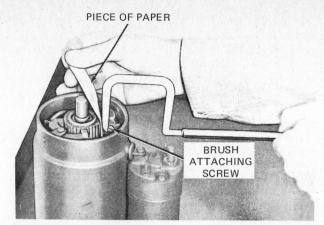

PIECE OF PAPER

BRUSH ATTACHING SCREW

Fig. 41–23 Measuring brush-spring tension. (J. I. Case Co., Components Div.)

them. Clean their contact surfaces with an 80-grit sandpaper. File off the corners of the brushes to ensure a smoothly gliding brush contact and faster brush seating.

Reinstall the brushes in their holders, but do not tighten the screws at this point. Install the armature to the commutator end frame. Align the brushes to the commutator. Now tighten the screws securely. Keep the brushes in alignment while the screws are being tightened.

Cranking-motor brushes must have a seating area of at least 25 percent, and the seating area must be in the center of the brush. A brush seating on either side of the center will cause a change in commutation and arcing.

To seat the brushes, cut a strip of 80-grit sandpaper the width of the commutator. Place it between the brushes and commutator with the cutting side toward the brushes. Then slide the sandpaper from underneath the brushes in the direction in which the armature is turning. Repeat this procedure until the brushes seat properly against the commutator. Make certain you follow the contour of the commutator.

Testing Brush-Spring Tension With the armature in the end plate, check the brush-spring tension by placing a small piece of paper under one of the sets of brushes. Hook a spring scale (calibrated in ounces [grams]) onto the brush holder (see Fig. 41–23). Pull on the spring scale until you can slide the paper out. At the same time, note the reading on the scale. It should coincide with the recommended specification.

Bushing Replacement When you replace armature bushings, do not substitute some other type of bushing. Motor bushing material, for example, is made from a special alloy quite different from that of bushings generally.

When driving bushings in or out, use the correct size bushing tool to prevent damage to the bushing or its bore. To remove a bushing which has a closed-off bushing bore from the commutator end frame, tap a thread into the bushing. Place a hex bolt in the threaded hole and, with a hammer puller, pull out the bushing.

NOTE Check all newly installed bushings to ensure that the armature rotates freely in them.

Assembling Cranking Motor Install the armature to the commutator end frame, but make certain the brushes do not come out of alignment.

NOTE Some cranking motors use a leather washer between the commutator end frame and the commutator to prevent the armature from overspeeding after the pinion is disengaged from the ring gear.

Slide the field frame over the armature. Align the commutator end frame with the prestamped center-punch marks, and install a few bolts to hold it in place. Install and secure the intermediate bearing. Assemble the Dyer drive on the armature shaft. Make certain that the tangs are pointing away from the pinion. Install and align the drive housing. Then tighten all the bolts. To be sure the armature will rotate, use a prybar and force the pinion to rotate. Install the solenoid switch and connect the linkage with the actuating mechanism.

Pinion-Clearance Check Cranking motors employing an inertia cranking-motor drive, do not require a pinion-clearance check as do cranking-motor drives using a solenoid switch. For the latter, a pinion-clearance check is essential. When inadequately adjusted, the pinion can cause a hammering effect on the end of the drive housing. This may break the housing or cause the armature to bounce back and break the commutator end frame. When properly adjusted, the pinion-clearance check assures effective pinion-to-ring-gear meshing and correct closing sequence of the cranking-motor circuit.

The electrical components and connections required to perform a pinion-clearance test are shown in Fig. 41–24. The connections for all cranking motors using a solenoid switch are the same, but the clearances and the points where measurements should be made vary between the different types of drives. For instance, when checking the Dyer-drive pinion travel, the clearance between the shift sleeve and the pinion must be measured. When checking a heavy-duty sprag clutch drive, however, the clearance is measured between the drive housing and the pinion.

To adjust the pinion for clearance on a Dyer drive and on certain sprag drives, turn the threaded solenoid plunger haft in or out. To adjust a sprag clutch drive having a serrated shift lever linkage, loosen the bolts and slide the linkage backward or forward to arrive at the correct clearance. On other sprag clutch drives, loosen the solenoid-switch mounting bolts, and then slide the solenoid switch forward or backward to arrive at proper clearance.

Cranking-Motor No-Load Test To test the cranking-motor efficiency, you should perform a no-load test. To make this test, connect the batteries, test instruments, and carbon pile (variable resistor) as shown in Fig. 41–25, or alternatively, connect the carbon pile across the batteries. The batteries must be fully charged and in good chemical and mechanical condition. Use an rpm indicator (tachometer) to measure the armature speed. Before you begin the test, close the master switch and, with the carbon pile, adjust the resistance to arrive at the desired specified voltage. Measure the rpm, note the current draw, and observe the brush action.

Inadequate Cranking-Motor Performance When a cranking motor has a low free speed and a high current draw, the problem may be a short-circuited armature, grounded armature or field coils, bent armature shaft, one or more loose pole shoes, or worn bearings.

When the cranking motor has low free speed and a low current draw, it could be due to a poor connection, defective leads, dirty commutator, open armature loops, open field coils, short brushes, broken brush spring(s), broken brush link, or high mica. When the cranking motor has a high current draw

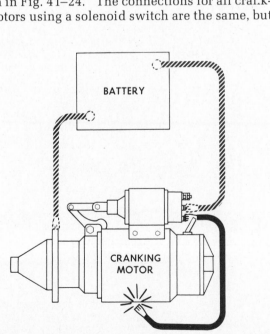

Fig. 41–24 Connections for checking pinion clearance. (*United Delco-AC Products, General Motors of Canada, Ltd.*)

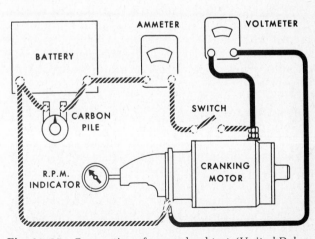

Fig. 41–25 Connections for a no-load test. (*United Delco-AC Products, General Motors of Canada, Ltd.*)

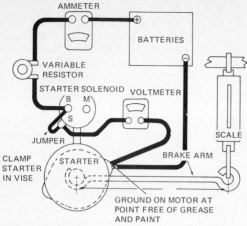

Fig. 41-26 Connections for a lock-torque test. *(J. I. Case Co., Components Div.)*

and either fails to rotate or rotates very slowly, it could be due to a grounded motor terminal, direct ground of the field, seized bearings, or a broken or badly twisted armature shaft.

Cranking-Motor Lock-Torque Test   The purpose of a lock-torque test is to prove the cranking motor's capability to develop the specified torque required to crank the engine.   Some manufacturers give no lock-torque test specifications.   They assume that if a cranking motor performs to specification during a no-load test, it will also produce the required torque.

To make a cranking-motor lock-torque test, securely mount the cranking motor in a test stand or a vise.   Connect the instruments and variable resistors as shown in Fig. 41-26.   Lock a 1-ft [30.42-cm] brake arm to the pinion while manually holding the pinion in the engaged position.   Connect a spring scale to the brake arm, but take care that the brake arm does not slip off the pinion when the current is applied to the motor.   Draw the specified current by varying the resistance, and then read the spring scale.   A 1-ft [30.42-cm] brake arm directly indicates foot-pounds [kilogram-meters] (1 ft·lb = 0.1383 kg·m).

Series-Parallel Switch   A series-parallel switch is used when 24 V are required for cranking when the charging system is only a 12-V system.   When energized, this switch connects two or four batteries in series to make 24 V available for the cranking-motor circuit.   When deenergized, it connects the batteries in parallel to 12 V.

Some series-parallel switches are combined with a magnetic switch (see Fig. 41-27).   The combined solenoid-operated series-parallel and magnetic switch has four heavy terminals and four heavy contacts, two spring-loaded disks, two small contact sets, and a solenoid.   The solenoid coil has two small terminals.   Terminal 6 is the ground connection, and terminal 7 is the starter switch terminal connection.   Terminal 5 is connected to the second ammeter to show the charging rate of the B battery (if used) and is the ground connection of the B battery (when the batteries are connected in parallel).   The

two small sets of contacts open or close the charging circuit to the B battery with the same plunger and rod which actuate the large disks.   The cranking motor (terminal) is connected to terminal 4, and the cranking motor is grounded at the A battery terminal post.   The positive post of the A battery is connected to terminal 1.   The negative side of the B battery is connected to terminal 2.   The positive battery post is connected to terminal 3.

SWITCH ACTION AND CURRENT FLOW WHEN CRANKING   As the operator closes the starting circuit, the solenoid coil is energized and pulls the plunger and rod inward.   The disks bridge the contacts between terminals 1 and 2 and between terminals 3 and 4 (Fig. 41-27).   This action connects the negative terminal of the B battery with the positive terminal of the A battery, and the positive terminal of the B battery with the cranking motor.   This closes the cranking-motor circuit to permit cranking to take place.   The ground side of the cranking motor is always connected to the negative terminal of the A battery.

SWITCH ACTION AND CURRENT FLOW WHEN CHARGING   As the operator opens the starting circuit, the solenoid coil deenergizes and the return spring forces the disk away from its contacts.   At the same time the two small sets of contacts are being closed (Fig. 41-28).   This action opens the cranking circuit and connects the battery "in parallel" with the charging circuit.   Current from the generator flows through the first ammeter to the series-parallel switch terminal 1.   Here the current flow splits.   The positive post of the A battery to ground becomes one circuit.   The other circuit includes termi-

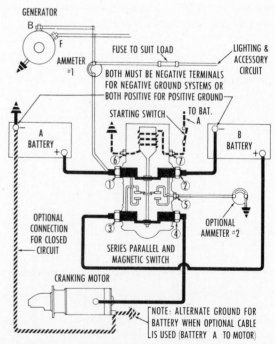

Fig. 41-27 Combined series-parallel and magnetic switch action and current flow when cranking. *(United Delco-AC Products, General Motors of Canada, Ltd.)*

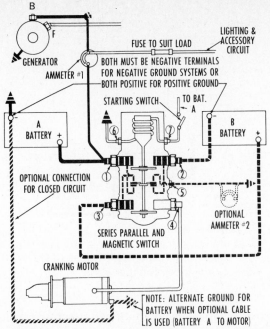

Fig. 41–28 Combined series-parallel and magnetic switch action and current flow when charging. *(United Delco-AC Products, General Motors of Canada, Ltd.)*

nal 1 to the first small contact set, from there to terminal 3, onto the B battery positive terminal, through the battery, out of the negative battery terminal, to terminal 2, to the second small contact set, and out of terminal 5 to the second ammeter, and then to ground.

NOTE Thinner gauge wire is used for this circuit and it has to pass over eight more connections than the first circuit. The first circuit uses battery cables as conductors, and it has only two connections.

Troubleshooting Series—Parallel Switches Ninety percent of all troubles caused by a series-parallel switch start with the undercharged condition of the B batteries. The most expedient way to prevent this situation is to:

1. Keep all connections clean and tight.
2. Keep the top of the batteries dry.
3. Replace batteries with those of the same make.
4. Use batteries of the same capacity, age, and chemical and mechanical condition.
5. Interchange A batteries with B batteries periodically to avoid undercharged conditions and to prolong battery life.

To check the charging circuit of the B batteries for high resistance, start the engine and run it about 1500 rpm. Connect a voltmeter across the B batteries. Let us assume that the voltage reading is 14.8 V. Connect a jumper cable (not a jumper wire) from terminal 1 of the switch to the positive post of the B batteries. The voltmeter reading should not increase more than 0.3 V. When a higher resistance is present, the series-parallel switch must be serviced.

Testing Series-Parallel Switch The service procedure is similar to that for servicing a solenoid switch with the exception that more contacts have to be serviced. To check the large contact set for continuity, energize the solenoid and connect an ohmmeter to terminals 1 and 2 and then to terminals 3 and 4. The ohmmeter should show no resistance. De-energize the solenoid and connect an ohmmeter to terminals 1 and 3 and then to terminals 2 and 5. In each case the ohmmeter reading should show no resistance.

12-V Charging System, 24-V without Series-Parallel Switch As today's diesel engines have ever larger displacements, the conventional series-parallel switches are no longer able to carry the high current demand placed on them. This has brought manufacturers to design fully solid-state switching devices that can do the work of a series-parallel switch. In addition, these devices prevent over- or undercharging of the batteries and thereby provide a more efficient starting voltage. One of these devices is shown in Fig. 41–29. This solid-state switching device does not require any additional components such as a transformer or rectifier [with the exception of the dual-voltage control (DUVAC) unit]. Any installed generator is suitable. Also note that the batteries are connected in series and that both batteries can be used for starting. However, battery B is used only for starting whereas battery A is connected to the electrical system of the vehicle, and all vehicle loads (including voltage regulator) are taken only from battery A.

OPERATION SEQUENCE When the engine is not operating, the controlled rectifier is open, preventing a current flow from the battery to the generator. When the operator energizes the cranking-motor activating circuit, the solenoid switch connects the cranking motor with the batteries and 24 V become available for cranking.

After the generator starts to operate and after it exceeds the cemf of both batteries (24 V), current begins to flow into the batteries.

Because the batteries are connected in series and the negative terminal of the A battery is grounded,

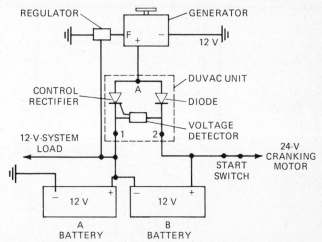

Fig. 41–29 Wiring diagram of the dual-voltage control.

both batteries are being charged. Current flows from the generator to terminal A of the DUVAC unit through the diode (out of terminal 2) to the positive post of the B battery, then to the A battery, and finally to ground. The current output of the generator increases until the A battery voltage rises to the value at which the voltage regulator is adjusted, and then regulates in its normal manner. Since the voltage regulator senses the voltage of the A battery, the B battery becomes fully charged first. As soon as the B battery becomes fully charged, the voltage detector in the control unit switches the control rectifier on, and the current can now only flow into the A battery. The B battery, therefore, is now out of the charging circuit and is only used again when the engine is to be restarted.

30-SI/TR Series Generator

Another method of carrying high current for the 24-V starting system without using a series-parallel switch is to use a generator that can supply 24 V and 12 V simultaneously. This is accomplished by modifying an ac generator (30-SI) and the electrical connection to the cranking motor (see Fig. 41-30). The generator has a 12- and 24-V battery terminal. The 12-V generator terminal is connected to the positive side of the two parallel-connected 12-V A batteries. The 24-V generator is connected to the positive side of the two (also parallel) 12-V B batteries. The entire vehicle load (including the circuit of the magnetic switch that activates the cranking motor) is taken only from the A batteries. The B batteries are used only for cranking. The batteries are permanently connected in series with the cranking motor. **NOTE** Additionally, a magnetic switch is used to activate the solenoid switch of the cranking motor.

CURRENT FLOW WHEN STARTING When the starter switch is closed, current (12 V) flows from the A batteries to the magnetic switch, through the coil winding and from there to ground, causing the switch to close. Closure of the magnetic switch energizes the solenoid switch of the cranking motor. This action causes the disk to bridge the battery terminal and the cranking-motor terminal, resulting in a current flow (24 V) into the cranking motor. When the starter

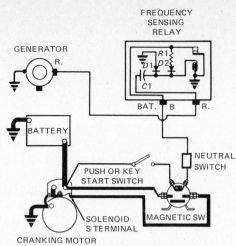

Fig. 41-31 Wiring diagram of a frequency-sensing relay when the generator is charging. *(United Delco-AC Products, General Motors of Canada, Ltd.)*

switch is opened, it deenergizes the magnetic switch. The magnetic switch opens the electric circuit of the solenoid switch and, as a result, the cranking motor is deenergized.

Frequency-sensing Relay Some large cranking-motor circuits utilize a frequency-sensing relay, as shown in Fig. 41-31, to protect the cranking motor. This relay automatically opens the starting circuit as soon as the generator begins to charge.

ACTION AND CURRENT FLOW WHEN STARTING As the operator closes the push or key starter switch, current flows to the pull-in winding of the magnetic switch, to the sensing relay B terminal, and over the closed contact points to ground. The magnetic switch pulls the disk and closes the starting circuit, energizing the solenoid switch. Cranking then takes place. As the engine runs, the generator generates voltage. Current present at the R terminal of the generator is directed to the R terminal of the relay and from there to the coil windings and to ground. The coil is energized and the magnetic pull pulls the armature and opens the points. With the points open, the coil of the magnetic switch loses its ground and the return spring opens the disk. This opens the starting circuit of the solenoid switch. Reengagement is not possible as long as the generator is charging.

Questions

1. Refer to Fig. 41-3. Why is the conductor loop forced to rotate?

2. Why are conductor loops placed in predesigned grooves in the armature?

3. Why is a jumper wire used between a pair of live brushes?

4. List three things that could cause a neutral switch to become ineffective.

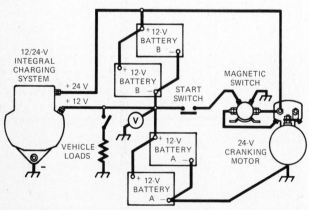

Fig. 41-30 12/24-V integral charging system. *(United Delco-AC Products, General Motors of Canada, Ltd.)*

5. List the three functions of a cranking-motor drive.

6. Describe the movement of the components during the engagement and disengagement of a heavy-duty sprag clutch.

7. List several conditions which will prevent a Dyer drive from engaging properly.

8. When is it necessary to undercut the mica?

9. Explain how to test the armature for a grounded circuit, an open circuit, and a short circuit.

10. When using a series-parallel switch, why is it recommended that the A and the B batteries be consistent in: capacity, size, age, and manufacturing source.

11. Why is it recommended that the position of the two and four batteries be periodically switched?

12. To find the voltage drop between the battery and starter switch, you would connect (a) an ammeter between the grounded post and the starter switch; (b) a voltmeter from the ungrounded battery post to the live side of the starter switch; or (c) a voltmeter from the starter switch to ground.

13. The field circuit of a dc motor should not be opened, especially if the motor is operating without a load, because the motor armature will become (a) heated, (b) short-circuited, (c) uncontrollable, or (d) dirty.

14. The direction of current flow in a dc motor is changed by means of the (a) commutator segments, (b) field windings, (c) brushes, or (d) field pole.

15. The rotating armature coil of a dc motor generates an electromotive force which (a) increases applied voltage, (b) decreases applied voltage, (c) opposes applied voltage, or (d) supports applied voltage.

16. If a cranking motor has excessive draw, which would be the likely trouble (a) badly worn brushes, (b) badly worn bushings, (c) battery cables too light, or (d) excessive air gap at pole shoes.

17. The most common cause of thrown armature windings is (a) overheating of armature, (b) grounding of armature, or (c) overspeeding of armature.

18. Burned commutator bars usually indicate (a) short-circuited armature windings, (b) open-circuit armature windings, or (c) grounded armature windings.

19. When checking a cranking motor on the tester, low free speed and high current draw with low torque may result from (a) open field, open armature, or high internal resistance; (b) grounded armature, weak brush springs, or short-circuited fields; or (c) internal ground, short-circuited armature, or worn bearings.

20. When checking a cranking motor on the tester, failure to operate with no current draw may result from (a) open or grounded fields, short-circuited armature, or worn bearings; (b) open field or armature, weak brush springs, or worn brushes; or (c) low commutator mica, grounded armature, or worn bearings.

21. When checking a cranking motor on the tester, failure to operate with a high current draw may result from (a) open or shorted field, high commutator mica, or weak brush springs; (b) jammed engine bearings, stuck piston or rings, or grounded armature; or (c) internal ground, frozen motor bearings, or grounded armature.

22. A starter may jam because of a (a) loose housing, (b) chipped starter gear, (c) worn ring gear, or (d) any of the above.

23. What are the advantages of a solid-state switching device over a charging system using a series-parallel switch?

24. Why is a 12-V magnetic switch used to energize the solenoid switch of the cranking-motor system shown in Fig. 41–32?

25. A cranking-motor circuit using a frequency sensing relay fails to energize the cranking motor after the operator has activated the control circuit. List the causes within the activated circuit that prevent the cranking-motor solenoid switch from energizing.

UNIT 42

Air and Hydraulic Starting Systems

Comparison of Starting Systems The air and hydraulic starting systems have become increasingly popular because of improvements in their design. Torque has increased noticeably, while the need for servicing has been reduced greatly.

Compared with an electric starting system, the air starting system is much less complex. It requires only an air tank, two control valves, a cranking motor, and the complementary hoses. A hydraulic system requires a hydraulic reservoir, a hand pump,

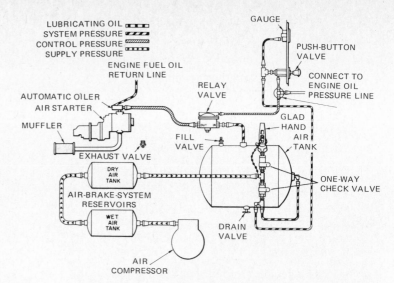

LUBRICATING OIL
SYSTEM PRESSURE
CONTROL PRESSURE
SUPPLY PRESSURE

ENGINE FUEL OIL
RETURN LINE

GAUGE

PUSH-BUTTON
VALVE

CONNECT TO
ENGINE OIL
PRESSURE LINE

AUTOMATIC OILER
AIR STARTER

RELAY
VALVE

MUFFLER

GLAD
HAND
AIR
TANK

FILL
VALVE

EXHAUST VALVE

DRY
AIR
TANK

ONE-WAY
CHECK VALVE

AIR-BRAKE-SYSTEM
RESERVOIRS

WET
AIR
TANK

DRAIN
VALVE

AIR
COMPRESSOR

Fig. 42–1 Schematic drawing of remote-control air starting system. *(Roosa Master Stanadyne/Hartford Div.)*

an engine-driven pump, an accumulator, a hydraulic cranking motor, and hydraulic pressure lines.

A complete air starting system, including the air tank, weighs as little as 150 lb [68.1 kg] while a hydraulic system weighs approximately 225 lb [102.2 kg]. An electric starting system, including the batteries, weighs approximately 400 lb [181.6 kg]. Cost and maintenance features of the air starting system are also more favorable, amounting to about one-third that of the hydraulic or electric systems. The hydraulic and the air starting systems are also more reliable than the electric starting system.

The air starting system has a slight advantage over the hydraulic system in regard to cranking time, but both these systems have a cranking speed of about 100 rpm more than an electric starting system. An air starting system with a standard 6 ft³ [0.1699 m³] air tank will crank an average diesel for 9 seconds at 70°F [21.1°C]. At the same temperature a hydraulic system with a standard accumulator will only crank a diesel engine for 3 seconds.

NOTE Cranking speed and not cranking time is the main essential for starting a diesel engine rather than the repeated number of times the engine will start. An electric starting system may crank the engine over for a longer period of time, but the cranking speed will gradually decrease. Additionally, with lower ambient temperature, the cranking speed may reduce to one-quarter of that of the air or hydraulic starters, which retain their speed notwithstanding adverse weather conditions.

Since air and electric starters and their components can be serviced in any shop having hand tools, the required service and test equipment is minimal. A hydraulic starting system, however, requires close-tolerance parts which cannot normally be serviced in the average shop.

AIR STARTING SYSTEM

Components A remote-control air starting system is illustrated in Fig. 42–1. Note that it has its own

reservoir (air tank). This is true of all air starting systems. The air tank is connected through a one-way check valve to the reservoirs of the air-brake system and to the relay valve. The relay valve is connected to the air starter by means of No. 16 or larger air hose. A No. 6 air hose connects the push-button valve to the air tank, to the relay valve, and in some cases to the automatic oiler. The automatic oiler is connected to the air motor by a hydraulic hose.

Push-Button Valve A push-button valve, as shown in Fig. 42–2, is used to activate the airflow to the air motor. It is a simple on-off air valve. When the operator pushes the knob, the push-button valve closes the exhaust valve and opens the inlet valve. Reservoir air is then directed to the relay valve and, in some cases, to the automatic oiler. The air pressure activates the automatic oiler and the relay valve.

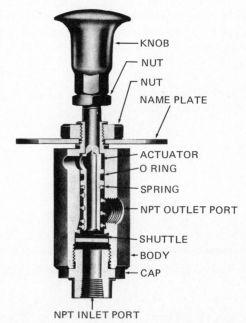

KNOB
NUT
NUT
NAME PLATE
ACTUATOR
O RING
SPRING
NPT OUTLET PORT
SHUTTLE
BODY
CAP
NPT INLET PORT

Fig. 42–2 Cutaway view of a push-button valve. *(Roosa Master Stanadyne/Hartford Div.)*

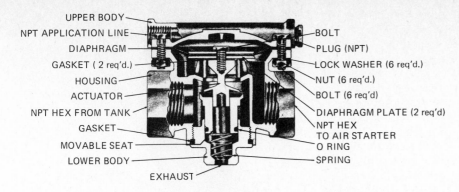

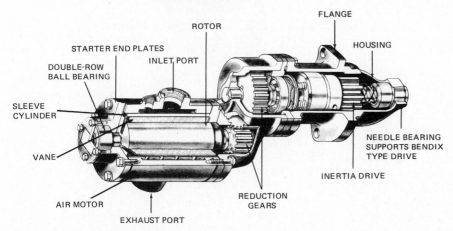

Fig. 42–3 Relay valve. *(Roosa Master Stanadyne/Hartford Div.)*

Fig. 42–4 Air cranking motor. *(Roosa Master Stanadyne/Hartford Div.)*

When the operator releases the knob, the valve closes the inlet valve, the spring opens the exhaust valve, and the compressed air is exhausted from the relay valve into the atmosphere.

Push-Button Valve Trouble   When the valve fails to operate or air exhausts during application (when air is applied or when the valve is at release position), the problem could be a damaged O ring, or the seat and valve of the inlet or exhaust valves could be damaged.

Relay Valve   A relay valve is a remote-control valve. It supplies air direct from the air reservoir to the air motor when the push-button valve directs air onto the diaphragm (Fig. 42–3). Air pressure forces the diaphragm downward, closes the exhaust valve, and opens the inlet valve. Air from the tank is directed to the inlet port of the air motor. When the push-button valve exhausts the pilot pressure from the diaphragm, the reservoir pressure and the return spring push the diaphragm upward. This closes the inlet valve and the spring force opens the exhaust valve.

Relay-Valve Trouble   A relay valve will cause very few problems if the air system is kept free of water and oil. Water and oil can cause the valve to stick, reduce airflow, and in time, damage the inlet valve, exhaust valve, and seats. When the relay valve requires servicing, dismantle the assembly and clean the valve internally and externally. Then install a parts replacement kit.

NOTE   Use only the recommended lubricant on all moving components.

Air Cranking Motor   The cranking motor consists of three major parts: an air motor, a reduction gear, and a friction-type inertia drive (Fig. 42–4). The rotating components of the air motor are well supported by ball bearings or roller bearings. A sleeve cylinder is pressed into the housing to decrease the wear and to achieve effective vane sealing. The large inlet and exhaust ports reduce air velocity to maintain air pressure. The reduction gear reduces the high motor speed and increases the torque.

Cranking-Motor Operation   Compressed air from the relay valve enters the inlet port and is forced into one or two pockets created by the rotor, vanes, end covers, and cylinder. Air pressure forces the rotor to rotate. This opens the adjacent pocket which then fills with compressed air. At the same time air exhausts from one other pocket as the vane passes by the exhaust slot. This action continues until the airflow from the relay valve ceases to flow. A muffler is screwed into the exhaust port to reduce the noise created by exhausting air (Fig. 42–1).

Air-Motor Trouble   Most air-motor troubles can be traced to the automatic oiler. Malfunction in this unit increase wear of the vanes, cylinder, end plates, and bearings. If detected early, efficiency can be restored by replacing the vanes and, if necessary, honing the cylinder and refinishing the end plates.

Unfortunately, when trouble in the oiler is assessed too late, bearing failure will already have caused the internal components to misalign, resulting in irreparable damage to the end covers, vanes, and rotor. The cylinder, which will also have been damaged, must be replaced.

While the reduction gear is seldom cause for concern when the friction clutch of the cranking-motor drive is correctly adjusted, incorrect adjustment can result in early bearing and gear failure.

Emergency Air Supply   Without compressed air, the air cranking motor will not function; therefore, an emergency air supply must always be available.

The most expedient way to replenish the air-starter reservoir, if you are in the shop, is to connect the reservoir to the shop's compressed air supply. Without the convenience of the compressed air, you must connect the emergency glad hand of another vehicle to the emergency glad hand of the stalled vehicle. (A glad hand is a fast-breakaway coupling used to connect the service line and emergency air line with the trailer.

Another method of assuring a supply of emergency air is the installation of an auxiliary air cartridge breech. It can be built into the reservoir or have separate connections. The breech is a steel container connected to the air motor or reservoir into which an air-pack cartridge is secured. When set off in the breech by a standard percussion primer (which may actuate mechanically or electrically), the cartridge releases high-pressure gases directly to the motor or into the reservoir. A safety valve protects the reservoir or motor from overpressurizing.

Hydraulic Starting System

Components   An air starting system relies on compressed air to crank its engine. A hydraulic starting system relies on hydraulic energy.

1. Reservoir
2. Hydraulic governor
3. Two-way check valve
4. Hand pump with gauge
5. Control valve
6. Engine driven pump
7. Accumulator

Fig. 42–5 Hydraulic starting system. *(American Bosch-Ambac Industries, Inc.)*

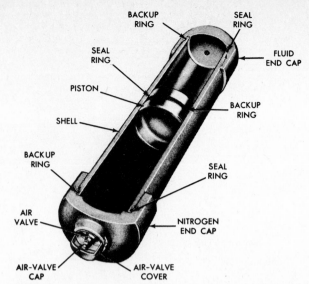

Fig. 42–6 Cross-sectional view of an accumulator. *(GMC Detroit Diesel Allison Div.)*

The components of one of many different hydraulic starting systems and the manner in which they are connected are illustrated in Fig. 42–5. The reservoir of a hydraulic starting system stores and cleans the hydraulic fluid. It is connected to the inlet ports of the hand- and engine-driven pumps. The return lines from the hydraulic motor and the engine-driven pump also are connected to the reservoir.

The accumulator, which stores hydraulic fluid under pressure, consists of a steel shell, or cylinder, closed by two end caps (Fig. 42–6). Inside the shell is a movable piston which divides the cylinder. The end cap on the hydraulic side (fluid end cap) is connected to the directional valve, and to the hand- and engine-driven pumps. The nitrogen end cap has an air valve through which the accumulator is charged with nitrogen at a pressure of about 1500 psi [105.4 kg/cm²]. Nitrogen is used instead of compressed air because it will not explode if seal leakage permits oil to infiltrate to the nitrogen side of the accumulator.

The engine-driven pump (Fig. 42–5) can be of the gear, vane, or piston type. Each can be driven by belts or directly from the engine. When the engine runs, the hydraulic pump forces fluid against the nitrogen pressure into the accumulator. When the preset pressure is reached, the hydraulic pump unloads. When the pressure is reduced, the pump again pumps fluid into the accumulator.

The hand pump of a hydraulic starting system is a double-action piston pump. It is used to pump fluid into the accumulator for initial cranking when the accumulator has exhausted all fluid.

Hydraulic Motors and Directional Valves
Hydraulic motors used for cranking motors are of the fixed displacement vane, swash-plate, or bent-axis piston types (Fig. 42–7). Except for the hydraulic motor, all cranking motors use the same inertia-type friction clutch. Their displacement per revolution ranges from 0.50 to 3.50 in³ [0.0081 to 0.057 l], de-

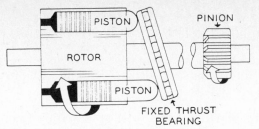

Fig. 42–7 Bent-axis piston-type hydraulic cranking motor. (*American Bosch-Ambac Industries, Inc.*)

pending on engine displacement. The directional valve, which directs oil to the motor when activated, may be bolted directly to the motor or it may be connected between the accumulator and the cranking motor.

Action and Fluid Flow when Cranking Assuming the accumulator is under pressure, the same pressure is also present at the directional spool valve. As the operator directly or indirectly moves the spool valve, oil enters the inlet port of the motor. When a seven-piston swash-plate motor is used, three pistons receive oil, one piston is in neutral, and three pistons are open to the outlet slot and the reservoir. The three receiving pistons are forced against the thrust bearing of the fixed-angle swash plate. They can only move outward by sliding along the swash plate until fully extended. In so doing, they cause the cylinder block to rotate and with it the shaft and motor drive. At the point where one piston begins to retract, the cylinder opening slides over the outlet-valve slot located in the valve plate, and oil pressure is released to the reservoir. At the same time one piston is fully retracted and its cylinder opening is over a solid area in the valve plate. However, the adjacent cylinder has just passed the inlet-valve slot and oil pressure is forced onto the piston. When the operator positions the spool valve to neutral, the oil flow and cranking cease.

Troubleshooting Hydraulic Starting System Brief cranking time or low rpm indicate that there is low oil pressure, or a malfunction in the directional valve, or that the cranking motor, motor drive, or engine are seized.

Check the engine first. A diesel engine should start (normal ambient temperature) within two revolutions. If it does not, check for the following:

1. Low accumulator pressure
2. A worn pump
3. Low oil level
4. A plugged oil screen or filter
5. Air in the system
6. External leakage
7. A restricted spool-valve movement
8. A damaged or worn motor

Low accumulator pressure caused by a damaged or worn air check valve or low nitrogen pressure can also reduce cranking efficiency.

Questions

1. State the purpose of the relay valve.

2. What is the purpose of the gear train in the cranking motor?

3. Explain the action in the relay valve when air from the push-button valve enters the chamber above the diaphragm.

4. What would prevent an air starter motor from operating? What would cause an air starter motor to operate at too slow a speed?

5. Which components, if faulty, could cause a continuous air supply to the air motor?

6. List several reasons why the cranking speed of a hydraulic starter could be reduced to an ineffective level.

7. What could prevent a hydro-starting motor from turning over?

8. Trace the oil flow and explain in detail the operation of each component of the starting system during the cranking period.

UNIT 43

Cold-Weather Starting Aids

Since some diesel engines will not start easily at temperatures of 25 to 50°F [−3.89 to 10°C], various devices are available to facilitate ignition. The most effective under these adverse temperatures are the cylinder-block coolant heater and the engine-oil heater. Because they keep the engine components warm, they also minimize wear during starting and during warm-up.

Glow Plugs Although a glow plug commonly is used as a starting aid because of its effectiveness and low price, it unfortunately requires the battery power needed for cranking.

Many different glow plugs are available. Some are placed in the intake manifold, while others are placed in the precombustion chamber (Fig. 43–1). The heating element of all glow plugs heats the sur-

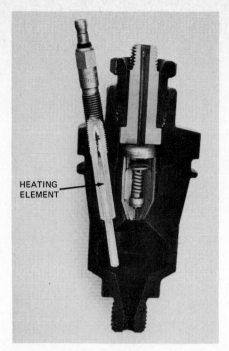

Fig. 43–1 Sectional view of a glow plug installed in a precombustion chamber.

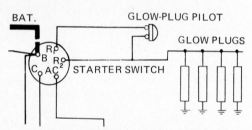

Fig. 43–2 Glow-plug parallel circuit.

Fig. 43–3 Testing a glow plug with an ohmmeter.

rounding air when the electric circuit is closed. When the engine is cranked, the heated air is drawn into the cylinder, giving the compressed air a higher temperature for ignition.

When more than one glow plug is used, they may be connected in series or in parallel. A parallel glow-plug circuit is shown in Fig. 43–2. It consists of a three-way starter switch, and four glow plugs. The serviceability of a glow plug can be determined by connecting an ohmmeter, as shown in Fig. 43–3, and comparing the resistance reading with the specified resistance.

Fluid Starting Aids Just as there are many different types of glow plugs, there are many different fluid starting aids. They all use the same principle, that is, supplying highly combustible ingredients to the air drawn in by the engine.

One fluid starting method is the ether aerosol can method. It requires two people, one to start the engine and the other to spray the ether. This method is very unsafe especially if too much ether is used without being properly atomized. If ether is not properly atomized, or if it is unevenly distributed, it can cause severe damage to the engine.

If you are obliged to resort to the ether method for starting, remember to crank the engine first. At an appreciable distance from the air intake, spray a moderate quantity of ether in its direction. This will allow the air to mix thoroughly with the ether before it is drawn into the engine. If you do not do this, the high-power starting fluid will explode too rapidly and excessive damage to the engine may result.

Another fluid starting method that is equally unsafe uses a predetermined volume of ether, in capsule form, that is placed in a sealed container. The container is connected with a piston-type hand pump to the intake manifold or to the air-intake system. Sometimes a spray nozzle is used to aid in atomization.

If you must use this type of starting aid, remember to crank the engine before breaking the capsule with the piercing pin. Use the hand pump to pump starting fluid into the manifold.

QUICK START A safer method, with greater protection for the engine, is the use of a fluid starting aid called *Quick Start*. A metered quantity of highly volatile fluid is atomized into the manifold before the engine is cranked. The valve and the atomizer can be selected to correspond to a particular engine model so that a safe quantity of starting fluid is metered and the atomization is determined precisely. When using this starting aid, activate the valve two to three times, and then crank the engine over.

To ensure a quick starting system at all times, periodically clean the orifice of the atomizer, and lubricate any movable valve components.

Questions

1. Outline briefly how to prepare and use a fluid starting aid.

2. Why is Quick Start safer than ether fluid starting methods?

3. Why are glow plugs the safest starting aid?

4. How do you test the serviceability of a glow plug?

Charging System (Generators)

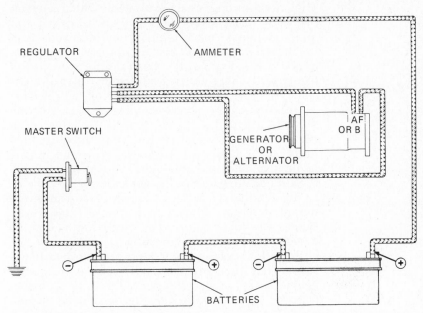

REGULATOR

AMMETER

MASTER SWITCH

GENERATOR OR ALTERNATOR

AF OR B

BATTERIES

Fig. 44–1 Schematic drawing of a typical charging system. *(Allis-Chalmers Engine Div.)*

Components and Function The charging system consists of a battery (or batteries), a belt-driven generator (ac or dc), a regulator, and the complementary wiring (Fig. 44–1). The function of the charging system is to supply the electrical system with current and to maintain the batteries in a full state of charge.

Generators A generator is a mechanical device that converts mechanical energy supplied by the engine into electrical energy. It furnishes the battery and the electrical system with direct current.

Today's ac generators (called *alternators*) convert alternating current to direct current through rectification, whereas earlier (dc) generators converted alternating current to direct current through brush commutation.

The ac generator has gradually replaced the dc generator because:

1. It can produce a high current at lower engine speed.
2. It can rotate in either direction.
3. It can rotate safely, without destruction, at high speed.
4. Its output voltage and current can be controlled easily.
5. It has a long service life.

Electromagnetic Induction The largest source of produced electric energy is through electromagnetic induction. Its history goes back to 1831 when Michael Faraday proved that voltage (emf) is induced in a conductor if the conductor is moved between the poles of a magnet or if the magnet is moved over the conductor. The electromotive force (emf) obtained in this way is said to be an *induced emf* and

the process is called *electromagnetic induction*. He also proved that the conductor, when moved through a magnetic field from left to right or from right to left, changes the induced emf polarity (when the magnetic poles maintain the same position). When the conductor is moved parallel to the magnetic lines of force, no voltage is induced. Faraday also proved that the polarity of the induced voltage changes when polarity of the magnetic field changes but the current flow in the conductor maintains its direction.

NOTE When the conductor moves across the magnetic field or the magnetic field moves over the conductor, voltage is induced. However, it must have a complete circuit or there is no current flow.

Alternating Electromagnetic Induction A simple one-loop (coil) generator is shown in Fig. 44–2. When the magnet rotates, the south and the north poles approach the conductor loop, and voltage is induced. Voltage gradually increases to its maximum, that is, when the south pole is under the conductor and the north pole is over it. With the magnet rotating as shown in Fig. 44–2, the direction of the induced voltage, which causes a current to flow, is determined by the right-hand rule. As the poles move away from the conductor, the induced voltage decreases and current ceases to flow when the poles arrive just at the middle of the conductor loop. At this point the poles again approach the conductor loop in a reverse polarity. The induced voltage reverses and gradually rises to its maximum when the north pole is under the conductor loop and the south pole is over it (bottom, Fig. 44–2). Further rotation decreases the induced voltage to zero, when the poles return to the middle of the conductor loop.

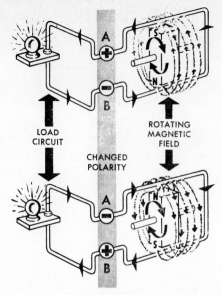

Fig. 44–2 Alternating electromagnetic induction. *(United Delco-AC Products, General Motors of Canada, Ltd.)*

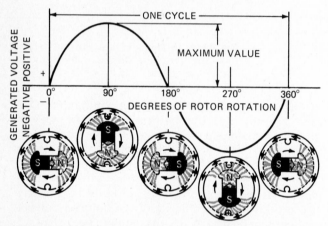

Fig. 44–3 Generated voltage during one cycle (one revolution). *(United Delco-AC Products, General Motors of Canada, Ltd.)*

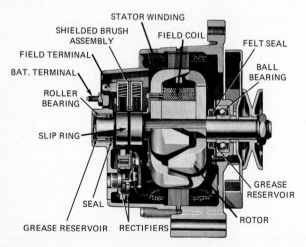

Fig. 44–4 Sectional view of a standard-duty generator. *(United Delco-AC Products, General Motors of Canada, Ltd.)*

Polarity again reverses, and the cycle is repeated (Fig. 44–3).

The induced voltage (emf) which periodically reverses its polarity (direction) is known as *alternating*

electromotive force and the resulting current flow is known as *alternating current*.

Generator Construction   There are many ac generator designs to complement the different electrical requirements of today's intricate electrical systems. They must accommodate equipment which operates over long distances or short stop-and-go service, or which requires special high-current output.

All generators have certain similarities in construction (see Fig. 44–4). Their magnetic fields (rotors) rotate between the stationary conductors (stators). Their rotors are mounted on ball, needle, or roller bearings. One brush carries current from the field terminal to the slip ring. The other brush completes the circuit to ground or to the other field terminal. The stator windings are wound on the laminated stator frame which, in most cases, is the housing or the frame of the generator.

The six diodes (rectifiers) used for rectification are mounted in the slip-ring end frame (Fig. 44–5). They are connected with the stator winding and the battery terminal of the generator.

Rotors of brushless generators are different in design. Here the rotor consists of two permanently magnetized pole pieces, (one south, the other north) fastened together with iron that cannot be magnetized (Fig. 44–6). The field coil, wound over an iron core, is fastened to the generator housing and does not rotate. When assembled it fits inside the rotor. **NOTE** The rotor is the only rotating component in this type of generator.

FACTORS AFFECTING OUTPUT VOLTAGE   The output voltage of a generator will increase when the speed of the rotor is increased, when the strength of the magnetic field is increased, or when the number of turns of wire in the stator winding are increased.

Inductive Reactance   Ac generators are electrically designed so that maximum current output will not exceed the output rating of the generator. To cite an example, a 12-V 60-A generator will produce approximately 66 A with a maximum voltage of about 19 V. As current flow increases in the stator windings, the magnetic lines of force cut across the adjacent winding. The direction in which the lines of force cut the adjacent conductor induces a voltage opposite to that of the applied voltage, limiting the current flow in the stator winding (Fig. 44–7). The inductive opposition to current flow is called *inductive reactance*.

Field Circuit   Generators are built with an electromagnet rather than with a permanent magnet because the former creates a stronger magnetic field. This magnetic field is used to vary the generator voltage. When increasing or decreasing the field current, the magnetic field strength is varied to safeguard the electrical components against high voltage. The unit which regulates the field current is called a *voltage regulator*.

The field current of a generator can be connected in two ways, as shown in Fig. 44–8. When con-

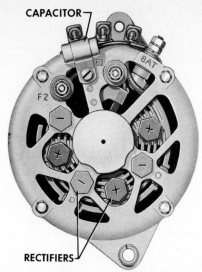

Fig. 44–5 End view of the slip ring and frame. (*United Delco-AC Products, General Motors of Canada, Ltd.*)

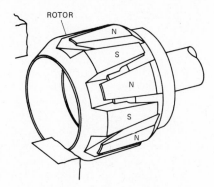

Fig. 44–6 Rotor used in a brushless generator.

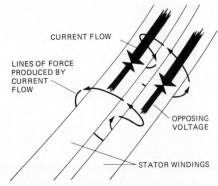

Fig. 44–7 Inductive reactance. (*United Delco-AC Products, General Motors of Canada, Ltd.*)

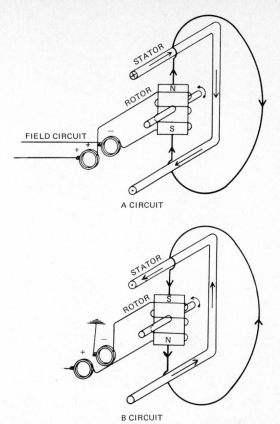

Fig. 44–8 Top, A-circuit generator; bottom, B-circuit generator.

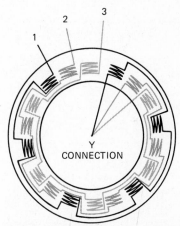

Fig. 44–9 Schematic illustration of a three-phase stator assembly, showing the position of the 18 coils.

nected as a B circuit, the field is grounded in the generator. When connected as an A circuit, the field is grounded in the regulator. In each case, both brushes are insulated. Neither circuit connection (A or B) changes the generator output or voltage.

Generator Stator Since a generator with a single loop has a low induced alternating voltage, many coils are placed over the changing magnetic field to increase the induced alternating voltage. Three independent stator windings (three phases) are used. Each independent phase is formed into as many coils as there are north or south poles. For example, a 12-pole rotor has each stator phase wound into six coils and spaced 60° from each other in the iron core

of the stator frame. Therefore, 18 coils are used for a three-phase 12-pole rotor (Fig. 44–9). A change in the rotor poles also changes the number of coils wound into the stator.

Phase As the rotor rotates 360°, an alternating voltage is induced in each winding. This causes alternating current to flow which is called a *phase*. Each phase coil is placed in the stator frame electrically 120° apart from the other. The three phases are connected to each other by a Y or delta connection. Each winding is connected to one negative and one positive diode (Figs. 44–10 and 44–11).

Cycle and Frequencies When the rotor has rotated 360° (one revolution), it has completed one cycle and has produced one positive and one negative sine

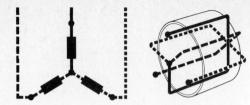

Fig. 44–10 Three-phase Y connection.

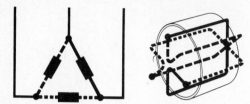

Fig. 44–11 Three-phase delta connection.

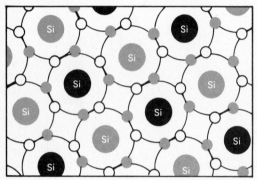

Fig. 44–12 Covalent bonding of silicon atoms. *(United Delco-AC Products, General Motors of Canada, Ltd.)*

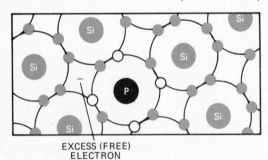

EXCESS (FREE) ELECTRON

Fig. 44–13 Covalent bonding with excess (free) electron. *(United Delco-AC Products, General Motors of Canada, Ltd.)*

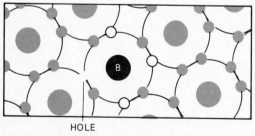

HOLE

Fig. 44–14 Hole. *(United Delco-AC Products, General Motors of Canada, Ltd.)*

wave. One half cycle (one sine wave) is called an *alternation*. The number of cycles generated per second is called *frequency*.

Semiconductors The alternating current produced by a three-phase generator must be rectified to direct current since direct current is required to operate the

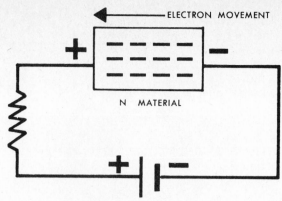

ELECTRON MOVEMENT

N MATERIAL

Fig. 44–15 Electron movement through N material. *(United Delco-AC Products, General Motors of Canada, Ltd.)*

electric units and charge the battery. This rectification is achieved through semiconductors known as *diodes*. A diode may also be used as a safety switch and as a resistor.

To understand how a diode can perform these various functions, you must look at its design. As outlined in the section on electricity (Unit 36), a conductor must have less than four electrons in its valence ring (outer orbit), and an insulator must have more than four electrons in its valence ring.

An element which has four electrons in its valence ring is neither a good conductor nor a good insulator. Four such elements are known; however, only two of these (silicon and germanium), in a purity of 1 billion to 1, are used to make N and P (negative and positive) type material. The element, silicon, is more popular because it can withstand higher temperature. When purified and prepared for doping (adding), the silicon is covalent bonded, which increases the number of electrons in the valence ring. Each atom then has eight electrons and becomes a good insulator (Fig. 44–12).

Doping Semiconductor Materials By doping (adding) the silicon material with a controlled portion of such elements as phosphorus or antimony (both of which have five electrons in their valence rings), covalent bonding will occur, as illustrated in Fig. 44–13, with one electron left over. This free electron becomes an electron carrier and can be made to move easily through the material. Material of such construction is called *negative or N-type material*, that is, material having an excess of electrons.

When the silicon material is doped with atoms of an element such as boron or indium, which have three electrons in their valence rings, the covalent bond is incomplete and a void is created. This void is referred to as a *hole* (Fig. 44–14). The hole is really a missing electron. Since an electron is a negative charge of electricity, the hole is considered a positive charge of electricity. By robbing an electron from the adjacent atom, the hole can be made to move as easily through the material as the free electron through the N material. Material of this construction is called *positive or P-type material* (material having a deficiency of electrons).

When either material is connected to a circuit with a power source of sufficient voltage, the N- and

P-type materials obey the simple law of opposite polarity attracting and like polarity repelling. Electrons will flow through the N-type material (as shown in Fig. 44–15) toward the positive terminal to the power source. Additional electrons flow out of the negative terminal to reenter the N-type material.

Similarly, when the P-type material is connected to the circuit shown in Fig. 44–16, the holes move to the negative terminal. There they will meet with the electrons from the power source and recombine.

Some of the electrons enter the P material, fill the holes and cause an imbalance within the semiconductor. The electrons are discharged forcibly at the positive terminal creating new holes which act as current carriers. This movement continues until the circuit is opened.

NOTE The hole movement is only in the semiconductor material. Electrons flow through the entire circuit.

Diode Construction By a fusion process, manufacturers join two small slices of N- and P-type material together. The point where they join is called a *junction* (Fig. 44–17). Because unlike charges attract, a few electrons near the junction drift into the P material and fill the holes, thereby nullifying them as current carriers. A few holes drift into the N material, recombine with the electrons, and eliminate them as current carriers. Equilibrium is achieved when no holes or electrons can move across the (PN) junction, as shown in Fig. 44–17.

This junction rests in the copper case (Fig. 44–18). The stem with its S-shaped conductor is welded to the PN junction. The S-shaped conductor prevents the stem seal as well as the electrical connection from breaking when the diode expands and retracts during operation. A dessicant is placed inside the case to absorb any moisture. The stem is glass-sealed and the top is bonded or welded to the case. The case makes up one side of the electrical connection and the stem is the other. When using a PN diode, the case is connected to the N material. When using an NP diode, the P-type material is connected to the case.

Diode Action When a diode is connected in the circuit as shown in Fig. 44–19 (forward bias), the positive terminal of the battery connects to the P material and the negative terminal connects to the N material. The holes in the P material repel and the electrons in the N material repel from their terminals and move toward the junction. Usually only a fraction of the voltage is required to cause the holes and the electrons to move through the junction.

As soon as a hole enters the N material and recombines, an electron enters the P material. This forces an electron to leave the P material, create a new hole, and a new electron to enter the N material.

When the battery is connected in reverse bias (reverse polarity), the negative terminal of the battery connects to the P material and the positive terminal connects to the N material. The holes move toward the negative terminal and the electrons move

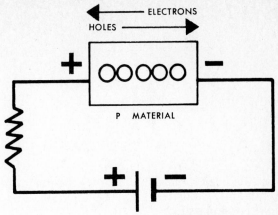

Fig. 44–16 Electron and hole movement through P material. *(United Delco-AC Products, General Motors of Canada, Ltd.)*

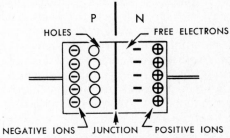

Fig. 44–17 Diode after the fusion process. *(United Delco-AC Products, General Motors of Canada, Ltd.)*

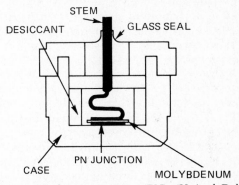

Fig. 44–18 Diode construction (PN). *(United Delco-AC Products, General Motors of Canada, Ltd.)*

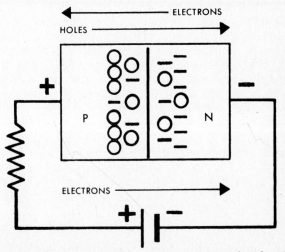

Fig. 44–19 Forward-bias connection. *(United Delco-AC Products, General Motors of Canada, Ltd.)*

CURRENT
FLOW

NO
CURRENT
FLOW

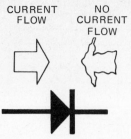

Fig. 44–20 Diode symbols. *(United Delco-AC Products, General Motors of Canada, Ltd.)*

toward the positive terminal causing the junction to void itself of current carriers. The junction becomes a high-resistance area.

REVERSE-BIAS VOLTAGE Some diodes will allow a very small current to flow in reverse direction. However, when the reverse voltage is increased, say by using a 110-V tester or by using a battery charger with a selected voltage higher than the system voltage, the maximum reverse voltage may cause the diode to overheat and finally to destroy itself. When using a battery charger with a selected current flow higher than the maximum system charging current, the diode may also overheat and destroy itself. This destruction occurs because the covalent-bond structure breaks down, and reverse current can flow freely through the diode. (The symbol used for a diode and for identification of the direction in which current is allowed to flow is shown in Fig. 44–20.)

Generators, regulators, and other electrical components use positive and negative diodes. They are identified according to their polarity by one of the methods shown in Table 44–1.

Diode Rectification To change the induced ac current of the ac generator into direct current, diodes are used. The diodes rectify (change) the alternating electric current into direct current by reversing the direction of alternate impulses. To utilize the full capacity of the generator, six diodes are needed instead of three, since three diodes give only half-phase rectification. One sine wave of each phase is not used since the sine wave cannot reach the battery.

To utilize each sine wave, each phase is connected to one positive and one negative diode (full-phase rectification). When connected in this sequence, six sine waves are rectified with each revolution of the rotor, and direct current flows into the battery. Current flow through each of the six sine waves is only for a fraction of a second during each stage of rectification. When the rotor passes under phases A and B, voltage is induced and current flows from junction C through the stator windings to junction B, through the positive diodes to the battery, from the battery to the negative diode, and to junction C (Fig. 44–21). The opposing voltage in the winding at junction A is in balance and, for a brief moment, in neutral.

Upon each rectification, the rotor poles reach the position where one phase winding is in neutral. When the rotor has turned 60°, current flows irre-

Table 44–1 DIODE IDENTIFICATION

Ground polarity of battery	Identification on diode in grounded heat sink	Insulated heat sink
Negative	−, black, right-hand, thread	+, red, left-hand, thread
Positive	+, red, left-hand, thread	−, black, right-hand thread

versibly through the phase winding at junction A to junction B, to a positive diode in the insulated heat sink, and to the battery. In this stage, the phase winding at junction B is in neutral. The electric circuit is completed through the negative diode junction A. The rectified voltage is by no means without pulsation, but for all practical purposes, it is smooth enough to be considered a nonpulsating dc voltage.

The current flow in a delta-connected stator is different from that in a Y-connected stator. In a Y-connected stator, current is only conducted through two stator windings, whereas a stator connected in a delta circuit conducts current through all three stator windings (see Fig. 44–22). The result is a higher current flow in the delta circuit. For example, say a generator with a Y-connected stator produces 60 A at 3000 rpm with a field current flow of about 2.5 A. If the same generator stator is changed to a delta connection and is run at the same rpm and with the same field current, the output would be 103.8 A. This is because a delta-connected stator produces 0.73 A more than a Y-connected stator. (60 A × 0.73 = 43.8 A and 60 A + 43.8 A = 103.8 A.)

Generator Cooling A generator, like any device which converts energy, produces heat. A generator also picks up heat through radiation from the engine. This heat must be removed to protect the insulation, soldering, and diodes. Heat shields sometimes are installed to reduce radiation from the engine.

The cooling methods employed vary with generator design. Generators not exposed to dirt, dust, and road splash, for example, generally have no external cooling fins. The fan is mounted on the rotor shaft either behind the pulley or inside the drive end frame. In either case the cooling air is drawn from the slip-ring side through the generator and out through the opening of the drive end frame. The temperature of the air used for cooling should not exceed 150°F [66°C].

Generators exposed to dust and road splash usually are sealed and cooled by external cooling fins. Cool air from the fan mounted at the regulator side of the generator blows air over the cooling fins.

Heat Sink When a generator is operating, the current flows through the diodes and causes heat. Semiconductors (diodes) lose their resistance to an opposing current flow when overheated. To ensure a safe diode operating temperature, three positive diodes are placed in one heat sink and three negative diodes are placed in another heat sink (Fig. 44–23).

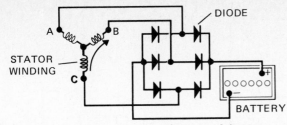

Fig. 44–21 One sine-wave rectification (delta-connected stator). *(United Delco-AC Products, General Motors of Canada, Ltd.)*

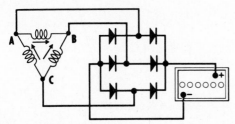

Fig. 44–22 Current flow in a delta-connected stator. *(United Delco-AC Products, General Motors of Canada, Ltd.)*

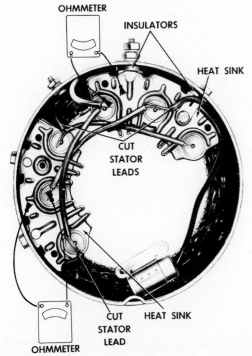

Fig. 44–23 Dual heat-sink testing diode. *(GMC Detroit Diesel Allison Div.)*

Another method of ensuring safe diode operating temperature is to place one set of diodes in a heat sink and the other set of diodes in the slip-ring end frame. The cool air drawn through the generator cools the diodes to a safe operating temperature.   **NOTE**   One heat sink always is connected with the battery (insulated) terminal of the generator, and the other is grounded in the generator.

When a two-wire system is used, it is also insulated and connected to the ground terminal of the generator. Two heat sinks are used to make the generator more versatile. It is then more suitable for a two-wire system and can accommodate either battery polarity.

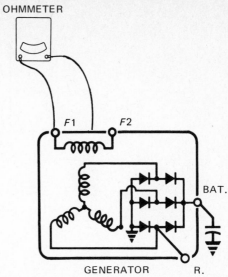

Fig. 44–24 Determining A or B circuit generator.

NOTE The heat sink connected with the battery terminal *must* have the same polarity as the live side of the battery.

CHARGING-SYSTEM SERVICE

Generator Testing Prior to Removal   When the engine cranking speed is not adequate, the first place to inspect is the charging system.   First, test the batteries.   Check the cables and wire connections, and then make a generator output test to determine whether the trouble is in the generator or in the regulator.

Before you proceed with the generator output test, you must verify the type of generator field circuit.   Is it connected as an A or a B circuit?   This information is contained in the service manual.   If no service manual is available follow this general procedure: disconnect the field wire(s) at the generator and connect an ohmmeter as shown in Fig. 44–24.

If the ohmmeter shows continuity, the field circuit is wired as a B circuit.   If the ohmmeter shows no continuity, the field circuit is wired as an A circuit.

After determining the type of circuit, connect the test instruments as shown in Fig. 44–25.

NOTE When the generator is wired as an A circuit, you must also connect a jumper from one field terminal to ground the field circuit.

An example of a generator test specification is given in Table 44–2.

After the instruments have been connected, run the generator at an output of about 15 A for about 15 minutes to warm up the components.   Then run the generator at about 760 rpm, and slowly remove the resistance to feed the field circuit.   The generator should show a current flow.   Increase the generator speed to 4100 rpm.   Adjust the field rheostat so that the field current flow is between 2.0 and 2.5 A. Check the generator speed constantly to maintain the recommended rpm.

Adjust the load rheostat, if necessary, to obtain the desired output (amperage).   Check readings of

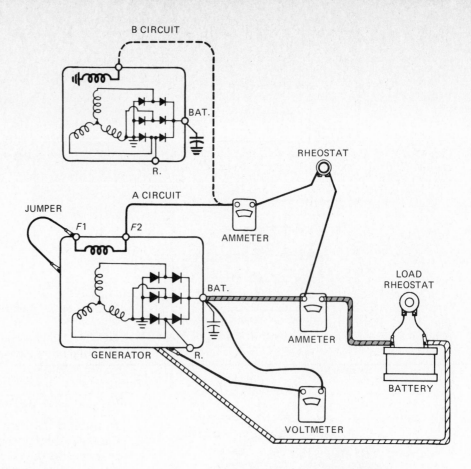

Fig. 44–25 Connections for testing A and B circuit generators.

Table 44–2 GENERATOR TEST SPECIFICATION

Vendor	Amperes at 15 V	Watts at 15 V	Field current, A at 12 V 75°F [23APC]	Cut-in-speed (alternator rpm)	Rated output speed (alternator rpm)	Belt width, in [mm]	Belt tension, lb [kg]	Brushes			
								Number used	Original length, in [mm]	Wear limit, in [mm]	Spring tension, oz [g]
Alternator	60	840	2.0–2.5	760	3300 cold 4100 hot	³⁄₈ [9.53]	80–110 [36.29–49.94]	2	½ [12.70]	⁹⁄₃₂ [7.14]	8–10 [226.8–283.5]

the rpm, field current, voltmeter, and output ammeter. All gauges should indicate specified value or higher. If the current output is lower or the field current is higher, or the voltage is higher, the generator must be disassembled and serviced.

Testing Generator with Oscilloscope If your shop has an electronic engine tester (an oscilloscope), use it! It is an excellent instrument with which to locate the source of generator trouble.

Connect the leads according to the manufacturer's recommendation. Set the instrument selectors to the proper positions. Run the generator at the recommended speed, using the correct field current, voltage, and ampere output. The wave pattern on the oscilloscope will indicate the source of the generator trouble.

Refer to Fig. 44–26. The generator, when tested at 3500 rpm showed 15.2 V, 30 A (Fig. 44–26a). When one positive diode during the test was open, grounded, or short-circuited, the current dropped 3 A because one sine wave wasn't used (Fig. 44–26b

and c). The same ampere drop was registered when one negative diode was open, short-circuited or grounded. The same tests were made at 1000 rpm, and the generator output dropped from 10 to 9 A.

With the generator running at the same speed and output, one stator winding was open. The output dropped to 22 A (Fig. 44–26d). When one stator winding was short-circuited, the output dropped to only 24 A (Fig. 44–26e). However, when the stator was grounded, the output dropped to 18 A (Fig. 44–26f). At the same time as these tests were performed the wave patterns throughout the tests were photographed. An open, grounded, or short-circuited stator at low generator speed (1000 rpm) did not change the generator output. When the R terminal of the generator was grounded, the output dropped to 18 A because two sine waves were directed to ground.

Disassembling Generator When generator components require servicing, the generator should be dis-

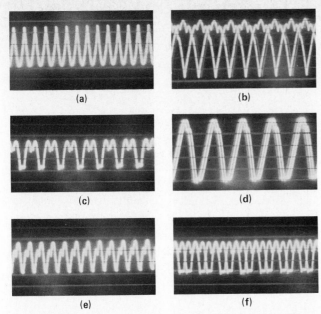

Fig. 44–26 Wave patterns on an oscilloscope. (a) Three-phase output; (b) negative or positive diode open; (c) short-circuited or grounded positive or negative diode; (d) one phase open; (e) one phase short-circuited (f) grounded stator winding.

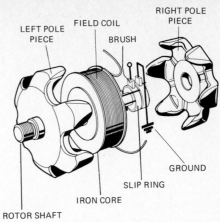

Fig. 44–27 Rotor construction. (*United Delco-AC Products, General Motors of Canada, Ltd.*)

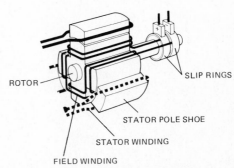

Fig. 44–28 Four-pole rotor. (*United Delco-AC Products, General Motors of Canada, Ltd.*)

assembled completely. Because there are so many varying generator designs, it is not possible to state a common disassembly procedure. However, the following are some typical steps to be taken and precautions to be noted:

1. Always clean the assembly thoroughly with compressed air.
2. Remove the "through" bolts and use a screwdriver (through the provided slots) to separate the drive end and slip-ring end frame from the stator assembly.
3. After disassembly is complete, be sure to protect the bearings to prevent their becoming dirty or damaged.
4. Take care when you remove the drive end frame from the rotor shaft. Use a proper puller to prevent damaging the pulley, the end plate, or the rotor.

Rotors The rotors of an ac generator vary in design from a 2- to 16-pole piece construction. Most rotor assemblies include a laminated iron core over which the field coil is wound. The coil ends are connected electrically to two slip rings. Two interlacing finger-type iron pole pieces are spaced evenly over the field coil and splined onto the rotor shaft (Fig. 44–27). A brush rides on each of the two slip rings connecting the field coil to the voltage regulator and to ground. When the coil is energized, the ends of the coil form a north and a south pole. The magnetic lines of force cause the left-rotor pole piece to become north and the right-rotor pole piece to become south. This creates alternating north and south poles which become six north and six south poles.

The rotor construction of high-output generators differs from that shown in Fig. 44–27. They have heavy gauge wire or copper ribbon wound onto individual pole shoes to create two south and two north poles (Fig. 44–28).

Servicing Rotor To service a rotor, first check the rotor for worn bearing surfaces and for a bent shaft. Either defect will reduce generator output since either will bring the rotor closer to the stator, reducing the magnetic field strength and, subsequently, the output. Make certain the pole pieces are tight on the shaft and equally spaced because looseness reduces the magnetic field strength.

Check the slip rings. When the rings are rough, brush contact is reduced, and this lowers the field current and the output. If they are rough or out-of-round, the slip rings must be turned in a lathe to within 0.002 in [0.054 mm] of dial-indicator reading. The surface must be polished to a mirror finish with a 400-grain (or finer) polish cloth (see Fig. 44–29).

Testing Field Coil for Grounded, Open, or Short Circuit When you test the field coil for a grounded circuit, hold one test probe of the ohmmeter against the rotor shaft or pole piece and the other against either of the slip rings (Fig. 44–30). If there is no grounded circuit, the ohmmeter will not move from its infinite position.

When you test the field coil for an open or short circuit, hold one probe against each slip ring. If there is no open or short circuit, the ohmmeter should indicate the resistance specified in the service manual.

Although the above methods are acceptable under certain circumstances, the current draw test is more reliable. The procedure is as follows: Connect the

Fig. 44–29 Resurfacing the slip ring.

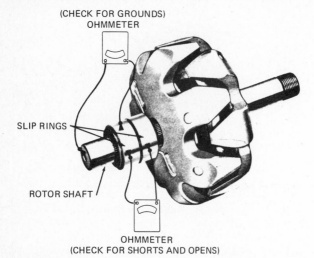

Fig. 44–30 Checking the rotor for grounded, short, and open circuits. *(United Delco-AC Products, General Motors of Canada, Ltd.)*

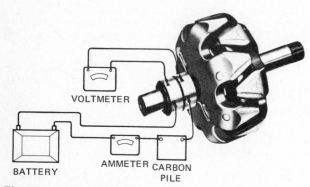

Fig. 44–31 Checking the rotor for short and open circuits by the current draw test. *(United Delco-AC Products, General Motors of Canada, Ltd.)*

rheostat or carbon pile, ammeter, battery, and voltmeter to the rotor as shown in Fig. 44–31.

NOTE Before you make the last battery connection, set the rheostat to its maximum resistance and connect the leads to the sides of the slip rings to prevent damage to the brush contact surfaces. Slowly reduce the resistance of the rheostat until the voltmeter reaches specified value. The ammeter at this time should indicate the specified amperage. If the ampere reading is above specification, the coil is

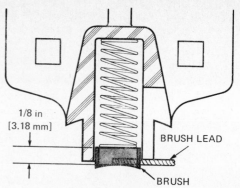

Fig. 44–32 Brush-holder assembly. *(United Delco-AC Products, General Motors of Canada, Ltd.)*

short-circuited. If it is below specification, the coil has high resistance. In either instance a replacement is necessary.

Bearing and Seal Replacement If it should become necessary to replace the bearing of the drive end frame, remove the pulley shaft nut according to instructions in your service manual. Next, remove the pulley and fan from the shaft. Use an arbor press to push the rotor shaft from the drive end frame. Remove the retainer plate and push out the bearing and seal from the end frame.

To remove the bearing and seal from the slip-ring end frame, first push out the plug. With the correct adapter, press out the bearing and the seal (from the outside to the inside). The adapter used must fit precisely within the bearing bore.

CAUTION Adequately support the end frame to prevent it from being damaged when pressing the bearing in or out. To safeguard the end frame from damage, some manufacturers recommend heating the end frame in an oven to 250°F [121°C] when installing bearings in the slip ring.

Generator Brush-Holder Assembly One of many different brush-holder assemblies is illustrated in Fig. 44–32. They all serve the same purposes: to position and hold the brush in sliding contact with the slip rings, to insulate the brush from the generator frame, and to connect the brush with the regulator and/or with ground.

As illustrated, the brushes are small and the springs used are light since the brushes carry only field current.

Brush Replacement In view of the many differently designed generators and brush-holder assemblies there is no single procedure governing the replacement of brushes. In the absence of defined procedure, use your common sense.

Brushes should be replaced when the generator is serviced or when they are worn beyond the specified length. (Use a small steel ruler to measure the wear length.)

NOTE Discoloration of a brush-holder assembly indicates overheating through a short or grounded circuit. You should, therefore, check the slip ring, brush spring, insulation, and field coil to determine the source of the short or grounded circuit.

Take care when inserting the brushes because they can be broken easily. Make sure that they can move freely within the holders once inserted. After you place the brush-holder assembly in position, use an ohmmeter to determine if the terminals have continuity and to ensure that the assembly is not grounded. When the generator is connected as a B circuit, one brush must be grounded.

Testing Stator The stator should be examined in detail. Begin by checking the stator leads and the point or points where the stator windings are soldered to each other. Check the laminated iron frame and the coils for looseness. If stator windings are discolored, they are short-circuited or grounded, and the stator should be replaced.

To test the stator winding for an open circuit, connect the test probe of an ohmmeter or test light between *A* and *B*, between *A* and *C*, and between *B* and *C* (Fig. 44–33). If the test lamp fails to light during any of these tests, or if the ohmmeter fails to show continuity, there is an open winding. The stator must then be replaced.

To test the stator for a grounded circuit, connect the test probe between *A* and the iron core, then between *B* and the iron core, and between *C* and the iron core. The test lamp should not light and the ohmmeter should show an infinity reading when the stator winding is not grounded.

If the generator output is low and the above tests have not revealed the problem, there may be a short-circuited stator winding or perhaps a short circuit between two stator windings. Since neither of these defects can readily be detected, the stator windings must be disconnected and the resistance of each winding measured.

An equally reliable test procedure is to reassemble the generator and run it at a constant speed of about 1200 rpm. With a voltmeter, measure the voltage across each phase winding. If the three voltage readings are not essentially the same when the current flow is steady, the trouble is surely in the stator.

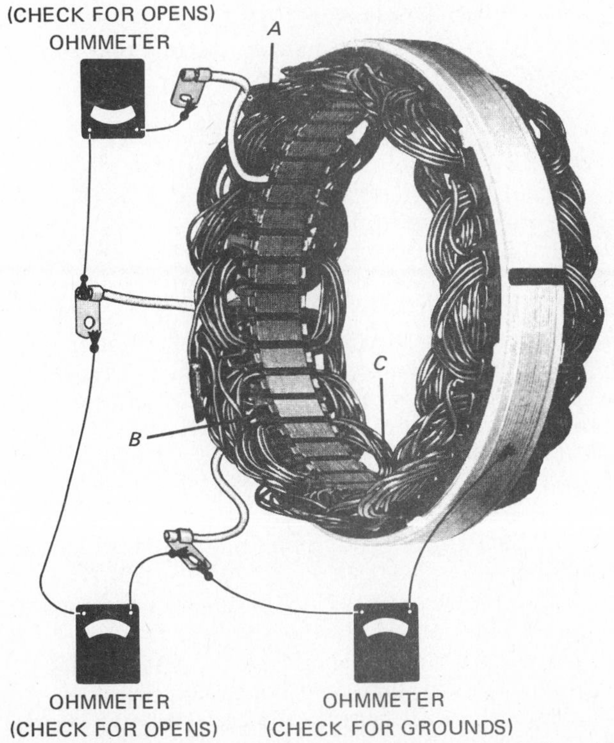

Testing Diode To test the diode, remove the diode lead from the battery terminal and connect one ohmmeter test probe to the diode lead and the other test probe to the diode case (Fig. 44–34). Take an ohmmeter reading, reverse the test probe, and again read the ohmmeter. If both ohmmeter readings are low, the diode is short-circuited, but if both readings are high, the diode has an open circuit. If one reading shows high resistance and the other reading shows low resistance, the diode is reusable.

To test the diodes in the heat sink or in the slip-ring end frame, you must remove the stator winding leads from the diodes since it is impossible to determine which diode is faulty when one is grounded, short-circuited, or open. The current of the test light or ohmmeter always goes through the damaged diode. When the diode has an open circuit, the test current will go through one of the other diodes.

CAUTION When soldering (or unsoldering) the stator lead to the diode, use a pair of pliers as a heat insulator to prevent overheating the diode.

Diode Replacement When you replace a diode, be careful not to damage the heat sink or the new diode. Use the special tools manufactured for this purpose.

Support the heat sink and then press the damaged diode out and the new diode in. Never "drive" them in or out.

When you replace a diode which is threaded into the heat sink, remember that negative diodes have right-hand threads and positive diodes have left-hand threads.

To ease the removal of a diode, cut the diode leads with a diagonal cutter. Immerse the heat sink in hot water or place it in an oven and heat it to about 150°F [66°C]. Coat the threads of the new diode with a silicon grease, and torque it to specification.

(CHECK FOR OPENS)
OHMMETER

A

C

B

OHMMETER
(CHECK FOR OPENS)

OHMMETER
(CHECK FOR GROUNDS)

Fig. 44–33 Checking stator windings. (*United Delco-AC Products, General Motors of Canada, Ltd.*)

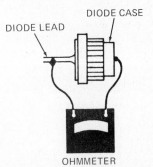

DIODE CASE

DIODE LEAD

OHMMETER

Fig. 44–34 Testing a diode using an ohmmeter. (*United Delco-AC Products, General Motors of Canada, Ltd.*)

Fig. 44–35 Soldering a diode lead.

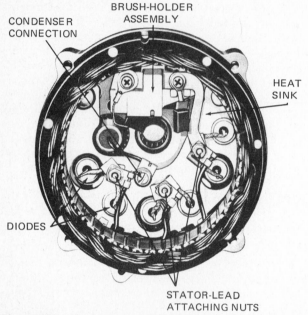

CONDENSER CONNECTION

BRUSH-HOLDER ASSEMBLY

HEAT SINK

DIODES

STATOR-LEAD ATTACHING NUTS

Fig. 44–36 Slip ring and frame showing diodes, stator, and condenser connections. *(United Delco-AC Products, General Motors of Canada, Ltd.)*

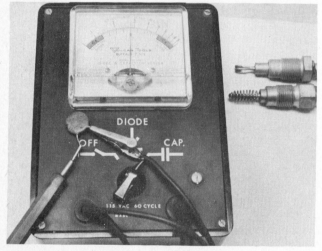

Fig. 44–37 Testing the condenser.

The silicon grease and the proper torque ensure a good diode circuit.

Always test the newly installed diode before resoldering. Solder the leads in a position where they do not interfere with the other components. Use a pair of pliers as a heat sink to avoid damaging the diode (Fig. 44–35).

Reduced Output Due to Heat-Sink Failure Any deterioration of the heat sink or components related to the heat sink can affect generator output. For example, although a broken heat sink does not instantly affect the output of a generator, it can cause the temperature to rise beyond the safe limit of the diode and, therefore, destroy the diode in the heat sink. Poor or broken insulation between the heat sink and the slip-ring end frame can send current partly to ground, reducing the current flow to the battery. The same is true when the battery terminal is partly grounded. An ohmmeter can be used to check the heat sink and battery terminal for a short or grounded circuit and for continuity.

Transit Condenser and Diodes Most generators have either a condenser or diode to protect the rectified diodes against high transit voltage and to aid in radio noise suppression (Fig. 44–36). This diode or condenser should, therefore, be checked. When either is short-circuited, part of the generator output goes to ground. When either is grounded, all generator output goes to ground. However, when the condenser or diode has an open circuit, no current output is lost because there is no circuit to ground.

To test the condenser for series resistance, capacity, and leakage, use a condenser tester as shown in Fig. 44–37. When you test for series resistance, the meter reading should be in the dark bar at the right end of the scale. When you test for leakage, the reading should be in the dark bar at the left end of the scale. To test the capacity of the condenser, take a meter reading and compare it with your specification. It should match the microfarad reading.

Reassembling Generator All generators must be protected in the same manner during reassembly. Some of the points to be noted are:

1. Fill the drive-end-frame bearing to the proper level with only the recommended grease.
2. Make certain the lip seal or felt seals are returned to the correct position.
3. Using the appropriate adapters, press the collar onto the shaft, then the drive end frame, and then the other collar.
4. Place the fan and pulley on the shaft after making certain that the blades are not bent or twisted.
5. Tighten the shaft nut to the recommended torque, following the procedure applicable to removing the shaft nut.
6. Assemble the stator and slip-ring end frame. Lift the brushes into position. Hold them in position by inserting a through bolt or a wire in the holes provided.
7. Clean the bearing surface of the rotor shaft on the slip-ring side before sliding the assembly over the rotor and onto the drive end frame. **CAUTION** If a lip seal is used, take care not to damage it.
8. Align the end frames. Install and tighten the through bolts. At the same time make sure the rotor turns freely.

9. Since the reservoir can be filled with grease either before or after assembly, depending on generator design, it should be checked so that the grease is not forgotten.

10. Remove the wire or bolt which holds the brushes away from the slip ring.

After a generator has been serviced, its performance must be tested. This can be done on a generator test bench or after installation but before being connected into the charging circuit.

Questions

1. Describe how the rotor (rotating field) is constructed and how the field is excited.

2. What would a discolored stator core indicate?

3. Explain why a 60-A ac generator reaches its limitations of approximately 68 A and 19 V?

4. What is the major difference in the current flow between rectification of a Y and a delta generator?

5. Why must the diodes be disconnected from the stator leads before testing?

6. What is the advantage of having either a delta-wound or Y-wound stator winding?

7. Why is it important to know which diodes are in the heat sink?

8. Why can an ac generator be driven in either direction and operate at a higher speed?

9. Outline the procedure of a generator output test.

10. What is the difference between an A and a B circuit?

11. Three things which increase the voltage being induced in the stator are: increasing the strength of the magnetic field, increasing the speed of rotor rotation, and (a) increasing regulator setting; (b) increasing speed of stator rotation; or (c) increasing number of conductors in stator.

12. In many ac generators, defective field windings require (a) stator replacement, (b) regulator replacement, or (c) rotor replacement.

13. The three components which require a check of an ac generator that produces no output are the (a) armature, rotor, and transistors; (b) stator, field windings, and armature; or (c) stator, fields, and diodes.

14. To make an alternator output test of a B circuit (a) disconnect the regulator and connect terminal F to the battery terminal with a jumper wire; (b) install a new regulator to check the alternator; (c) disconnect the regulator and connect terminal F to ground; or (d) connect terminal F to ground at the regulator.

UNIT 45

Regulators

REGULATOR COMPONENTS

To protect the battery and the accessories against high voltage, the generator voltage must be controlled. This is done by using a voltage regulator which varies the current flow to the rotating field (rotor). Although vibrating points and transistorized regulators are still used, they have been replaced on diesel engines by transistor regulators.

A transistor regulator consists primarily of resistors, capacitors (condensers), diodes, and transistors. It is a complete static unit which controls the generator voltage. It is durable and efficient. It safely allows a high field-current flow, and it has a longer service life than the vibrating contact regulator. An equally important feature is the ease with which it can be tested, adjusted, and serviced.

Resistors Resistors used in transistor regulators vary in material and shape (Fig. 45-1). They reduce

current flow (calculable by Ohm's law), stabilize voltage, and reduce voltage (also calculable by Ohm's law). [See Ohm's Law (Unit 36).]

Capacitors Capacitors, also called condensers, may be used to store electricity and to equalize sys-

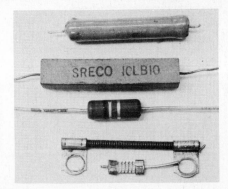

Fig. 45-1 Typical resistors.

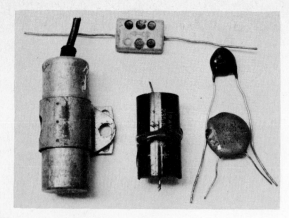

Fig. 45-2 Typical capacitors.

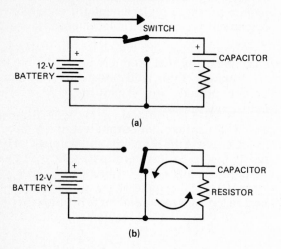

(a)

(b)

Fig. 45-3 Switch action and current flow of a capacitor when (a) charging and (b) discharging. (*United Delco-AC Products, General Motors of Canada, Ltd.*)

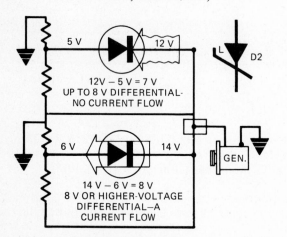

Fig. 45-4 Zener diode symbols. (*United Delco-AC Products, General Motors of Canada, Ltd.*)

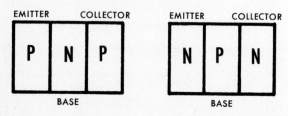

Fig. 45-5 Transistor construction. (*United Delco-AC Products, General Motors of Canada, Ltd.*)

tem voltage which varies when the transistor is switched on or off (Fig. 45-2). They are also used to assist in switching a transistor on and off and to prevent excessive heat from developing.

A capacitor, when placed in a circuit as shown in Fig. 45-3a, will receive a charge when the switch is closed. The capacitor voltage will increase from zero to battery voltage, at which time the current flow stops. When the switch is reversed, the capacitor will discharge until all stored energy is dissipated by the resistor (Fig. 45-3b).

Zener Diodes   A zener diode is an electric switch that reacts to a predetermined value of system voltage to switch a transistor on and off. The symbols used to identify the zener diode are shown in Fig. 45-4.

This specially designed diode can safely conduct current flow in reverse bias when the reverse voltage reaches a value of, say, 8 V. When the reverse voltage is below 8 V, the diode will not conduct current. This factor has been made possible by doping the silicon with more atoms having three electrons in their outer orbit when P-type material is required or by doping the silicon with more atoms having five electrons in their outer orbit when N-type material is required. The number of atoms controls the reverse voltage flow. Unlike a conventional diode, this diode will not overheat and destroy itself.

Transistors   A transistor is somewhat of an electrical switch and at the same time, an amplifier. However, it only allows a larger current flow (from the emitter to collector) when the emitter-to-base circuit is complete.

A transistor is made from a PN or NP junction by fusing one or more P- or N-type materials to it (see Fig. 45-5). The left P or the left N material is called the *emitter*; the N or P material at the right is the *collector*. The base is usually the transistor case and the emitter and collector stems from the wire leads. The emitter-to-base diode is always in forward bias; the base-to-collector diode is always in reverse bias; the base and collector are always of the same polarity. Both the PNP and NPN transistors perform in essentially the same manner but their polarity is different.

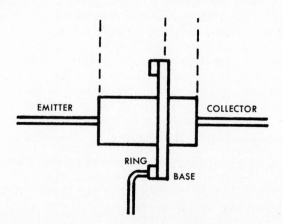

Fig. 45-6 Transistor potential. (*United Delco-AC Products, General Motors of Canada, Ltd.*)

The unusual operating characteristics of a transistor are based on the potential difference between the base and the collector, and the base and the emitter. When the base potential is closer to the collector potential, as shown in Fig. 45–6, the ratio of current flow between the emitter-base and the emitter-collector is higher, and the ratio will decrease by moving the base potential toward the emitter. When the emitter and collector potential are equal, there is no current flow.

Transistor Operation   When a diode is placed in a simplified field circuit, and the circuit is completed by closing the switch, or through a voltage relay or zener diode, field current will flow (Ohm's law) from the positive terminal of the battery, through the field coil, diode, and switch, to the negative terminal of the battery.

When you replace one diode with a PNP transistor, the current flow remains the same but something rather unexpected happens: Refer to Fig. 45–7. From the illustration you can see that 5 A enter the emitter, 4.8 A leave the collector, and a current of only 0.2 A flows in the emitter-base circuit.

The holes created in the emitter by the departing electrons, travel directly into the collector due to velocity, and some holes at the emitter-base junction combine with electrons and neutralize. However, the number which combine is quite small because of the base potential which is usually less than 5 percent of the emitter-collector circuit.

Transistor Testing   A multitude of transistors are available, varying in size and appearance. The procedure by which they are tested remains the same.

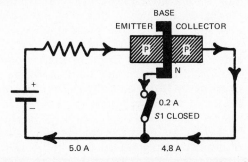

Fig. 45–7   Current flow when a transistor is placed in the circuit. *(United Delco-AC Products, General Motors of Canada, Ltd.)*

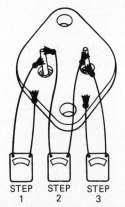

Fig. 45–8   Testing a transistor. *(United Delco-AC Products, General Motors of Canada, Ltd.)*

NOTE   Ohmmeters sometimes differ in resistance and in their battery's state of charge. As a result, two ohmmeters used under identical situations may show different readings.

Testing a transistor actually involves testing two diodes. When the ohmmeter is connected as shown in step 1 in Fig. 45–8, you are testing the emitter-collector circuit. There should be continuity and upon reversing the ohmmeter leads, there should be a high resistance reading. If both readings show high resistance, the emitter-collector circuit is open. If both readings show low resistance, there is a short circuit.

To determine the condition of the PN junction, connect the ohmmeter as shown in step 2 of Fig. 45–8. This tests the emitter-base circuit (the PN junction). The ohmmeter should indicate low resistance and, when the leads are reversed, a high resistance should be indicated. If both ohmmeter readings show high resistance, the circuit is open. If both readings show low resistance, there is a short circuit.

To determine the condition of the NP junction, connect the ohmmeter as shown in step 3 in Fig. 45–8. This will test the base-collector circuit. The ohmmeter should indicate low resistance and, when the leads are reversed, a high resistance should be indicated. If both readings are low, there is a short circuit. If both readings are high, the circuit is open.

Transistor Regulators   Transistor regulators are superior to the transistorized regulators in many ways. They are fully static units and are not sensitive to vibration. They allow better voltage control since they can switch a transistor on and off as often as 1000 times per second. They can be adjusted quickly and easily, are better protected from external damage, and are less affected by heat. Of prime importance to the mechanic is the fact that they are easy to test and service.

The basic components and the internal electrical connections are similar in all models of transistor regulators. However, the value of the resistors, diodes, transistors, and condensers must be changed when the charging voltage and/or battery polarity changes.

The basic circuit which connects the battery to the field coil is shown in Fig. 45–9. When the transistor is switched on, maximum field current flows from the emitter-collector through the field coil to ground. The generator will then produce maximum output.

While the switch is closed, the battery is connected to the emitter and the base is connected over resistor $R1$ to ground. Since resistor $R1$ has a resistance of 520 Ω, the emitter-base current is very low.

To switch the transistor $TR1$ off, thereby cutting off the field current flow, transistor $TR\,2$ and diode $D1$ are placed in the circuit (see Fig. 45–10). The emitter of $TR2$ is connected to the battery. The $TR2$ collector is connected to the base of the transistor $TR1$, and the base of $TR2$ is connected to ground. Diode $D1$ is placed between the emitter connection of $TR1$ and $TR2$.

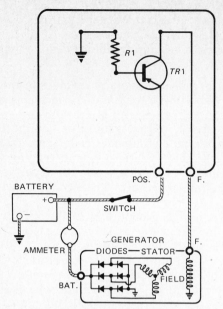

Fig. 45–9 Action when transistor TR1 is switched on. (*United Delco-AC Products, General Motors of Canada, Ltd.*)

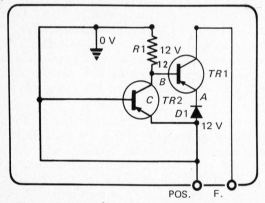

Fig. 45–10 Action when transistor TR1 is switched off. (*United Delco-AC Products, General Motors of Canada, Ltd.*)

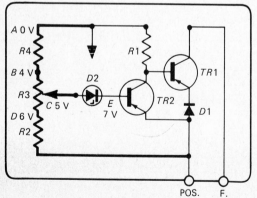

Fig. 45–11 Zener diode blocking the current flow. (*United Delco-AC Products, General Motors of Canada, Ltd.*)

Diode $D1$ is used to prevent a flow of reverse current and to reduce the emitter voltage to $TR1$ by about 1 V. When the battery voltage is 12 V, the potential at the emitter, point A in Fig. 45–10, and the base of $TR1$, point B, is 11 V, when nullifying the resistance of the transistor.

However, battery voltage, 12 V, is present at the emitter and collector of $TR2$, point C, because the base is connected to ground. Therefore, 12 V is present at the base of $TR1$. The higher potential robs the base circuit of the transistor $TR1$. With no base circuit, transistor $TR1$ is switched off, and no current flows to the field.

To switch $TR1$ on, and allow field current to flow to the field coil. A zener diode is placed in the base circuit of transistor $TR2$ and one circuit with three resistors is added from the battery over $R2$, $R3$, and $R4$ to ground (Fig. 45–11).

The base circuit of $TR2$ is now connected to a variable resistor (potentiometer) $R3$. When measuring the voltage at points A, B, C, and D, the reading would be 6 V at D, 4 V at B, 0 V at A, and 5 V at point C (the variable resistor), as shown in Fig. 45–11.

Since the breakdown voltage of the zener diode is 8 V, the diode will not conduct current because there are 12 V from the battery opposing 5 V from the variable resistor. This leaves a potential of only 7 V at point E. With no base circuit, the transistor $TR2$ is switched off, allowing the base circuit of $TR1$ to complete its circuit to ground. As a result, transistor $TR1$ is switched on and the full field current flows through the field coil.

As the generator voltage rises, so does the voltage within the regulator. As the charging voltage reaches say, 14.1 V, the voltage at point E will be 8.1 V; at point C, 6 V; at points F and G, 13.1 V; and at point H, 14.1 V (see Fig. 45–12). With 8.1 V at the zener diode, the diode conducts current, causing the transistor $TR2$ to switch on, $TR1$ to switch off, and the charging voltage to decrease. As soon as the charging voltage drops below 14 V, the zener diode blocks the current flow because of the 8 V present only at point E, and transistor $TR1$ switches on again. As stated earlier, the switching cycle may occur as often as 1000 times per second.

To protect the transistor from the high inductive-current flow which may come from switches, solenoids, or from the induced voltage of the field coil when the transistor is switched off, diodes $D3$ and $D4$ are placed in the regulator circuit (Fig. 45–13). Diode $D3$ diverts the inductive current created by the reversed induced voltage in the field coil back to the field coil, thereby protecting the transistor. The transistor's suppression diode $D4$ diverts any other induced voltage (called *transient voltage*) that would allow damaging current flow into the regulator to divert back to its original source.

Since the charging voltage increases with increased battery temperature, it is necessary to reduce the charging voltage when the battery is hot in order to protect it against overcharging. It is also necessary to increase the voltage when the battery is cold to prevent it from undercharging.

The special resistor $R5$ decreases in resistance with an increase in temperature, thereby reducing the potential at the variable resistor. This allows the zener diode to conduct current at a slightly lower voltage thereby lowering the charging voltage.

Resistor $R6$ prevents the occurrence of small current leakage through the emitter-collector due to

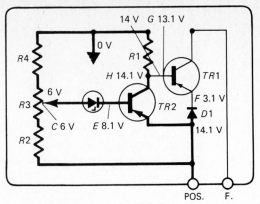

Fig. 45–12 Zener diode conducting current (regulator voltage increasing).

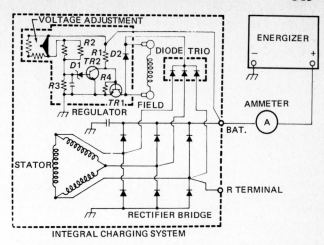

Fig. 45–14 Wiring diagram of a 27-SI Series integral charging system. (*United Delco-AC Products, General Motors of Canada, Ltd.*)

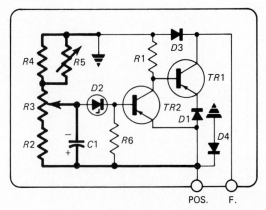

Fig. 45–13 Added components to protect the regulators. (*United Delco-AC Products, General Motors of Canada*)

high temperature. The filter capacitor stabilizes the system voltage at the resistors as the voltage varies from maximum to nearly zero while transistor TR1 switches off and on.

To speed up transistor switching and, as a result, reduce the amount of heat developing within the regulator, capacitor C2 is placed in the circuit.

When transistor TR1 is switched off, capacitor C2 lowers the voltage at the potentiometer of the zener diode by 0.1 V. This brings the voltage across the zener diode to 8.2 V and causes the transistors to quickly switch on and off. As the charging voltage across the zener diode decreases to 8 V, the transistor shuts off, and transistor TR1 conducts current to the field coil. Capacitor C2 becomes charged, and because of this increased voltage at the potentiometer side of the zener diode, the zener diode shuts off.

Integral Charging System Generators that have the regulator internally mounted are called *self-rectifying generators* or *integral charging systems*. Delco-Remy generators of this design use an A field circuit and when the field circuit is not externally energized, they have a residual magnetized rotor. When the rotor is rotating, ac voltage is induced in the stator winding and the current then flows through the diode trio into the regulator as well as from the battery terminal into the regulator (see Fig. 45–14). Current then flows through resistor R1 to forward bias, to the emitter base of TR1, to ground, and back to the ground diode in the rectifier bridge

to the stator, completing the circuit. Transistor TR1 switches on, and current can then flow from the diode trio into the field coil, from there to TR1 (collector-emitter), to the grounded rectifier bridge and to the stator winding. At the same time, depending on the position of the voltage adjustment, there is a current flow from the battery terminal, through the voltage regulator resistors, through the two parallel resistors R2, and then through R3 to ground. As the voltage between R2 and R3 increases to the value where the zener diode conducts (that is, at 8.1 V), TR2 suddenly conducts current from emitter to base to ground. This causes TR1 to switch off, thus limiting the output voltage. As soon as the voltage drops at the battery terminal of the generator, the zener diode acts as an open switch, forcing TR2 to switch off and resulting in TR1 switching on.

The external voltage adjustment consists of a square plug in which terminals can be inserted in one of the four positions to increase or decrease the voltage that acts on the zener diode.

Self-rectifying Brushless Generators Self-rectifying brushless generators are used when operating conditions are subject to dust, water, etc., or when an explosionproof generator is required. One type is shown in Fig. 45–15. The rotor is permanently magnetized, and it is the only rotating component over the stationary field. Note the unusual location of the fan and regulator.

To increase generator output, the stator is commonly wired in delta connection rather than in Y connection and a specially designed output terminal is used. The transistor regulators and the rectifying diodes are mounted at the rear of the generator. Some regulators are similar to those previously mentioned and some are sealed and cannot be serviced. The electrical connections of the sealed type are different from those of other regulators (see Fig. 45–16). Note that although each stator phase is connected over a diode to the regulator, from that point on the electrical connections differ.

The regulator shown in Fig. 45–16 has an additional transistor, TR3, and the transistors and diodes used have NPN- and NP-type junctions. It also has

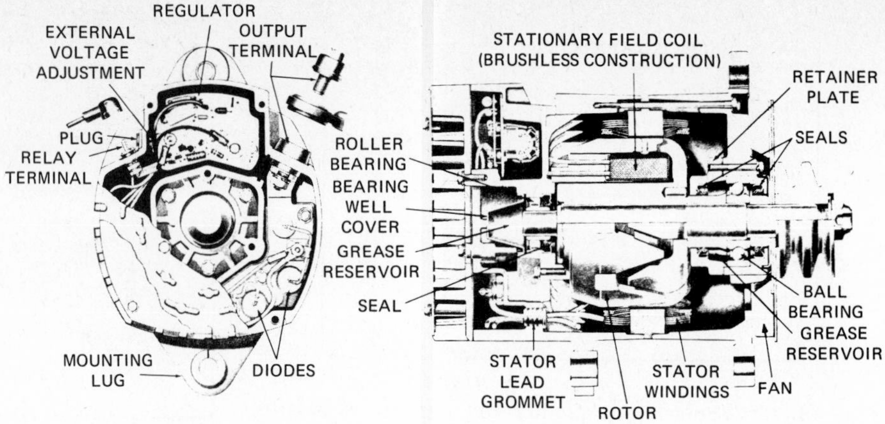

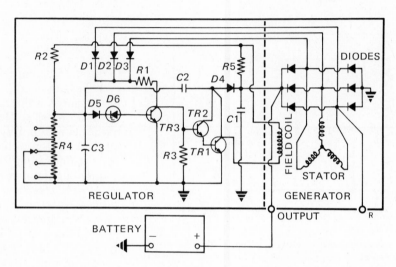

Fig. 45–15 Sectional views of a brushless generator having an integral charging system. *(United Delco-AC Products, General Motors of Canada, Ltd.)*

Fig. 45–16 Sealed-type regulator circuit. *(United Delco-AC Products, General Motors of Canada, Ltd.)*

an additional diode, $D5$, which is connected between the zener diode and resistors $R2$ and $R4$.

Generator and Regulator Operation When the permanently magnetized rotor rotates, an alternating voltage is induced in the stator winding which is rectified by the three negative and three positive diodes, and dc current flows into the battery. The rectified current of each phase winding also flows over diodes $D1$, $D2$, $D3$ into the regulator to resistor $R1$, to the collector of resistor $TR3$, and to the resistor $R3$ to ground (see Fig. 45–16). The transistor $TR3$ is not switched on because the low voltage allows zener diode $D6$ and diode $D5$ to block the base circuit. However, transistors $TR2$ and $TR1$ are switched on because current can now flow over both emitter bases to ground.

With both transistors switched on, current from the output terminal of the generator supplies current to the regulator over resistor $R5$ to the field coil and transistor $TR1$ (collector elements) to ground. Output current also flows from resistor $R5$ to resistors $R2$ and $R4$ to ground. As charging voltage increases, the voltage impressed across resistor $R4$ is also impressed across diode $D5$ and zener diode $D6$.

When the breakdown voltage is reached, transistor $TR3$ switches on because the emitter-base circuit to ground is completed. This causes $TR2$ and $TR1$ to shut off since current now flows over the lower resistance circuit from resistor $R1$, transistor $TR3$

(collector-emitter) to ground, robbing the current flow from transistor $TR2$. The field current flow stops. As system voltage decreases, diodes $D5$ and $D6$ stop conducting current, and transistor $TR3$ shuts off. This cycle repeats many times per second to maintain present generator voltage. The capacitors $C1$, $C2$, and $C3$, and diode $D4$ perform the same function as the capacitors and diodes in the regulator previously outlined.

REGULATOR SERVICE

Voltage Regulator Adjustment Most charging-system trouble is drawn to the operator's attention by a slow cranking speed or by an overcharged battery. When trying to locate a charging system's malfunction, start your check and tests with the battery. Next, check and measure the voltage drop of the battery, generator, and the regulator connections. If none of these checks have disclosed the problem, perform a generator output test.

To check regulator voltage, connect a voltmeter from the generator's B terminal to ground. Run the engine at half-throttle for about 15 minutes. (You may have to use a carbon pile across the battery to obtain a current flow of about 10 A.) After a 15-minute interval, measure the temperature by placing a thermometer close to the regulator cover. Take a voltage reading and compare it with the suggested voltage setting (Table 45–1) for the measured temperature.

Table 45–1 SUGGESTED VOLTAGE SETTING RANGE

Ambient temperature, °C (°F)	Voltage setting range, V
18.3 (65)	13.5–15.5
29.4 (85)	13.3–15.3
40.5 (105)	13.1–15.1
51.5 (125)	12.9–14.9
63.0 (145)	12.8–14.8

If the voltage setting is within the recommended specified range but the battery was undercharged prior to testing, increase the setting by 0.3 V. If the battery was overcharged, decrease it by the same amount.

To change the charging voltage, remove the external plug and, with a screwdriver, turn the potentiometer. When the regulator adjustment is internal, stop the engine. Remove the regulator, remove the bottom cover of the regulator, and then make the adjustment. **NOTE** Always recheck the voltmeter setting after an adjustment.

Generator Output Tests and Voltage Adjustment Before you attempt any voltage adjustment or output test, make some preliminary checks and tests to verify the condition of the charging system. If the circuit connection and the battery appear to function satisfactorily, start the generator output test by connecting the test instruments as shown in Fig. 45–17.

NOTE The battery or batteries must be fully charged when making a voltage regulator adjustment. Note also, that some generator terminals are of special design and need special connections.

Open the carbon pile and run the engine at recommended speed for a short period of time to stabilize the temperature. Check the voltage; it should be within operating range (see Table 45–1).

If the charging voltage is below operating range, increase engine speed to high idle. You may have to use a carbon pile to obtain maximum current output.

If the current output is less than specified, but within 5 A of suggested maximum, adjust the voltage regulator to come within the range of voltage setting. To make a regulator adjustment, remove the plug, and turn the adjusting screw one or two notches clockwise to raise the voltage.

When the output is less than specified or the voltage cannot be adjusted to show a satisfactory charging voltage, remove the generator end cover and check the field leads, stator leads, and regulator lead. Make certain their connections are tight. When the connections are tight and there is still no change in current output or voltage, the generator must be disassembled so that the stator, the diodes, and the field coil can be checked.

If the charging voltage is higher than recommended or if the battery is overcharged, turn the adjusting screw one or two notches counterclockwise to lower the charging voltage.

If the voltage cannot be adjusted, recheck the battery. Also check the regulator connection. If repeating the charging voltage test yields a charging voltage which is still too high, check the field coil. If the field coil checks out satisfactorily, replace the regulator, repeat the test, and make your adjustment.

Field-Coil Test To check for an open and a short-circuited field-coil circuit, connect both ohmmeter leads to the field-coil leads as shown in Fig. 45–18. The ohmmeter needle should indicate the resistance of the field coil.

To check the field coil for a grounded circuit, connect an ohmmeter lead to one field-coil lead and the other ohmmeter lead to ground. The ohmmeter needle should show maximum (infinite) resistance.

There is another method used to check the field coil for a short circuit. Connect the battery, ammeter, and a rheostat in series with the field coil. Then connect a voltmeter across the field coil. Slowly reduce the rheostat resistance until the voltmeter indicates specified voltage, and then read the ammeter. The ammeter reading should come within specified value. If the reading is above the specified amperage, the field coil must be replaced.

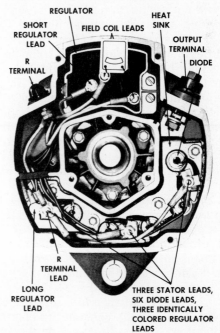

Fig. 45–18 Checking for open or shorted field-coil circuit. (*United Delco-AC Products, General Motors of Canada, Ltd.*)

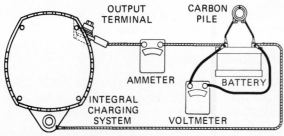

Fig. 45–17 Generator output check. (*United Delco-AC Products, General Motors of Canada, Ltd.*)

Remagnetizing Rotor When a generator is being disassembled, it is often demagnetized in the process. Therefore, it should be remagnetized. To remagnetize the rotor, connect the battery lead to the generator B terminal. Using a jumper wire from the generator terminal, momentarily touch the R terminal. Current will flow through the regulator and field coil, to ground, magnetizing the pole pieces to their correct polarity.

30 SI/TR Series To provide two different voltages from one generator (Delco 30 SI) (12 V for charging and 24 V for cranking), a TR unit transformer rectifier has been added. The unit consists of three transformers and a rectifying bridge.

Electrically, each stator output phase is connected to the primary coil of one transformer. The secondary transformer winding is connected to the rectifying bridge of the TR unit and from there connected with the 24-V terminal of the generator.

30 SI/TR Series Operation When the generator is operating, the voltage regulator reacts to the cemf of the A batteries and regulates in its normal manner (between 13 to 15 V). The transformer steps up the regulated voltage of the voltage regulator to 26 to 30 V, and a current flow between 5 and 15 A flows to the rectifying bridge, is rectified to direct current, and from there flows to the 24-V generator terminal.

30 SI/TR Generator Service When either set of batteries is overcharged or undercharged, check the batteries for defective battery cables and connections, for wiring defects, and for loose drive belts. Then check the generator output voltage at the 12- and 24-V terminals just as you would test a conventional generator.

If the output voltage is above or below specification and will not test out to specification, remove the battery ground cable and then remove the TR unit. Remove the primary transformer leads from the 30 SI rectifier bridge, and the single lead from the 30 SI rectifier-bridge heat sink, and retighten the nuts. The generator will then be a regular 30 SI generator. The generator output test can then be made by connecting an ammeter in series with the 12-V battery terminal of the generator and with the disconnected battery wire and a carbon pile across the batteries. If the current output does not come within 10 A of the rated output, the generator must be serviced.

When the generator checks out satisfactorily, check the rectifying bridge of the TR unit in the same manner as previously outlined (Fig. 45–19). If still no defects are found, install the TR unit to the generator. Connect an ammeter in series with the 24-V generator terminal and the removed 24-V wires. Connect a voltmeter from the 24-V terminal to ground. Connect a carbon pile across the battery. You can then operate the generator to obtain rated maximum output. The voltmeter should read twice the regulated 12-V voltage and the ammeter should show not less than 5 A. If the ammeter shows less than 5 A under this test condition, it indicates that

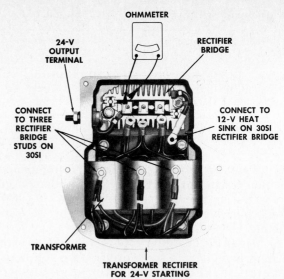

Fig. 45–19 Ohmmeter checks of the rectifier bridge diodes in the transistor unit. *(United Delco-AC Products, General Motors of Canada, Ltd.)*

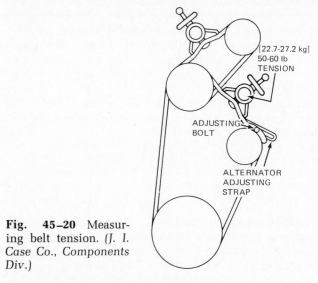

Fig. 45–20 Measuring belt tension. *(J. I. Case Co., Components Div.)*

Table 45–2 CORRECT BELT TENSION

Belt width, in [mm]	Deflection per foot of span, in [mm]
½ [12.700]	¹³/₃₂ [10.3187]
¹¹/₁₆ [17.4625]	¹³/₃₂ [10.3187]
¾ [19.0500]	⁷/₁₆ [11.1125]
⅞ [22.2250]	½ [12.7000]
1 [25.4000]	⁹/₁₆ [14.2875]

one (or more) transformer is damaged and it must be replaced.

Installing Alternator Before you install an alternator, check the sheaves of the generator and the drive. Check the drive belt(s).

NOTE The drive belts are in matched sets. Do not replace only one belt, replace the entire set. Make certain the battery ground cable is disconnected.

When you install the generator, use the correct washers and bolts. Fit it snugly to its mounting brackets.

GROUNDED HEAT SINK

INSULATED HEAT SINK

DIODE TRIO

REGULATOR

*TR*1 COLLECTOR

Fig. 45–21 Special connections when testing an integral charging system for maximum output.

Check the pulley alignment. Any misalignment must not exceed $^1/_{16}$ in [1.59 mm] for each 1 ft [30.48 cm] between pulley centers.

Install belt(s), but do not roll or pry belts over pulleys. Once the belts are installed, tighten them and lock the adjustment. Then measure the belt tension, as shown in Fig. 45–20, or with your index finger (in the middle of the free belt span), push the belt straight downward. Correct belt tension will be within the specified value shown in Table 45–2.

After you have established the correct belt tension, torque all mounting bolts to the specified values. Connect the battery (and R terminal wire, if used) to the generator. Connect the battery ground cable.

NOTE If the drive belts are new, check the tension 1 hour or so after operation because they tend to loosen (stretch) slightly and may need to be readjusted. Recheck them again after 8 hours of operation, and thereafter at least every 5000 miles [8045 km] or 150 hours of operation. Do not use any kind of belt dressing.

Belt Cleaning Remove oil or grease from the drive belts as soon as detected so that it will not penetrate them. Remove the belts from the engine. Clean both sheaves and belts with a nonflammable solvent or cleaner. Occasionally soap and water may be used, but it is not as effective as a solvent cleaner.

CAUTION Never attempt to clean belts when the engine is running and never use a flammable solvent.

Integral-Charging-System Service When testing an integral charging system for its maximum output, connect the same instrument as you did when testing the maximum output of a conventional generator. However, to achieve a maximum output, you must bypass the regulator and at the same time ground the generator brush which is connected to the collector of *TR*1 (see Fig. 45–21).

The service checks for the rectifier bridge, rotor, stator, and brushes of an integral charging system are the same as for a conventional generator except that you also must test the diode trio and the voltage regulator.

To test the three diodes in the diode trio, remove the diode trio from the generator. Connect one ohmmeter lead to the single connector and the other ohmmeter lead to one of the three phase connectors, then reverse the ohmmeter leads. A good diode should have one high and one low ohmmeter reading. Now repeat the ohmmeter test with the other two diodes.

The regulator must be checked with a regulator tester, but note that the regulator is used with an A field circuit generator.

Questions

1. Why are resistors used in transistor regulators?

2. List the three functions of a (condensor) capacitor.

3. What is the major difference between an N and a P material?

4. Explain why a diode conducts current in only one direction.

5. Explain how to test a transistor for an open and a short circuit.

6. Explain how to test a diode for an open and a short circuit.

7. Explain how the test instruments are connected when performing a generator output test for a generator with an integral charging system.

8. Outline the procedure to check the field coil of a brushless generator.

9. Refer to Fig. 45–16. Would the generator output be affected when one diode has a short circuit?

10. A precaution to be observed when soldering diode leads is (a) to avoid overheating during soldering; (b) to heat high enough to get a good connection in the diode; or (c) not to damage the insulation.

11. A major advantage of the transistor regulator is that the transistor (a) permits a lower field current; (b) permits a higher field current; or (c) eliminates the need for cutout relay.

12. List the order in which the field current flows through the generator shown in Fig. 45–16 when the battery is in a low-charge state.

13. A transistor permits a current to flow when the base circuit is (a) open, (b) connected to field, (c) grounded, or (d) carries current.

14. The diode in an ac generator performs two functions. One is to rectify, the other is (a) to act as a cutout relay; (b) to change the ac current to dc current; or (c) to change the current flow.

15. What are the advantages of an integral charging system over a conventional charging system (that is, where the voltage regulator is not part of the generator)?

16. In order to determine if the generator or regulator of a Delco integral charging system is faulty, which test must you make? How is the test made?

17. Why is the voltage first stepped up to 24 V and then rectified to direct current (30 SI/TR Series generator)?

18. How would you increase the output voltage of the generator shown in Fig. 45–15 if the voltage regulator were set too low?

SECTION 6
Shop Equipment

Hand Tools

It is almost impossible to quickly and accurately diagnose and repair a component that is not functioning correctly if you do not understand how the unit operates or what the law is that governs its behavior. It is also necessary for you to know about hand tools and shop tools. It may be possible for you to haphazardly repair a component with the first available tool, but only the correctly designed and selected tool will ensure that the job is well done.

Selection of Hand Tools The most important advice you can be given at the beginning of your career is to purchase top quality tools. These are made from good steel and manufactured to precision. Special consideration is given to balance so that the tool will feel comfortable in your hand. Since you, the mechanic, must work with your tools daily, balance is of major importance.

The initial cost of a complete set of tools is high but the accompanying warranty ensures satisfaction and many years of service. It is much better to start with a few carefully selected tools. These will take care of your most common needs. Later you can gradually build up a complete high-quality set.

Care of Tools Your tools should always be cleaned before you return them to your toolbox. Ideally, each type of tool should have its own tray or drawer. This not only protects the tools but also allows you to locate them quickly.

Always place the most commonly used tools in the most accessible box drawer. Make certain that your tool box is large, sturdy, and if possible, equipped with casters. Hammer handles should always be tight and of the proper length. Your chisels, punches, and screwdrivers should be sharpened and dressed at all times. Be proud of your tools; they are your bread and butter!

Punches A punch is a forged steel tool which has any one of many shapes at one end (Fig. 46–1). It is used to drive a pin, bolt, or other fastener in and out of a hole.

The working end may be sharp or blunt, tapered or cylindrical, or it may have a cutting edge. Punches are classified according to their specific uses. The *starter punch*, for example, is used to start an object moving, whereas the *drift punch* is used to complete the job. A *center punch* is used to make a center before drilling or to show alignment of components. An *alignment punch* is used to align holes and thereby facilitate insertion of bolts, etc., or to line up parts for assembly work. A *hole punch* is used to make holes in gasket material.

A *prybar* can be defined as a forged steel tool tapered at one end, flat and angled at the other, and tempered over its total length. A prybar can be used as an alignment punch on one side while the other side can be used as a lever to move, remove, or lift components.

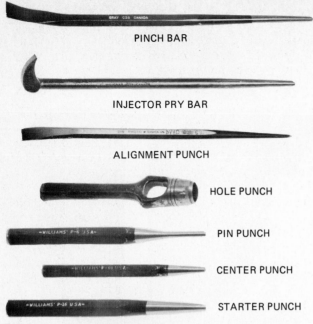

Fig. 46-1 Punches.

Fig. 46-2 Files.

Files One of the oldest tools is the file. Every mechanic should have an assortment of files to fit, prepare, or shape a component when restoring its serviceability. This does not mean that as a mechanic you must possess one of each of the very many files used in your trade, but you should at least acquire the files shown in Fig. 46-2.

A file is a tool with cutting ridges made from hardened steel for forming or smoothing surfaces. It is usually used on metal. In the trade a file is described by its shape or its specific function, that is, flat file, mill file, round file, etc. Each of these files is made in bastard, single, or double cut. Bastard files are the least used as they are too coarse for filing many surfaces, especially the faces of flanges. Double or single-cut files are more applicable.

Pointers for Selecting and Using Files
1. Be certain to choose the appropriate file. The incorrect file may ruin the component.
2. Choose a file which is fitted with a handle so that the point of the tang will not pierce your hand.
3. Grasp the file firmly. Stand balanced with your left foot forward so that your right hand is square and parallel to the object being filed.
4. Push the file down and forward with a gentle force. On the return stroke, lift the file from the object to prevent excessive damage to the file cut.
5. Clean the file frequently with a file cleaner or a piece of wood to produce a smooth surface.
6. Fill the cut with chalk. This prevents particles from sticking in the file cut and increases the effectiveness of the file.

Screwdrivers The screwdriver is often misused by many tradesmen because of its capacity to substitute as a prybar or chisel. Every mechanic should have a variety of screwdrivers varying in type and size to accommodate the many different screw recesses. The screwdrivers most popular in the diesel mechanics trade are shown in Fig. 46-3. It is advisable to have a similar set of screwdrivers for the socket drive to permit working in tight quarters.

Pointers for Using Screwdrivers
1. Keep screwdrivers dressed. This can be done by grinding and dressing with a file (see Fig. 46-4).
2. Use a screwdriver only for the purpose for which it was designed. Do not use it as a scraper, chisel, or prybar.
3. Select the proper size screwdriver (appropriate to the screw) to avoid damage to the screw head. This will also ensure troublefree removal of the screw, or when tightening, will help to achieve the proper torque.
4. Hold the components so that the screwdriver cannot slip and pierce your hand or body.
5. When you work on electrical equipment, make certain that the ground cable is disconnected. If the cable cannot be disconnected, be sure to use an *insulated* screwdriver.
6. When loosening a tight screw, use a strong downward force. Using both hands, apply a sharp twisting effort. A sharp blow with a hammer on the head of the screw will also help to loosen it, as will the application of penetrating oil or moderate head.

Hacksaws The hacksaw is seldom used by diesel mechanics, but it is still indispensable. When buy-

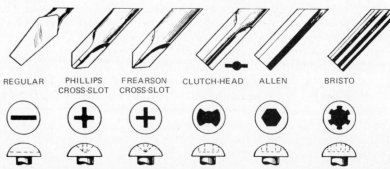

Fig. 46-3 Screwdrivers.   STANDARD TYPES OF SCREWDRIVER TIPS AND SCREW RECESSES

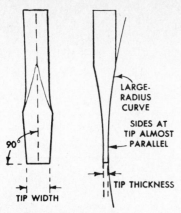

Fig. 46–4 Dressing a screwdriver.

ing a hacksaw, select one with a 12-in [30.5-cm] fixed frame. This will give you a stable blade when sawing and thereby reduce the number of blade replacements. Keep spare blades with 18, 24, and 32 teeth per inch so that you will have the proper blade when needed. Buy good quality blades. They may be more expensive, but they will last much longer.

Pointers for Using Hacksaws

1. Carefully select the blade according to the job to be done. A properly selected blade will have at least two teeth in contact with the surface being sawed and will be coarse enough to give good chip clearance. Be sure the teeth point away from the handle.

2. Test the metal with the first or last teeth of the hacksaw blade, otherwise you may inadvertently ruin the entire set of teeth.

3. Use your thumb to guide the blade until the cut is started at the desired location. Apply sufficient pressure so that the blade begins to cut immediately.

4. Use the entire length of the hacksaw blade and enough pressure to ensure effectiveness.

5. When the kerf is deep enough to guide the blade, the saw should be lifted or raised lightly on the back stroke to extend the life of the teeth.

6. If you break a hacksaw blade, start a new cut. Since a new blade has more set than a worn one, continuing on the original cut could jam or break the blade.

7. Protect the hacksaw and blades so that the teeth will not be damaged.

In addition to the ordinary saw it is handy to have a jab saw or junior hacksaw. These can reach into areas where no other saw can operate.

Pliers   As a diesel engine mechanic without pliers, you would be almost helpless. Furthermore, you need all the types shown in Fig. 46–5. At times you will require specially designed pliers if you are to hold, cut, bend, or remove various snap rings and hose clamps. The name of the pliers will usually indicate to you their shape and specific use.

Pointers for Using Pliers

1. Select pliers to suit the job.

2. Do not use pliers to cut spring steel or steel harder than the pliers themselves, otherwise the jaws may be permanently damaged.

3. Never use pliers to tighten bolts, nuts, or fittings.

4. Never use pliers as a jumper cable.

Hammers   Hammers differ in type and style throughout the trades. The most commonly used hammers in the diesel mechanics trade are the ball-peen, plastic, bronze, or lead-tip hammers.

How to Use Hammers   Hammers are rated and described according to their shape, their material, and the weight of the head without the handle. Al-

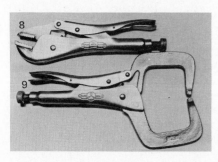
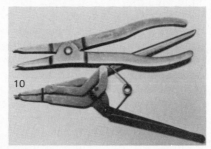
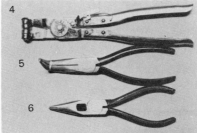
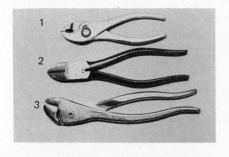
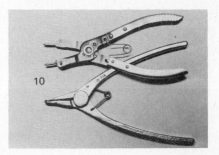
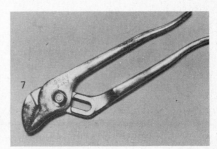

1. Combination pliers
2. Diagonal cutter
3. Angle-nose pliers
4. Hose-clamp pliers
5. Bent-needle-nose pliers
6. Needle-nose pliers
7. Interlock-joint pliers
8. Vise grip
9. Vise-grip clamp
10. Various snap-ring pliers

Fig. 46–5 Pliers.

though the hammer is a simple tool, there is a right and a wrong way to use it:

1. The hammer should be gripped closely to the end of the handle to increase the length of the lever arm. This increases the force of the blow.
2. The hammer should be parallel to the object being struck to ensure contact with its full face. This distributes the force of the blow over the entire surface of the hammer face, minimizing damage to the edges of the hammer and to the object being struck. The novice has a tendency to grip the hammer too close to the head. This is known as *choking* and reduces the force of the blow.
3. Hammer handles must always be tight in the head to prevent the head from flying off and endangering other workers.

The taper in the hammer head, known as the *eye*, tapers from the center in both directions. The handle is made more secure by driving in a steel wedge or wedges. Care must be taken to ensure that the wedges remain tight. If they come out the head will become loose.

When working in dry or heated areas, hammer handles have a tendency to dry out and become loose. This may be corrected by periodically immersing the hammerhead in water overnight.

Never use the heel of the hammer handle to bump or drive components into position because this can split the handle. Do not use another hammer as a punch. Such unorthodox procedures may result in the loss of your eyesight. The peen end of hammers is used for riveting and tapping out bolt holes when making a gasket.

Sledge hammers are used for heavier work. They too come in different weights and shapes. They are described as single-jack and double-jack sledge hammers.

Cold Chisels A cold chisel can be defined as a forged steel tool with a hardened wedge-shaped cutting edge at one end. Although hand-operated chisels are still the most common, there is an increasing number of pneumatically and electrically operated chisels appearing on the market.

The size of a chisel is determined by the distance across the cutting edge; manufacturers have not yet standardized on lengths. A cold chisel must be hard enough to cut metal without its edge breaking down. At the same time, it must be capable of withstanding heavy hammer blows. To meet both these requirements, only a short portion of the wedge end is hardened and tempered.

Chisel Cutting Angle The cutting angle of a flat chisel is determined by the strength of the material to be cut. For general-purpose work, an angle of 60° to 70°, is preferred. With softer metal (such as copper or brass) the angle may be increased to 45°.

When you grind the chisel, take great care to prevent the point from overheating because that will remove the hardness of the cutting edge.

Other types of cold chisels handy to have in your toolbox are the cape, the round-nose cape, and the

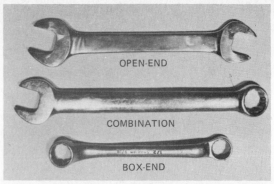
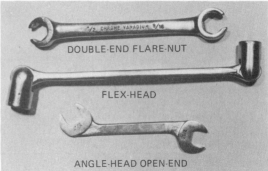

Fig. 46–6 Wrenches.

diamond-point chisels. These are used to cut narrow grooves of various shapes.

Combination, Box-End, and Open-End Wrenches One wrench suitable for almost any job is the combination wrench (see Fig. 46–6). You should have wrenches varying in size from $1/4$ to $1^1/2$ in [6 to 39 mm]. These wrenches give you the advantage of having a 15° box-end wrench on one side and a 15° angle open-end on the other side. Seldom used are the box-end, open-end, angle-head open-end, or ratcheting box-end wrenches. However, you should have a set of double-end, single-hex, flank-drive flare-nut wrenches when working on fuel-injection equipment. When your work requires adjustment of fuel-injection equipment, you must also have a tappet wrench set because standard-type open-end wrenches are too thick.

Another wrench that is used often is the flex-head wrench. It can be used at various angles in small areas.

Pointers for Using Combination and Other Wrenches
1. Use your wrenches to tighten or loosen bolts, nuts, and fittings, and for nothing else!
2. Select the wrench that is right for the job.
3. Hold the wrench parallel to the nut, bolt, or fitting to be tightened or loosened to prevent the wrench from slipping.
4. When you use an open-end wrench, make certain that the longest and widest side carries the torque, otherwise the wrench may spread and slip.
5. Do not use an extension or pipe to increase torque. It may either slip or break the wrench.

Socket Wrenches Without a complete socket wrench set in $3/8$-and $1/2$-in drives, your mechanic's toolbox is not complete. Also if your work includes the larger size bolts and nuts, then you must invest in a ³/₄-in drive.

Socket wrenches of various designs are among the greatest time- and work-saving tools you can have in your possession (Fig. 46–7). Ninety percent of all tightening and loosening of nuts and bolts, etc., can be made with a socket wrench or socket adapter. Learn to use this tool to its fullest advantage.

How to Use Socket Wrenches First use a speed, or sliding T handle to spin the bolt or nut tight. Then use the ratchet, flex handle or the sliding T handle (with or without ratchet adapter) to tighten the nut or bolt (Figs. 46–8 and 46–9). If you have to work at an angle, use either the flex handle or a universal joint or swivel socket. Take advantage of the extension bar to swing the handle freely, but avoid too long an extension. Make use of the various socket attachments to improve and speed up your work (Fig. 46–10).

Pointers for Using Socket Wrenches
1. Select the correct size and most suitable handle.
2. Avoid using a socket as a driver or a handle as a prybar.
3. Use an impact socket and extension when using an impact wrench.
4. Support the socket to secure engagement, and when possible, pull on the handle or wrench.
5. When pushing is the only possible method to loosen or tighten the nut or bolt, push with the palm of your hand.
6. Work with care. The slip of a wrench can cause serious injury.

Torque Wrenches Either you or your employer must have a torque wrench because every bolt or nut on a diesel engine must be torqued to the manufacturer's exact specification. As you know, torque is based on the principle of the lever, that is, distance times force equals the movement or torque about a point measured at 90° to the direction of force.

A torque wrench is a tool which measures the resistance to turning (referred to as torquing). This tool is comparable to a micrometer, dial indicator, or any other accurate measuring device.

TYPES OF TORQUE WRENCHES There are many types of torque wrenches in varying shapes. Some are direct reading and others have a sensory signaling mechanism that warns the mechanic the moment the predetermined torque is reached. Sensory torque wrenches also incorporate a direct-reading scale and, therefore, may be used as either direct-reading or signaling models.

Torque wrenches are calibrated to read in pound-inches, pound-feet, or for special applications, in kilogram-meters. Your employer should have a torque wrench with a combination scale in pound-feet and kilogram-meters, varying in torque range.

TORQUE-WRENCH DESIGN AND OPERATION The torque wrench is an adaptation of the principle of Hooke's law. This law states that, within the boundary of

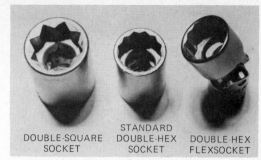

Fig. 46–7 Socket wrenches.

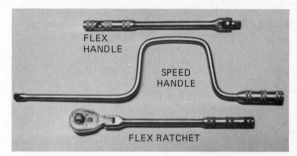

Fig. 46–8 Handles and ratchet for socket wrenches.

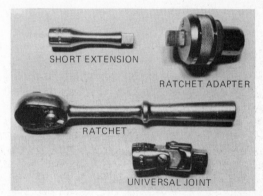

Fig. 46–9 Handles and adapters for socket wrenches.

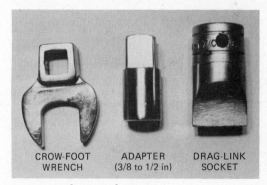

Fig. 46–10 Socket attachments.

the norm, steel will bend in proportion to the amount of load it carries. In other words, a 1-lb [0.45-kg] load creates a 1-in [25.4-mm] deformation of a steel spring. This increases to a 2-in [50.8-mm] deformation with a 2-lb [0.90-kg] load, and so on. This 1 in-to-1 lb ratio law remains constant until the stress in the spring has exceeded the norm, the norm being bound by the elastic limit of the spring. Past this point, the spring attains a permanent set. Even complete removal of the load will not bring back the original shape.

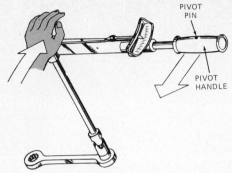

Fig. 46-11 Proper method of supporting a torque wrench. *(International Harvester Co.)*

The spring itself (measuring element) can be a single beam, double beam, or coil. It is attached between the drive square and the pivot handle a precise length in relation to the scale. The handle is pivoted, in some cases, to ensure a fixed length and to maintain a 90° angle to the centerline of the torque wrench. The pointer is attached to the socket-drive square and is thereby directly or indirectly connected to the bolt or nut. Because the scale is attached to the handle, it bends with the measuring element when force is applied. It moves underneath the pointer and thereby permits a direct torque reading.

How to Use Torque Wrenches Proper use of a torque wrench is of vital importance when a specific torque is to be achieved. Many small mistakes may be made when using a torque wrench with the result that:

1. A connection may remain loose or may become loose.
2. Loss of water, oil, or entrance of air and dirt may occur.
3. Flanges and rod bearings may be destroyed because of overtorquing and undertorquing.
4. Wear of components is accelerated.

Hold the pivot handle so that as you pull on it, the handle will float on the pivot pin (Fig. 46-11). The handle end may contact the yoke in the process of pulling it, but as long as you do not alter its position, this will not affect the accuracy. However, an error in torque will result if you change the 90° angle of direction of force or if you change the distance by extending the handle.

As an example, if you have changed the lever length from 12 to 10 in, or from 12 to 14, but continued to apply the same force (100 lb, for instance), the proper torque-wrench reading would be

$$T = 12 \times 100 = 1200 \, \text{lb} \cdot \text{in}$$

$$T = \frac{12 \times 100}{12} = 100 \, \text{lb} \cdot \text{ft}$$

When altered to 10 or 14 in, the torque would be

$$T = \frac{10 \times 100}{12} = \frac{500}{6} = 83.6 \, \text{lb} \cdot \text{ft}$$

$$T = \frac{14 \times 100}{12} = \frac{700}{6} = 116.6 \, \text{lb} \cdot \text{ft}$$

When the centerline of the torque wrench has been extended by the use of an extension or an adapter (Fig. 46-12), the following formulas will apply:

$$T = \frac{TS \times (L1 + L2)}{L1} = \frac{100 \times (12 + 6)}{12} = 150 \, \text{lb} \cdot \text{ft}$$

where T = torque, lb·ft
  TS = torque seale, lb
  $L1$ = torque-wrench frame, in
  $L2$ = extension or adapter, in

Or if you know the recommended torque is 100 lb·ft, the torque-wrench scale reading may be found by using the following formula:

$$TS = \frac{T \times L1}{(L1 + L2)} = \frac{100 \times 12}{(12 + 6)} = 66.66 \, \text{lb} \cdot \text{ft}$$

Pointers for Using Torque Wrenches
1. By selecting the appropriate size and range torque wrench, the accuracy of the torque-wrench reading is improved. A good rule of thumb is to select a torque wrench with an adequate capacity for a working range within the central two quarters of the scale.
2. When the torque range is limited to 100 lb [45.36 kg], and you are required to torque a nut to 150 lb·ft [20.70 kg·m], use an extension of the same length as the torque wrench. This gives you double lever length and, therefore, your torque scale also doubles to 200 lb·ft [27.60 kg·m]. (See Fig. 46-12.)
3. Clean the threads of both components thoroughly, and restore the threads when necessary. Remove water, oil, or dirt for the holes.
4. Check the manufacturer's specifications to determine whether a antiseize lubricant should be used or whether a dry torque is recommended.
5. When the rundown resistance is high due to a self-locking device or a damaged thread, take the resistance measurement at the last rotation and add it to the specified torque.

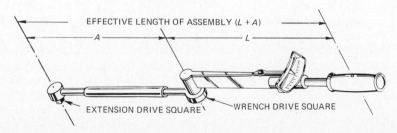

Fig. 46-12 Making a torque wrench multipurpose and multirange. *(International Harvester Co.)*

6. Follow the manufacturer's recommended sequence when tightening. First, torque each bolt to half of the desired torque. Then tighten the bolts to the three-quarter point. Finally, torque each bolt to the specified torque. The important point here, of course, is not to completely torque a bolt in one movement. If sequence procedures have not been outlined by the manufacturer, then you should start tightening from the center in a circular or diagonal pattern to the outside.

7. To check the torque, use the breakaway method. This means you should loosen the fastener and take a reading. The difference between this reading and your applied torque should be about 10 percent.

8. When retorquing, loosen the fastener, and then torque to specification. This method is important because a torque reading from a static position of the fastener is usually higher than the torque of the original tightening.

9. When using a torque multiplier, multiply the ratio by the scale reading to determine the torque of the fastener.

10. It is advisable to rest your palm on the drive square to equalize the force you apply to the handle. Failure to support the head end of the torque wrench will result in the socket rocking off the fastener.

Hex-Head and Multispline Wrenches   As a diesel engine mechanic, you will require wrenches in both inch and metric sizes to loosen or tighten set screws and cap screws. Both sizes are used on fuel-injection equipment (Fig. 46–13).

Pipe and Adjustable Wrenches   Both of these wrenches are useful tools, but they should be used only for the purposes for which they were designed. The pipe wrench, for instance, was designed to hold or loosen round stock. It should never be used on bolts, studs, or nuts. An adjustable (crescent) wrench should only be used to loosen a bolt, nut, or fitting when no other wrench fits.

Pointers for Using Pipe Wrenches

1. Select the proper size pipe wrench for the job.
2. Do not use a pipe over the handle for additional leverage as the stress may damage your wrench.
3. Use only pipe wrenches with sharp jaws.
4. Adjust the wrench so that the object to be torqued is held by the center area of the jaws.

5. To prevent lost motion or slippage at the beginning of the turn, press the adjustable jaw against the object. This ensures that the wrench will bite and hold at once.

Pointers for Using Adjustable Wrenches

1. Use an adjustable wrench only when other wrenches are not available. Be prepared to have this wrench slip.
2. Adjust its jaws to the fastener.
3. Apply the force so that the fixed jaw will carry most of the torque.

Stud Removers   Special tools have been designed to remove studs quickly and easily. The most common stud removers are shown in Fig. 46–14. Note that the collet type, when properly applied, will not bend or damage the thread or stud, whereas the wedge type may do so.

How to Use Collet-Type Stud Removers

1. Screw the tapered collet onto the stud.
2. Slide the housing over the collet and tighten the nut. This will draw the collet into the housing and lock the puller tightly onto the stud.
3. Use a wrench, preferably an impact wrench, to remove the stud.

How to Use Wedge-Type Stud Removers

1. Slip the wrench over the stud and push the wedge into the bore until it contacts the stud. As the wrench is turned, friction forces the wedge tight against the stud, holding the stud extractor to the stud bolt.
2. To install the stud, follow the same procedure, but push the wedge in from the other side. When no stud removal tools are available, simply screw two nuts onto the stud and jam them tight. Then use a wrench on the lower nut for removal. To install the stud use the upper nut. Make sure the nuts are tight to prevent slippage and resulting thread damage.

Removing Broken-off Bolts, Studs, or Screws   A simple but not always successful way of removing a stud broken above or below the surface is to use a center punch and hammer. Gentle tapping will force the broken object out by turning it in coun-

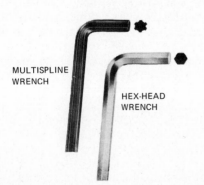

MULTISPLINE WRENCH

HEX-HEAD WRENCH

Fig. 46–13 Wrenches used to tighten and loosen set screws and cap screws.

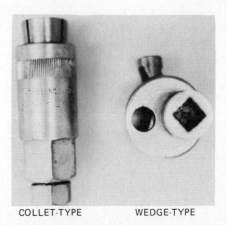

COLLET-TYPE   WEDGE-TYPE

Fig. 46–14 Typical stud removers.

Fig. 46–15 Taper-bit screw extractors.

FINE
LEFT-HAND
TWIST   SQUARE
COARSE
LEFT-HAND
TWIST

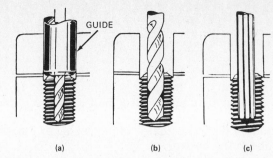

GUIDE

(a)   (b)   (c)

Fig. 46–16 Procedure for extracting a stud that breaks below the surface. (a) Center-punch stud and drill pilot hole; (b) enlarge hole; (c) insert extractor.

CAUTION Do not blow out a broken stud, tap, etc., with a cutting torch because particles of the molten metal will fuse with the threads.

terclockwise direction. Once a part of the object extends above the surface, take hold of it with a pair of pliers or vise grip and work it back and forth to loosen it. Removal is sometimes aided by a hard hammer blow on the object or by the application of penetrating oil or moderate heat.

Some other methods of removing a bolt, stud, or screw when the pliers cannot be used are: (1) weld a nut to the broken bolt or screw, (2) cut a slot with a hacksaw and then use a screwdriver to remove the bolt, (3) file the protruding part so that an adjustable wrench can be used to remove the bolts.

Many different types of extractors are on the market to remove bolts, screws, taps, or pipes which have broken off at or below the surface of a piece of work (Fig. 46–15). Although some are more effective than others, they all have one common factor: in every case, a hole must be drilled into the object and the extractor then forced or threaded into the hole.

Pointers for Removing Broken-off Bolts, Studs, or Screws

1. Blow the hole or surface clean. Use goggles to protect your eyes. It is a good practice to apply penetrating oil to the broken studs, etc., and allow it to soak into the threads before attempting to remove them.

2. Center-punch the object and drill a pilot hole. If the object is broken below the surface, be sure to use the correct size guide to ensure centering of the hole (Fig. 46–16).

3. Enlarge the hole sufficiently to insert the extractor tool. It is wise to drill the hole over the total length of the object to reduce stress on the threads.

4. When using extractors similar to those shown in Fig. 46–15, you must drive the extractors into the drill hole, and then turn them counterclockwise to remove the threaded object.

5. If you use the extractor shown in Fig. 46–16, you must tap the extractor lightly after you have inserted it into the drilled hole, and then turn it counterclockwise to remove the threaded object.

Tap Extractors Many times during your employment as a mechanic you may break a tap. It is possible to remove the broken tap by heating it to take out the temper. After this, the procedure previously outlined should be followed.

Work carefully when restoring the threads with a tap because the tap or threads are easily damaged.

Twist Drills Although twist drills are always to be found in the shop tool crib, you can save yourself many steps, and much time, by having a small kit of your own. There are many different designs and sizes, but those most useful to the diesel mechanic are the high-speed conventional twist drills of the straight- or tapered-hank design.

The value of any tool or instrument largely depends on your own performance and the manner in which you have prepared the tool. Any twist drill can cut effectively if you have sharpened it to suit the material to be drilled. You must also adjust the speed at which it will operate according to the size of the drill and the material to be drilled. Proper and adequate lubrication is another essential factor.

Drill Sizes Standard drill sizes may be either in inches or millimeters. Taper-shank drills in fractional inch sizes run from ⅛ to 1¾-in diameter by increments of sixty-fourths, to 2¼-in diameter by increments of thirty-seconds, and to 3½-in diameter by increments of sixteenths. Straight-shank drills run in increments of sixty-fourths up to 1¼-in diameter, in increments of thirty-seconds up to 1½ in, and in increments of sixteenths up to 2-in diameter.

Sizes smaller than ⅛-in diameter are not regularly made with tapered shanks but rather with straight shanks. They are made in so-called "jobbers" lengths ranging from 1/64 to ½-in diameter in increments of sixty-fourths. Jobbers drills are shorter than regular drills. This lessens the risk of breakage.

Both straight and tapered shanks are available in letter size drills from A (0.234-in diameter) to Z (0.413-in diameter). Straight-shank drills are available in wire gauge or number sizes from No. 80 to No. 1 (0.0135 to 0.2280-in diameter).

Metric-size drills are available in 0.75 to 77-mm diameter [0.295 to 2.9921 in].

Drill Sharpening The importance of correct repointing or sharpening of twist drills cannot be too strongly emphasized. A drill, no matter how well designed and heat-treated, will fail in its performance if point-ground incorrectly. It is safe to say that by far the largest percentage of failures experienced in drilling is due to incorrect repointing of drills.

Generally speaking, it is not possible to sharpen and drill by hand with sufficient accuracy. To get the maximum efficiency, drills should be resharpened with properly designed drill-grinding machines. This is especially true of drills over ⅜ in [9.52 mm] in diameter. You should, however, by using only a drill gauge, be able to hand-sharpen a drill ½ in [12.7 mm] in diameter and under as proficiently as any machine, once you gain experience.

Procedure for Drill Sharpening Use a soft, medium-grain grinding wheel for pointing drills. Before the drill is ground, the wheel should be dressed on the face to ensure that it is flat and sharp. Drills should be ground on the face of the wheel, not on the side. All point grinding should be done dry. If during grinding the drill is accidentally allowed to get too hot, it should never be cooled in cold water. It should be allowed to cool of its own accord. A sudden cooling is almost sure to produce grinding checks.

When you are repointing a drill, four features must be carefully considered: point angle, lip relief, cutting edges, and web thinning.

Drill Point Angle The proper point angle, as measured from the cutting edges to the axis of the drill, depends on the material being drilled. A point angle of 118° (Fig. 46–17a), has been found to be satisfactory for the average class of work. In general, it can be said that the harder the material, the greater the point angle should be. For example, a point angle of 136° is recommended for manganese steel (Fig. 46–17b), while for wood, fiber, and similar materials, a point angle of 60° is recommended (Fig. 46–17c). However, in the case of a smaller drill it has been found from years of experience that, in sizes of ¼ in [6.35 mm] and smaller, a 136° angle is more satisfactory for production work. Thus, this drill is furnished with this point, unless otherwise specified.

Lip Relief (Clearance) In order that the points may have effective cutting edges, the heel, or part of the drill point behind each cutting edge, is removed. This lip clearance may vary from 12 to 15°. If the clearance angle is greater, it weakens the cutting edge. If the clearance angle is smaller, there may not be sufficient clearance to allow the cutting edge to enter the work. This may split the drill up the center. The angle of clearance at the center or chisel point should be about 130° (Fig. 46–18).

Cutting Edges The two cutting edges of a drill must make exactly the same angle with the centerline of the drill. In other words, while the point angle may vary a trifle one way or the other, the variation must be uniform in regard to both cutting edges. A difference in angle will cause one edge to take a larger cut and, consequently, put an unequal strain on the two cutting edges. If the cutting edges are of unequal length, the result is that the chisel

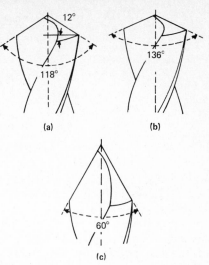

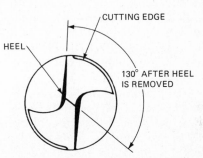

Fig. 46–17 Twist-drill point angles for (a) the average class of work, (b) manganese steel; (c) wood, fiber, and similar materials.

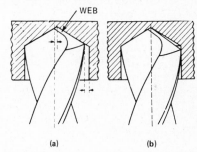

Fig. 46–18 Lip relief (top view) before removing the heel.

Fig. 46–19 (a) Before web thinning and (b) after web thinning.

point is off-center even though the point angle may be uniform on both edges. This condition will cause the drill to cut oversize.

Web Thinning As the web of a drill usually increases in thickness toward the shank, a web-thinning operation becomes necessary when the drill has been shortened by repeated grindings. This operation is essential in order to minimize the pressure required to make the drill penetrate. This point-thinning operation must be carried out equally on both sides of the web in order to ensure that the web will be central. Care must also be taken to see that the point thinning is not carried too far up the web, thereby weakening the drill. As a general rule, a web thickness at point of approximately one-eighth the drill diameter is recommended. Figure 46–19b shows a drill with the web thinned correctly.

Table 46-1 CUTTING FLUIDS

Metal	Drilling	Reaming	Tapping
Mild and medium steel	Soluble	Lard oil	Lard oil
Tool steel	Soluble	Lard oil	Lard oil
Brass and bronze	Dry	Soluble	Dry
Copper	Soluble	Soluble	Soluble
Aluminum	Kerosene	Lard oil	Soluble
Cast iron	Dry	Dry	Lard oil

Note Soluble oil is generally used 1 part oil to 40 parts water.

Drill Speeds The speed of a drill is the rate at which the periphery of the drill moves in relation to the workpiece, expressed in surface feet per minute (sfm). For convenience, this sfm is best converted to revolutions per minute (rpm). The correct rpm of a twist drill depends upon: the diameter of the drill, the kind of steel used in the manufacture of the drill, the kind of material being drilled, the quality of the hole desired, the way in which the workpiece is set up or held, and the quality of the machine itself.

To calculate the rpm of a drill, use the formula

$$\text{rpm} = \frac{4 \times \text{cutting speed}}{\text{diameter of drill}}$$

Recommended Cutting Fluids for Drilling, Reaming, and Tapping Perhaps even more important than the rpm of a twist drill is the cooling of the drill. If the drill is not properly cooled with the recommended cooling fluid, the cutting edge of the drill will wear faster and the hole surface will harden. Table 46-1 lists recommended cutting fluids to be used with various metals.

Reamers The reamers which you will use for servicing a diesel engine are of the straight, tapered, and adjustable design (Fig. 46-20). They are a multiple-tooth cutting tool and are used for hole sizing or smoothing. With this tool the diameter of an existing hole can be increased by 0.003 to 0.030 in [0.0762 to 0.762 mm].

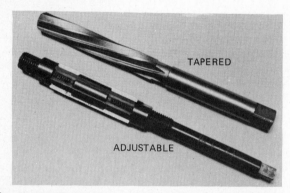

Fig. 46-20 Reamers.

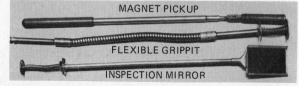

Fig. 46-21 Mechanic's helpers.

Reaming does not improve the angular alignment of drilled holes, rather it tends to follow the direction of the existing hole. It is, therefore, essential to make certain that the hole is located accurately before reaming. The drill should be well sharpened to prevent it from wandering.

Pointers for Using Reamers
1. Select the reamer to suit the job. When an adjustable reamer is used, adjust it to the drilled hole size *plus* 0.002 in [0.05 mm].
2. Lubricate the hole with cutting oil and insert the reamer.
3. Maintain a 90° angle to the hole. Gently turn the reamer clockwise. *Never* turn the reamer counterclockwise because this will chip the edges of the reamer. Make only small cuts from 0.001 to 0.002 in [0.025 to 0.05 mm].
5. Store your reamers in a container or some well-protected place when they are not in use.

Hand Cleaning Tools The first step in service maintenance procedure is to superficially clean the components to be serviced. This speeds up the repair job and prevents foreign matter from entering the components during disassembly. It is important to clean components thoroughly after disassembly so that they can be examined visually to determine the cause of failure.

Pointers for Using Cleaning Tools
1. Use only approved cleaning solvent (never gasoline) and keep all containers labeled. Keep the solvent tank covered when not in use.
2. When cleaning with caustic solution, use rubber gloves.
3. Always protect your eyes, especially when using hand- or power-operated wire brushes. Special protection shields are available for this purpose.
4. When using a scraper to remove gasket material (or any type of dirt) always keep your free hand *behind* the scraper.
5. Take care when using compressed air to remove foreign material, because it has great impact force.

Mechanic's Helpers In every vocation there are certain "tricks of the trade." In mechanics, speed and performance are very often aided by miscellaneous gadgets known as *mechanic's helpers*. For instance, suppose you lose a bolt in an engine and know only vaguely where it has dropped. With such helpers as an inspection mirror and a magnet pickup or flexible grippit, you can retrieve the bolt easily (Fig. 46-21). A mechanic's stethoscope can save time when trying to locate or diagnose any abnormal engine sound, noise, or knock. The screw starter is another useful device which will start a screw when your hand fails.

Questions

1. What factors should be considered when purchasing a set of tools?

2. What safety precautions must be taken when using a punch or chisel?

3. Name the three most common punches and state their applications.

4. Why should you *never* use a file without a handle?

5. Why should you lift the file on the return stroke?

6. Describe how to dress a ½-in regular screwdriver.

7. What safety precaution must be taken when using a screwdriver?

8. Why should you start a new cut after a hacksaw blade breaks?

9. What must you consider when selecting a hacksaw blade?

10. List a number of safety precautions which should be taken when using a hammer.

11. Describe how to restore the cutting edge of a chisel.

12. When is a speed handle used?

13. List a number of occasions when a pipe wrench is required.

14. Describe how to remove a stud bolt with a wedge-type stud remover.

15. Describe how to sharpen a twist drill to drill a hole into a broken grade 8 stud bolt.

16. Which wrench is used on a hollow-head screw: (*a*) open, (*b*) box, (*c*) torque, or (*d*) Allen?

17. Wrench sizes are governed by the (*a*) thread length, (b) length of the bolt, (*c*) width of the bolt head across the flats, or (*d*) width of the bolt head across the corners.

18. Diamond-point chisels are used to (*a*) cut V-shaped grooves, (*b*) cut round-shaped grooves, (*c*) cut square-shaped grooves, or (*d*) cut hardened nuts and bolts.

19. The number of threads per inch on a hex bolt can be determined by a (*a*) radius gauge, (*b*) screw thread micrometer, (*c*) thread pitch gauge, or (*d*) acme screw thread gauge.

UNIT 47

Shop Tools

The range and variety of shop tools available to you depends, of course, on the shop in which you are working. However, there are some tools which are standard in all service shops.

Steam Cleaner　The most efficient way to clean assembled engines or components is with a steam cleaner—dry cleaning is not nearly as effective. All steam cleaners, whether portable or stationary, operate on the same principle. Water, mixed with a cleaning solution, is forced under controlled pressure through a heating coil. It then passes into a special hose and out through a handle-type nozzle. The heating coil is located in an open furnace where it is heated by an oil or gas burner. By controlling the burner flame (and in some steam cleaners, the water flow), the desired temperature is regulated to produce a wet steam.

Pointers For Using Steam Cleaners

1. Steam-clean in an isolated area where there is good ventilation because steam will cause oxidation. Protect all components which could be damaged by the steam. Close all openings to prevent the entry of water or foreign matter.
2. Make certain that any electrical equipment is grounded properly.
3. Check supplies of fuel, water, and cleaning solution before igniting the steam-cleaning unit.
4. Wear safety gloves and goggles.
5. Make sure that the object to be cleaned is raised off the ground to ensure better cleaning.
6. Ensure that the water flow from the nozzle is free of air before you ignite the fuel, otherwise steam pockets will form in the heating coil. Steam pockets can prevent water from flowing and cause damage to the coil.
7. Adjust the flame heat to the desired steam pressure by controlling the fuel flow.
8. Vary the nozzle distance to achieve the greatest cleaning action.
9. When the object is clean, shut off the supply of solution first and, a short time later, shut off the fuel valve.
10. Do not try to wash the object with cold water because this will increase the possibility of corrosion.
11. Do not shut down the steam cleaner until cool water leaves the nozzle. This will help to prevent damage to the heating coil.
12. When you are done, clean the work area and store the equipment carefully, especially the hose and electrical wires.

Hot Tank　Another method of cleaning is the hot tank method. This method is increasing in popularity due to its effectiveness. The cleaning solution is simply a mixture of water and caustic soda which is

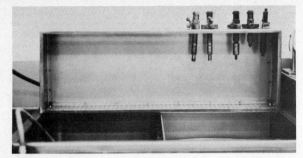

Fig. 47–1 Two-sectional cold tank with built-in pump and filter element.

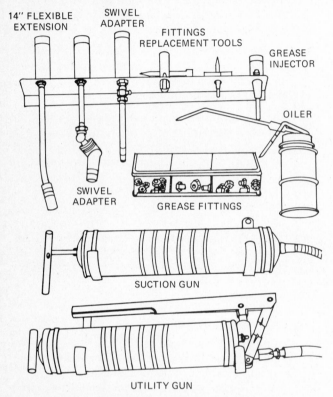

Fig. 47–2 Standard lubrication equipment.

heated to a temperature of 200°F [93°C]. There are many hot-tank cleaning solutions on the market, but close attention should be paid to manufacturers' warnings. Some solutions are irritating to the skin and some are very poisonous. Eye protection must *always* be worn.

Pointers for Using Hot Tanks

1. Follow the manufacturer's operating instructions regarding the solution to be used.
2. Protect your hands with rubber gloves. Shield your face.
3. Lower the components gently into the tank to avoid splashing.
4. After the component has been removed from the hot tank, wash it thoroughly to remove all the solution.
5. Finally, the clean component must be protected from surface oxidation. Either immerse it in an oil solution or use an air-operated spray oiler.

Cold Tank As a temporary measure any container and cleaning solvent may be used to clean compo-

nents. However, inferior containers can, in time, contaminate the solution and reduce its cleaning power.

There are many well-designed cold-tank cleaners available which guarantee an excellent cleaning job. An example of one is shown in Fig. 47–1. This cold tank consists of two separate compartments which hold 6 gal [22.71 l] of solution. One side can be used as a soaking tank. This cleaner also has a filter element, hydraulic pump, and a spray nozzle to clean areas which are difficult to reach.

Pointers for Using Cold Tanks

1. Use a recommended solvent with a high flash point. Never smoke or approach the tank with an open flame or spark.
2. Clean parts or components with a scraper or wire brush before placing them in the solution.
3. Keep the solution and tank clean. Change the filter and the fluid regularly to prevent damage to the pump and electric motor.
4. Keep the tank lid closed when not in use to prevent accidental explosion.
5. Protect your hands with rubber gloves or wash your hands thoroughly after the cleaning job is completed.
6. Always wear eye protection.

Sonic Cleaner A sonic cleaner is ideal for cleaning smaller parts and components such as the injection equipment.

Lubrication- and Fuel-Supply Tools The type of lubrication-supply tools used will depend on the type of service your workshop supports. However, the equipment illustrated in Fig. 47–2 is standard.

SPRING AND PISTON/TYPE OILER Suction guns and grease guns are tools used in the lubrication of component parts during assembly. Suction guns are used for both draining and refilling filters, gear cases, and cylinders. Grease guns may be hand- or power-operated. Various grease guns with the necessary adapters are shown in Fig. 47–2.

Pointers for Using Grease Guns

1. Select the correct coupler to complement the grease fittings.
2. Use an extension when necessary to ensure proper coupling fitting.
3. Clean and examine the grease fitting before you push the coupling on.
4. Check the specifications for both low- and high-pressure grease guns to prevent overgreasing.
5. When you remove the coupling, try to use a straight pull. Do not twist because this will damage the coupling tang and seal.
6. Wipe off the excess grease.

Lubrication and Fuel Dispenser Time will be saved and cleanliness will be ensured if the service shop has a separate dispenser for each liquid used. The dispensers may be hand- or power-operated.

When using air-operated lubrication- or fuel-supply tools, be sure to use a pressure regulator, moisture-separator, lubricator, and nylon retrocoil airhose.

Be sure to set the regulator to the recommended pressure to avoid damage to valve, piston, and seals.

Glass Blaster Some workshops use a glass-blasting cleaning tool to quickly remove rust, paint scale, or any other surface coating. This tool operates on the principle of air, controlled by a valve, flowing into and around a venturi. Small glass beads or pearls are drawn into the stream of air and forced through a specially designed blasting gun against the surface to be cleaned. By regulating the airflow, the amount of air and glass pearls impinging on the surface is readily controlled.

Pointers for Using Glass Blasters
1. Make certain that the air pressure is adjusted to the specified pressure and that the component to be cleaned is free of dirt and is dry.
2. Be sure that the recycling filter is clean.
3. Use masking tape to protect all surfaces which are not to be cleaned.
4. Do not accidentally spray the glass window in the cabinet. Such action will damage the surface of the glass and reduce your vision.
5. Upon completion of the cleaning operation, make sure that all the glass dust is removed from the components.

Hydraulic Jacks Statistics indicate that most accidents in a service shop are related to lifting, pushing, or pulling. In the majority of instances these accidents could have been avoided if normal safety rules had been obeyed. The following common sense rules should always apply:

1. Do not play Hercules. Lift or carry no more than safety boards recommend.
2. When you lift a load, make certain your back is straight and your legs are bent. Let the strength of your legs help you lift the weight.
3. Make sure that the load you are about to carry does not obstruct your vision.
4. Do not twist your body while lifting or carrying a load (move your feet in a straight line position).
5. When you lift sharp-cornered objects, use work gloves to protect your hands. It is wise to remove rings from your fingers to prevent them from getting caught on some protrusion and possibly ripping off your finger.

HAND, FLOOR, AND PORTABLE JACKS Mechanically, hydraulically, and air-operated jacks are designed to make lifting easier. You should be aware of their full capacity. It is even more important that you know the recommended procedures for using them.

Pointers for Using Hydraulic Jacks
1. Always clean the floor and surrounding area before you use a floor jack or portable crane.
2. Select the jack which suits the particular lifting situation.
3. In the interest of safety, position the jack in the correct location. For instance, never place a jack under an oil pan, radiator, clutch housing, or any other area which could cause component damage or unsafe lifting conditions.
4. Place the jack square to the component to be lifted and use a suitable saddle or head to ensure safe lifting.
5. Lift only as high as necessary and support with safety stands or cross blocking. *Never* work on or under an object which is supported only by a jack, crane, or any other lifting device.
6. Before you place safety stands under the object, make certain that the stands can hold and support the load. In some cases it is wise to place wood between the jack or stand and the object to be lifted. This will increase friction and prevent slippage.
7. Never remove the jack or lift an object from the block until you have checked that the work area is clear.
8. When using a portable crane, block and tackle, or chain hoist, lower the load as much as possible before moving it into position.
9. Never leave the load hanging unprotected while you are lowering it into position. It may release accidentally.
10. Always use the correct type of lifting cable, chain, or manilla rope. Never use a synthetic rope.

Hydraulic Presses A floor or portable hydraulic press is essential to any service shop because it is so versatile (Fig. 47–3). It can be a real time-saver when gears or hubs are to be removed. Also, the controlled force on the object being pushed or pulled is evenly distributed, thereby eliminating shock load and vibration.

Pointers for Using Hydraulic Presses
1. Before you use a hydraulic press, clean the surrounding area of any obstructions.
2. Adjust the table to suitable working height. Secure it firmly and make certain that there is slack in the winch cables.
3. Support the object being pulled or pushed by using a pulling attachment or accessory. Remember that the center of the hydraulic ram must be in the center of the object to be pulled or pushed.
4. Screw out the extension and bring it in contact with the object. Apply moderate pressure and check the alignment.
5. Bearings, gears, or any other object which can break during press action should be wrapped with a cloth in order to protect you in the event of breakage.
6. In order to prevent damage to pressed out object (which may eject with great force), let it fall into a pail of sawdust or onto a slab of wood.

Pullers and Attachments Nearly all service shops possess some type of puller set with attachments. Figure 47–3 shows some of the more common pullers. Never substitute a punch, hammer, or prybar for a puller. By using the wrong tool, the job will take longer and you may damage the component.

Tool manufacturers have designed a great variety of pullers and attachments to take care of the three basic pulling problems: pulling gear-bearing pulleys, etc., out of shafts (either to replace them or to

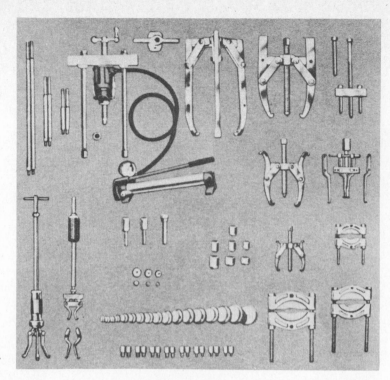

Fig. 47–3 Typical pullers and attachments. *(Owatonna Tool Co., Tools and Equipment Div.)*

get at other worn-out parts); pulling press-fitted internal bearing cups, retainers, or oil seals out of a housing; and pulling shafts out of a housing.

One of the major difficulties when removing or installing press-fitted parts is getting a firm grip on the parts themselves. Precision gears, bearings, races, etc., whether very small or very large, are delicate pieces of machined steel. They must be gripped in a manner that will not damage them, yet the grip must be firm enough to prevent the puller from slipping off. Unfortunately these parts are often located in places which are extremely difficult to reach.

The proper grip is not the only aspect of pulling. Force is equally important. Not only do you need enough force to remove press-fitted parts that may be frozen or rusted in place, but the force must be applied properly. This means that the pull must be perfectly straight. If a gear, bearing, or bearing cup is cocked by an improperly applied pulling force, it will be damaged during removal if it can be removed at all!

Some basic methods of gripping and pulling bearings, gears, pulleys, etc., are shown in Fig. 47–4. Figure 47–5 shows how to grip and pull bearing races and retainer seals, and Fig. 47–6 illustrates how to grip and pull shafts.

How to Select the Right Puller

1. Analyze the area of resistance, or area of press fit. It can vary greatly between seemingly similar jobs.

2. Select a pulling screw which is at least half as large in diameter as the shaft of the object to be pulled. When using a hydraulic puller the maximum force exerted, in tons, should be 7 to 10 times the diameter of the shaft in inches (see Table 47–1).

3. Select the proper shaft protector or step plate. Determine which puller type is best for getting a grip

on the part. Perhaps a combination of pullers may be required.

4. The puller you select must have a reach equal to, or larger than, the corresponding dimension of the object to be pulled.

5. Measure the width of the part to be pulled. It will determine the required spread of the puller. A

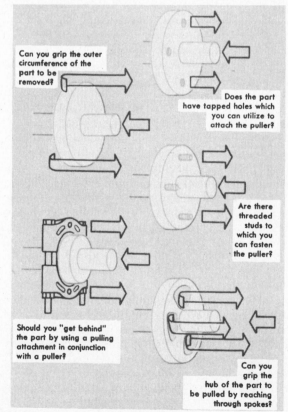

Can you grip the outer circumference of the part to be removed?

Does the part have tapped holes which you can utilize to attach the puller?

Are there threaded studs to which you can fasten the puller?

Should you "get behind" the part by using a pulling attachment in conjunction with a puller?

Can you grip the hub of the part to be pulled by reaching through spokes?

Fig. 47–4 Pointers for gripping and pulling gears, bearings, wheels, etc. *(Owatonna Tool Co., Tools and Equipment Div.)*

Table 47-1 HYDRAULIC-PULLER FORCE RELATED TO SHAFT DIAMETER

Shaft diameter, in	OTC hydraulic-puller ton range
0–2	17½
2–3½	30
3½–5½	50
5½–10	100

puller with proper reach and spread will usually be strong enough. When in doubt, always use the next largest size. A larger size may be needed for rusted parts or when the area of resistance is large.

Pointers for Using Pullers
1. Make sure you have as much room as possible when preparing a large pulling job.
2. Lubricate the pulling screw and the center of the step plate or shaft protector to reduce friction.
3. Check for alignment and grip after the slack has been taken up.
4. When necessary, use penetrating oil or moderate heat for rusted parts or when the area of resistance is large.

CAUTION Do not overlook the possibility that the object to be pulled could eject with a "bang" from the shaft or bearing.

Bushing Inserter and Removing Tools
Bushing inserter and removing tools are essential to every diesel service shop to ensure proper bushing removal and installation. These can also be used for the installation of oil seals and many other press-fitted components.

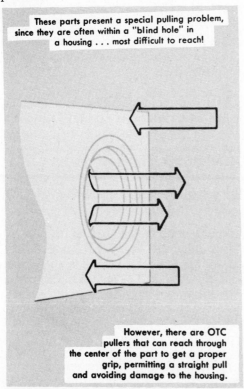

These parts present a special pulling problem, since they are often within a "blind hole" in a housing . . . most difficult to reach!

However, there are OTC pullers that can reach through the center of the part to get a proper grip, permitting a straight pull and avoiding damage to the housing.

Fig. 47-5 Pointers for gripping and pulling bearing races, retainers, seals, etc. *(Owatonna Tool Co., Tools and Equipment Div.)*

Electric Hand and Bench Drills
Although it is daily routine to use a hand or bench drill in conjunction with a twist drill to drill holes, do you realize how often you use them for other purposes? They may be used as the power source for the wire brush or the rotary file, and also as hole-cutter saws. Bench and slow-speed hand drills can also be used as the power source to rotate reamers, taps, and dies. Learn to use these tools to their maximum capacity.

Pointers for Using Hand and Bench Drills
1. Secure the object to be drilled so that it cannot rotate with the drill, if the drill should grab. A vise should be used whenever possible.
2. Make sure that your work clothes are buttoned up so that they cannot be caught in the drill.
3. Make sure that you use a three-prong plug connector. Remember, *never* use power tools if your feet are wet.
4. Make sure that the twist drill is sharpened properly and the hole to be drilled is center-punched.
5. Whenever possible select the correct speed. Alternatively, you can start with a slow speed and light feed and gradually increase both until the drill forms the correct chips. Usually the most efficient operation results from a heavy feed and moderate speed.
6. Use cutting oil to prevent surface hardening. This is also useful in cooling the drill. Cooling the

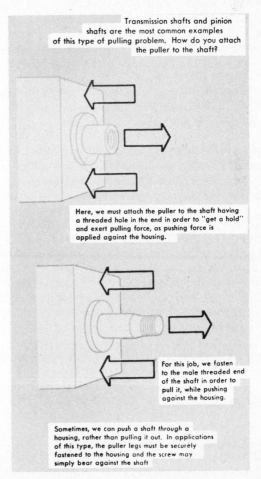

Transmission shafts and pinion shafts are the most common examples of this type of pulling problem. How do you attach the puller to the shaft?

Here, we must attach the puller to the shaft having a threaded hole in the end in order to "get a hold" and exert pulling force, as pushing force is applied against the housing.

For this job, we fasten to the male threaded end of the shaft in order to pull it, while pushing against the housing.

Sometimes, we can push a shaft *through* a housing, rather than pulling it out. In applications of this type, the puller legs must be securely fastened to the housing and the screw may simply bear against the shaft

Fig. 47-6 Pointers for gripping and pulling shafts. *(Owatonna Tool Co., Tools and Equipment Div.)*

drill makes the cutting operation easier and helps to retain the cutting edge. The use of a cutting fluid will improve the hole finish and the cooler chips will minimize the danger of burns. Use kerosene, kerosene and lard oil, or soluble oil when drilling aluminum bronze. For cast iron, use soluble oil or air pressure. When drilling holes in steel, tool steel, or steel forgings, use soluble oil or mineral lard oil.

Hand and Floor Bench Grinders and Hand Sanders Grinding or sanding is the cutting action of thousands of sharp abrasive grains on the face of the grinding wheel or disk against the object being ground. The abrasive grains actually cut chips out of the component being ground.

Selection of Grinding Wheels Grinding wheels and sanding disks come in many different sizes and shapes to meet a multitude of demands. Your concern, however, is with an abrasive grain which suits the various types of metals you will be handling. The two types of abrasive grains are: aluminum oxide and silicon carbide.

Grinding wheels are identified by a color code. The regular aluminium oxide grinding wheel is red. It is acceptable for grinding all types of steel. The special aluminium oxide wheel is white. Although it is not as tough as the regular grinding wheel, it is used for grinding tools and high-speed steel. The regular silicon carbide is green and is used for grinding cast iron, brass, copper and aluminium.

The object being ground determines the grain size to be used. It is advisable to have one coarse and one medium grinding wheel for the bench grinder. You can grind most soft materials with the coarse grinding wheel and most hard materials with the medium grinding wheel.

Testing and Mounting Grinding Wheels Before mounting a grinding wheel, test the wheel for damage by inserting (through the center hole) a screwdriver and tapping the wheel every 45° around its circumference. A sound, undamaged wheel will give a clear metallic tone. If it is cracked, there will be a dull sound.

When mounting the wheel, make sure that the flanges are undercut and that the wheel goes easily onto the spindle. Do not force the wheel onto the spindle, instead, scrape the bushing hole to a larger size. After installation or after the wheel becomes loaded, glazed or out-of-round, the wheel should be dressed with a star wheel dresser. Work the dresser freehand, but always use the work rest to support the dresser. Adjust the work rest as close as possible to the wheel. The maximum distance should not exceed ⅛ in [3.18 mm]. This will prevent the work from being caught between the wheel and the rest.

Pointers for Using Grinders or Sanders
1. Wear safety goggles at all times when performing any operation with a sander or grinder.
2. Make sure that all safety equipment (such as guards) is in place.

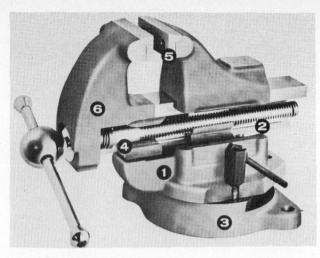

1. Heavy horn extension
2. Buttress thread on steel screw
3. Nonslip swivel base
4. Spring screw fastener
5. Jaw facing
6. Front jaw

Fig. 47–7 Bench vise.

3. Hold the work securely. Do not let the work hammer or hit against the grinding wheel because the wheel may explode.
4. Keep your shirt sleeves buttoned or rolled up. Avoid all personal contact with the moving wheel.
5. Stand to one side when starting the grinder. Allow the wheel to reach full rpm before using it.
6. Do not grind on the side of the wheel unless it was manufactured for that purpose.

Bench Vise Each mechanic should have a working area and a vise if unnecessary steps are to be saved. The vise (Fig. 47–7) is the proverbial workhorse. Without it, many service jobs would be very difficult or even impossible.

Pointers for Using Bench Vises
1. Keep the vise clean and securely supported on the workbench. Check that the jaws are tightened.
2. Do not expose the jaws of the vise to excessive heat. *Never* use the vise as an anvil.
3. Do not use an extension over the vise handle. Do not use a hammer to increase jaw force. The latter can break the spindle, nut, or jaws.

Impact Wrenches Impact wrenches should be standard equipment in every service shop as time- and labor-saving devices. Whether they are air- or electrically operated will depend on the type of service the workshop has to perform. Many mechanics do not use the impact wrench to its fullest advantage. You will find it a most useful tool even if you only use it to tighten a few bolts or nuts.

Pointers for Using Impact Wrenches
1. Keep this tool in good operating condition by using an airline filter and oiler.
2. When possible, use a recoil airhose for safer and easier operating and handling.

3. Use only impact sockets and extension.

4. When using an air-impact wrench, adjust the air-valve regulator to achieve the desired torque.

Questions

1. List the safety precautions you must take when using a steam cleaner, a hot tank, and a solvent cleaner.

2. List the safety precautions you must take when using a grease gun.

3. What safety precautions must you take when using a floor jack, a chain hoist, and a portable crane hoist?

4. Describe how to operate a frame shop press. What safety precautions must be taken while operating it?

5. What factors govern the selection of the correct puller?

6. List the safety precautions you must take when using a hand drill, a bench drill, a hand sander or grinder, and a bench grinder.

7. Give the color code of the grinding wheel which is best suited for aluminium and tool steel.

8. List the reasons why a new grinding wheel should be tested before being installed; a new grinding wheel should be dressed after installation.

9. List the circumstances under which you should *not* use the aid of a vise.

10. List the safety precautions to be taken when using an impact wrench.

UNIT 48

Measuring Tools

All service and repair jobs require analysis, diagnosis and measurement, to determine the cause of the trouble. After disassembly, a regular series of steps are repeated to determine each component's serviceability. When you reassemble and install components to the engine, you must make exact measurements to ensure proper operation.

It is obvious then, that knowing how to use and read the various measuring tools is of vital importance. All measuring tools used by you (with the exception of the protractor) measure distance in inches or thousandths of an inch [millimeters or hundredths of a millimeter]. It is essential for you to be able to interpret both measuring scales.

Selection and Care of Measuring Tools Earlier in this textbook it was suggested that for long-range accuracy and performance you should buy only high-quality tools. This statement is even more relevant when buying measuring tools. Also, careful handling and proper storage cannot be overemphasized. Measuring tools are delicate instruments and their precision is quickly affected by mishandling or improper storage. Therefore, keep all tools that are not in use lightly oiled. A padded box or cloth wrapping will help minimize exposure to atmospheric air, thereby preventing oxidation.

Some measuring tools are not precision instruments, but they are still to be listed among the essential tools for your toolbox. The main items are the 6- or 12-in steel rule, straightedge, feeler gauge, divider, and inside and outside calipers. You

should also have access to spring scales, screw pitch gauges, and a pocket knife.

Reducing Common Fractions Dimensions used on engine or fuel-injection systems or assembly operations are given in terms of common fractions or decimal fractions. As it is often necessary to change a common fraction into a decimal fraction or vice versa, you should memorize the following formula: divide the numerator by the denominator.

For example, to change $3/8$ to a decimal fraction:

- Divide the numerator 3 by the denominator 8.
- Place a decimal point after the 3.
- Locate the decimal point in the quotient.
- Add as many zeros as are needed to obtain a quotient which can be rounded-off to the required number of decimal places.
- Divide. The result is the decimal fraction equivalent of the common fraction $3/8$.

$$\begin{array}{r} 0.375 \\ 8\overline{)3.000} \\ \underline{2\ 4} \\ 60 \\ \underline{56} \\ 40 \end{array}$$

Reducing Decimal Fractions to Common Fractions

- Write the number after the decimal point as the numerator of a common fraction.
- Write the denominator as 1 with the same number of zeros after it as there are digits to the right of the decimal point.
- Reduce resulting fraction to lowest terms.

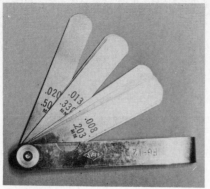

Fig. 48–1 Feeler gauge marked in hundredths of a milli-meter and thousandths of an inch.

For example, to change 0.3125 to a common fraction:

• Write number after the decimal point as the numerator: 3125.
• Determine the denominator: 1 + 4 zeros = 10000.
• Reduce the fraction to lowest terms:
3125/10000 = 5/16.

6-in Steel Rules The steel rule should have English measurements in thirty-second and sixty-fourth graduations on one side and metric measurements in half-millimeter graduations on the other side.

Straightedge As its name implies, a straightedge is used to check the surface of the flywheel, cylinder head, or a manifold flange for flatness. Handle this tool with care. Do not use it for any other purpose than that for which it was designed.

Feeler Gauge A feeler gauge or thickness gauge is a fixed distance measuring gauge with its thickness marked in thousandths of an inch or hundredths of a millimeter. It is wise to have a set of no-go feeler gauges as well as a standard set and a millimeter set (Fig. 48–1).

Screw Pitch Gauge A most useful tool to quickly determine the pitch of various threads is the screw pitch gauge. Each of its folding leaves has teeth corresponding to a definite pitch. By matching the teeth with the thread on the work, the correct pitch can be read directly from the leaf.

Calipers and Dividers Calipers are used to measure chambers, cavities, bores, grooves, flanges, or shafts (Fig. 48–2). The distance between the caliper points is then measured with a rule. For even greater precision, measurements can be made with a micrometer (see the section on Micrometers later in this unit). A divider is different from a caliper in that it has pointed legs and therefore is a handy tool for layout work, surface measurements, etc. (Fig. 48–3).

Protractor A combination square, combination set, and bevel protractor are among the most useful tools you can have if your work demands the mea-

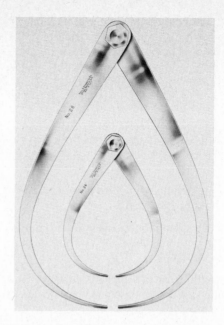

Fig. 48–2 Outside calipers. *(The L. S. Starrett Co.)*

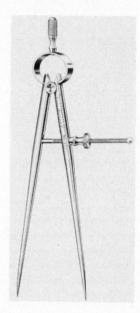

Fig. 48–3 Divider. *(The L. S. Starrett Co.)*

suring of levels and angles during engine installation.

How to Use Protractors To measure an object for its horizontal installation, hold the square head on the surface to be measured. The air bubble in the spirit-filled glass tube will indicate if the object is parallel to the horizontal. To find the exact measurement of this angle in degrees, lay the protractor on the surface to be measured and then turn the degree ring so that the air bubble is in the middle of the two graduation marks. The graduation scale on the protractor head will show the tilt of the object in degrees and minutes.

Spring Scale When deciding whether to purchase a spring scale, remember that a graduation in pounds and ounces [kilograms] is useful when measuring torque or the resistance of a spring.

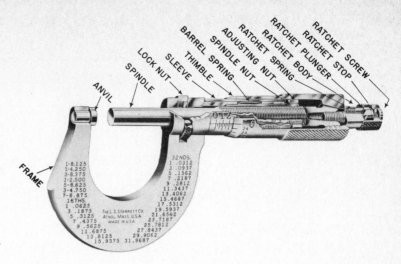

Fig. 48–4 Cutaway view of a micrometer. *(The L. S. Starrett Co.)*

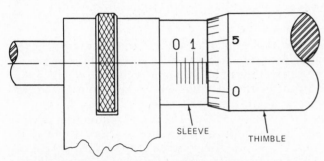

Fig. 48–5 Reading a micrometer graduated in thousandths of an inch. *(The L. S. Starrett Co.)*

Precision Measuring Tools The most common adjustments to components are in measurements of thousandths or ten-thousandths of an inch [millimeters or hundredths of a millimeter]. The instruments with which you can measure such precision are the micrometer, the dial indicator and the vernier caliper.

Micrometers All micrometers have an accurately ground screw which rotates in a fixed nut (spindle nut in Fig. 48–4). This opens and closes the distance between two measuring faces on the ends of the anvil and spindle or between the spindle and the object, or between one fixed and one movable spindle. The pitch of the screw is $^{1}/_{40}$ in or 40 threads per inch. With one complete revolution of the thimble, the spindle will move outward or inward precisely 0.025 in. The sleeve is divided by vertical lines into 40 equal parts that correspond to the numbers of threads on the spindle; each vertical line designates 0.025 in. For faster and easier reading, the manufacturers of micrometers have increased the vertical lines by three different lengths. Every fourth line, which is the longest line, designates 0.100 in. For example, the line mark "1" represents 0.100 in, and the line mark "2" represents 0.200 in. Every short line represents 0.025 in, and every middle length line represents 0.050 in.

As you can see from Fig. 48–4, the thimble is bevelled. It is divided into 25 equal parts with each line representing 0.001 in. Rotating the thimble from one of these lines to the next line will move the screw and spindle longitudinally 0.001 in.

For speedier reading, every fifth graduation (on the thimble) has a longer line and is numbered 0, 5, 10, and so on. Therefore, when the thimble is rotated from No. 0 to No. 15, the spindle has moved 0.015 in or, when the thimble has made a complete revolution, the spindle has moved 0.025 in.

How to Read Outside Micrometers Graduated in Thousandths of an Inch (0.001 in) To read the micrometer graduated in thousandths of an inch (0.001 in), multiply the number of vertical divisions visible on the sleeve (past the 0 mark) by 0.025. Add to this the number on the thimble which coincides with the longitudinal line on the sleeve. As an example see Fig. 48–5:

- Seven lines are visible. Multiplying 7 by 0.025 equals 0.175 in.
- The third line of the thimble coincides with the longitudinal line on the sleeve. Therefore, multiplying 3 by 0.001 equals 0.003 in.
- Adding 0.175 to 0.003 equals 0.178 in.

There is a second method to determine the micrometer reading shown in Fig. 48–5:

- Read the last number on the sleeve: 0.100 in.
- Multiply each additional line shown on the sleeve by 0.025: $3 \times 0.025 = 0.075$ in.
- Read the number which coincides with the longitudinal line on the sleeve: 3 = 0.003 in.
- The total reading is $0.100 + 0.075 + 0.003 = 0.178$ in.

How to Read Outside Micrometers Graduated in Ten-thousandths of an Inch (0.0001 in) Micrometers graduated in 0.0001 in are similar to those which are graduated in 0.001 in except they have a vernier graduation scale on the sleeve (Fig. 48–6). The vernier scale consists of 10 divisions on the sleeve, which occupy the same space as 9 divisions on the thimble. Therefore the difference between the width of one of the 10 spaces on the vernier scale and one of the 9 spaces on the thimble is one-tenth of a division on the thimble, or one-tenth of 0.001 in which is 0.0001 in.

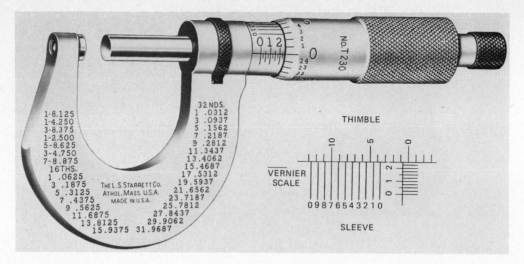

Fig. 48–6 Reading a micrometer graduated in ten-thousandths of an inch. *(The L. S. Starrett Co.)*

To read a micrometer graduation of 0.0001 in, you must first obtain the 0.001-in reading, then check which line on the vernier scale coincides with the line on the thimble. If it is the line "1," add 0.0001 in. If it is the line marked "2," add 0.0002 in. If it is line marked "3," add 0.0003 in, and so on.

In Fig. 48–6, the number 2 line on the sleeve is visible, representing 0.200 in. Two additional lines are visible, each representing 0.025 in, totaling 0.050 in. The longitudinal line on the sleeve lies between 0 and 1 on the thimble, indicating 0.000 in. **NOTE** It must also be added when it reads more. The seventh line on the vernier scale coincides with a line on the thimble representing 0.0007 in. The total equals 0.2507 in.

How to Read Micrometers Graduated in Hundredths of a Millimeter (0.01 mm) The design of a micrometer graduated in 0.01 mm is similar to the micrometer graduated in thousandths of an inch. The only differences are: (1) The spindle screw has a 0.5 mm pitch and, therefore, one complete revolution of the thimble moves the screw and spindle 0.5 mm. (2) The thimble is graduated equally in 50 divisions with every fifth line numbered from 0, 5, 10, to 50. Therefore, each graduation on the thimble is equal to one-fiftieth of $1/2$ mm or 0.01 mm. Two complete turns of the thimble will move the spindle 1 mm.

The procedure for reading a metric micrometer is the same as for reading a micrometer graduated in thousandths of an inch. For example, to read a metric micrometer, you add the total reading in millimeters visible on the sleeve to the reading in hundredths of a millimeter indicated by the graduation on the thimble which coincides with the longitudinal line on the sleeve.

Refer to Fig. 48–7. The 5-mm graduation is visible, thereby representing 5 mm. There is one additional 0.5 mm line visible, representing $1/2$ mm. Line 28 on the thimble coincides with the longitudinal line on the sleeve. Each line represents 0.01 mm and the total is 0.28 mm. Therefore, the aggregate total reading for this example is 5.78 mm.

Inside Micrometers As the name suggests, the inside micrometer is capable of measuring distance between two inside surfaces. The inside micrometer has the same micrometer head as the outside micrometer and, therefore, is read in the same manner.

Micrometer Depth Gauge The micrometer depth gauge is used to measure the distance between two surfaces. You will notice from Fig. 48–8 that the micrometer head is the same as the inside and outside micrometers. However, the anvil is moved and enlarged to a precision-ground and lapped base. The numbers on the sleeve start from the right and the graduation numbers on the thimble start from below the 0 mark.

How to Read Micrometer Depth Gauges Graduated in Thousandths of an Inch (0.001 in) When the thimble is turned in a clockwise direction so that the bevel edge reaches the left 0 mark of the sleeve and the thimble 0 mark is in longitudinal line on the sleeve, the rod is then precisely 1, 2, 3 in, etc. (depending on the rod length) from the base. Conversely, when the thimble is turned counterclockwise so that the bevel edge of the thimble reaches the right 0 mark of the sleeve, and the thimble 0 mark is in longitudinal line on the sleeve, the rod is then precisely flush with the base or 1, 2 in, etc., depending on the rod length used. This, of course, is assuming that the micrometer is properly adjusted.

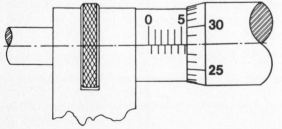

Fig. 48–7 Reading a micrometer in hundredths of a millimeter. *(The L. S. Starrett Co.)*

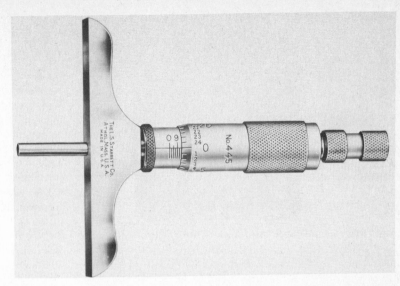

Fig. 48–8 Micrometer depth gauge. *(The L. S. Starrett Co.)*

When you turn the thimble one complete revolution in a clockwise direction, the rod will extend exactly 0.025 in. When you turn five complete revolutions, the rod will extend exactly 0.125 in.

To read a micrometer depth gauge in 0.001 in, multiply the last concealed division on the vertical scale by 0.025 in and add to this the number of 0.001 in indicated by the lines on the thimble which coincide with the longitudinal line of the sleeve.

Refer to Fig. 48–8. The concealed number on the sleeve is 8, which represents 0.800 in. There are two additional lines not visible (each line representing 0.025 in); 2 × 0.025 = 0.050 in. The 0 on the thimble coincides with the longitudinal line on the sleeve, which means no additional thousandths of an inch measured. Therefore the micrometer reading is 0.800 + 0.050 + 0.000 = 0.850 in.

Pointers for Using Micrometers
1. Check the micrometer for accuracy by using the gauge stock provided for that purpose.
2. Clean the surfaces to be measured as well as the anvil and spindle before attempting any measurement.
3. Balance the micrometer with two or three fingers so that it is parallel to the object to be measured. Use your other hand to lightly support the micrometer head.
4. Do not force the spindle against the object to be measured. The micrometer or object should slide through without restriction.
5. Use the ratchet stop to ensure proper force. Move the object to be measured back and forth between the anvil and spindle when making the final ratchet stop adjustment. Then lock the spindle so that the spindle cannot move out of adjustment.
6. When using a micrometer depth gauge, use only two fingers across the base so that you can feel when the rod contacts the surface to be measured.

Dial Indicators Dial indicators are used to measure distance of movement in thousandths, half-thousandths, or hundred thousandths of an inch or to measure a millimeter or hundredths of a milli-meter (Fig. 48–9). Your workshop should have in-dicators scaled in both inches and millimeters for measuring such things as the movement of timing gears, fuel-injection pumps, plungers, the straightness of a valve stem, or the travel of the fuel rack. The face of the dial indicator should be of the balanced type, that is, both positive and negative graduations. The measuring range should be at least 1 in. A dial indicator without an attachment can seldom be used to measure movement unless it is to check the roundness of bores or of cylinders. Many mounting attachments and contact points of various designs are needed to install the dial indicator correctly in order to make a precise measurement.

Special Dial Indicators Special dial indicators do not differ from any other dial indicators except that they are mounted on their own specially designed base to suit some particular purpose.

CYLINDER AND OUT-OF-ROUND GAUGES Cylinder gauges and out-of-round gauges are ideal measuring

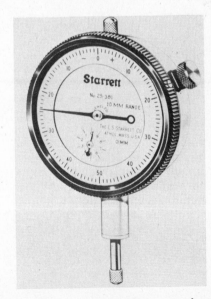

Fig. 48–9 Balance-type dial indicator graduated in hundredths of a millimeter. *(The L. S. Starrett Co.)*

instruments when your service work includes reboring or regrinding cylinders and connecting rods or rebuilding unit injectors. They readily determine taper, out-of-roundness, or concentricity of the nozzle tip (Fig. 48–10).

DIAL DEPTH GAUGE The dial depth gauge is a useful tool to have when servicing fuel-injection pumps because it can quickly and accurately measure plunger lift, head clearance, or fuel rack travel. As you can see from Fig. 48–11, the dial face is of the continuous type.

Pointers for Using Dial Indicators
1. Make certain that you have selected the best mounting attachment and a suitable contact point.
2. Make certain that the dial indicator and base are firmly attached and that the measuring rod is parallel to the object being measured.

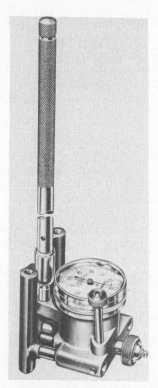

Fig. 48–10 Cylinder gauge. *(The L. S. Starrett Co.)*

SPINDLE → ← MEASURING ROD

Fig. 48–11 Dial depth gauge.

3. Check that the rod, spindle, and dial needle move freely and that the dial range is adjusted so that the full movement of the object can be measured.
4. Always turn the dial face so that the zero line is placed over the dial needle.
5. Take at least two readings. Move or turn the object against its stop or end of travel to ensure the accuracy of measurement.
6. Store the measuring tool in a protective case. Do not oil the spindle or gear mechanism.
7. When you measure with a cylinder gauge, always lock the locking mechanism before you move the tool from the cylinder because this clamps the adjustable contact points. These points are synchronized with the indicator needle. Remeasure with a micrometer the exact diameter you have measured with the cylinder gauge.

Vernier Calipers The vernier caliper differs from the vernier micrometer both in construction and in its principle of operation. The reading on the vernier caliper is not obtained by the relationship between the pitch of a screw and the movement of the thimble as it is on the vernier micrometer. Instead, the vernier caliper's legs are slid into position and accurately adjusted for measurement by means of a fine screw which moves a sliding leg on a beam or bar. The measurement is then determined by adding the reading on the beam and the graduation on the vernier scale (Fig. 48–12).

Vernier calipers are designed to measure inside and outside dimensions in thousandths of an inch or hundredths of a millimeter. By varying the standard design, the vernier caliper can also be made to measure height and depth.

The beam of the tool is graduated in inches, and each is divided into 20 or 40 equal parts representing $\frac{1}{20}$ or $\frac{1}{40}$ in, respectively, (or equal to 0.050 or 0.025 in).

Every second or fourth division is longer and is numbered to represent 0.100 in. The vernier scale has 50 or 25 divisions which correspond with either 49 or 24 divisions on the beam.

How to Read a 50-Division Calipers If you set the tool so that the 0 line on the vernier scale coincides with the 0 line on the bar, the line to the right of the 0 of the vernier will differ from the line to the right of the 0 on the bar by 0.001 in, the second line by 0.002 in, and so on. The difference will continue to increase by 0.001 in for each division until the line 50 on the vernier scale coincides with line 49 on the bar. When you read the tool, note how many inches and how many 0.100 in and 0.050 in that the 0 mark on the vernier scale is from the 0 mark on the bar. Then note the number of divisions on the vernier scale from the 0 to a line which exactly coincides with a line on the bar.

For example, refer to Fig. 48–13. An outside measurement has been made and the measurement is locked in. The two legs are exactly 1.464 in apart because the 0 mark on the vernier scale is past the 1.000-in graduation line and 9 additional divisions to the right are visible. (Multiply 9 × 0.050 = 0.450

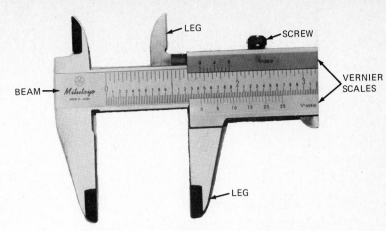

Fig. 48–12 Vernier caliper graduated in 1/25 and 1/16 in.

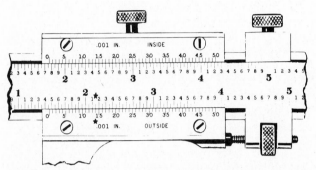

Fig. 48–13 Reading a vernier caliper. (*The L. S. Starrett Co.*)

in.) The fourteenth line on the vernier scale coincides with a line on the bar which equals 0.014 in. When you add the reading on the bar to the vernier reading, you will obtain the exact measurement: 1.000 + 0.450 + 0.014 = 1.464 in.

How to Read a 25-Division Metric Vernier Calipers The difference between the English and metric vernier caliper is that the bar on the metric vernier is graduated in centimeters, millimeters, and half-millimeters. The vernier gives a final reading of ⅕ mm (equal to 0.02 mm). Apart from this, the same principles apply when reading metric measurements as when reading English measurements.

For example, the result of a measurement made with a metric vernier caliper is shown in Fig. 48–14. The 0 mark on the vernier scale is past the 40-mm graduation mark on the bar. Also, three divisions are visible to the right: 3 × 0.5 mm = 1.5

mm. The ninth line on the vernier scale coincides with a line on the bar. Therefore, 9 × 0.02 mm = 0.18 mm. When you add the reading of the bar to the vernier reading, you will have the exact measurement: 40.00 + 1.50 + 0.18 = 41.68 mm.

Pointers for Using Vernier Calipers
1. Clean the legs and check the tool for accuracy.
2. Clean the objects to be measured and slide the movable leg close to the surface. Lock it with the right screw to the bar and have light tension on the left screw.
3. Be certain that the legs are parallel to the object to be measured and do not force the legs together.
4. Adjust the leg with adequate force so that the object just sits between the two legs. Then lock the reading in.

Hole and Telescopic Gauges Hole and telescopic gauges are used to determine hole size or to check a hole for taper or out-ot-roundness (Fig. 48–15). The tool itself only fixes the distance between the two surfaces. A micrometer or a vernier caliper is needed to measure the distance.

How to Use Hole and Telescopic Gauges Hole gauges are used to measure openings up to a ½-in diameter, whereas telescopic gauges will measure up to a 6-in diameter. Each gauge has the ability to measure over a fixed distance, for example, hole gauge from 0.300 to 0.400 in and telescopic gauge from 0.700 to 1.250 in.

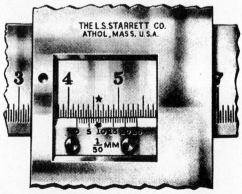

Fig. 48–14 Reading a metric 25-division vernier. (*The L. S. Starrett Co.*)

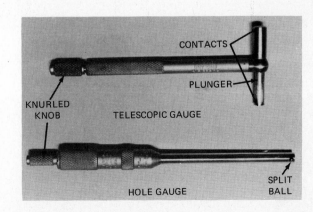

Fig. 48–15 Telescopic and hole gauges.

To use a hole gauge, insert the gauge into the hole and adjust the split ball by means of the knurled knob so that the two sides of the split ball contact the surface lightly. With the telescopic gauge, the contact plunger expands to the hole size. When you feel the plunger touch the sides of the hole, lock the plunger by turning the knurled knob. Withdraw the gauge, and with a micrometer or vernier caliper, measure the distance across the two contact points.

Speed Indicator A speed indicator (hand tachometer) is used to measure the rotating speed of a shaft in either direction. The outer dial plate is graduated from 0 to 100, reading from left to right, or vice versa. Each graduation represents one revolution of the spindle. Therefore, one complete revolution of the large dial indicates 100 revolutions. The small dial has 50 groove lines on its face, each representing 100 revolutions. Therefore, one complete revolution of the small dial indicates 5000 revolutions.

How to Use Speed Indicators Select a tip suitable for the desired application and push it gently onto the spindle. Set the raised point on the inner dial to 0. Position the indicator tip on the rotating object. (The rotating inner dial can be prevented from turning by firmly pressing on the raised knob.) Release the knob, and for the count of 1 minute, allow the rotating disk to turn to record the number of revolutions. Read the inner and outer dials.

Questions

1. What line on the thimble of a micrometer will be aligned with the zero reference line when the reading of the micrometer is 2.071 in?

2. What is the fractional equivalent of 3.750 in expressed in its lowest terms?

3. What will be the finished size of a crankshaft bearing journal when ground 0.020 in undersize if the original size was 2.0625 in?

4. Which of the following can be used to measure inside diameters: outside micrometer, telescope gauge and outside micrometer, measuring hole gauge, or small-hole gauge and ruler.

5. Which of the following turns when the thimble of a micrometer is turned: hub, spindle, anvil, or sleeve spindle nut.

6. Which of the following is the analogous term for step-type feeler gauges: angle step gauges, no-go gauges, spark plug gauges, or go and no-go gauges.

7. List the things you would do to take care of your measuring tools.

8. What are the decimal equivalents of $11/32$, $13/64$, and $9/16$?

9. For what measurements can a feeler gauge be used?

10. Explain how to measure the shaft diameter of a turbocharger.

11. Explain how to measure a camshaft-bearing bore using an inside micrometer.

12. Explain how to measure the depth of a gear-pump housing, using a micrometer depth gauge and by using a dial-indicator depth gauge.

13. List the steps you must take to mount or fasten a dial indicator to measure the runout of a flywheel or the end play of a shaft.

14. What carelessness during installation or mounting would lead to a false reading of the dial indicator?

15. What measuring advantage does a vernier caliper have over a micrometer?

16. Explain how to measure a valve guide using a hole gauge and a micrometer.

UNIT 49

Fasteners, Taps, and Dies

As you may be aware, there are only three ways to join metal to metal: by welding (soldering), by riveting, and by a screw thread. The great advantage of the latter method is that the pieces can be taken apart and reassembled without damage.

You know, of course, that nuts have internal threads while bolts, studs, and screws have external threads. Bolts and screws are similar except that a bolt requires a nut whereas a screw is turned into a threaded hole. Stud bolts are screws with threads on both ends.

Classification and Identification of Bolts and Screws The dimension and design of almost every bolt and screw meet SAE standards. There are a number of ways to categorize the different bolts and screws. One way is by the design of the head of the bolt or screw (see Fig. 49–1).

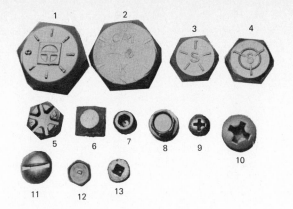

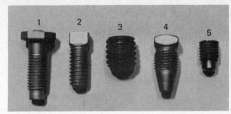

1. Hexagon-head dog point
2. Square-head cup point
3. Slotted-headless oval point
4. Square-head cone point
5. Headless-cone-point internal hexagon

Fig. 49–2 Setscrews.

1. Hex bolt—exceeds SAE grades, tensile strength 180,000 psi [12,664 kg/cm²]
2. Hex bolt—grade 8, tensile strength 150,000 psi [10,545 kg/cm²]
3. Hex bolt—grade 7, tensile strength 133,000 psi [9,349 kg/cm²]
4. Hex bolt—grade 5, tensile strength 120,000 psi [8,436 kg/cm²]
5. Self-locking hex bolt
6. Square-head bolt
7. Socket-head bolt
8. Hexagon-washer-head bolt
9. Round-head machine screw
10. Oval-head-trim machine screw
11. Thrust-head machine and tapping screw
12. Hexagon-reset-head machine and tapping screw
13. 100° flat-head machine and tapping screw

Fig. 49–1 Bolt and screw head designs.

NOTE Because you cannot screw an English bolt into a metric threaded hole or vice versa, straight metric conversions of bolts, screws, nuts, threads, etc., are not always useful and will not be listed in this unit. However, both the English and metric systems of measurement use the same method of categorizing bolts and screws.

Special Screws One of many special screws is the set screw. It is used to hold collars and gears, etc., in position. As you can see from Fig. 49–2, the head and point designs vary to suit specific applications.

Other special bolts and screws are shown in Fig. 49–3. These specially designed heads and shanks have an exact alignment and, therefore, ensure interference-free connections and fastenings.

The reverse head screw, round head, step bolt, countersunk, square neck, ripped neck, and cartridge bolts are used where appearance and smoothness are important.

Self-tapping Screws Self-tapping screws, with various head and thread designs, are most commonly used on sheet metal work. Because of the thread design, tapping screws cut their own threads into a drilled or punched-out hole.

Size of Bolts and Screws Another way in which to identify a bolt or screw is by its size. The size of a bolt or screw is measured by the bolt diameter. The diameter of hex-head bolts, cap screws, and studs is

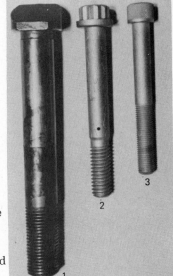

1. Connecting-rod bolt
2. Rocker-shaft bolt (note oilhole)
3. Socket-head screw

Fig. 49–3 Bolts and screws.

measured in fractions of an inch starting with ¼ in and increasing by sixteenths up to ⅞ in. From ⅞ in, it increases to 1½ in, in ⅛-in increments. Bolts and screws of less than ¼-in diameter are classified by a numbering system from 0 to 14, for example, No. 0 = 0.001-in diameter and No. 14 = 0.024-in diameter. Some bolts and screws are numbered from 0 to 24, and these range from 0.060 to 0.371 in.

Bolt and Screw Length Bolt and screw lengths are measured from the extreme end to the bearing surface of the head. Bolts and screws of diameters up to 2½ in increase in length by ¼ in. From 2½ to 6 in, they increase in length by ½ in.

Bolt and Screw Thread Series and Pitch The thread series and pitch are directly related: the thread series governs the pitch. *Pitch* is defined as the distance from a point on a screw thread to a corresponding point on the next thread, measured parallel to the axis. The pitch of a bolt or screw can be measured with a thread gauge by comparing the various blades which have different pitches with the threads of the bolt or screw.

The *thread series* of a bolt or screw identifies the number of threads per inch which are equal to the number of pitches per inch. Although there are about 24 different thread series used in industry, you need only be concerned with those most commonly

Table 49–1 ENGLISH BOLT AND SCREW THREAD SERIES AND PITCH

Nominal size of wrench, in	Basic diameter, in	Nominal size of screw and bolt, in	(U)NC	(U)NF	(U)NEF	(U)N8	(U)N12	(U)16	NPTF size, in	NPSF threads per inch
	0.060	0		80					1/16	27
	0.073	1	64	72					1/8	27
	0.086	2	56	64					1/4	18
	0.099	3	48	56					3/8	18
	0.112	4	40	48					1/2	14
	0.112	4	32*	36*					3/4	14
	0.125	5	40	44					1	11½
	0.138	6	32	40					1¼	11½
	0.164	8	32	36					1½	11½
	0.190	10	24	32					2	11½
	0.216	12	24	28	32				2½	8
	0.242	14	20*	24*					3	8
3/16	0.1875	3/16	24*	32*						
7/16	0.2500	1/4	20	28	32					
1/2	0.3125	5/16	18	24	32					
9/16	0.3750	3/8	16	24	32					
5/8	0.4375	7/16	14	20	28					
3/4	0.5000	1/2	13	20	28		12			
13/16	0.5625	9/16	12	18	24		12			
15/16	0.6250	5/8	11	18	24		12			
1 1/16	0.6875	11/16	11*	16*			12			
1 1/8	0.7500	3/4	10	16	20		12	16		
1 5/16	0.8750	7/8	9	14	20		12	16		
1 1/2	0.1000	1	8	12	20	8	12	16		
1 11/16	1.1250	1 1/8	7	12	18	8	12	16		
1 7/8	1.250	1 1/4	7	12	18	8	12	16		
2 1/16	1.3750	1 3/8	6	12	18	8	12	16		
2 1/4	1.5000	1 1/2	6	12	18	8	12	16		
2 1/2	1.750	1 3/4	5			8	12	16		
2 5/8	2.000	2	4½			8	12	16		

Note: *Denotes nonstandard.

used on diesel engines or related machines (see Table 49–1):

1. (U)NC (American National Coarse Thread Series)
2. (U)NF (American National Fine Thread Series)
3. (U)NEF (American National Extra Fine Thread Series)
4. N (American National, 8-, 12-, and 16-Pitch Series)
5. NPGF (American Standard Tapered Pipe (Dryseal)

There is a difference between the American National Coarse Thread Series and the Unified Standard. Figures or letters on the bolt head are used to identify Unified Standard bolts. In the Unified System of screw thread limits, the term *Class of Fits* once applied to the assembly characteristics of mated parts. The term *Class of Thread* now indicates only the limits of one component and does not imply that the mating part need necessarily be made to the corresponding class of limits. In fact, any class of internal thread may be used to assemble with any class of external thread so long as the assembly meets the requirements of the end product.

Thread Classes and Grade Marking The Unified Screw Thread System includes three classes of external threads, 1A, 2A, 3A, and three classes of internal threads, 1B, 2B, 3B. Associated with these three classes will sometimes be found two of the old American National classes 2 and 3, formerly called *fits*.

It is obvious then that when ordering screws and bolts from the parts department or tool room, you should specify the head design, the bolt size, the bolt length, the thread series, and the bolt grade. The class or fit will most commonly be 2A or 2B unless you require the bolt, screw, or nut specially designed for a particular purpose.

Metric Threads Most fuel-injection equipment is manufactured with the international metric 60° thread. The identification of a metric thread is simple because the diameter and the pitch are used to identify the bolt, screw, or nut.

As you can see from Table 49–2, most bolts, screws, or studs are produced in three pitches which are about equal to (U)NC, (U)NFC, (U)NEF.

Washers Lock washers are able to hold the nut and bolt in position because the energy of the spring or

Table 49-2 METRIC BOLT AND SCREW THREAD SIZE AND PITCH

Size, mm	Pitch, mm	Size, mm	Pitch, mm
1.6	0.35	6	0.8, 0.75, 1.00
1.7	0.35	8	0.75, 1.00, 1.25
1.8	0.35	9	1.00, 1.25
2.0	0.4	10	1.00, 1.25, 1.5
2.2	0.45	12	1.00, 1.25, 1.5
2.3	0.4	13	1.75
2.5	0.45	14	1.00, 1.25, 1.5
2.6	0.45	16	1.5
3.0	0.5	18	1.5, 2.00, 2.5
3.5	0.6	20	1.5
4	0.5, 0.7		
5	0.5, 0.75, 1.00		

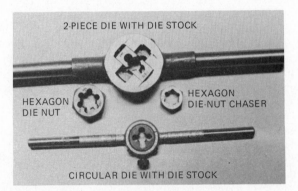

Fig. 49-4 Dies used to cut external threads.

the multitooth lock washer bites into the surface when the screw or bolt is tightened. Although flat washers serve many purposes, their main functions are to act as a seal or to allow more even distribution of the bolt-head force.

Pointers for Using Bolts, Studs, and Screws
1. Select the proper thread, series, size, length and class. If any one of these factors is incorrect, the bolt, stud, or screw will fail as a fastener.
2. Under all circumstances tighten the bolt to a precise torque and sequence.
3. Use the correct lock washer and flat washer.

Causes of Bolt or Cap-Screw Failure Bolts and screws, like any other components, can eventually become ineffective due to varying circumstances. Listed below are some of the causes of their ineffectiveness and/or deterioration.

1. Excessive residual tension (adds to the stress on the bolt or screw)
2. Improper torque, improper grade, or improper installation (creates stress changes which in time lead to fatigue and failure)
3. Exposure to extreme heat (reduces tensile strength of bolts)
4. Improper mating of fastener components (causes decarburization and stripping of the nuts and bolts)
5. Cracked or dished washers (cause the fasteners to fail)
6. Improper material, heat treatment, or excessive hardness (cause the head to pop off or the nut to crack)
7. Excessive vibration (causes fracture failure)

Dies The dies shown in Fig. 49-4 are used to cut external threads. When it is necessary to cut large pitch threads, adjustable (split) or two-piece dies are recommended to ensure a clean cut.

How to Cut External Threads
1. Select an adjustable die which has the desired dimension and thread series. For instance, it might be ½ in (U)NF.
2. Insert the die in the die stock with its full thread toward the die-stock shoulder. Enlarge the diameter with the adjustable screw. Then use the set-screw to lock the die into the die stock.
3. Make certain that the round stock is of ½-in dimension and that the end is chamfered about 4°. This will ensure easier starting.
4. Use the lubricant recommended for drilling.
5. Cut the thread by turning the die stock. While turning, push down with a moderate force. Make certain that you maintain a 90° angle to the round stock.
6. Cut the thread by turning the die stock at slow speed. Keep the die stock well lubricated. Do not reverse rotation because this may break the thread. If it becomes necessary to turn backward, only turn the width of one land, then start cutting again.
7. After you have cut the desired thread length, remove and adjust the die, then repeat the cutting procedure. Check the thread fit with a nut or die to make sure that it has the proper tolerance.

Restoring External Threads There are many ways in which an external thread can be damaged. To restore a damaged thread, you can use a rethreading die or a thread file (see Fig. 49-5). However, in both cases you are limited to a 7 or 8 thread series,

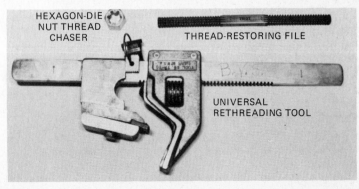

Fig. 49-5 Tools used to restore external threads.

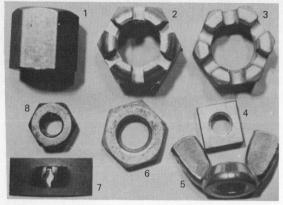

1. Hex high nut
2. Hex castle nut
3. Hex slotted nut
4. Square machine
   screw nut
5. Wing nut
6. Hex flat jam nut
7. Spring nut
8. Standard hex nut

Fig. 49–6 Nuts.

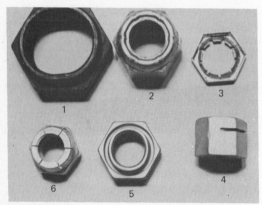

1. Hex collar
2. Nylon collar
3. Palnut locknut
4. Distorted thread
5. Tensilock nut
6. Marsden nut

Fig. 49–7 Self-locking nuts.

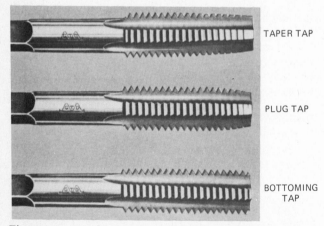

TAPER TAP

PLUG TAP

BOTTOMING TAP

Fig. 49–8 Hand taps. *(Greenfield Tap and Die Corp.)*

whereas the universal rethreading tool shown in Fig 49–5 is not limited to size or to thread series.

Nuts and Internal Threads   Nuts of various thread series, size, grade, class, and design are used with bolts and studs. Some of the most common nuts which require lock washers are shown in Fig. 49–6.

Palnut locknuts, and jamnuts are used to lock a nut to a bolt or stud after the first nut has been properly positioned or tightened. Slotted or castle nuts are used where thread friction and a lock washer are not sufficient to prevent the nut from coming loose. A cotter pin properly installed will prevent the nut from coming loose.

Self-locking nuts are widely used (see Fig. 49–7). They are designed principally to increase thread friction. This is done (1) by using a nylon insert or nylon collar, (2) by distorting the threads by cutting vertical slots in the upper one-third of the nut and shrinking its dimension slightly, or (3) by flattening the upper one-third of the nut to a slightly smaller dimension than the lower two-thirds of the nut (see Fig. 49–8).

Nut Grade   The standard grades are Nos. 2, 5, or 8. A grade marking is only required for a No. 5 or No. 8 nut. Grade No. 5 is identified by a dotted point on one corner and a radial line at the corner located 120° clockwise from the corner with the dot. Grade No. 8 nut is marked with a dot at one corner and a radial line at the corner located 60° counterclockwise from the corner with the dot.

Hand Taps   As a minimum requirement you should understand the differences between the standard hand taps you will be using when servicing diesel engine components. Hand taps used for cutting English and metric threads are supplied in sets of three of equal size. The set includes a taper tap, plug tap, and bottoming tap, 1 to 1 ½ threads (Fig. 49–8).

Series taps are used to tap deep holes. They may be either open or blind. They are similar in general dimension to taper, plug, and bottoming hand taps, but they differ from the latter in that each tap cuts only a certain percentage of the thread. This divides the tapping strain among the three taps and lessens the possibility of tap breakage.

Pointers for Hand Tapping   Tapping a hole follows the drilling of that hole. Three types of holes can be drilled into an object, as shown in Fig. 49–9. If possible, avoid drilling a blind bottoming hole because it is difficult to tap.

When hand tapping a hole, you should consider a drill size which will give you 60 to 70 percent thread depth. You should be aware that thread depth is a very important factor in tap breakage. A bolt inserted in an ordinary nut which has only 50 percent thread will break before stripping the threads. Remember, also, that a 100 percent thread depth is only 5 percent greater in strength than a 75 percent depth.

1. Use a tap drill size chart (see Conversion Tables) or the following formula to determine theoretical maximum and minimum drill sizes for average conditions. For example, ¼ in (U)NC thread:

Maximum drill size basic thread outside diameter =

$$0.250 - \frac{3}{8} \times \text{No. of thread}$$

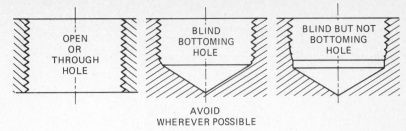

OPEN OR THROUGH HOLE

BLIND BOTTOMING HOLE

BLIND BUT NOT BOTTOMING HOLE

AVOID WHEREVER POSSIBLE

Fig. 49–9 Types of holes that can be drilled in tapping. *(Greenfield Tap and Die Corp.)*

Maximum drill size =

$$\frac{0.250 - 3}{8 \times 20} = 0.250 - 0.01875 = 0.2312 \text{ in}$$

Maximum drill size basic thread outside diameter =

$$0.250 - \frac{1}{2N}$$

Minimum drill size =

$$0.250 - \frac{1}{2 \times 20} = 0.250 - 0.025 = 0.225 \text{ in}$$

2. Make certain that the drill is properly sharpened. A dull drill can create an egg-shaped hole which will result in an uneven thread depth (Fig. 49–10).
3. Dull drills or the lack of lubricant can cause surface hardening. As a result you will find it hard to start the tap or you will round over the threads. The tap may even break altogether (Fig. 49–10).
4. Blow out the drilling chips from the bottom of the hole. **CAUTION** Use goggles.
5. Use the taper or No. 1 tap to ensure a straight start.
6. Lubricate the area you are cutting with the proper lubricant. [See Recommended Cutting Fluids for Drilling, Reaming, and Tapping (Unit 46).]
7. Select a tap wrench which will hold the tap securely.
8. Using both hands, turn the tap in a clockwise direction as far as possible. Periodically remove the tap to clear out the swarf (metallic particles) and repeat the procedure.
9. When it is necessary to tap in both a clockwise and counterclockwise direction to break the chips, do not reverse the tap more than the width of one land.
10. Frequently clear out the chips when tapping a bottom hole. This will prevent tap breakage caused by the impact of the tap on the chips.
11. When you tap a bottom hole, fill the hole with lard. This will force the chips out as the tap is turned in.
12. When you tap with a taper-pipe tap, use a pipe reamer before threading. Remember that the diameter of the tap increases at a rate of ¾ in/ft.

Rivets Except for the friction clutch or for sheet metal work, the rivet is very seldom used as a fastener on diesel engine components. Nevertheless, you should be thoroughly familiar with the various types of rivets and methods of riveting.

THIS CONDITION . . . . CAN CAUSE THIS TROUBLE!

Hole out of round. "Egg shape" because of improperly ground drill.	Threads do not extend all the way around hole or are of uneven depth at different points.
Hole too small. Drilled to give more than 75% height of thread.	Uses excessive power, tap breakage high. If much too small tap will ream hole instead of threading.
Surface of hole "work hardened" by using dull drill.	Too hard for tap to start, may chip or round over threads or break tap.
Drilling chips left in bottom of hole.	Packed Chips will shatter tap.
Punched hole in thin sheet metal, too small.	Flare on underside tends to load tap on withdrawal.

Fig. 49–10 How faulty drilling can cause tapping troubles. *(Greenfield Tap and Die Corp.)*

Methods of Riveting Some of the more common riveting methods are shown in Figs. 49–11 and 49–12. Although there is no established rule regarding the most suitable rivet, it is wise to use solid rivets when maximum strength is needed. Semitubular rivets can be used if tensile or fatigue strength is not a major concern.

Impact riveting refers to the use of a hardened die and a buck. By striking the die repetitively with a hammer, the head is formed (Fig. 49–12).

Squeeze riveting is similar to impact riveting except that it is essential to apply a steady force on the die to form the head.

Staking riveting refers to deforming the rivet material at one or more points with a sharp tool in order to force the metal at these points tight against the mating parts (Fig. 49–11).

Clinch riveting is the most popular method used on diesel engines. This method of riveting involves the forcing of a die against the opening end of a semitubular rivet to form a star- or roll-clinch cap against the materi___ ___ fastened (Fig. 49–12).

ENDS OF RIVETS

RIVETING BURR

Fig. 49-11 Staking method of riveting.

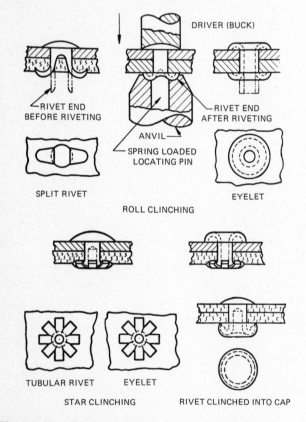

DRIVER (BUCK)

RIVET END BEFORE RIVETING

RIVET END AFTER RIVETING

ANVIL

SPRING LOADED LOCATING PIN

SPLIT RIVET

EYELET

ROLL CLINCHING

TUBULAR RIVET   EYELET

STAR CLINCHING

RIVET CLINCHED INTO CAP

Fig. 49-12 Clinch riveting.

Pointers for Riveting

1. Select the most suitable rivet and material for the job.
2. The rivet should protrude by the amount of its diameter. It should fit in the hole snugly.
3. Keep the two joining surfaces tightly together at all times.

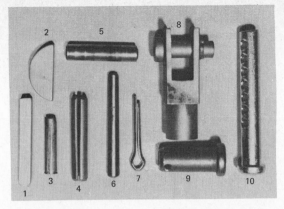

1. Flat key
2. Woodruff key
3. Groove tapered pin
4. Spring-type pin (roll pin)
5. Press-fit pin
6. Tapered pin
7. Cotter pine
8. Rod end
9. Clevis pin
10. Universal clevis pin

Fig. 49-13 Locking devices.

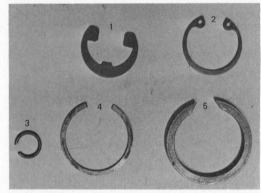

1. External-finger snap ring
2. Internal snap ring
3. External snap ring
4. Internal snap ring
5. External snap ring

Fig. 49-14 Snap rings.

4. For soft or resilient materials, use the die which forms a star clinch.
5. Do not use a split rivet with clutch facings.
6. When possible, change every rivet so that the heads alternate from side to side.

Locking Devices   Locking devices used to lock, hold, or position a pulley or gear are: woodruff keys, splines, pins, snap rings, metal lock plates, and lock wires (Figs. 49-13 and 49-14).

Woodruff keys or square keys are used to position a component to a shaft or tapered shaft when moderate strength is required. For greater strength, or when it is necessary for a shaft gear or pulley to move lengthwise, multisplines are used.

The most common pins and snap rings you will encounter are those shown in Figs. 49-13 and 49-14. Each is used under special circumstances to hold the component in position.

Metal Wire Locks   A metal lock, or wire lock, is used when a self-locking cap screw or a nut is inadequate. Both methods will prevent the cap screw or bolt from loosening. The correct locking method for installing flat metal locks is shown in Fig. 49–15.

CAUTION Never reuse a metal lock.

Pointers for Using Different Locking Devices

1. Always use a new cotter pin of the proper length and dimension. Use special cotter keys for engine clutches and connecting rods.
2. Use new self-locking pins to prevent premature failure.
3. Check splines, keys, and keyways for wear.
4. Make certain that the setscrew is seated and locked properly.
5. Do not substitute an internal snap ring for an external one.
6. When you install a snap ring, make certain that its diameter and thickness are the correct size and that it is inserted properly in the groove.

Questions

1. What is the difference between a hex bolt and a screw?

2. What does a thread gauge measure?

3. Between what two points is the length of a bolt measured?

4. Which types of thread series are most commonly used on diesel engines?

5. Name two major causes for bolt or cap screw fatigue and failure.

6. What is the difference between a die and a tap?

7. When it becomes necessary while cutting a thread to turn the die backwards, why must it be turned only the width of one land?

8. Name the standard grades of nuts.

9. Why should you never reuse a self-locking nut?

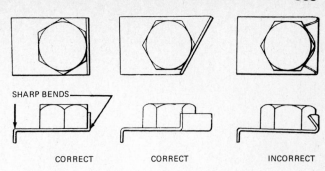

Fig. 49–15 Installing flat metal locks.

10. Before drilling a hole for tapping, what must be considered?

11. When fastening by riveting, what must you know in order to select the correct rivet?

12. Why, in some instances, is a lock plate used instead of a lock washer?

13. When a nut is tightened, it tends to (a) expand its threads; (b) pull inward at the center; (c) bulge outward at the center; or (d) none of the above.

14. The strain on a loose bolt is (a) the same as when it is tight; (b) less than when it is tight; (c) greater than when it is tight; or (d) chiefly on the nut.

15. What is the official name for coarse thread? (a) SAE, (b) unified coarse, (c) unified national coarse, or (d) NF.

16. A die is a special type of tool that (a) is easily damaged; (b) repairs and makes male threads; (c) should be used only by an expert; or (d) makes threads stronger than new.

17. A headless fastening device threaded at both ends is a (a) screw, (b) bolt, (c) stud, or (d) pin.

18. Which nut is secured with a cotter pin? (a) slotted nut, (b) acorn nut, (c) Palnut locknut, or (d) hex nut.

SECTION 7

Break-in, Troubleshooting, and Tuneup

UNIT 50

Checks and Adjustments Before Starting the Engine

Need for Break-in Tests   A new or rebuilt engine must be operated for a certain period of time (break-in) under various speeds and load conditions to remove high spots from the moving components and allow mating surfaces to acquire a full seat.   During this time possible errors in the assembly, in the torque value of the mounting bolts, or in the adjustment can be detected and promptly corrected.   The engine should also be tested and adjusted *after* the break-in period to ensure faultless performance.

Preparation Prior to the Break-in Run   To prepare the engine for its break-in run, secure and align it to the dynamometer test stand, then proceed:

1.   Connect the dynamometer drive shaft strictly in accordance with the manufacturer's instructions.
2.   Connect the exhaust and intake piping to the engine.
3.   Connect the water supply and return hoses to the cooling tower.
4.   If necessary, connect the throttle and shutoff lever.
5.   Connect the batteries but make certain you have corrected battery voltage and ground polarity.
6.   Depending on dynamometer cooling arrangement, you may have to remove the thermostat.
7.   Connect a mercury manometer to the intake manifold or to the air box (to measure turbocharger boost pressure or blower pre

8.   Connect a pryrometer to the exhaust manifold (to measure the exhaust temperature).   Connect a water manometer to the exhaust-manifold outlet (to measure the exhaust back pressure).
9.   Connect a temperature gauge to the inlet water manifold (to measure the coolant temperature).
10.   Connect a pressure gauge to the main oil gallery (to measure the oil pressure).   Position a temperature gauge as recommended by the engine manufacturer (to measure the oil temperature).
11.   Connect a water manometer to the crankcase in accordance with the procedure recommended by the engine manufacturer (to measure piston blow-by).
12.   Connect a fuel-pressure gauge, or a flowmeter, to the fuel manifold (to measure fuel pressure and/or flow.
13.   Connect the fuel supply and return line to the engine (Fig. 50–1).

Before you start the engine, proceed as follows:

1.   Check the coolant-system hose connection, and adjust coolant flow if necessary.
2.   Check the drive belts and make any adjustments that are required.
3.   Adjust the valves as required.   Check the injection pump timing, the injector timing, or the injector adjustment, if necessary.
4.   Fill the crankcase to the correct level, with the oil specified by the engine manufacturer.   Connect a priming pump to the recommended oil manifold.

Fig. 50–1 Smokemeter connected to the engine.

5. Lubricate the cranking motor and alternator bushings or bearings, if required.

6. Prime the fuel-injection system when necessary.

7. If the engine is equipped with a turbocharger, remove the oil-inlet line and lubricate the bearings according to the engine manufacturer's recommendation.

8. Connect the ground battery cable.

9. Pressurize (prime) the lubrication system to about 40 psi [2.8 kg/cm²] using a power primer or a hand primer. Then set the throttle to the no-fuel position. If the throttle is equipped with a decompression lever, place it in the on position. Crank the engine for a minimum of 20 seconds while maintaining (with the priming pump) a minimum pressure of about 20 psi [1.4 kg/cm²].

10. If not previously done, open the coolant supply to the engine and to the dynamometer as recommended by the manufacturer. Then check all plugs, tubings, fittings, hoses, and connections for coolant leaks.

11. Disconnect the turbocharger oil-return line and set the throttle at idle speed. **NOTE** If the engine is equipped with a compression release, pull the release while cranking the engine for a short period of time. Release the compression lever, and then continue cranking the engine until it starts.

12. If the engine fails to start within 30 seconds although the cranking speed reaches 150 to 200 rpm, check the fuel supply, the transfer pump, or the manifold pressure. If they are not faulty, bleed the low- and high-pressure system. If a glow plug circuit is used, check the wiring and the glow plugs.

13. Once the engine starts, check the turbocharger return-oil drain. Oil should flow from the hose in the time specified in the engine service manual. When oil flows within the specified time, reconnect the oil drain hose. If oil does not flow, then stop the engine and correct the cause of failure.

Break-in Run Each engine manufacturer gives dynamometer test specifications or charts which present concise instructions regarding rpm, load, temperatures, and pressures. Nevertheless, there are break-in procedures which are common to all engines. For instance, when the engine is operated initially, make certain there is no load on the engine, that there is instant coolant circulation, and that the oil are at operating temperature, stop the engine.

Then check and, if necessary, adjust the valve clearance and injector adjustment.

On Detroit Diesel engines you must make a complete sequence of tuneup checks and adjustments. The oil pressure should be checked repeatedly and it should remain constant at all loads and speeds. The oil temperature must not exceed 225°F [107°C] and the cooling temperature must not exceed 200°F [90°C] or drop below 160°F [71°C]. After each run-in phase or step, the oil level must be checked.

NOTE Never shut the engine down immediately after a test run; first allow it to cool down.

Constantly check the blow-by reading, bearing in mind that the blow-by increases with increased load and speed (Fig. 50–2). Nevertheless, it must remain within the manufacturer's specification. If the

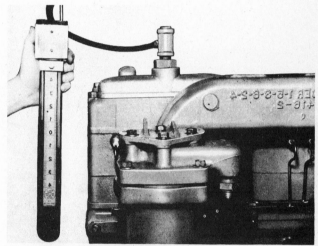

Fig. 50–2 Checking blow-by.

Fig. 50–3 Checking boost pressure.

blow-by is higher than specified at this phase of run-in, the engine speed and load must be reduced for a period of time and then brought back to the original setting. When the engine is operating at the recommended rpm and has reached about 90 percent of its power, and the blow-by is at (or below) specification (about 2 in [50.8 mm] of water), measure the exhaust smoke. Always follow engine and smokemeter manufacturers' instructions when making this measurement. When the exhaust smoke exceeds the emission specification, check the following: injection pump timing, automatic advance operation, injector timing or injector adjustment, governor and valve adjustment, blower or turbocharger boost pressure (Fig. 50–3, p. 387), operation of the aftercooler, and aneroid or fuel ratio control.

NOTE Naturally aspirated engines lose about 3 percent of their power for each 1000 ft [304.8 m] above sea level. All engines lose about 1 percent for each 10°F [6°C] above 60°F [16°C] ambient temperature.

Questions

1. Why is it essential to thoroughly prepare an engine prior to starting it up?

2. Which check must you make immediately after the engine has been started and is running at say, 900 rpm?

3. Why should you *not* operate the engine at idle speed during the break-in run?

4. List the checks and adjustments you must make after the engine reaches operating temperature.

5. Why should an engine be allowed to cool down before it is shut off?

UNIT 51

Troubleshooting and Tuneup

Tuneup It cannot be overemphasized that over 90 percent of engine problems and the causes of engine performance complaints can be prevented by a good maintenance program which includes engine tuneup at the correct interval. An engine *tuneup* means checking the entire engine and its supporting systems, checking various adjustments, and comparing them against the service manual specifications and, where necessary, making readjustments.

NOTE A tuneup can only be made when the engine is mechanically sound.

Troubleshooting A methodical approach to troubleshooting is required if you are to solve the problems which you will face daily. In order to relate troubleshooting to as many different fuel-injection systems and engines as possible, only generalizations will be made in this unit.

Locating component failure, or quickly isolating the cause of such engine problems as overheating, misfiring, and smoking excessively, or determining why the engine is hard to start, etc., requires a thorough knowledge of the basic operating principles of the following systems: cooling system, lubrication system, intake system, exhaust system, fuel-injection system, and the accessory systems (cranking and charging). Good comprehension of the engine's construction and design is essential. In addition to theoretical knowledge, the practical skill to test and measure engine performance, rpm, temperature, and pressure, etc., and the ability to evaluate results against service manual specifications is essential.

Diagnosis The first step in solving engine problems is to narrow them down to only those systems which might be involved, determine the cause of the problem, and then decide how it can be prevented in the future.

Begin your evaluation by asking the operator the following questions:

1. When was the trouble first noticeable?
2. Was this complaint or a similar problem recorded at any time?
3. Was the trouble noticeable at all speeds and load?
4. Did any unusual noise precede the trouble?
5. Has the cooling temperature increased?
7. When was the last maintenance check?
8. When was the last oil and filter change?
9. When was the cooling system last checked?
10. How does the engine respond to acceleration or deceleration?
11. Is there excessive exhaust smoke?
12. Does the engine start easily?
13. Does the engine hunt or surge at idle, high idle, or full load?

A wealth of information is available to you if the engine has a service record sheet to indicate which components or parts have been serviced or replaced.

Finally, a personal inspection should be made. During this appraisal you should check to determine if the engine is clean or dirty or if there are any external oil, coolant, or fuel leaks. Check the condition of the coolant, the fuel hoses or the tubings, the drive belts and pulleys. Determine if any components have been added which are not standard.

After gathering all data possible, analyze your findings in relation to the operating principles of the components concerned.

Engine Will Not Start   When an engine fails to start (this is one of the most common complaints), it is often because the cranking speed is too low or because the ambient temperature is very low. Regardless of these two factors, first check the electrical system (see Units 40, 41, 43, and 44). When the ambient temperature is very low, always use No. 1 fuel and engine oil of the correct viscosity. Other factors which affect starting are referred to later in this unit.

Air-Intake Restriction and Poor Compression   When the air-intake system is restricted due to a plugged air cleaner, or when the engine has low compression because of a damaged or worn blower or turbocharger, worn or damaged rings and cylinder sleeves, or misadjusted or burned valves, you will notice the following:
- The engine is hard to start.
- It has lost power.
- The engine temperature is below normal.
- The exhaust smoke has increased in density.
- The consumption of oil and fuel has increased.

Early formation of carbon on the valves and pistons is likely to occur, fuel may have diluted the engine oil, and the engine may run rough or vibrate excessively.

To check for a restricted air-intake system, connect a water manometer to the inlet manifold. Then operate the engine at high idle. The manometer reading must not exceed specification.

To check the condition of the blower or turbocharger, connect a mercury manometer to the air box or intake manifold. Operate the engine at the specified speed and load. Then compare the manometer reading with your service manual specification. (See the sections on blowers and turbochargers in Unit 19.)

To determine piston-ring wear, check the blow-by by connecting a water manometer to the crankcase. Operate the engine at specified speed (load). Compare the blow-by reading shown on the manometer against that specified in the service manual. If the blow-by reading is within specification, check the cylinder compression. (Compression check procedure varies somewhat among engine manufacturers.)

One recommended procedure is as follows:

1. Bring the engine to operating temperature.
2. Remove the injection line and the injector from the first cylinder head to be checked.
3. Clean the injector bore or sleeve with the recommended tool, but make certain that no carbon remains in the combustion space.
4. Crank the engine over for a few revolutions to blow out the carbon.

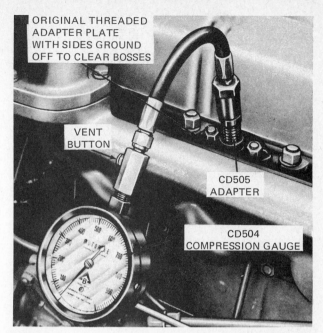

Fig. 51–1 Checking cylinder compression. (*J. I. Case Co., Components Div.*)

5. Place a new gasket in the bore, and then install and tighten the correct adapter with the gauge as shown in Fig. 51-1.
6. Operate the engine on three, five, or seven cylinders, at recommended rpm, until the pressure gauge stablizes. Then read and compare the compression pressure with the specification shown in your service manual.
7. Check the remaining cylinders. The pressure variation should not be greater than 50 psi [3.4 kg/cm²].

Air-Intake Leaks   When unfiltered air enters the cylinders, the engine may show a loss in power because of increased wear on the piston rings and the cylinder sleeve. Piston-ring wear causes poor compression, hard starting, and an increase in exhaust smoke density. (Effects of an air leak are more noticeable when a turbocharger is used.)

When the air leak occurs before the compressor, the compressor wheel will show dirt and excessive wear. If the air leak occurs after the compressor, the boost pressure will be lower and a hissing noise may be audible. Examine the compressor wheel and measure the boost pressure. To locate the leak, pressurize the intake system as explained in the section on air cleaners (Unit 18).

High Exhaust Back Pressure   When the exhaust back pressure is high, the engine temperature will increase since more exhaust heat must be carried away by the cooling system. Furthermore, the engine power will decrease because of incomplete scavenging, the exhaust smoke will become denser, and the engine will probably run rough at idle speed. Other components that are affected due to high exhaust back pressure are the valves, injectors, and pistons. These components will show exces-

Fig. 51–2 Checking the exhaust pressure with a water manometer.

sive carbon buildup and the lubricant will become contaminated.

To locate the cause of high exhaust back pressure, install a water manometer to the exhaust manifold, and measure the back pressure. If it is within specification, check the valve adjustment and/or the valve timing (Fig. 51-2).

High Oil Consumption When the engine burns lubricant excessively (after the run-in phase), you will notice that the oil becomes dirty early, the engine temperature increases, and the exhaust smoke is of a bluish color.

To locate the cause of high oil consumption, first check the dipstick. If an incorrect dip stick is used, it may result in too high an oil level. Check the oil pressure. If it is higher than specified because of a malfunctioning relief valve, excessive oil is thrown onto the cylinder wall. The oil cannot be controlled by the piston rings and, therefore, enters the combustion chamber. The result is the same when the incorrect grade or quality of oil is used.

Check the cooling system and the drive belt. Either may have caused high engine temperature. Measure the blow-by and cylinder compression and, of course, check for external oil leaks.

Check the crankcase breathers. When they are plugged, the crankcase pressure increases, forcing the oil out through the gaskets and seals, and sometimes past the piston rings.

If none of the above tests have revealed the cause of high oil consumption, then check the valve guides, the rocker shaft position, the main bearings, the connecting rod, and the camshaft bearings for excessive clearance.

Low Oil-Pressure Gauge Reading When the oil-pressure reading is lower than specified and the engine is noisy or knocks (see Engine Noise and Knocks later in his unit), instantly stop the engine to prevent additional damage.

Immediately check the crankcase oil to determine if it is diluted. To ensure that the low pressure was not due to a damaged pressure gauge, recheck the oil pressure using a master gauge. If the oil pressure is actually low, drain the oil and remove the filters. Then check the filters and the crankcase for metal particles. If there are no metal particles present, check the oil pump and the pressure relief valve.

Crankcase Oil Dilution

DIESEL FUEL When diesel fuel has diluted the engine oil, any one or more of the following may result: (1) a cylinder may misfire, (2) the engine may run rough, (3) the exhaust smoke may increase in density, or (4) the oil pressure may drop.

Misfiring can be caused by a sticky injector, the nozzle valve not seating properly, an incorrect spray pattern, an enlarged orifice, or not high enough compression.

Test for a malfunctioning injector. (The various procedures are described in Units 27, 31, and 32.) Measure cylinder compression and blow-by. Check the transfer-pump seal.

When the engine does not show any of the four symptoms outlined above, ask the operator when the oil and filters were last changed. Ask if the engine was operated while the engine was below recommended temperature or if it operated at idle speed over a long period of time.

When working on a Detroit Diesel engine, check the fuel jumper lines. They may be loose or broken. When working on a Cummins diesel engine check the injector O rings for damage.

SLUDGE OR COOLANT When coolant has diluted the engine oil, you may notice that the engine temperature increases and the oil has become creamy. However, before you conclude that coolant has leaked into the crankcase, make sure the problem is not condensation.

Excessive condensation can be caused by operating the engine at a low ambient temperature, operating the engine at low engine temperature, or running the engine at idle speed for a long period of time.

Engine oil can be diluted by coolant when there is a cracked cylinder head or block (due to overheating or freezing) or when there is a leaking cylinder-head gasket or a leaking cylinder-sleeve O ring.

To locate any of these leaks, remove the oil pan and clean the crankcase. Then lay a piece of white paper under the engine and allow it to stand overnight. The location of the coolant on the paper will indicate the area inside the block which you must examine.

If this method is unsuccessful in spotting cracks or the source of a leak, remove the water pump and the radiator hoses from the engine and seal the openings. Install an airhose with a pressure regulator to the cylinder block. Fill the engine with a strong solution of antifreeze. Operate the engine until the operating temperature is about 180°F [82°C], then pressurize the coolant system to about 40 psi [2.8 kg/cm²]. Drain the oil and remove the oil pan. Then again check the cylinder block and crankcase for coolant leaks.

Engine Noise and Knock   The source of engine noise or knocks may or may not be easy to trace. The conditions under which the noise or knock is more easily found are when the engine is noticeably misfiring or operating irregularly, when there is excessive exhaust smoke, or when the coolant pump, compressor, or generator is noisy.   Fairly obvious sources of knocks are due to incorrect timing, the engine overheating, or the incorrect fuel being used.

More difficult to trace is the source of noise or knock due to piston "slap," broken piston rings or a broken piston pin, a damaged camshaft bearing or follower.   It is also difficult to locate the source of noise when the engine has one main or connecting-rod bearing excessively worn or damaged.

To locate a noise or knock caused by engine misfiring, loosen the high-pressure fitting or hold the injector down.   If there is no change in the noise, then that particular cylinder is not at fault.

Following are some further hints which will help you locate engine knock:

- A loose wrist pin gives a sharp metallic rap at low speed and decreases as temperature increases.
- A worn main bearing causes a dull knock when accelerating under load, in contrast to a worn connecting-rod bearing which causes a noticeable dull knock at light load or idle speed and decreases at full load.
- A worn piston makes a lighter sound than a worn connecting-rod bearing, and the noise may disappear as the engine warms up.
- A broken piston ring and/or land will give a sharp clicking noise throughout all speeds and load.

When the timing gears have too much backlash or the gears are loose on their shafts, they usually make a rattling noise.   Timing gears with reduced backlash will whine.

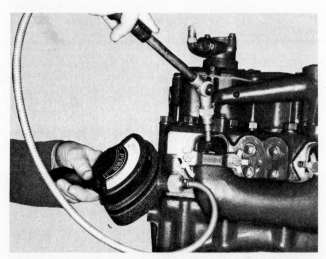

Fig. 51–3   Using a contact pyrometer to measure exhaust temperature of each cylinder. *(Cummins Engine Co., Inc.)*

Excessive Engine Vibration   Determining the cause of engine vibration can be a time-consuming task unless the vibration is due to a misfiring cylinder, a damaged vibration damper, or loose or broken engine mounts or accessories.

However, when engine vibration is the result of accessories that vibrate at various frequencies (depending on their own malfunction) or because of their location, weight, and operating speed, a precise test instrument is necessary to measure the vibration and thereby determine its source.

Engine Overheating   When an engine overheats, you may notice that it loses power (because it inhales less air by volume), and ignites too early.   It may run rough, vibrate excessively, or knock.   The density of the exhaust may have increased.

Many malfunctions or misadjustments can contribute to overheating an engine.   To find the cause, proceed as follows:

1.   Check the radiator hoses and coolant filter.
2.   Check the adjustment and the operation of the thermostat fan and shutterstat.
3.   Inspect the exterior of the engine for excessive dirt, grease, etc., and/or noticeable leaks.
4.   Check the coolant level and the oil level.
5.   If the radiator core is plugged or scaled, clean it and the engine (see Unit 20, Cooling System).
6.   Check for airflow restriction in the radiator. Check the radiator's pressure cap.   The fan shroud should be positioned properly so that no vacuum pockets can build up.
7.   Make certain the fan blades are not bent or broken.   When a clutch-type fan drive is used, check its operation and adjustment.
8.   Check the drive belts for correct tension.

To check the cooling system for aeration, install a liquid eye into the top radiator hose.   Operate the engine at high idle, and then check for air bubbles. Remove and test the thermostat if necessary (see Unit 20, "Cooling System").

When the injection system is the cause of the high temperature, the exhaust smoke is dense.   Check the high: idle speed, the governor-to-fuel-rack adjustment, the injector adjustment, injector timing, or the fuel-injection-pump timing.

Using a contact pyrometer, measure the exhaust temperature of each cylinder to determine those which vary from the average exhaust temperature (See Fig. 51–3).

To check the injection-pump timing and/or the operation of the automatic advance, use a diesel timing light.   Install the specified transducer into the injection line (either at the injection pump or at the injector, depending on the transducer number) and make the necessary electrical connections (see Fig. 51–4).   Before you start the engine, clean the timing marks and be sure you have the manufacturer's timing specifications at hand.

Start the engine and operate it at idle speed. Aim the flashing timing light at the timing marks. Synchronize the timing marks by using the knurled

Fig. 51–4 Installed diesel-engine timing light.

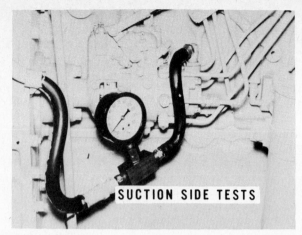

Fig. 51–5 Vacuum gauge installed to test fuel-inlet restriction. (International Harvester Co.)

knob on the timing light. Read from the timing light scale the advance of the engine at idle speed. Increase the engine speed to the specified rpm to check the remaining advance points. (See the sections on injector timing and adjustment in Units 31 and 32.)

Engine Has Lost Power or Is Not Developing Full Power A reduction in engine power is usually coincidental with hard starting, increased exhaust smoke density, increased engine temperature, engine misfiring, or rough operating. Power loss or underdeveloped engine power are constant problems. Both usually are traced to the fuel-injection system. The fuel-injection system, however, is not always the source of the trouble. Many other problems may be the source of the trouble. Therefore, check and measure the intake system for restriction and the exhaust system for excessive back pressure. Use a smokemeter or a smoke chart to measure the density of the exhaust smoke (see Caution, p. 393). Check and measure the turbocharger or blower boost pressure. Check for a leaking manifold. Check the cylinder compression, valve adjustment, and/or valve timing. Check for a leaking cylinder head or gasket.

When the reduced power is suspected to be the result of damaged, worn, or misadjusted fuel-injection system components, your theory can quickly be proved or disproved. Check first for external fuel leaks. If none exist, check for restricted fuel hoses or tubings, damaged fittings, incorrect fuel-tank installation, or a malfunctioning shutoff valve. The fuel-tank filter and/or filler cap should also be checked. To check the fuel-inlet restriction and transfer-pump pressure, install a pressure gauge to the manifold or immediately after the transfer pump, and install a vacuum gauge before the transfer pump (Fig. 51–5). Then operate the engine. The inlet restriction should not exceed 8 in Hg [20.32 cmHg], and the transfer-pump pressure must be within specification.

To check for the presence of air in the fuel during engine operation, install a liquid eye into the fuel line immediately before the transfer pump (to check the inlet side). If no air can be seen, install the liquid eye before the injection pump or manifold (to check the transfer pump and secondary filter). If the point of air entry occurs in any of these areas, it will be visible through the liquid eye.

When you check a Detroit Diesel engine for fuel-inlet restriction and the condition of the transfer-pump pressure, remove the return line at a suitable place after the restriction orifice and install a pressure gauge to the manifold. Then operate the engine at the specified rpm. Measure the return flow and the manifold pressure. While measuring the return flow, check for the presence of air in the return fuel.

If the cause of engine power loss has not yet been discovered, check the injection-pump timing or the injector timing or adjustment. Check the adjustment of the governor and the control rack travel, the torque spring, or the stop-plate adjustment. If the engine is equipped with either an aneroid or fuel-ratio control, check its adjustment.

When the problem is still unsolved, check the opening pressure of the injectors. This can be done by removing the injectors from the engine and then testing them on the injector tester, or by connecting a suitable tester to the injector as shown in Fig. 51–6.

Engine Has Too Much Power Excessive power is a fairly rare complaint, but when it is reported you may notice any or all of the following: the engine rpm is above specification, the exhaust temperature is higher than normal, or the exhaust smoke is very dense.

To locate the cause of excessive power, check the governor-to-fuel-rack adjustment, the adjustment of the maximum-fuel stop, and the manifold boost pressure. (The wrong turbocharger may have been installed.)

Diesel Exhaust Smoke A qualified mechanic can, by analyzing the exhaust smoke of a diesel engine, quickly evaluate the engine's performance and de-

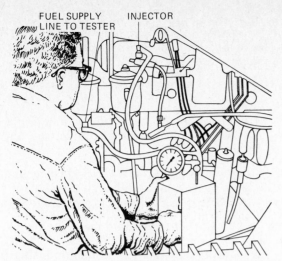

FUEL SUPPLY
LINE TO TESTER   INJECTOR

Fig. 51–6 Testing injector opening pressure while the injector is installed. *(International Harvester Co.)*

termine the source of trouble. Evaluation of engine trouble through the exhaust smoke should never be attempted when the engine is not at operating temperature or when the ambient temperature is below normal. The combustion chamber under these conditions is not hot enough to completely burn all the fuel injected because the time lag is increased. This increased time lag causes late ignition which, depending on combustion temperature, produces grey to white exhaust smoke. It is advisable to test drive the unit, to put a load on the engine using a chassy dynamometer, or to put a load on the engine using the torque converter of the power shift transmission so that you can analyze the exhaust smoke under all speeds and load.

Diesel exhaust smoke may be divided into three major groups: (1) liquid or cold exhaust smoke, which can range from grey to white, (2) hot or dense exhaust smoke which can range from 20 to 100 percent density (see Fig. 51–7), and (3) blue exhaust smoke.

LIQUID EXHAUST SMOKE When an engine at operating temperature emits grey to white exhaust smoke, this indicates that part of the fuel in the combustion chamber has not been ignited. This may be caused by low compression as a result of broken piston rings or lands, leaking valves, or misadjusted valves. To determine the cause, check cylinder pressure, blow-by, and valve adjustment. If the engine is misfiring, check the injector and/or the injector nozzles. There may be a leaking fuel nozzle, the opening pressure may be low, or the orifices may be enlarged. The injection timing may be late or the injectors may be misadjusted. Grey or white exhaust smoke also re-

sults from too low a fuel cetane rating, or from an excessive coolant leak into the combustion chamber.

HOT OR DENSE EXHAUST SMOKE Hot (dense) exhaust smoke indicates that the injected fuel has not burned completely. This may be due to reduced air in the cylinders or from overfueling.

CAUTION Never guess the density (percentage) of the exhaust smoke. Measure it with a smoke-meter or compare it with a smoke scale (Fig. 51–7).

The smoke scale is divided into five groups ranging from 20 to 100 percent in density and is consistent with the Hartridge smokemeter scale. Although 40 percent density is permissible, it is near the upper range of the acceptable limit of emission control standards. If the exhaust becomes denser than 40 percent check the intake system for restriction, and check the exhaust system for excessive back pressure. Check the turbocharger or blower boost pressure, the valve adjustment, the cylinder compression, and the blow-by. Also check the coolant system and the oil level since both affect engine temperature.

When the excessive exhaust density is due to overfueling or early timing, check one or more of the following:

1. The adjustment of the aneroid or fuel-ratio control.
2. The injector timing, injector adjustment, or injection-pump timing.
3. Check the governor and the adjustment of the fuel control rack.
4. If necessary, check the opening pressure and the spray pattern of the injectors. Poor spray pattern and/or low opening pressure will reduce atomization and cause an increase in exhaust smoke density.

BLUE EXHAUST SMOKE When the smoke emitted is a mixture of blue and white (occasionally it may be completely blue), it indicates that the engine burns excessive oil.

When the exhaust is bluish, check the crankcase oil and the oil level. The oil may be too light for the ambient temperature or its level may be too high. In either case, excessive oil is thrown onto the cylinder walls and the piston rings cannot control it. Worn main bearings and connecting-rod bearings as well as excessive oil pressure can also cause oil to pass by the piston rings.

Check the cylinder compression and blow-by. Do not forget to check the crankcase breather be-

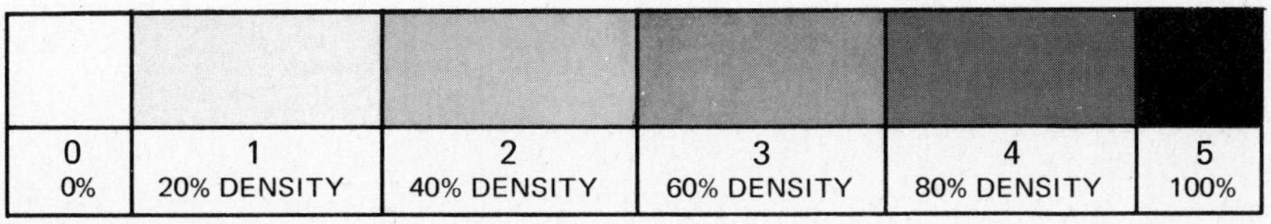

0	1	2	3	4	5
0%	20% DENSITY	40% DENSITY	60% DENSITY	80% DENSITY	100%

Fig. 51–7 Smoke chart. *(American Bosch-Ambac Industries, Inc.)*

cause a plugged breather can increase the crankcase pressure and force oil into the combustion chamber.

When an oil-bath cleaner is used, check its level. It may be too high, thus permitting oil to be drawn into the cylinders. Also, check the valve guides and valve seals for excessive wear. Worn guides or seals can permit oil to be drawn into the cylinder on the intake stroke.

Questions

1. How would you define the term *tuneup*?

2. List several circumstances which would call for a tuneup.

3. Why is it necessary to have a good basic knowledge of the fundamental operating principles of the various systems?

4. When servicing an engine, why should you seek information, from the operator and from the engine service record?

5. What components would you check if the engine were hard to start but the cranking speed, compression, and timing checked out satisfactorily?

6. What symptoms would be noticeable if the camshaft gear were timed late by one tooth?

7. If the compressor on a Cummins engine were not timed, how would this affect the engine operation?

8. What components would you check when an engine overheats even though the coolant system checked out satisfactorily or the injection-system timing checked out satisfactorily?

9. List the components within the fuel-injection system of a Caterpillar compact-housing fuel pump which, if defective, could cause a reduction in power output.

10. Why, in your opinion, does an engine emit "cold smoke" prior to its reaching operating temperature?

CONVERSION TABLES

Well over 95 percent of the world's population already uses metric units of measurement or is converting to them. The tremendous advantage of the metric system lies in its simplicity and universality: All relationships between the various units of the metric system work in powers of 10 and unified symbols are used for each unit instead of a conglomerate of abbreviations.

Since many informative service manuals are still in use wherein the specifications are written in English units of measurement, the basic conversion charts have been included in this textbook to help you cope during this time of transition. These charts will provide you with a fast and easy means of converting from English to metric units of measurement, or vice versa.

For example, if you want to know the displacement in liters or cubic centimeters of a diesel engine having a displacement of 600 in^3 (cubic inches), select the capacity conversion table. Refer to the cubic inches entry in the left-hand column. The comparable amount in liters (l) and in cubic centimeters (cm^3) is shown to the right. If 1 in^3 equals 16.39 cm^3, then 600 in^3 equals 600 × 16.39 = 9834 cm^3. Likewise, if 1 in^3 equals 0.0164 l, then 600 in^3 equals 600 × 0.0164 = 9.84 l.

AREA CONVERSION

Unit	in²	ft²	yd²	mi²	cm²	dm²	m²
1 in²	1	0.0069	.....	....	6.452	0.06452	0.00064
1 ft²	144	1	0.1111	0.0092	929	9.29	0.0929
1 yd²	1296	9	1	0.0288	8361	83.61	0.8361
1 mi²	....	27,878,400	....	1	....	....	2,589,998
1 cm²	0.155	0.0017	.....	...	1	0.01	0.0001
1 dm²	15.5	0.1076	0.01196	...	100	1	0.01
1 m²	1550	10.76	1.196	...	10,000	100	1

BRITISH THERMAL UNITS (Btu)

1 Btu = 778.3 foot-pounds (ft · lb)
1 Btu = 107.5 meter-kilograms (m · kg)
1 Btu = 0.003931 horsepower-hour (hp · h)
1 Btu = 0.002928 kilowatthour (kWh)

CAPACITY CONVERSION

Unit	in³	ft³	U.S. liquid	U.S. gal	Imp. qt	Imp. gal	cm³	dm³	m³	l
1 in³	1	....	0.01732	....	0.01442	...	16.39	0.01639	......	0.0164
1 ft³	1728	1	29.92	7.481	24.92	6.229	.....	28.32	0.02832	28.32
1 yd³	46,656	27	807.9	202	672.8	168.2	.....	764.6	0.7646	764.6
1 U.S. liquid qt	57.75	0.03342	1	0.25	0.8327	0.2082	946.4	0.9464		
1 U.S. gal	231	0.1337	4	1	3.331	0.8327	3785	3.785	0.003785	3.785
1 Imp. qt	69.36	0.04014	1.201	0.3002	1	0.25	1136	1.136		
1 Imp. gal	277.4	0.1605	4.804	1.201	4	1	4546	4.546	0.004546	4.546
1 cm³	0.06102	....	.....	......	.....	....	1	0.001	0.000001	0.001
1 dm³	61.02	0.03531	1.057	0.2642	0.88	0.22	1000	1	0.001	1
1 m³	61,023	35.31	1057	264.2	880	220	....	1000	1	1000
1 l	61.02	0.03531	1.057	0.2642	0.88	0.22	1000	1	0.001	1

LENGTH CONVERSION

Unit	in	ft	yd	mi (statute)	nmi	mm	cm	m	km	μm
1 in	1	0.08333	0.02778	....	....	25.4	2.54	0.0254	....	25.400
1 ft	12	1	0.3333	....	....	304.8	30.48	0.30488	....	
1 yd	36	3	1	....	....	914.4	91.44	0.9144	....	
1 mi (statute)	63,360	5280	1760	1	0.8684	....	....	1609.3	1.609	
1 nmi	72,960	6080	2027	1.152	1	....	....	1853.3	1.853	
1 mm	0.03937	3281×10^3	....	....	....	1	0.1	0.001	....	1,000
1 cm	0.3937	4.716	14.148	....	....	10	1	100	....	10,000
1 m	39.37	3.281	1.094	....	....	1000	100	1	0.001	
1 km	39,370	3281	1093.6	0.6214	0.5396	10^6	100,000	1000	1	
1μm	0.000039	....	....	....	....	0.01	0.0001	0.000001	....	1

PRESSURE CONVERSION

Unit	inH₂O	cmH₂O	inHg	cmHg	psi	kg/cm²	atm
1 inH₂O	1	2.54	0.0735	1.866	0.0361		
1 inHg	13.6	34.544	1	2.54	0.491	0.0345	0.0334
1 psi	27.7	70.104	2.036	5.171	1	0.0703	0.068
1 kg/cm²	393.73	1000.0	28.96	73.55	14.22	1	0.9678
1 atm	407.19	1033.0	29.92	75.96	14.70	1.033	1

WEIGHT CONVERSION

Unit	oz	lb	g	kg
oz	1	0.0625	28.35	0.02835
lb	16	1	453.59	0.45359
g	0.03527	....	1	0.001
kg	35.27	2.2046	1000	1

TORQUE CONVERSION

Unit	lb · ft	kg · m
1 lb · ft	1	0.1383
1 kg · m	7.233	1

POWER CONVERSION

Unit	hp	hp (metric)	ft · lb/s	kg · m/s	kW	W	Btu/min
hp	1	1.014	550	76.04	0.746	746	42.4
hp (metric)	0.986	1	542.5	75.00	0.736	736	41.8
ft · lb/s	.....	.....	1	0.30488	....	0.0226	0.001285
kg · m/s	.....	.....	3.281	1	....	0.0741	0.0042
kW	1.341	1.360	737.28	102.00	1	1000	56.8
W	0.00134	0.00136	0.737	0.102	0.001	1	17.6
Btu/min	0.0236	0.0239	12.96	3.939	0.0176	17.6	1

PRESSURE VERSUS THE BOILING POINTS OF WATER
(VARIOUS ALTITUDES)

Altitude		psi	inHg	kg/cm²	Boiling point of water	
ft	in				°F	°C
Sea level		14.69	29.92	1.032	212	100
1000	304.8	14.16	28.86	0.995	210.1	99
2000	609.6	13.66	27.82	0.960	208.3	98
3000	914.4	13.16	26.81	0.925	206.5	97
4000	1219.4	12.68	25.84	0.891	204.6	95.9
5000	1524.0	12.22	24.89	0.859	202.8	94.9
6000	1828.8	11.77	23.98	0.827	201.0	94.1
7000	2133.6	11.33	23.09	0.796	199.3	93.0
8000	2438.4	10.91	22.22	0.766	197.4	91.9
9000	2743.2	10.50	21.38	0.738	195.7	91
10,000	3048.0	10.10	20.58	0.701	194.0	90

FUEL CONSUMPTION FORMULAS

Fuel consumption (lb/hr) = specific fuel consumption [lb/(bhp)/(h)] × bhp

$$\text{Fuel consumption (gal/h)} = \frac{\text{specific fuel consumption [lb/(bhp)/(h)]} \times \text{bhp}}{\text{fuel specific weight (lb/gal)}}$$

Fuel specific weight (lb/gal) = fuel specific gravity × 8.34 lb

$$\text{Specific fuel consumption [lb/(bhp)/(h)]} = \frac{\text{fuel consumption (gal/h)} \times \text{fuel specific weight (lb/gal)}}{\text{bhp}}$$

$$\text{Specific fuel consumption [kg/(mhp)/(h)]} = \frac{\text{specific fuel consumption [lb/(bhp)/(h)]}}{2.24}$$

TEMPERATURE CONVERSION

°F = 9/5 × °C + 32
°C = 5/9 × (°F − 32)

°F	°C	°F	°C	°F	°C	°F	°C	°F	°C
−60	−50	+30	0	120	50	210	100	3500	2000
−70								3000	
−80	−60	+20		110		200		2500	1500
−90			−10		40	190	90	2000	
−100	−70	+10		100					1000
	−80							1500	900
−120		0		90		180			800
	−90		−20		30		80		700
−140	−100					170			600
−160		−10		80				1000	500
−180	−120					160	70	900	
−200		−20	−30	70	20			800	400
	−140							700	300
−250	−160	−30		60		150		600	
	−180							500	
−300	−200	−40	−40	50	10	140	60	400	200
−350									180
	−250	−50		40		130		300	160
−400								250	140
−459.67	−273.75	−50	−50	32	0	122	50		120
									100

BRAKE HORSEPOWER FORMULA

$$bhp = \frac{torque\ (ft \cdot lb) \times engine\ rpm}{5252}$$

CONVERSION FOR TEST STANDARDS

	Altitude, ft [m]		Barometric pressure, inHg [kg/cm²]		Ambient temperature, °F [°C]		Air density, %	Brake power, lb·ft/s [kg·m/s]	
SAE	500	[152.4]	29.38	[746.25]	85	[29.4]	50	550.0	[76.04]
British	500	[152.4]	29.38	[746.25]	85	[29.4]	50	550.0	[76.04]
Din (German)	984	[300.0]	28.88	[733.55]	68	[20.0]	60	542.5	[75.00]
Metric	984	[300.0]	28.88	[733.55]	68	[20.0]	60	542.5	[75.00]
Sea level	Sea level	Sea level	29.92	[759.97]	60	[15.55]	Dry air	*550.0	[*76.04]

* Standard ambient condition.

DECIMAL EQUIVALENTS OF MILLIMETERS

mm	in	mm	in	mm	in	mm	in	mm	in
0.01	0.00039	0.41	0.01614	0.81	0.03189	21	0.82677	61	2.40157
0.02	0.00079	0.42	0.01654	0.82	0.03228	22	0.86614	62	2.44094
0.03	0.00118	0.43	0.01693	0.83	0.03268	23	0.90551	63	2.48031
0.04	0.00157	0.44	0.01732	0.84	0.03307	24	0.94488	64	2.51968
0.05	0.00197	0.45	0.01772	0.85	0.03346	25	0.98425	65	2.55905
0.06	0.00236	0.46	0.01811	0.86	0.03386	26	1.02362	66	2.59842
0.07	0.00276	0.47	0.01850	0.87	0.03425	27	1.06299	67	2.63779
0.08	0.00315	0.48	0.01890	0.88	0.03465	28	1.10236	68	2.67716
0.09	0.00354	0.49	0.01929	0.89	0.03504	29	1.14173	69	2.71653
0.10	0.00394	0.50	0.01969	0.90	0.03543	30	1.18110	70	2.75590
0.11	0.00433	0.51	0.02008	0.91	0.03583	31	1.22047	71	2.79527
0.12	0.00472	0.52	0.02047	0.92	0.03622	32	1.25984	72	2.83464
0.13	0.00512	0.53	0.02087	0.93	0.03661	33	1.29921	73	2.87401
0.14	0.00551	0.54	0.02126	0.94	0.03701	34	1.33858	74	2.91338
0.15	0.00591	0.55	0.02165	0.95	0.03740	35	1.37795	75	2.95275
0.16	0.00630	0.56	0.02205	0.96	0.03780	36	1.41732	76	2.99212
0.17	0.00669	0.57	0.02244	0.97	0.03819	37	1.45669	77	3.03149
0.18	0.00709	0.58	0.02283	0.98	0.03858	38	1.49606	78	3.07086
0.19	0.00748	0.59	0.02323	0.99	0.03898	39	1.53543	79	3.11023
0.20	0.00787	0.60	0.02362	1.00	0.03937	40	1.57480	80	3.14960
0.21	0.00827	0.61	0.02402	1	0.03937	41	1.61417	81	3.18897
0.22	0.00866	0.62	0.02441	2	0.07874	42	1.65354	82	3.22834
0.23	0.00906	0.63	0.02480	3	0.11811	43	1.69291	83	3.26771
0.24	0.00945	0.64	0.02520	4	0.15748	44	1.73228	84	3.30708
0.25	0.00984	0.65	0.02559	5	0.19685	45	1.77165	85	3.34645
0.26	0.01024	0.66	0.02598	6	0.23622	46	1.81102	86	3.38582
0.27	0.01063	0.67	0.02638	7	0.27559	47	1.85039	87	3.42519
0.28	0.01102	0.68	0.02677	8	0.31496	48	1.88976	88	3.46456
0.29	0.01142	0.69	0.02717	9	0.35433	49	1.92913	89	3.50393
0.30	0.01181	0.70	0.02756	10	0.39370	50	1.96850	90	3.54330
0.31	0.01220	0.71	0.02795	11	0.43307	51	2.00787	91	3.58267
0.32	0.01260	0.72	0.02835	12	0.47244	52	2.04724	92	3.62204
0.33	0.01299	0.73	0.02874	13	0.51181	53	2.08661	93	3.66141
0.34	0.01339	0.74	0.02913	14	0.55118	54	2.12598	94	3.70078
0.35	0.01378	0.75	0.02953	15	0.59055	55	2.16535	95	3.74015
0.36	0.01417	0.76	0.02992	16	0.62992	56	2.20472	96	3.77952
0.37	0.01457	0.77	0.03032	17	0.66929	57	2.24409	97	3.81889
0.38	0.01496	0.78	0.03071	18	0.70866	58	2.28346	98	3.85826
0.39	0.01535	0.79	0.03110	19	0.74803	59	2.32283	99	3.89763
0.40	0.01575	0.80	0.03150	20	0.78740	60	2.36220	100	3.93700

SOURCE: The L.S. Starrett Co.

TUBE OD SIZES

in	mm
1/8	3.18
3/16	4.76
1/4	6.35
5/16	7.94
3/8	9.52
13/32	10.32
1/2	12.70
5/8	15.88
3/4	19.05
7/8	22.22
1	25.40
1 1/8	28.58
1 1/4	31.75
1 1/2	38.10
1 3/4	44.45
2	50.80

TAP DRILL SIZES FOR FRACTIONAL SIZE THREADS—75% DEPTH THREAD, IN—AMERICAN NATIONAL THREAD FORM

Tap size	Threads per inch	Diameter hole	Drill
1/16	72	0.049	3/64
1/16	64	0.047	3/64
1/16	60	0.046	56
5/64	72	0.065	52
5/64	64	0.063	1/16
5/64	60	0.062	1/16
5/64	56	0.061	53
3/32	60	0.077	5/64
3/32	56	0.076	48
3/32	50	0.074	49
3/32	48	0.073	49
7/64	56	0.092	42
7/64	50	0.090	43
7/64	48	0.089	43
1/8	48	0.105	36
1/8	40	0.101	38
1/8	36	0.098	40
1/8	32	0.095	3/32
9/64	40	0.116	32
9/64	36	0.114	33
9/64	32	0.110	35
5/32	40	0.132	30
5/32	36	0.129	30
5/32	32	0.126	1/8
11/64	36	0.145	27
11/64	32	0.141	9/64
3/16	36	0.161	20
3/16	32	0.157	22
3/16	30	0.155	23
3/16	24	0.147	26
13/64	32	0.173	17
13/64	30	0.171	11/64
13/64	24	0.163	20
7/32	32	0.188	12
7/32	28	0.184	13
7/32	24	0.178	16
15/64	32	0.204	6
15/64	28	0.200	8
15/64	21	0.194	10
1/4	32	0.220	7/32
1/4	28	0.215	3
1/4	27	0.214	3
1/4	24	0.209	4
1/4	20	0.201	7
5/16	32	0.282	9/32
5/16	27	0.276	J
5/16	24	0.272	I
5/16	20	0.264	17/64
5/16	18	0.258	F
3/8	27	0.339	R
3/8	24	0.334	Q
3/8	20	0.326	21/64
3/8	16	0.314	5/16
7/16	27	0.401	Y
7/16	24	0.397	X
7/16	20	0.389	25/64
7/16	14	0.368	U
1/2	27	0.464	15/32
1/2	24	0.460	00/64

Tap size	Threads per inch	Diameter hole	Drill
1/2	20	0.451	29/64
1/2	13	0.425	27/64
1/2	12	0.419	27/64
9/16	27	0.526	17/32
9/16	18	0.508	33/64
9/16	12	0.481	31/64
5/8	27	0.589	19/32
5/8	18	0.571	00/64
5/8	12	0.544	35/64
5/8	11	0.536	17/32
11/16	16	0.627	5/8
11/16	11	0.599	00/32
3/4	27	0.714	00/32
3/4	16	0.689	11/16
3/4	12	0.669	43/64
3/4	10	0.653	21/32
15/16	12	0.731	47/64
13/16	10	0.715	00/32
7/8	27	0.839	27/32
7/8	18	0.821	00/64
7/8	14	0.805	13/16
7/8	12	0.794	51/64
7/8	9	0.767	49/64
15/16	12	0.856	55/64
15/16	9	0.829	53/64
1	27	0.964	31/32
1	14	0.930	15/16
1	12	0.919	00/64
1	8	0.878	7/8
1 1/16	8	0.941	15/16
1 1/8	12	1.044	1 3/64
1 1/8	7	0.986	63/64
1 3/16	7	1.048	1 8/64
1 1/4	12	1.169	1 11/64
1 1/4	7	1.111	1 7/64
1 5/16	7	1.173	1 11/64
1 3/8	12	1.294	1 19/64
1 3/8	6	1.213	1 7/32
1 1/2	12	1.419	1 27/64
1 1/2	6	1.338	1 11/32
1 5/8	5 1/2	1.448	1 29/64
1 3/4	5	1.555	1 9/16
1 7/8	5	1.680	1 11/16
2	4 1/2	1.783	1 00/00
2 1/8	4 1/2	1.909	1 00/00
2 1/4	4 1/2	2.034	2 1/32
2 3/8	4	2.131	2 1/8
2 1/2	4	2.256	2 1/4
2 5/8	4	2.381	2 3/8
2 3/4	4	2.505	2 1/2
2 7/8	3 1/2	2.597	2 00/32
3	3 1/2	2.722	2 00/32
3 1/8	3 1/2	2.847	2 00/32
3 1/4	3 1/2	2.972	2 00/32
3 3/8	3 1/4	3.075	3 1/16
3 1/2	3 1/4	3.200	3 3/16
3 5/8	3 1/4	3.325	3 5/16
3 3/4	3	3.425	3 7/16
4	3	3.675	3 11/16

TAP DRILL SIZES FOR MACHINE SCREW THREADS—75% DEPTH OF THREAD

A bolt inserted in an ordinary nut, which has only one-half of a full depth of thread, will break before stripping the thread. Also a full depth of thread, while very difficult to obtain, is only about 5% stronger than a 75% depth.

These tables give the exact size of the hole, expressed in decimals, that will produce a 75% depth of thread, and also the nearest regular stock drill to this size. Holes produced by these drills are considered close enough for any commercial tapping.

$$\text{Diameter of tap} - \frac{0.974}{\text{no. threads per inch}} = \text{diameter of hole}$$

Tap size	Threads per inch	Diameter hole	Drill	Tap size	Threads per inch	Diameter hole	Drill
0	80	0.048	3/64	10	32	0.160	21
1	72	0.060	53	10	30	0.158	22
1	64	0.058	53	10	28	0.155	23
2	64	0.071	50	10	24	0.149	25
2	56	0.069	50	12	28	0.181	14
3	56	0.082	45	12	24	0.175	16
3	48	0.079	47	14	24	0.201	7
4	48	0.092	42	14	20	0.193	10
4	40	0.088	43	16	22	0.224	2
4	36	0.085	44	16	20	0.219	7/32
5	44	0.103	37	16	18	0.214	3
5	40	0.101	38	18	20	0.245	D
5	36	0.098	40	18	18	0.240	B
6	40	0.114	33	20	20	0.271	I
6	36	0.111	34	20	18	0.266	17/64
6	32	0.108	36	22	18	0.292	L
7	36	0.124	1/8	22	16	0.285	9/32
7	32	0.121	31	24	18	0.318	O
7	30	0.119	31	24	16	0.311	5/16
8	36	0.137	29	26	16	0.337	R
8	32	0.134	29	26	14	0.328	21/64
8	30	0.132	30	28	16	0.363	23/64
9	32	0.147	26	28	14	0.354	T
9	30	0.145	27	30	16	0.389	25/64
9	24	0.136	29	30	14	0.380	V

SOURCE: The L. S. Starrett Co.

ABDC Abbreviation for *after bottom dead center.*

Absolute pressure Gauge pressure plus atmospheric pressure.

Absolute Temperature The temperature measured using absolute zero as a reference. Absolute zero is −273.16°C, or [459.69°F] and is the lowest point of temperature known.

Acceleration The rate of increase of velocity per time unit (i.e., seconds, minutes, hours).

Accumulator A device used for storing liquid under pressure (sometimes used to smooth out pressure surges in a hydraulic system).

Actuator A device which uses fluid power to produce mechanical force and motion.

Additive A matter which is added to improve fuel.

Advance To set the timing of the injection pump or injectors for an earlier injection.

Aftercooler A device used on turbocharged engines to cool air which has undergone compression.

Air bind The presence of air in a pump or pipes which prevents the delivery of liquid.

Air bleeder A device used to remove air from a hydraulic system. Types include a needle valve, capillary tubing to the reservoir, and a bleed plug.

Air cleaner A device (filter) for removing unwanted solid impurities from the air before it enters the intake manifold.

Air compressor A device used to increase air pressure.

Air/fuel ratio The ratio (by weight or by volume) between air and fuel.

Air gap The distance between two components.

Air pollution Contamination of the earth's atmosphere by pollutants such as smoke, harmful gases, etc.

Air starting valve A valve which admits compressed air to the air starter for starting purposes.

Align To bring two or more components of a unit into the correct positions with respect to one another.

Alloy A mixture of two or more different metals, usually to produce improved characteristics.

Allowance The difference between the minimum and the maximum dimensions of proper functioning.

Alnico magnet A magnet composed of aluminum (Al), nickel (Ni), and cobalt (Co).

Alternator An electromechanical device which produces ac current.

Alternating current (ac) An electric current that changes polarity.

Ambient temperature Surrounding air temperature.

Ammeter An instrument used to measure the rate of current flow in amperes.

Ampere (A) A unit of measurement defined as the current that one volt can send through one ohm resistance.

Ampere-hour capacity A measurement of the battery capacity to deliver a specified current over a specified length of time.

Angle Inclination of two lines to each other.

Angularity Having or being at an angle.

Aneroid A pressure-measuring device containing no liquid.

Anneal To toughen metals by heating and then cooling.

Annular In the form of an annulus; ring-shaped.

Annulus A figure bounded by concentric circles or cylinders (e.g., a washer, ring, sleeve, etc.).

Antifreeze A chemical added to the coolant in order to lower its freezing point.

API Abbreviation for *American Petroleum Institute.*

API gravity Gravity expressed in units of standard API (hydrometer).

Arc Portion of a curved line or of the circumference of a circle.

Arcing Electrons leaping the gap between the negative and the positive poles.

Armature The movable part of a relay, regulator, or horn or the rotating part of a generator or starter.

Asbestos A heat-resistant and nonburning organic mineral.

Aspirate To breathe (to draw out gas by suction).

ATDC Abbreviation for *after top dead center.*

Atmosphere The mass or blanket of gases surrounding the earth.

Atmospheric pressure (Barometric pressure) The pressure exerted by the atmosphere, averaging 14.7 psi at sea level with a decrease of approximately 1/2 lb per 1000 ft of altitude gained.

Atom The smallest particle of an element.

Atomizer A device which disperses liquid (e.g., fuel) into fine particles (pulverized spray).

Attrition Wearing down by rubbing or by friction; abrasion.

Automatic valve A valve assisted by a spring, which is opened by a difference of pressure acting in one direction and closed by a difference in pressure acting in the opposite direction.

Auxiliary An aid to the main device which may only be used occasionally.

Babbit An antifriction metal used to line bearings, thereby reducing the friction of the moving components.

Backlash The distance (play) between two movable components.

Back pressure A pressure exerted contrary to the pressure producing the main flow.

Baffle A device which slows down or diverts the flow of gases, liquids, sound, etc.

Balanced valve A valve in which the fluid pressure is equal on both sides (i.e., the opening and closing directions).

Ball bearing A bearing using steel balls as its rolling element between the inner and outer ring (race).

Ball check valve A valve consisting of a ball held against a ground seat by a spring. It is used to check the flow or to limit the pressure.

Barometer An instrument which measures atmospheric pressure.

Basic size The theoretical or nominal standard size from which all variations are made.

Battery An electrochemical device that produces electric current.

Battery charging (See *Charge*).

BBDC Abbreviation for *before bottom dead center.*

BDC Abbreviation for *bottom dead center.*

Bearing The contacting surface on which a revolving part rests.

Bearing clearance The distance between the shaft and the bearing surface.

Bell housing (Clutch housing) The metal covering around the clutch or torque converter assembly.

Bendix-type starter drive (Inertia starter drive) A type of starter drive that causes the gear to engage when the armature starts rotating and to automatically disengage when it stops.

Bernoulli's principle Given a fluid flowing through a tube, any constriction or narrowing of the tube will create an increase in that fluid's velocity and a decrease in pressure.

Blow-by Exhaust gases which escape past the piston rings.

Blower A low-pressure air pump, usually of one rotary or centrifugal type.

Bond The holding together of different parts.

Boiling point The temperature at which a liquid begins to boil.

Bore A cylinder, hole, or the inside diameter of the cylinder or hole.

Bore diameter The diameter of a hole or cylinder.

Boring Enlarging the cylinders by cutting or honing them to a specified size.

Boring bar (cylinder) A tool used to machine the cylinders to a specific size.

Bosch metering system A metering system with a helical groove in the plunger which covers or uncovers ports in the pump barrel.

Bound electrons The inner-orbit electrons around the nucleus of the atom.

Boyle's law The absolute pressure which a given quantity of gas at constant temperature exerts against the walls of the containing vessel is inversely proportional to the volume occupied.

Brake horsepower (bhp) The usable power delivered by the engine.

Brake mean effective pressure (bmep) Mean effective pressure acting on the piston which would result in the given brake horsepower output, if there were no losses due to friction, cooling, and exhaustion. Equal to mean indicated pressure times mechanical efficiency.

Brake thermal efficiency Ratio of power output in the form of brake horsepower to equivalent power input in the form of heat from fuel.

Brazing The fastening of two pieces of metal together by heating the edges and then melting brass or bronze on the area.

Breather pipe A pipe opening into the crankcase to assist ventilation.

Brinnel hardness The surface hardness of a metal, alloy, or similar material according to J. A. Brinnell's method of measurement. A metal's surface is struck at a given force by a rigid steel ball of given diameter and the indentation is measured.

British gallon (Imperial gallon) A gallon measurement of 277.4 in³.

British thermal unit (Btu) Approximate definition: The amount of heat required to raise one pound of water one degree Farenheit. Exact definition: 1/180 the amount of heat required to raise one pound of water from freezing to boiling at a standard atmospheric pressure.

Brush The pieces of carbon or copper that make a sliding contact against the commutator or slip rings.

BTDC Abbreviation for *before top dead center.*

Buoyancy The upward or lifting force exerted on a body by a fluid.

Burnish To polish or shine a surface with a hard, smooth object.

Bushing A metallic or synthetic lining for a hole which reduces or prevents abrasion between components.

Butane A hydrocarbon gas formed synthetically by the action of zinc or ethyl iodide. This gas becomes a liquid when under pressure.

Butterfly valve A valve in the venturi to control the airflow.

Bypass filter An oil filter that only filters a portion of the oil flowing through the engine lubrication system.

Bypass valve A valve that opens when the set pressure is exceeded. This allows the fluid to pass through an alternative channel.

Cage A housing in which a valve operates and seats.

Calibrate To make an adjustment to a meter or other instrument so that it will indicate accurately its input.

Calipers A tool for measuring diameter, usually having curved legs and resembling a pair of compasses.

Calorie The amount of heat required to raise one gram of water from 17 to 18°C.

Calorific value The amount of heat produced by burning one pound of fuel. (See *heating value.*)

Cam A rotating component of irregular shape. It is used to change the direction of the motion of another part moving against it, e.g., rotary into reciprocating or variable motion.

Cam follower (Valve lifter) A part which is held in contact with the cam and to which the cam motion is imparted and transmitted to the pushrod.

Cam-ground A piston that is ground slightly oval but becomes round when heated.

Cam nose That portion of the cam that holds the valve wide open. It is the high point of the cam.

Camshaft A shaft with cam lobes.

Camshaft gear The gear that is fastened to the camshaft.

Capacitor An arrangement of insulated conductors and dielectrics for the accumulation of an electric charge with small voltage output.

Carbon One of the nonmetallic elements constituting fuel and lubricating oil.

Carbon dioxide (CO₂) A colorless, odorless gas which results when carbon is burned completely.

Carbon monoxide (CO) A colorless, odorless, poisonous gas which results from the incomplete burning of carbon.

Carbon pile Carbon disks or plates capable of carrying high current.

Carbon tetrachloride A colorless liquid, the fumes of which are toxic. Used in fire extinguishers and for cleaning.

Carburizing To combine or add to the metal the element carbon for hardening purposes.

Case-harden To harden the outer surface of metal to a given case or shell depth, while leaving the inner portion soft to absorb shocks and allow bending. (See *Carburizing.*)

Cavitation The formation of cavities in the fluid due to excessive speed of the activator resulting in the loss of efficiency in the pump.

Cells (battery) The individual (separate) compartments in the battery which contain positive and negative plates suspended in electrolyte.

Cell connectors The lead straps connecting the cell groups.

Celsius (Centigrade) Thermometer scale on which the freezing temperature of water is 0°C and boiling temperature of water at atmospheric pressure is 100°C.

Center of gravity That point in a body at which all its mass can be concentrated without allowing the effect of gravity.

Centigrade thermometer A thermometer with a centigrade scale.

Centrifugal force A force exerted on a rotating object in a direction outward from the center of rotation.

Centrifugal governor A governor

which uses flyweight force to sense speed in order to control the amount of fuel supplied to the combustion chambers.

Centrifugal pump A pump using the centrifugal force produced by a rapidly rotating impeller to displace liquid.

Centrifuge A device with a rapidly rotating bowl which separates the impurities of a fluid by intense centrifugal force. It is one of the most efficient means known for purifying fuel and lubricating oils.

Cetane number The rating of a diesel fuel's ignition.

Chamfer (Taper lead) The taper at the thread end of a tap or the throat of a die, made by cutting away the crests of the first few threads. This distributes the work of cutting over several threads and acts as a guide in starting the tap or die. The chamfer is relieved to facilitate cutting. The tap is classed as taper, second, or bottoming, according to the length of chamfer, which approximates the following:

- Taper tap: 4° per side
- Second tap: 8° per side
- Bottoming tap: 23° per side
- Nut taps: Approximately 75 percent of the thread length

Charge (battery) To force current (electrons) into a battery to restore it to a full state of charge.

Charles' law The physical law which states that the rise of temperature in all gases produces the same increase in volume if the pressure remains constant.

Check valve A valve which permits only one direction of flow.

Chemical change A change which alters the composition of the molecules of a substance producing new substances with new properties.

Circuit (electric) A source of power. A path for current flow with one or more resistant units.

Circuitbreaker (lighting system) A device that opens the circuit when the current draw becomes excessive and closes the circuit when the current flow is reduced.

Circulating pump The term applied to cooling-oil pumps which give circulation of fluid.

Circumference The distance, or perimeter, of a circle. (π times the diameter.)

Clearance The space between two components.

Clearance volume The volume remaining above the piston when it is at TDC.

Closed cooling system A cooling system which is not exposed to the atmosphere.

Closed nozzle A fuel nozzle having a valve between the combustion chamber and the fuel chamber.

Clutch A device used to connect or disconnect the power input to the power output.

Clutch pilot bearing A small bushing, or ball bearing positioned in the crankshaft or flywheel.

Coil spring A spring-steel wire wound in a spiral pattern.

Color code Colored markings or wires to identify the different circuits.

Combustion The act or process during burning.

Combustion chamber The chamber in which combustion mainly occurs.

Combustion chamber volume The volume of the combustion chamber (when the piston is at TDC) measured in cubic centimeters.

Combustion cycle A series of thermodynamic processes through which the working gas passes to produce one power stroke. The full cycle is: intake, compression, power, and exhaust.

Commutator A number of copper bars connected to the armature windings but insulated from each other and from the armature.

Compound A combination of two or more elements that are mixed together.

Compressed air Air that at any pressure in excess of atmospheric pressure is considered to be compressed.

Compressibility The property of a substance (e.g., air) by virtue of which its density increases with increase in pressure.

Compression The process by which a confined gas is reduced in volume through the application of pressure.

Compression check A measurement of the compression of each cylinder at cranking speed or as recommended by the manufacturer.

Compression gauge A test instrument used to test the cylinder compression.

Compression ignition The ignition of fuel through the heat of compression.

Compression pressure Pressure in the combustion chamber at the end of the compression stroke, but without any of the fuel being burned.

Compression ratio The ratio between the total volume in the cylinder when the piston is at BDC and the volume remaining when the piston is at TDC.

Compression release A device to prevent the intake or exhaust valves from closing completely, thereby permitting the engine to be turned over without compression.

Compression ring The piston rings used to reduced combustion leakage to a minimum.

Compression stroke That stroke of the operating cycle during which air is compressed into a smaller space creating heat by molecular action.

Compressor A mechanical device to pump air, and thereby increase the pressure.

Concentric Having the same center.

Condensation The reduction of a vapor or gas to a liquid state.

Condense To reduce from gas or vapor to liquid.

Condenser An arrangement of insulated conductors and dielectrics for the accumulation of an electric charge.

Conduction The transmission of heat through matter without motion of the conducting body.

Conductor Any material or device forming a path for the flow of electrons.

Connecting rod The rod joining the piston with the crankshaft.

Connecting-rod bearing The bearing used in the connecting-rod bore.

Constant pressure combustion Combustion which occurs without a change in pressure. In an engine, this is obtained by a slower rate of burning than with constant volume combustion.

Contamination The presence of harmful foreign matter in a fluid or in air.

Contour Outline.

Contract To reduce; to make smaller.

Control To regulate or govern the function of a unit.

Controlled port scavenging Scavenging method using ports which are controlled by valves in addition to the power piston.

Convection The transfer of heat through a liquid by motion of its parts.

Conventional According to the most common or usual mode.

Converge To incline to or approach a certain point; to come together.

Convolution One full turn of a screw.

Coolant A liquid used as a cooling medium.

Cooling system The complete system for circulating coolant.

Core The central or innermost part of an object.

Corrode To eat or wear away gradually by chemical action.

Corrosion The slow destruction of material by chemical agents and electromechanical reactions.

Corrosive Having the ability to corrode.

Counterbalance A weight, usually attached to a moving component, that balances another weight.

Counterbore A cylindrical enlargement of the end of a cylinder bore or bore hole.

Counterelectromotive force (cemf) The electromotive force (voltage) that opposes the applied voltage.

Countersink To cut or shape a depression in an object so that the head of a screw may set flush or below the surface.

Counterweight Weights that are mounted on the crankshaft opposite each crank throw. These reduce the vibration caused by putting the crank in practical balance and also reduce bearing loads due to inertia of moving parts.

Coupling A device used to connect two components.

Crankcase The casing which surrounds the crankshaft.

Crankcase scavenging Scavenging method using the pumping action of the power piston in the crankcase to pump scavenging air.

Crankpin The portion of the crank throw attached to the connecting rod.

Crankshaft A rotating shaft for converting rotary motion into reciprocating motion.

Crankshaft gear The gear that is mounted to the crankshaft.

Crank throw One crankpin with its two webs (the amount of offset of the rod journal).

Crank web The portion of the crank throw between the crankpin and main journal. This makes up the offset.

Crest The top surface joining the two sides of a thread.

Crest clearance Defined on a screw form as the space between the top of a thread and the root of its mating thread.

Critical compression ratio Lowest compression ratio at which any particular fuel will ignite by compression under prescribed test procedure. The lower the critical compression ratio the bet-

ter ignition qualities the fuel has. (Gasoline engine, 4:1; oil engine, 7:1; diesel engine, 12.5:1.)

Critical speeds Speeds at which the frequency of the power strokes synchronize with the crankshaft's natural frequency or torsional damper. If the engine is operated at one of its critical speeds for any length of time, a broken crankshaft may result.

Crocus cloth A very fine abrasive polishing cloth.

Crowned A very slight curve in a surface (e.g., on a roller or raceway).

Crude oil Petroleum as it comes from the well (unrefined).

Crush A deliberate distortion of an engine's bearing shell to hold it in place during operation.

Current The flow of electrons passing through a conductor. Measured in amperes.

Cycle A series that repeats itself in the same sequence.

Cylinder The piston chamber of an engine.

Cylinder head The replaceable portion of the engine that seals the cylinder at the top. It often contains the valves, and in some cases, it is part of the combustion chamber.

Cylinder hone A tool used to bring the diameter of a cylinder to specification and at the same time smooth its surface.

Cylinder liner A sleeve which is inserted in the bores of the engine block which make up the cylinder wall.

Dead center Either of the two positions when the crank and connecting rod are in a straight line at the end of the stroke.

Deceleration Opposite of acceleration, that is, implying a slowing down instead of a speeding up. Also called *negative acceleration*.

Deflection Bending or movement away from the normal position, due to loading.

Deglazer A tool used to remove the glaze from cylinder walls.

Degree (circle) 1/360 of a circle.

Degree wheel A wheel marked in degrees to set the lifter height.

Depth of engagement The depth of a thread in contact with two mating parts measured radially. It is the radial distance by which their thread forms overlap each other.

Density The weight per unit volume of a substance.

Detergent A chemical with cleansing qualities added to the engine oil.

Detonation Burning of a portion of the fuel in the combustion chamber at a rate faster than desired (knocking).

Dial indicator (Dial gauge) A precision measuring instrument.

Diaphragm Any flexible dividing partition separating two compartments.

Die (thread) A thread-cutting tool.

Diesel engine An internal-combustion engine having fuel injected into the combustion chamber near the end of the combustion stroke. The fuel is ignited by the heat of compression only.

Diesel index A rating of fuel according to its ignition qualities. The higher the diesel index number, the better the ignition quality of the fuel.

Differential pressure fuel valve A closed fuel valve with a needle or spindle valve which seats onto the inner side of the orifices. The valve is lifted by fuel pressure.

Dilution Thinning, e.g., as when fuel mixes with the lubricant.

Diode A device which allows current to pass but only in one direction.

Dipstick A device to measure the quantity of oil in the reservoir.

Direct-cooled piston A piston which is cooled by the internal circulation of a liquid.

Direct current (dc) Current that flows in one direction only.

Directional control valve A valve which selectively directs or prevents flow to or from specific channels. Also referred to as selector valve, control valve, or transfer valve.

Discharge A draw of current from the battery.

Displacement In a single-acting engine, the volume swept by all pistons in making one stroke each. The displacement on one cylinder in cubic inches is the circular area (in square inches) times the stroke (in inches).

Distillation Heating a liquid and then condensing the vapors given off by the heating process.

Division plate A diaphragm surrounding the piston rod of crosshead-type engine, usually having a wiper ring to remove excess oil from the piston rod as it slides through. It separates the crankcase from the lower end of the cylinder.

Double acting An actuator producing work in both directions.

Double flare A flared end of the tubing having two wall thicknesses.

Dowel A pin, usually of circular shape like a cylinder, used to pin or fasten something in position temporarily or permanently.

Dribbling Unatomized fuel running from the fuel nozzle.

Drill A tool used to bore holes.

Drill press A fixed machine to drive a tool in rotary motion.

Drive-fit A fit between two components, whose tolerance is so small that the two parts must be pressed or driven together.

Drop-forged Formed by hammering or forced into shape by heat.

Dry cell (dry battery) A battery that uses no liquid electrolyte.

Dry-charged battery A battery in a precharged state but without electrolyte. The electrolyte is added when the battery is to be placed in service.

Dry sleeve A cylinder sleeve (liner) where the sleeve is supported over its entire length. The coolant does not touch the sleeve itself.

Dual valves Refers to cylinders having two valves performing one function, e.g., two intake valves, two exhaust valves.

Dynamic balance Condition when the weight mass of a revolving object is in the same plane as the centerline of the object.

Dynamic pressure The pressure of a fluid resulting from its motion, equal to one-half the fluid density times the fluid velocity squared. In incompressible flow, dynamic pressure is the difference between total pressure and static pressure.

Dynamometer A device for absorbing the power output of an engine and measuring torque or horsepower so that it can be computed into brake horsepower.

Eccentric Circles which do not have the same center.

Edge filter A filter which passes liquid between narrowly separated disks or wires.

Efficiency In general, the proportion of energy going into a machine which comes out in the desired form, or the proportion of the ideal which is realized.

Electromotive force (emf) Forces that move or tend to move electricity.

Electrolyte A solution of sulfuric acid and water.

Element (battery) A group of plates—negative and positive.

Emulsify To suspend oil in water in a mixture where the two do not easily separate.

End play The amount of axial movement in a shaft that is due to clearance in the bearings or bushings.

Energize To make active.

Energy Capacity for doing work.

Engine displacement The volume each piston displaces when it moves from BDC to TDC times the number of cylinders.

Erode To wear away.

Ethylene glycol A compound added to the cooling system to reduce the freezing point.

Evaporative cooling system A cooling system in which the heat finally passes to the atmosphere by evaporation. This system may be either open or closed.

Excess air Air present in the cylinder over and above that which is theoretically necessary to burn the fuel.

Excite To pass current through a coil or starter.

Exhaust gas The products of combustion in an internal-combustion engine.

Exhaust analyzer (Smoke meter) A test instrument used to measure the density of the exhaust smoke to determine the combustion efficiency.

Exhaust manifold A device which connects all the exhaust ports to one outlet.

Exhaust port The opening through which exhaust gas passes from the cylinder to the manifold.

Expansion ratio Ratio of the total volume when the piston is at BDC to the clearance volume when the piston is at TDC. (Nominally equal to compression ratio.)

Exhaust valve The valve which, when opened, allows the exhaust gas to leave the cylinder.

Eye bolt A bolt threaded at one end and bent to a loop at the other end.

Fahrenheit A designated temperature scale in which the freezing temperature of water is 32°F and boiling point 212°F (when under standard atmospheric pressure).

Fahrenheit thermometer A thermometer using a Fahrenheit scale.

Fatigue Deterioration of material caused by constant use.

Feeler gauge A strip of steel ground to a precise thickness used to check clearance.

Field The area affected by magnetic lines of force.

Field coil An insulated wire wound around an (iron) pole piece.

Fillet A curved joint between two straight surfaces.

Filter A device for cleaning or purifying fluid or air.

Finishing stone (Hone) A honing stone with a fine grid.

Fire point Lowest temperature at which an oil heated in standard apparatus will ignite and continue to burn.

Firing order The order in which the cylinders deliver their power stroke.

Firing pressure The highest pressure reached in the cylinder during combustion.

Fit The closeness of contact between machined components.

Fixed displacement pump A type of pump in which the volume of fluid per cycle cannot be varied.

Flange A metal part which is spread out like a rim; the action of working a piece or part spread out.

Flank (side or thread) The straight part of the thread which connects the crest with the root.

Flank angles The angle between a specified flank of a thread and the plane perpendicular to the axis (measured in an axial plane).

Flare To open or spread outwardly.

Flaring tool A tool used to form a flare on a tubing.

Flash point The temperature at which a substance, usually a fluid, will give off a vapor that will flash or burn momentarily when ignited.

Flat crank A crankshaft in which one of the bearing journals is not round.

Flow control valve A valve which is used to control the flow rate of fluid in a fluid power system.

Flowmeter An instrument used to measure the quantity or flow rate of a fluid in motion.

Fluctuating Wavering, unsteady, not constant.

Fluid A liquid, gas, or mixture thereof.

Fluid flow The stream or movement of a fluid; the rate of a fluid's movement.

Fluid power Power transmitted and controlled through the use of fluids, either liquids or gases, under pressure.

Flute The grooves of a tap that provide the cutting rake and chip clearance.

Flux (magnetic) Magnetic force.

Flyball governor Conventional type of centrifugal governor commonly called a mechanical governor.

Flywheel A device for storing energy in order to carry the piston over a compression and to minimize cyclical speed variations.

Flywheel ring gear A circular steel ring having gear teeth on the outer circumference.

Foot-pound (ft · lb) The amount of

work accomplished when a force of one pound produces a displacement of one foot.

Force The action of one body on another tending to change the state of motion of the body acted upon. Force is usually expressed in pounds [kilograms].

Force-feed lubrication A lubricating system in which oil is pumped to the desired points at a controlled rate by means of positive displacement pumps.

Forged Shaped with a hammer or machine.

Foundation The structure on which an engine is mounted. It performs one or more of the following functions: holds the engine in alignment with the driven machine; adds enough weight to the engine to minimize vibration; adds to rigidity of the bed plate.

Four-stroke cycle Cycle of events which is completed in four strokes of the piston, or two crankshaft revolutions.

Frame The main structural member of an engine.

Free electrons Electrons which are in the outer orbit of the atom's nucleus.

Free flow Flow which encounters negligible resistance.

Friction The resistance to motion due to the contact of two surfaces, moving relatively to each other.

Fuel mixture A ratio of fuel and air.

Fuel transfer pump A mechanical device used to transfer fuel from the tank to the injection pump.

Fuel valve A valve admitting fuel to the combustion chamber. In a more general sense, this term may also apply to any manual or automatic valve controlling flow of fuel.

Fulcrum The pivot point of a lever.

Full-floating piston pin A piston pin free to turn in the piston boss of the connecting-rod eye.

Full-flow oil filter All engine oil passes through this oil filter before entering the lubrication channels.

Gauge pressure Pressure above atmospheric pressure.

Gauge snubber A device installed in the fuel line to the pressure gauge used to dampen pressure surges and thus provide a steady reading. This helps protect the gauge.

Gallery Passageway inside a wall or casting.

Galvanic action When two dissimilar metals are immersed in certain solutions, particularly acid, electric current will flow from one to the other.

Gas Matter that has no definite form or volume, but instead tends to expand indefinitely.

Gasket A layer of material used between machined surfaces in order to seal them against leakage.

Gassing Hydrogen bubbles rising from the electrolyte when the battery is being charged.

Gate valve A common type of manually operated valve in which a sliding gate is used to obstruct the flow of fluid.

Gear-type pump A pump which uses the spaces between the adjacent teeth of gears for moving the liquid.

Generator An electromagnetic device used to generate electricity.

Gland A device to prevent the leakage of gas or liquid past a joint.

Glaze A smooth, glassy surface finish.

Glow plug A heater plug for the combustion chamber. It has a coil of resistance wire heated by a low-voltage current.

Governor A device for controlling the speed of a prime mover.

Gravity The force which tends to draw all bodies toward the center of the earth. The weight of a body is the resultant of all gravitational forces on the body.

Grid (battery) The lead frame to which the active material is affixed.

Grinding Removing metal from an object by means of a revolving abrasive wheel, disk or belt.

Grinding compound Abrasive for resurfacing valves, etc.

Ground (battery) The battery terminal that is connected to the engine or the framework.

Growler A test instrument used for testing the armature of a starter or generator for open, short, and grounded circuits.

Half-moon key A fastening device in a shape somewhat similar to a semicircle. (See Key.)

Heat A form of energy.

Heat exchanger A device used to cool by transferring heat.

Heating value Amount of heat produced by burning one pound of fuel.

Helical gear A gear wheel of a spiraling shape. (The teeth are cut across the face at an angle with the axis.)

Hone A tool with an abrasive stone used for removing metal.

Horsepower (hp) A unit of power equivalent to 33,000 foot-pounds of work per minute (English) or 75 kilogram-meters per second (metric). (See *Brake horsepower* and *Indicated horsepower*.)

Horsepower-hour (hp·h) A unit of energy equivalent to that expended in one horsepower applied for one hour. Equal to approximately 2545 Btu.

Hunting Alternate overspeeding and underspeeding of the engine caused by governor instability.

Hydraulics That branch of mechanics or engineering which deals with the action or use of liquids forced through tubes and orifices under pressure to operate various mechanisms.

Hydraulic governor A governor using fluid to operate the fuel control.

Hydrocarbon (HC) A compound of hydrogen and carbon. Petroleum and its derivatives are mixtures of various hydrocarbons.

Hydrogen One of the elements constituting fuel and lubrication oil.

Hydrometer A test instrument for determining the specific gravities of liquids.

Idling An engine running without load.

Ignition The start of combustion.

Ignition lag The time between start of injection and ignition.

Immersed To be completely under the surface of a fluid.

Impact wrench An air wrench or an electrically driven wrench.

Impeller A wheel or disk with fins.

Indicated horsepower (ihp) The power transmitted to the pistons by the gas in the cylinders.

Indicated thermal efficiency The ratio of indicated horsepower to equivalent power input in the form of heat from fuel.

Indicator An instrument for recording the variation of cylinder pressure during the cycle.

Indicator card A graphical record of the cylinder pressures made by an indicator.

Indirect cooled piston A piston cooled mainly by the conduction of heat through the cylinder walls.

Induction Using the magnetic field to impart electricity into an object which is not otherwise connected to the first ones.

Inertia That property of matter which causes it to tend to remain at rest if already motionless or to continue in the same straight line of motion if already moving.

Inhibitor Any substance which retards or prevents such chemical reactions as corrosion or oxidation.

Injection pump A high-variable-pressure pump delivering fuel into the combustion chamber.

Injection system The components necessary for delivering fuel to the combustion chamber in the correct quantity, at the correct time, and in a condition satisfactory for efficient burning.

Injector A device used to bring fuel into the combustion chamber.

In-line engine An engine in which all the cylinders are in a straight line.

Insert bearing A removable, precision-made bearing.

Insulator (electrical) A material that, under normal conditions, will not conduct electricity.

Intake manifold A connecting casting between the air filter or turbocharger and the port openings to the intake valves.

Intake valve The valve which when open allows air to enter into the cylinder.

Intercooler Heat exchanger for cooling the air between stages of compression.

Internal-combustion engine An engine that burns fuel within itself as a means of developing power.

Isochronous governor A governor having zero speed droop.

Jet A small hole in a carburetor passage to measure the flow of gasoline.

Jet cooling A method of passing cooling oil below the piston by means of a jet or nozzle.

Journal The portion of a shaft, crank, etc., which turns in a bearing.

Kelvin scale A temperature scale having the same size divisions as between Celsius degrees, but having the zero point at absolute zero.

Key A fastening device wherein two components each have a partially cut groove, and a single square of square is inserted in both to fasten them together.

Keyway The groove cut in a component to hold the key.

Kilometer (km) A metric measurement of length equal to 0.6214 mile.

Kilowatt (kw) A unit of power equal to 1000 watts.

Kilowatt hour (kwh) A unit of electric energy.

Kinetic energy The energy which an object has while in motion.

Knocking A sharp pounding sound occurring periodically in an engine.

Knurling A method of placing ridges in a surface, thereby forcing the areas between these ridges to rise.

Lag To slow down or get behind; time interval, as in *ignition lag*.

Laminar A layer of fluid.

Laminar flow A smooth flow in which no crossflow of fluid particles occurs.

Land The projecting part of a grooved surface; for example, that part of a piston on which the rings rest.

Lap (lapping) A method of refinishing (grinding and polishing) the surface of a component.

Letter drills Drills on which the size is designated by a letter.

Line A tube, pipe, or hose which is used as a conductor of fluid.

Liner The sleeve forming the cylinder bore in which the piston reciprocates.

Linkage A movable connection between two units.

Liquid Matter which has a definite volume but takes the shape of any container.

Liter A metric measurement of volume equal to 0.2642 gallon (U.S.).

Live wire A conductor which carries current.

Load The power that is being delivered by any power-producing device. The equipment that uses the power from the power-producing device.

Load factor The mean load carried by an engine, expressed in percent of its capacity.

Load Line A center line indicating the points of contact where the load passes within the bearing.

Load-line angle The angle of a load line with respect to the shaft center or bearing radial centerline.

Lobe The projecting part, usually rounded, on a rotating shaft.

Lubricant A substance to decrease the effects of friction, commonly a petroleum product (grease, oil, etc.).

Lubricator A mechanical oiler which feeds oil at a controlled rate.

Lug (engine) Condition when the engine is operating at or below its maximum torque speed.

Magnaflux A method used to check components for cracks.

Magnetic field The affected area of the magnetic lines of force.

Main bearing A bearing supporting the crankshaft on its axis.

Mandrel A mounting device for a stone, cutter, saw, etc.

Manual valve A valve which is opened, closed, or adjusted by hand.

Matter Any substance which occupies space and has weight. The three forms of matter are solids, liquids, and gases.

Mean effective pressure (mep) The calculated combustion in pounds per square inch (average) during the power stroke, minus the pounds per square inch (average) of the remaining three strokes.

Mean indicated pressure (mip) Net mean gas pressure acting on the piston to produce work.

Mechanical advantage The ratio of the resisting weight to the acting force. The distance through which the force is exerted divided by the distance the weight is raised.

Mechanical efficiency (1) The ratio of brake horsepower to indicated horsepower, or ratio of brake mean effective pressure to mean indicated pressure. (2) An engine's rating which indicates how much of the potential horsepower is wasted through friction within the moving parts of the engine.

Mechanical injection Mechanical force pressurizing the metered fuel and causing injection.

Mechanically operated valve A valve which is opened and closed at regular points in a cycle of events by mechanical means.

Metal fatigue When metal crystallizes and is in jeopardy of breaking due to vibration, twisting, bending, etc.

Metering fuel pump A fuel pump delivering a controlled amount of fuel per cycle.

Metric size Size of a component, part, etc., in metric units of measurement (e.g. meters, centimeters.

Micrometer (Mike) A precision measuring tool that is accurate to within one thousandth of an inch or one hundredth of an millimeter.

Micron (μm) One millionth of a meter or 0.000039 inch.

Millimeter (mm) One thousandth of a meter or 0.039370 inch.

Milling machine A machine used to remove metal, cut splices, gears, etc., by the rotation of its cutter or abrasive wheel.

Misfiring When the pressure of combustion of one or more cylinders is lower than the remaining cylinders, one or more cylinders have an earlier or later ignition than the others.

Mixed cycle Where fuel burns partly at constant volume and partly at constant pressure. Sometimes applied to the actual combustion cycle in most high-speed internal-combustion engines.

Molecule The smallest portion to which a substance may be reduced by subdivision and still retain its chemical identity.

Motor An actuator which converts fluid power or electric energy to rotary mechanical force and motion.

Multiviscosity oil An oil meeting SAE requirements.

Needle bearing A roller-type bearing in which the rollers are smaller in diameter than in length proportional to the race.

Negative terminal A terminal from which the current flows back to its source.

Neoprene A synthetic rubber highly resistant to oil, light, heat, and oxidation.

Neutron A neutral charged particle of an atom.

Newton's third law For every action there is an equal, opposite reaction.

Nitrogen oxide (NO_x) The combination of nitrogen and oxygen that occurs during the combustion process.

Nonferrous metals Any metals not containing iron.

North pole (magnet) The pole from which the lines of force emanate (thereafter entering the south pole).

Nozzle The component containing the fuel valve and having one or more orifices through which fuel is injected.

Number drills Drills on which the size is designated by a number.

Ohm (Ω) A unit of electrical resistance.

Ohmmeter An instrument for measuring the resistance in a circuit or unit in ohms.

Ohm's law The number of amperes flowing in a circuit is equal to the number of volts divided by the number of ohms.

Oil-bath air cleaner An air filter that utilizes a reservoir of oil to remove the impurities from the air before it enters the intake manifold or the compressor of the turbine.

Oil cooler A heat exchanger for lowering the temperature of oil.

Oil filter A device for removing impurities from oil.

Oil gallery A pipe-drilled or casted passage in the cylinder-head block and crankcase that is used to carry oil from the supply to an area requiring lubrication or cooling.

Oil pump A mechanical device to pump oil (under pressure) into the various oil galleries.

Oil pumping An engine condition wherein excessive oil passes by the piston rings and is burned during combustion.

Oil seal A mechanical device used to prevent oil leakage, usually past a shaft.

Oil slinger A special frame disk fastened to a revolving shaft. When the shaft rotates and oil contacts the disk, it is thrown outward away from the seal, and thus reduces the force on the seal lip.

Open circuit A circuit in which a wire is broken or disconnected.

Opposed piston engine An engine having two pistons operating in opposite ends of the same cylinder, compressing air between them.

Orderly turbulence Air motion which is controlled as to direction or velocity.

Orifice An aperture or opening.

Oscilloscope A device for recording wave forms on a fluorescent screen, proportional to the input voltage.

Oscillate To swing back and forth like a pendulum; to vibrate.

Output shaft The shaft which delivers the power.

Overhead camshaft A camshaft which is mounted above the cylinder head.

Overrunning clutch A clutch mechanism that transmits power in one direction only.

Overrunning-clutch starter drive A mechanical device that locks in one direction but turns freely in the opposite direction.

Overspeed governor A governor that shuts off the fuel or stops the engine only when excessive speed is reached.

Oversquare engine An engine that has a larger bore diameter than the length of its stroke.

Oxidation That process by which oxygen unites with some other substance causing rust or corrosion.

Packing A class of seal of flexible

material used to seal two parts which move in relation to each other.

Paper air cleaner An air filter with a special paper element through which the air is drawn.

Parallel circuit An electric circuit with two or more branch circuits. It is wired so as to allow current to flow through all branches at the same time.

Pascal's law Pressure applied anywhere to a body of confined fluid is transmitted undiminished to every portion of the surface of the containing vessel.

Peen The thin end of a hammer head (opposite to the face).

Penning Flattening the end of a rivet, etc., using the force of a hammer.

Penetrating oil A special oil that aids the removal of rusted parts.

Perforate To make full of holes.

Periphery The external boundary or circumference.

Petroleum An oil liquid mixture made up of numerous hydrocarbons chiefly of the paraffin series.

Phosphor-bronze A bearing material composed of tin, lead, and copper.

Physical change A change which does not alter the composition of the molecules of a substance.

Pilot shaft A shaft position in or through a hole of a component as a means of aligning the components.

Pilot valve A valve used to control the operation of another valve.

Pintle-type nozzle A closed-type nozzle having a projection on the end of the fuel valve which extends into the orifice when the valve is closed.

Pipe In diesel applications, that type of fluid line, the dimensions of which are designated by nominal (approximate) inside diameter.

Piston A cylindrical plug which slides up and down in the cylinder and which is connected to the connecting rod.

Piston boss The reinforced area around the piston-pin bore.

Piston displacement The volume of air moved or displaced by a piston when moved from BDC to TDC.

Piston head The portion of the piston above the top ring.

Piston lands That space of the piston between the ring grooves.

Piston pin (Wrist pin) A cylindrical alloy pin that passes through the piston bore and is used to connect the connecting rod to the piston.

Piston ring A split ring of the expansion type placed in a groove of the piston to seal the space between the piston and the wall.

Piston-ring end cap The clearance between the ends of the ring (when installed in the cylinder).

Piston-ring groove The grooves cut in the piston into which the piston rings are fitted.

Piston-ring side clearance The clearance between the sides of the ring and the ring lands.

Piston skirt The portion of the piston which is below the piston bore.

Piston speed The total distance traveled by each piston in one minute. The formula is: Piston speed-stroke (ft) $\times$ rpm $\times$ 2 or stroke (in) $\times$ rpm/6.

Pivot The pin or shaft on which a component moves.

Plate (battery) A flat, square, rigid body of lead peroxide or porous lead.

Play The movement between two components.

Plunger pump A pump which displaces fluid by means of a plunger.

Pneumatics That branch of physics pertaining to the pressure and flow of gases.

Polarity Can refer to the grounded battery terminal or to an electric circuit or to the north and south pole of a magnet.

Polarizing To develop polarization of the pole shoes in respect to battery polarity.

Polar timing diagram A graphic method of illustrating the events of an engine cycle with respect to crankshaft rotation.

Pole (magnet) Either end of a magnet.

Pole shoe A soft iron piece over which the field coil is placed.

Ports Openings in the cylinder block and cylinder head for the passage of oil and coolant. (Also exhaust-intake connection and valve openings.)

Port bridge The portion of a cylinder or liner between two exhaust or scavenging ports.

Port scavenging Introducing scavenging air through ports in the cylinder wall when they are uncovered by the power piston near the end of the power stroke.

Positive terminal The terminal which has a deficiency of electrons.

Potential energy The energy possessed by a substance because of its position, its condition, or its chemical composition.

Pour point The lowest tempera-

ture at which an oil will flow.

Power Rate of doing work.

Precison insert bearing A precision type of bearing consisting of an upper and lower shell.

Precombustion chamber A portion of the combustion chamber connected to the cylinder through a narrow throat. Fuel is injected into and is partly burned in the precombustion chamber. Heat released by this partial burning causes the contents of the precombustion chamber to be ejected into the cylinder with considerable turbulence.

Preloading Adjusting taper roller bearings so that the rollers are under mild pressure.

Press-fit (See *Drive-fit*.)

Pressure Force exerted per unit of area.

Pressure cap A special radiator cap with a pressure relief and vacuum valve.

Pressure differential The difference in pressure between any two points of a system or a component.

Pressure lubrication A lubricating system in which oil at a controlled pressure is brought to the desired point.

Pressure relief valve A valve that limits the maximum system pressure.

Printed circuit An electric circuit where the conductor is pressed or printed in or on an insulating material (panel) and at the same time is connected to the resistor, diodes, condenser, etc.

Prony brake A friction brake used for engine testing.

Proton The positively charged particle in the nucleus of an atom.

Prussian blue A blue pigment, obtainable in tubes, which is used to find high spots in a bearing.

Pulsate To move with rhythmical impulse.

Pulverize To reduce or become reduced to powder or dust.

Pump A device for moving fluids.

Pumping loss The power consumed by replacing exhaust gas in the cylinder with fresh air.

Pump scavenging Using a piston-type pump to pump scavenging air.

Push fit The part of the bearing that can be slid into place by hand if it is square with its mounting.

Push rod A rod used for transmitting cam motion to a rocker arm.

Pyrometer A temperature indicator used for comparing exhaust temperatures of the various cylinders.

Quench To cool heated steel or iron by thrusting it into water.

Quicksilver Metallic mercury.

Race (bearing) The inner or outer groove or channel bearing ring.

Raceway The surface of the groove or path which supports the balls or rollers of a bearing roll.

Radial Perpendicular to the shaft or bearing bore.

Radial clearance (Radial displacement) The clearance within the bearing and between the balls and races, perpendicular to the shaft.

Radial load A "round the shaft" load, that is, one that is perpendicular to the shaft through the bearing.

Radiator A heat exchanger in which cooling water gives up heat to the air without coming into direct contact with it.

Radius The distance from the center of a circle to its outer edge or the straight line extending from the center to the edge of a circle.

Ratio The numerical relationship between objects.

Rebore To bore out a cylinder to a size slightly larger than the original.

Reciprocating action A back-and-forth (alternating) movement.

Rectifier A device used to convert alternating current into direct current.

Regulator (electrical) An electromagnetic or electronic device used to control generator voltage.

Relay An electromagnetic switch which utilizes variation in the strength of an electric circuit to effect the operation of another circuit.

Relief valve An automatic valve which is held shut by a spring of correct strength. Excess pressure opens the valve and releases some of the gas or liquid. This valve is for protecting filters, air tanks, etc., from dangerous pressures.

Resistance (electrical) The opposition offered by a body when current passes through it.

Resistor A device placed in a circuit to lower the voltage, to reduce the current, or to stabilize the voltage.

Retard (injection timing) To set the timing so that injection occurs later than TDC or fewer degrees before TDC.

Reverse flush To pump water or a cleaning agent through the cooling system in the opposite direction to normal flow.

Rheostat A device to regulate current flow by varying the resistance in the circuit.

Ring expander A type of spring which is placed between the ring and ring groove to hold the ring with fixed force against the cylinder wall.

Ring groove A groove machined in the piston to receive the piston ring.

Ring job The service work on the piston and cylinder including the installation of new piston rings.

Rivet A soft-metal pin having a head at one end.

Rocker arm A first-class lever used to transmit the motion of the pushrod to the valve stem.

Rocker-arm shaft The shaft on which the rocker arms pivot.

Rockwell hardness A measurement of the degree of surface hardness of a given object.

Rod Refers to a connecting rod.

Roller bearing An antifriction bearing using straight (cupped or tapered) rollers spaced in an inner and outer ring.

Roller tappets (Roller lifters) Refers to valve lifters having a roller at one end which is in contact with the camshaft and is used to reduce friction.

Roots blower An air pump or blower similar in principle to a gear-type pump.

Rope brake A friction brake used for engine testing.

Rotary blower Any blower in which the pumping element follows rotary motion, the centrifugal blowers being the exception.

Roughing stone (Hone) A coarse honing stone.

Running-fit A machine fit with sufficient clearance to provide for expansion and lubrication.

SAE Abbreviation for Society of Automotive Engineers.

SAE horsepower (Rated horsepower) Formula to determine power: bore diameter^2 × number of cylinders/2.5 = hp

SAE viscosity numbers Simplified viscosity ratings of oil based on Saybolt viscosity.

Safety factor Providing strength beyond that needed as an extra margin of insurance against parts failure.

Safety valve (See *Relief valve*.)

Sand blast (Glass blast) A cleaning method using an air gun to force the sand at low pressure (about 150 psi) against the surface to be cleaned.

Saybolt viscosimeter A container with a calibrated outlet tube for determining the viscosity of liquids. (This method is now obsolete.)

Saybolt viscosity The number of seconds necessary for 60 milliliters of liquid to pass through the outlet tube of a Saybolt viscosimeter under standardized test conditions.

Scale Precipitated hardness (salts) from water.

Scavenging The displacement of exhaust gas from the cylinder by fresh air.

Scavenging air The air which is pumped into a cylinder to displace exhaust gas.

Scavenging blower A device for pumping scavenging air.

Scavenging pump A piston-type pump delivering scavenging air to an engine.

Scraper ring An oil-control ring.

Screw extractor A device used to remove broken bolts, screws, etc., from holes.

Sealed bearing A bearing which is lubricated and sealed at the factory and which cannot be lubricated during service.

Seat (rings) Rings fitted or seated properly against the cylinder wall.

Semiconductor An element which is neither a good conductor nor a good insulator.

Sediment Solid impurities in a liquid.

Semifloating piston pin A piston pin which is clamped either in the connecting rod or in the piston bosses.

Separator (battery) A porous insulation material placed between the positive and negative plates.

Series circuit An electric circuit wired so that the current must pass through one unit before it can pass through the other.

Series-parallel circuit A circuit with three or more resistance units in a combination of a series and a parallel circuit.

Shaft horsepower Power delivered at the engine crankshaft. This term is commonly used instead of brake horsepower to express output of large marine engines.

Shim Thin, flat pieces of brass or steel used to increase the distance between two components.

Short circuit A circuit whose resistance is reduced in power owing to one or more coil layers contacting one another.

Shrink-fit A fit between two components made by heating the outer component so that it will expand and fit over the inner component. As the outer component cools, it shrinks and thereby fits tight to the inner component.

Shroud The enclosure around the fan, engine, etc., which guides the air flow.

Shunt A parallel circuit where one resistance unit has its own ground.

Shunt winding A resistance coil with its own ground.

Shutoff valve A valve which opens and thereby stops the flow of a liquid, air, or gas.

Silencer A device for reducing the noise of intake or exhaust.

Single-acting cylinder An actuating cylinder in which one stroke is produced by pressurized fluid, and the other stroke is produced by some other force, such as gravity or spring tension.

Sludge Deposits inside the engine caused by dust, oil, and water being mixed together by the moving components.

Snap ring A fastening device in the form of a split ring that is snapped into a groove in a shaft or in a groove in a bore.

Sodium valve A valve designed to allow the stem and head to be partially filled with metallic sodium.

Solenoid An electrically magnetic device used to do work.

Specific gravity The ratio of the weight of a given volume of any substance to that of the same volume of water.

Spline The land between two grooves.

Spool valve A hydraulic directional control valve in which the direction of the fluid is controlled by the means of a grooved cylindrical shaft (spool).

Spur gear A toothed wheel having external radial teeth.

Squish area The area confined by the cylinder head and flat surface of the piston when on compression stroke.

Stability The resistance of a fluid to permanent change such as that caused by chemical reaction, temperature changes, etc.

Starting air Compressed air used for starting an engine.

Starting-air valve A valve which admits compressed starting air to the cylinder.

Static electricity Electricity at rest; pertaining to stationary charges.

Staybolt A stress bolt running diagonally upward from the bedplate to the opposite side of the frame.

Steady flow A flow in which the velocity components at any point in the fluid do not vary with time.

Stethoscope A device for conveying the sound of a body (engine noise) to the mechanic.

Streamline flow A nonturbulent flow, essentially fixed in pattern.

Stresses The forces to which parts are subjected.

Stroboscope (Timing Light) An instrument used to observe the periodic motion of injection visible only at certain points of its path.

Stroke One of a series of recurring movements of a piston or the distance of such movement.

Stroke-to-bore ratio The length of stroke divided by the diameter of bore.

Stud A rod having threads on both ends.

Stud puller A device used to remove or to install stud bolts.

Stuffing box A chamber having a manual adjustment device for sealing.

Suction The process of producing a pressure differential (partial vacuum).

Suction valve Often used interchangeably with intake valve.

Sulfur An undesirable element found in petroleum in amounts varying from a slight trace to 4 or 5 percent.

Sump A receptacle into which liquid drains.

Sump pump A pump which removes liquid from the sump.

Supercharger An air pump driven by the engine which fills the cylinders with a higher pressure than atmospheric pressure.

Supply line A line that conveys fluid from the reservoir to the pump.

Surge A momentary rise and fall of pressure or speed in a system or engine.

Synchronize To make two or more events or operations occur at the proper time with respect to each other.

Synchronous Happening at the same time.

Synthetic material A complex chemical compound which is artificially formed by the combining of two or more compounds or elements.

Tachometer An instrument indicating rotating speeds. Tachometers are sometimes used to indicate crankshaft rpm.

Tap A cutting tool used to cut threads in a bore. (See *Chamfer.*)

Tap and die set A set of cutting tools used to cut internal and external threads.

Tapered roller bearing (See *roller bearing.*)

Tappet The rocker arm.

Tappet noise The noise caused by excessive clearance between the valve stem and the rocker arm.

TDC Abbreviation for *top dead center.*

Temper The condition of a metal with regard to hardness achieved through heating and then suddenly cooling.

Temperature of compression The temperature of the compressed air charge in a power cylinder at the end of the compression stroke before combustion begins.

Temporary hardness Dissolved substances which precipitate out when water is heated.

Tension Stress applied on a material or body.

Terminal The connecting point (post) of a conductor.

Theory A scientific explanation tested by observations and experiments.

Thermal efficiency (See *Brake thermal efficiency* and *Indicated thermal efficiency.*)

Thermal expansion The increase in volume of a substance caused by temperature change.

Thermocouple The part of a pyrometer which consists of two dissimilar metal wires welded together at the inner end and held in a protective housing.

Thermometer An instrument for measuring temperature.

Thermostat A temperature-responsive mechanism used for controlling heating systems, cooling systems, etc., usually with the object of maintaining certain temperatures without further personal attention.

Throttling Reducing the engine speed (flow of fuel).

Through bolt Term usually applied to the stress rod passing through the engine frame to carry combustion stresses.

Throw The part of a crankshaft to which the connecting rod is fastened.

Thrust bearing (washer) A bearing or washer of bronze or steel which restrains endwise motion of a turning shaft, or withstands axial loads instead of radial loads as in common bearings.

Thrust load A load which pushes or reacts through the bearing in a direction parallel to the shaft.

Time lag of ignition (See *Ignition lag.*)

Timing gears Gears attached to the crankshaft, camshaft, idler shaft, or injection pump to provide a means to drive the camshaft and injection pump and to regulate the speed and performance.

Timing marks (injection) The marks located on the vibration damper or flywheel, used to check injection timing.

Tolerance A fractional allowance for variations from the specifications.

Torque A force or combination of forces that produces or tends to produce a twisting or rotary motion.

Torque wrench A wrench used to measure the turning force being applied.

Torsional vibration The vibration caused by twisting and untwisting a shaft.

Transfer pump A mechanical device for moving fuel from one tank to another or bringing fuel from the tank to the injection pump.

Troubleshooting The act of analyzing, testing, and measuring the engine to remedy the cause of trouble.

Tube cutter A tube-cutting tool having a sharp disk which is rotated around the tube.

Tubing That type of fluid line whose dimensions are designated by actual measured outside diameter.

Tuneup The act of checking, testing, measuring, repairing, and adjusting the engine components in order to bring the engine to peak efficiency.

Turbine A series of curved vanes mounted on a shaft and actuated by the action of a fluid or gas under pressure.

Turbocharger An exhaust-gas-driven turbine directly coupled with a compressor wheel.

Turbulence chamber A combustion chamber connected to the cylinder through a throat. Fuel is injected across the chamber and turbulence is produced in the chamber by the air entering during compression.

Twist drill (See *Drill.*)

Two-stage combustion Combustion occurring in two distinct steps such as in a precombustion chamber engine.

Two-stroke cycle Cycle of events which is complete in two strokes of the piston or one crankshaft revolution.

Uniflow scavenging Scavenging method in which air enters one end of the cylinder and exhaust leaves the opposite end.

Unit injector A combined fuel-injection pump and fuel nozzle.

U.S. gallon United States gallon (231 in^3).

Vacuum A pressure less than atmospheric pressure.

Vacuum gauge A gauge used to measure the amount of vacuum existing in a chamber or line.

Valve Any device or arrangement used to open or close an opening to permit or restrict the flow of a liquid, gas, or vapor.

Valve duration The time (measured in degrees of engine crankshaft rotation) that a valve remains open.

Valve float A condition where the valves are forced open due to valve-spring vibration or vibration speed.

Valve grinding Resurfacing the valve face by a special grinding machine.

Valve guide A hollow-sized shaft pressed into the cylinder head to keep the valve in proper alignment.

Valve keeper (Valve retainer) A device designed to lock the valve-spring retainer to the valve stem.

Valve lift The distance a valve moves from the fully closed to the fully open position.

Valve lifter (See *Cam follower*.)

Valve margin The distance between the edge of the valve and the edge of the face.

Valve oil seal A sealing device to prevent excess oil from entering the area between the stem and the valve guide.

Valve overlap The period of crankshaft rotation during which both the intake and exhaust valves are open. It is measured in degrees.

Valve rotator A mechanical device locked to the end of the valve stem which forces the valve to rotate about 5° with each rocker-arm action.

Valve seat The surface on which the valve face rests when closed.

Valve-seat insert A hardened steel ring inserted in the cylinder head to increase the wear resistance of the valve seat.

Valve timing The positioning of the camshaft (gear) to the crankshaft (gear) to ensure proper valve opening and closing.

Vaporization The process of converting a liquid into vapor.

Venturi A specially shaped tube with a small or constricted area used to increase velocity and reduce pressure.

Vibration damper A specially designed device mounted to the front of the crankshaft to reduce torsional vibration.

Viscosity The property of an oil by virtue of which it offers resistance to flow.

Viscosity index (VI) Oil decreases in viscosity as temperature changes. The measure of this rate of change of viscosity with temperature is called the *viscosity index* of the oil.

Volatile Evaporating readily at average temperature on exposure with air.

Volatility A measurement of the ease with which a liquid may be vaporized at relatively low temperature.

Volt (V) A unit of electromotive force that will move a current of one ampere through a resistance of one ohm.

Voltage Electrical potential expressed in volts.

Voltage drop Voltage loss due to added resistance caused by undersized wire, poor connection, etc.

Voltmeter A test instrument for measuring the voltage or voltage drop in an electric circuit.

Volume The amount of space within a given confined area.

Volumetric efficiency The difference between the volume of air drawn in on the intake stroke and the air mechanically entering the cylinder.

Water brake A device for engine testing in which the power is dissipated by churning water.

Water jacket The enclosure directing the flow of cooling water around the parts to be cooled.

Wet sleeve A cylinder sleeve which is about 70 per cent exposed to the coolant.

Yoke A link which connects two points.

Zener diode A diode that allows current to flow in reverse bias at the designed voltage.